CANCER: OXIDATIVE STRESS
AND DIETARY ANTIOXIDANTS

T0329415

CANCER: OXIDATIVE STRESS AND DIETARY ANTIOXIDANTS

VICTOR PREEDY

Kings College London, London, UK

Amsterdam • Boston • Heidelberg • London
New York • Oxford • Paris • San Diego
San Francisco • Singapore • Sydney • Tokyo
Academic Press is an imprint of Elsevier

Academic Press is an imprint of Elsevier
32 Jamestown Road, London NW1 7BY, UK
225 Wyman Street, Waltham, MA 02451, USA
525 B Street, Suite 1800, San Diego, CA 92101-4495, USA

First edition 2014

Copyright © 2014 Elsevier Inc. All rights reserved.

No part of this publication may be reproduced, stored in a retrieval system or transmitted in any form or by
any means electronic, mechanical, photocopying, recording or otherwise without the prior written permission
of the publisher

Permissions may be sought directly from Elsevier's Science & Technology Rights Department in Oxford, UK:
phone (+44) (0) 1865 843830; fax (+44) (0) 1865 853333; email: permissions@elsevier.com. Alternatively, visit
the Science and Technology Books website at www.elsevierdirect.com/rights for further information

Notices
No responsibility is assumed by the publisher for any injury and/or damage to persons or property as a
matter of products liability, negligence or otherwise, or from any use or operation of any methods, products,
instructions or ideas contained in the material herein.

Because of rapid advances in the medical sciences, in particular, independent verification of diagnoses and
drug dosages should be made

British Library Cataloguing-in-Publication Data
A catalogue record for this book is available from the British Library

Library of Congress Cataloging-in-Publication Data
A catalog record for this book is available from the Library of Congress

ISBN: 978-0-12-405205-5

For information on all Academic Press publications
visit our website at elsevierdirect.com

Typeset by TNQ Books and Journals
www.tnq.co.in

Printed and bound in United States of America

14 15 16 17   10 9 8 7 6 5 4 3 2 1

Working together
to grow libraries in
developing countries

www.elsevier.com • www.bookaid.org

Contents

2

ANTIOXIDANTS AND CANCER

Preface

In the past few decades there have been major advances in our understanding of the etiology of disease and its causative mechanisms. Increasingly it is becoming evident that free radicals are contributory agents: either to initiate or propagate the pathology or add to an overall imbalance. Furthermore, reduced dietary antioxidants can also lead to specific diseases and pre-clinical organ dysfunction. On the other hand, there is abundant evidence that dietary and other naturally occurring antioxidants can be used to prevent, ameliorate, or impede such diseases. The science of oxidative stress and free radical biology is rapidly advancing, and new approaches include the examination of polymorphism and molecular biology. The more traditional sciences associated with organ functionality continue to be explored but their practical or translational applications are now more sophisticated.

However, most textbooks on dietary antioxidants do not have material on the fundamental biology of free radicals, especially their molecular and cellular effects on pathology. They also fail to include material on the nutrients and foods that contain antioxidative activity. In contrast, most books on free radicals and organ disease have little or no text on the use of natural antioxidants.

The series **Oxidative Stress and Dietary Antioxidants** aims to address the aforementioned deficiencies in the knowledge base by combining in a single volume the science of oxidative stress and the putative therapeutic use of natural antioxidants in the diet, its food matrix, or plants. This is done in relation to a single organ, disease, or pathology. These include cancer, addictions, immunology, HIV, ageing, cognition, endocrinology, pregnancy and fetal growth, obesity, exercise, liver, kidney, lungs, reproductive organs, gastrointestinal tract, oral health, muscle, bone, heart, kidney, and the central nervous system.

In the present volume, **Cancer: Oxidative Stress and Dietary Antioxidants,** holistic information is imparted within a structured format of two main sections:

- Oxidative Stress and Cancer
- Antioxidants and Cancer

The first section on **Oxidative Stress and Cancer** covers the basic biology of oxidative stress from molecular biology to physiological pathology. Topics include cancer of the breast, prostate, lung, and stomach as well as viral carcinogenesis and oxidative DNA damage. The second section, **Antioxidants and Cancer**, covers cellular and molecular approaches, genetic polymorphisms, herbs and spices, dietary antioxidants, vitamins, n-3 polyunsaturated fatty acids, the Indian blackberry, black chokeberries, broccoli bioactives, fern extract, polyphenols, resveratrol and lycopene, curcumin and curcumin analogs, cocoa, green tea polyphenols, quercetin, capsaicin, tocotrienols, pterostilbene, and iron. More scientifically vigorous trials are needed to ascertain the reported properties of many of these antioxidants and their extracts as well as the properties of agents that increase oxidative stress (pro-oxidants) and the development of cancers.

The series is designed for dietitians and nutritionists, food scientists, as well as health care workers and research scientists. Contributions are from leading national and international experts including those from world renowned institutions.

Professor Victor R. Preedy,
King's College, London

List of Contributors

Lucas Aidukaitis, CNA Brigham Young University, Department of Microbiology and Molecular Biology, Provo, UT, USA

Jennifer L. Allensworth, PhD Department of Surgery, Division of Surgical Sciences and Duke Cancer Institute, Duke University Medical Center, Durham, NC, USA

B. Andallu, MSc, PhD Food Science and Nutrition Division, Sri Sathya Sai Institute of Higher Learning, Anantapur Campus, Anantapur, Andhra Pradesh, India

Farrukh Aqil, PhD James Graham Brown Cancer Center and Department of Medicine, University of Louisville, Louisville, KY, USA

Vipin Arora, M.Pharm, PhD University Institute of Pharmaceutical Sciences, UGC-CAS, Panjab University, Chandigarh, India

Khaled Aziz, MD Medical Scientist Training Program, Mayo Graduate School, Mayo Clinic, College of Medicine, Rochester, MN, USA

Yasutaka Baba, MD, PhD Department of Radiology, Graduate School of Medical and Dental Sciences, Kagoshima University, Kagoshima-city, Kagoshima, Japan

Yun-Jung Bae Department of Food and Nutritional Sciences, Hanbuk University, Gyeonggi, Korea

Ankita Baveja, M.Pharm University Institute of Pharmaceutical Sciences, UGC-CAS, Panjab University, Chandigarh, India

Marco Bisoffi, PhD Chapman University Schmid College of Science and Technology, Biological Sciences, Biochemistry and Molecular Biology, Orange, CA, USA and University of New Mexico Health Sciences Center, Department of Biochemistry and Molecular Biology, School of Medicine Albuquerque, NM, USA

Robert Burky, MS UCLA David Geffen School of Medicine, Department of Obstetrics and Gynecology, Los Angeles, CA, USA

David Bynum School of Natural Sciences, Fairleigh Dickinson University, Teaneck, NJ, USA

Gloria M. Calaf, PhD Instituto de Alta Investigación, Universidad de Tarapaca, Arica, Chile and Center for Radiological Research, Columbia University Medical Center, New York, NY, USA

Rosa A. Canuto, MD Department of Clinical and Biological Sciences, University of Turin, Turin, Italy

Maria G. Catalano, MD Department of Medical Sciences, University of Turin, Turin, Italy

Kanishka Chakraborty, MD Department of Internal Medicine, Division of Oncology, James H. Quillen College of Medicine, East Tennessee State University, Johnson City, TN, USA

Yin-Chiu Chen Institute of Environmental and Occupational Health Sciences, School of Medicine, National Yang-Ming University, Taipei, Taiwan

Rong-Jane Chen, PhD Department of Environmental and Occupational Health, National Cheng Kung University Medical College, Tainan, Taiwan, Graduate Institute of Clinical Medicine, Taipei Medical University, Taipei, Taiwan

Chin-Wen Chi, PhD Department of Medical Research and Education, Taipei Veterans General Hospital and Institute of Pharmacology, School of Medicine, National Yang-Ming University, Taipei, Taiwan

Kanwaljit Chopra, M.Pharm., PhD, MNASc University Institute of Pharmaceutical Sciences, UGC-CAS, Panjab University, Chandigarh, India

Raffaella Coccia, PhD Department of Biochemical Sciences, Sapienza University of Rome, Rome, Italy

Joshua Cohen, MD UCLA David Geffen School of Medicine, Department of Obstetrics and Gynecology, Los Angeles, CA, USA

Ana Cruz, MS UCLA David Geffen School of Medicine, Department of Obstetrics and Gynecology, Los Angeles, CA, USA

Sreemanti Das Cytogenetics and Molecular Biology Laboratory, Department of Zoology, University of Kalyani, Kalyani, India

Palika Datta, PhD Department of Biomedical Sciences and Cancer Biology Center, Texas Tech University of Health Sciences Center, Amarillo, TX, USA

Cristian Del Bo', MS Università degli Studi di Milano, DeFENS, Department of Food, Environmental and Nutritional Sciences, Division of Human Nutrition, Milan, Italy

Gayathri R. Devi, PhD Department of Surgery, Division of Surgical Sciences and Duke Cancer Institute, Duke University Medical Center, Durham, NC, USA

Myron K. Evans, MD Department of Surgery, Division of Surgical Sciences and Duke Cancer Institute, Duke University Medical Center, Durham, NC, USA

Maurizio Fadda Department of Clinical Nutrition, San Giovanni Battista Hospital, Turin, Italy

Alexandra M. Fajardo, PhD University of South Alabama Mitchell Cancer Institute, Mobile, AL, USA

Robin Farias-Eisner, MD, PhD UCLA David Geffen School of Medicine, Department of Obstetrics and Gynecology, Los Angeles, CA, USA

Concetta Finocchiaro Department of Clinical Nutrition, San Giovanni Battista Hospital, Turin, Italy

Cesira Foppoli CNR Institute of Molecular Biology and Pathology, Sapienza University of Rome, Rome, Italy

Alexandros G. Georgakilas, PhD Physics Department, School of Applied Mathematics and Physical Sciences, National Technical University of Athens (NTUA), Zografou 15780, Athens, Greece

Yolanda Gilaberte, MD, PhD Department of Dermatology, San Jorge Hospital, Huesca, Spain

Salvador Gonzalez, MD, PhD Dermatology Service, Memorial Sloan-Kettering Cancer Center, New York, NY, USA

Luis Goya, PhD Department of Metabolism and Nutrition, Instituto de Ciencia y Tecnologia de Alimentos y Nutricion (ICTAN-CSIC), Ciudad Universitaria, Madrid, Spain

Ramesh C. Gupta, PhD James Graham Brown Cancer Center and Department of Pharmacology and Toxicology, University of Louisville, Louisville, KY, USA

Chris Hamilton Brigham Young University, Department of Microbiology and Molecular Biology, Provo, UT, USA

Vasiliki I. Hatzi, PhD Laboratory of Health Physics & Environmental Health, Institute of Nuclear Technology & Radiation Protection, National Center for Scientific Research "Demokritos," 153 10 Aghia Paraskevi, Athens, Greece

Sadao Hayashi, MD, PhD Department of Radiology, Graduate School of Medical and Dental Sciences, Kagoshima University, Kagoshima-city, Kagoshima, Japan

Charles Hummel, MD, PhD UCLA David Geffen School of Medicine, Department of Obstetrics and Gynecology, Los Angeles, CA, USA

Jeyaprakash Jeyabalan James Graham Brown Cancer Center, University of Louisville, Louisville, KY, USA

Thwisha Joshi James Graham Brown Cancer Center, University of Louisville, Louisville, KY, USA

Wei Sheng Joshua Loke Inflammation and Infection Research Centre, Faculty of Medicine, University of New South Wales and Department of Respiratory Medicine, Prince of Wales Hospital, Randwick, Sydney NSW 2031, Australia

Angeles Juarranz, MD, PhD Biology Department, Sciences School, Universidad Autónoma de Madrid, Madrid, Spain

Daehee Kang, MD, PhD Department of Preventive Medicine, Seoul National University College of Medicine, Seoul, Republic of Korea

Anisur Rahman Khuda-Bukhsh, PhD Cytogenetics and Molecular Biology Laboratory, Department of Zoology, University of Kalyani, Kalyani, India

Koyamangalath Krishnan, MD, MRCP, FACP Department of Internal Medicine, Division of Oncology, James H. Quillen College of Medicine, East Tennessee State University, Johnson City, TN, USA

Anurag Kuhad, M.Pharm., Ph.D, MNASc University Institute of Pharmaceutical Sciences, UGC-CAS, Panjab University, Chandigarh, India

Sang-Ah Lee, PhD Department of Preventive Medicine, Kangwon National University School of Medicine, Gangwon-do, Republic of Korea

Craig R. Lewis, MMed, FRACP Department of Medical Oncology, Prince of Wales Hospital, Randwick, Sydney NSW 2031, Australia

Mann Ying Lim Inflammation and Infection Research Centre, Faculty of Medicine, University of New South Wales and Department of Respiratory Medicine, Prince of Wales Hospital, Randwick, Sydney NSW 2031, Australia

Pingguo Liu, MD Brigham Young University, Department of Microbiology and Molecular Biology, Provo, UT, USA

Marina Maggiora Department of Clinical and Biological Sciences, University of Turin, Turin, Italy

Olga A. Martin, PhD Department of Radiation Oncology & Laboratory of Molecular Radiation Biology, Peter MacCallum Cancer Centre, and The Sir Peter MacCallum Department of Oncology, University of Melbourne, Melbourne, VIC, Australia

María Angeles Martín, PhD Department of Metabolism and Nutrition, Instituto de Ciencia y Tecnologia de Alimentos y Nutricion (ICTAN-CSIC), Ciudad Universitaria, Madrid, Spain

Sudhir Mehrotra, PhD Department of Biochemistry, Lucknow University, Lucknow, India

Radha Munagala, PhD James Graham Brown Cancer Center and Department of Medicine, University of Louisville, Louisville, KY, USA

Giuliana Muzio, PhD Department of Clinical and Biological Sciences, University of Turin, Turin, Italy

Seiji Naito, MD, PhD Department of Urology, Graduate School of Medical Sciences, Kyushu University, Fukuoka, Japan

Masayuki Nakajo, MD Department of Radiology, Graduate School of Medical and Dental Sciences, Kagoshima University, Kagoshima-city, Kagoshima, Japan

Toshihiro Nishizawa Division of Gastroenterology, National Hospital Organization Tokyo Medical Center, Tokyo, Japan

Somaira Nowsheen Medical Scientist Training Program, Mayo Graduate School, Mayo Clinic, College of Medicine, Rochester, MN, USA

Kim O'Neill, PhD Brigham Young University, Department of Microbiology and Molecular Biology, Provo, UT, USA

Beata Olas, PhD Department of General Biochemistry, Faculty of Biology and Environmental Protection, University of Lodz, Lodz, Poland

Concepción Parrado, MD Department of Histology and Pathology, Faculty of Medicine, University of Málaga, Spain

Marzia Perluigi, PhD Department of Biochemical Sciences, Sapienza University of Rome, Rome, Italy

Neena Philips, PhD School of Natural Sciences, University College, Fairleigh Dickinson University, Teaneck, NJ, USA

Kartick C. Pramanik, PhD Department of Biomedical Sciences and Cancer Biology Center, Texas Tech University of Health Sciences Center, Amarillo, TX, USA

C.U. Rajeshwari Food Science and Nutrition Division, Sri Sathya Sai Institute of Higher Learning, Anantapur Campus, Anantapur, Andhra Pradesh, India

Sonia Ramos, PhD Department of Metabolism and Nutrition, Instituto de Ciencia y Tecnologia de Alimentos y Nutricion (ICTAN-CSIC), Ciudad Universitaria, Madrid, Spain

Victoria Palau Ramsauer, PhD Department of Pharmaceutical Sciences, Bill Gatton College of Pharmacy, East Tennessee State University, Johnson City, TN, USA

Patrizia Riso, PhD Università degli Studi di Milano, DeFENS, Department of Food, Environmental and Nutritional Sciences, Division of Human Nutrition, Milan, Italy

Richard Robison, PhD Brigham Young University, Department of Microbiology and Molecular Biology, Provo, UT, USA

Anand Kamal Sachdeva, M.Pharm University Institute of Pharmaceutical Sciences, UGC-CAS, Panjab University, Chandigarh, India

Santu Kumar Saha Cytogenetics and Molecular Biology Laboratory, Department of Zoology, University of Kalyani, Kalyani, India

Scott J. Sauer, PhD Department of Surgery, Division of Surgical Sciences and Duke Cancer Institute, Duke University Medical Center, Durham, NC, USA

Marina Schena, MD Department of Oncology - San Giovanni Battista Hospital, Turin, Italy

Masaki Shiota, MD, PhD Department of Urology, Graduate School of Medical Sciences, Kyushu University, Fukuoka, Japan

R.I. Shobha, MSc, MPhil Food Science and Nutrition Division, Sri Sathya Sai Institute of Higher Learning, Anantapur Campus, Anantapur, Andhra Pradesh, India

Inder P. Singh, MSc, PhD National Institute of Pharmaceutical Education and Research, Punjab, India

Prathistha Singh, M.Pharm University Institute of Pharmaceutical Sciences, UGC-CAS, Panjab University, Chandigarh, India

Halyna Siomyk, MSc School of Natural Sciences, Fairleigh Dickinson University, Teaneck, NJ, USA

Shankar Siva, MBBS, FRANZCR Department of Radiation Oncology, Peter MacCallum Cancer Centre and The Sir Peter MacCallum Department of Oncology, University of Melbourne, VIC, Australia

Shunro Sonoda, MD Southern Region Hospital, Makurazaki-city, Kagoshima, Japan

Sanjay K. Srivastava, PhD Department of Biomedical Sciences and Cancer Biology Center, Texas Tech University of Health Sciences Center, Amarillo, TX, USA

William Stone, PhD Department of Pediatrics, James H. Quillen College of Medicine, East Tennessee State University, Johnson City, TN, USA

Mi-Kyung Sung, PhD Department of Food and Nutrition, Sookmyung Women's University, Seoul, Korea

Ming-Ta Sung Department of Medical Research and Education, Taipei Veterans General Hospital, Taipei, Taiwan

Hidekazu Suzuki, MD, PhD, FACG Division of Gastroenterology and Hepatology, Department of Internal Medicine, Keio University School of Medicine, Tokyo, Japan

Paul S. Thomas, MD, FRCP Inflammation and Infection Research Centre, Faculty of Medicine, University of New South Wales and Department of Respiratory Medicine, Prince of Wales Hospital, Randwick, Sydney NSW 2031, Australia

Nanako Tosuji Kagoshima University, Kagoshima-city, Kagoshima, Japan

Stefano Vendrame, MS Università degli Studi di Milano, DeFENS, Department of Food, Environmental and Nutritional Sciences, Division of Human Nutrition, Milan, Italy, Department of Food Science and Human Nutrition, University of Maine, Orono, ME, USA

Ying-Jan Wang Department of Environmental and Occupational Health, National Cheng Kung University Medical College, Tainan, Taiwan

Matthew White, BS UCLA David Geffen School of Medicine, Department of Obstetrics and Gynecology, Los Angeles, CA, USA

Akira Yokomizo, MD, PhD Department of Urology, Graduate School of Medical Sciences, Kyushu University, Fukuoka, Japan

OXIDATIVE STRESS AND CANCER

The Role of Oxidative Stress in Breast Cancer

Gayathri R. Devi, Jennifer L. Allensworth, Myron K. Evans, Scott J. Sauer

Department of Surgery, Division of Surgical Sciences and Duke Cancer Institute, Duke University Medical Center, Durham, NC, USA

List of Abbreviations

8-OHdG 8-Hydroxydeoxyguanosine
γ-GCL Gamma-glutamylcysteine synthetase
γ-GCS Gamma-glutathione synthase
Akt Protein kinase B
AP-1 Activator protein-1
ARE Antioxidant response element
Bax Bcl-2-associated X protein
Bcl-2 B-cell lymphoma 2
Bcl-xl B-cell lymphoma-extra large
BRCA1 Breast cancer susceptibility gene 1
BSO Buthionine sulfoximine
CA9 Carbonic anhydrase IX
c-Abl Abelson murine leukemia viral oncogene homolog
CD Conjugated dienes
cFLIP Cellular FLICE inhibitory protein
cIAP Cellular inhibitor of apoptosis
COX2 Cyclo-oxygenase 2
CYP Cytochrome P450
CYP1B1 CYP family 1, subfamily B, polypeptide 1
CYP1A1/2 CYP family 1, subfamily A, polypeptide 1/2
DNMT DNA methyltranserase
EGFR Epidermal growth factor receptor
eNOS Endothelial nitric oxide synthases
EPO Erythropoietin
ER Estrogen receptor
ERE Estrogen response elements
ERK Extracellular-signal-regulated kinase
GLUT1 Glucose transporter 1
GPCR G-protein coupled receptor
GPx Glutathione peroxidase
GSH Glutathione
GSK3β Glycogen synthase kinase 3β
GSR Glutathione reductase
GSS Glutathione synthetase
GSSG Glutathione disulfide
GST Glutathioine S-transferase
GSTP Glutathione-S-transferase P1
HDAC1 Histone deacetylase 1
HER2 Human epidermal growth factor receptor 2
HIF-1 Hypoxia inducible factor 1
HO-1 Heme oxygenase 1
HRE Hypoxia response elements
IGFBP5 Insulin-like growth factor-binding protein 5

IHC Immunohistochemistry
IKK IκB Kinase
IL-6 Interleukin-6
iNOS Inducible nitric oxide synthases
Keap1 Kelch-like ECH-associated protein 1
LDH-A Lactate dehydrogenase-A
LOH Loss of heterozygosity
LOOH Lipid hydroperoxides
MAPK Mitogen-activated protein kinase
MAPKKK MAPK kinase kinase
MDA Malondialdehyde
MMP Matrix metalloproteinase
mETC Mitochondrial electron transport chain
NDRG1 N-Myc downstream-regulated 1
NF-κB Nuclear factor kappa-light-chain-enhancer of activated B cells
nNOS Neuronal nitric oxide synthases
NOS Nitric oxide synthases
NQO1 NADH quinone oxidoreductase 1
Nrf2 Nuclear factor (erythroid-derived 2)-like 2
NSAID Nonsteroidal anti-inflammatory drug
PGE2 Prostaglandin 2
PI3K Phosphatidylinositide 3-kinase
PR Progesterone receptor
Prx Peroxiredoxins
PTEN Phosphatase and tensin homolog
Ras Rat sarcoma protein
RNS Reactive nitrogen species
ROS Reactive oxygen species
RTK Receptor tyrosine kinase
SERM Selective estrogen receptor modulator
siRNA Small interfering RNA
Smac(Diablo) Second mitochondria-derived activator of caspases
SOD Superoxide dismutase
Src Sarcoma protein
TAK1 Transforming growth factor-β activated kinase-1
tBHQ tert-Butylhydroquinone
TERT Telomerase reverse transcriptase protein
TNBC Triple negative breast cancer
TNFα Tumor necrosis factor α
TRAF TNF receptor associated factor
TRAIL TNF-related apoptosis-inducing ligand
Trx Thioredoxins
VEGF Vascular endothelial growth factor
XIAP X-linked inhibitor of apoptosis protein

© 2014 Elsevier Inc. All rights reserved.

INTRODUCTION

Redox Homeostasis: ROS Production and Elimination

Reactive species, also termed oxidants, are byproducts of key aerobic cellular processes of respiration, metabolism, and the mitochondrial electron transport chain (mETC),[1,2] and are removed continuously by an array of antioxidant mechanisms. These species include reactive oxygen species (ROS) and reactive nitrogen species (RNS). ROS are mainly comprised of neutral molecules (H_2O_2), radicals (hydroxyl radicals), and ions (superoxide).[3] On the other hand, nitric oxide, the main form of RNS in the cell, is produced by a family of enzymes (nitric oxide synthases, NOSs) that include iNOS (inducible), eNOS (endothelial), and nNOS (neuronal).[4] ROS can also be produced at somewhat low levels in response to the activation of certain signaling pathways, such as the epidermal growth factor receptor (EGFR) pathway.[5] Activation of these pathways has been shown to be important for proliferation, as well as the oncogenic and metastatic potential of cancer cells. Extracellular sources of ROS include tobacco, smoke, drugs, xenobiotics, radiation, and high levels of heat, most of which either activate a stress response or directly damage cellular components leading to ROS production.[6]

Cells have natural defense systems against ROS that consist of antioxidant enzymes and scavengers. Some of these antioxidants are produced inside cells and the human body, mostly falling into the enzymatic category, as they are predominantly protein in nature. These proteins include the superoxide dismutase (SOD) enzymes (which have differential subcellular localization and dismute superoxide to H_2O_2), glutathione peroxidase (GPx) and catalase (both of which clear peroxide), thioredoxins (Trxs; reduce oxidized proteins), and glutathione synthetase (GSS; synthesizes glutathione [GSH], an important antioxidant), among others.[1,7] Antioxidant scavengers are mostly obtained from nutritional sources and include ascorbic acid (vitamin A), tocopherol (vitamin E), polyphenols, carotenoids, and uric acid.[8]

Therefore, a fine balance exists between the levels of ROS and antioxidants within the cell. Oxidative stress occurs when the level of ROS exceeds the cellular antioxidant capacity either due to increased ROS production and/or impairment of the antioxidant capacity of the cells.[1] This stress promotes damage to key cellular structures including DNA, proteins, and lipids, which play a pivotal role in the development of multiple types of cancer.[9] Expression of oncogenes (e.g., Ras, myc, telomerase) and loss of tumor suppressor genes (p53, p21, PTEN) can also increase ROS, leading to senescence or escape from apoptosis.[10–12] Oxidative stress can cause arrest or induction of transcription, activation of signaling pathways, and genomic instability, which are all hallmarks of cancer (including breast cancer) and are key factors that modulate cancer cell proliferation, evasion of apoptosis, angiogenesis, and metastasis.[9]

OXIDATIVE STRESS AS A DRIVER OF BREAST CANCER DEVELOPMENT AND PROGRESSION

ROS as Second Messengers in Breast Cancer

The role of oxidative stress in the etiology of breast cancer is supported by multiple lines of evidence.[13] Although ROS are generally thought of as damaging to cells due to their ability to induce oxidative stress at high concentrations, low levels of ROS are actually essential to normal cell function. This is in part due to the fact that ROS can act as second messengers in signaling cascades that are vital for cellular responses to external stimuli. To be characterized as a second messenger, the molecule must (1) exhibit concentration control at the level of synthesis and removal, (2) exhibit effector molecule specificity, and (3) take part in a reversible signaling interaction.[14] Due to very high reactivities that preclude substrate specificity, superoxides, hydroxyl radicals, and singlet oxygen are not considered to be second messengers of signaling.[15] However, the enzymatic production and degradation of H_2O_2, along with its preferential reactivity with protein thiols, which are reversibly oxidized, allows for its characterization as a second messenger.[15] It is known that levels of ROS are often upregulated in cancer cells, and their role in promoting certain signaling cascades is likely one reason that this adaptation is advantageous.[16] Figure 1.1 summarizes the well-characterized signaling effects of ROS in breast cancer and highlights the role of ROS in regulating growth factor receptor signaling, epithelial-mesenchymal transition, and stem cell-like phenotype in breast cancer.

Mutation and Inactivation of Antioxidants

As mentioned earlier there exists a balance between ROS and cellular antioxidants, and alterations in the genes that encode certain antioxidants are associated with increased proliferation and progression of cancer.[16] These alterations in antioxidant genes can be either gain of function or loss of function depending on the cell type, gene, and function of that gene in the context of cancer. In human breast cancer patients there is a multitude of contrasting data; however, it has been posited that during progression of cancer, low levels of SOD2 lead to increased ROS and a significant accumulation of mutations, while in late stages SOD2 is increased to combat ROS and promote carcinogenesis.[17] Robinson

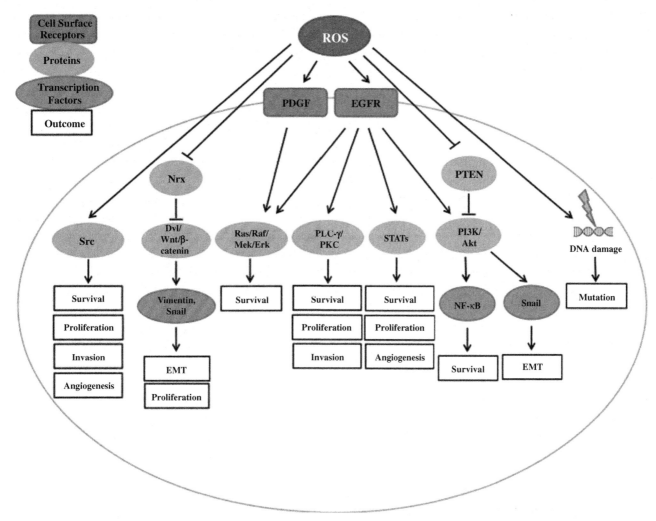

FIGURE 1.1 ROS Drives Oncogenesis. ROS promotes the development of breast cancer through activation of cell signaling pathways that promote survival, proliferation, invasion, angiogenesis, and EMT including Src,[40] Wnt/β-catenin,[70] Ras/Raf/Mek/ERK,[71] EGFR,[72] and PI3K/Akt;[39] additionally, DNA damage promotes the acquisition of mutations.[46]

et al. reported an inverse correlation between GPx levels and cancer progression.[18] A loss of heterozygosity (LOH) on chromosome 3p, where the GPx gene is located, has been frequently found in breast cancer.[19] It was previously reported that GPx expression inversely correlates with estrogen receptor status in breast cancer cell lines; however, a follow-up study using additional cell lines weakened this proposed correlation, and ER status is therefore not considered a good surrogate marker for GPx expression.[20]

Tumor Hypoxia, HIF-1, and Oxidative Stress in Breast Cancer

Due to immature vascularization, areas of solid breast tumors often have inadequate blood supply. The level of tumor oxygenation in breast tumors is typically half that of normal tissues, with 30 to 40% of breast cancer tissue having a quarter of the amount of oxygenation as normal

tissue.[21] Further, hypoxia can induce a quiescent state in tumor cells, which makes them less sensitive to chemotherapy designed to target rapidly dividing cells.[21] One of the most important mediators of the cellular hypoxic response is the transcription factor hypoxia inducible factor 1 (HIF-1). HIF-1 expression has been correlated with aggressive breast cancer and poor response to treatment.[22] Through changes in HIF-1 levels, hypoxic gene activation can occur through HIF-1 binding to hypoxia response elements (HREs).[21] This upregulation of HIF-1 has been shown to increase breast cancer cell proliferation and p53 accumulation,[23] and correlate with an increase in oxidative stress and production of vascular endothelial growth factor (VEGF) through the HRE, inducing production of blood vessel growth within the tumor and heightening the risk for metastasis.[24] Like VEGF, increased expression of another angiogenesis protein, erythropoietin, in breast cancer is HIF-1 dependent.[25] Additionally, the breast cancer-related oncogene

HER2 has been shown to promote metastasis during hypoxia by increasing resistance to anoikis through a HIF-1-mediated mechanism.[26]

Further, regulation of tumor metabolism in hypoxic cancer cells is mediated predominantly by HIF-1, wherein glucose transporter 1 (GLUT1), a protein that facilitates cellular glucose uptake, is increased in a HIF-1-dependent manner in breast cancer, increasing the dependence of the cell on glycolysis for energy.[27] Expression of mRNA levels of lactate dehydrogenase-A (*LDH-A*), involved in the glycolytic pathway and under hypoxic control through HIF-1, was reported to be markedly increased in breast cancer, along with a modest increase in activity.[28] HIF-1 has also been shown to increase levels of mRNA for *CA9* (carbonic anhydrase IX), which is involved in proliferation and the neutralization of hypoxia-induced pH increases through increased glycolysis, *NDRG1* (N-myc downstream-regulated 1), a stress-related gene involved with differentiation and *IGFBP5* (insulin-like growth factor-binding protein 5), in breast cancers.[29]

Inflammation and Oxidative Stress in Breast Cancer

The presence of persistent free radicals during oxidative stress leads to induction of a chronic inflammatory response and evidence of this inflammation in breast tissue has been found through the increase in levels of tumor necrosis factor α (TNFα) due to infiltrating macrophages, dysregulated interleukin-6 (IL-6) production and upregulation of inflammatory enzymes such as cyclo-oxygenase 2 (COX2).[30] These features lead to poor prognosis and drug response, as well as increased metastasis.

Increased COX2 production in breast cancer leads to higher production of prostaglandin 2 (PGE2). This stimulation of PGE2 production can then also lead to downstream signaling through MAPK, Src, and Akt pathways, as well as VEGF and HIF-1α.[31] Particularly interesting is the effect of PGE2 on HIF1α, as the hypoxic environment of the tumor can give rise to inflammation and inflammation can affect HIF1α, thereby providing a potential link for cross-talk between hypoxia and inflammation in the tumor. PGE2-induced activation of these pathways results in the progression of breast cancer.[32] Further evidence of COX2's role in breast cancer is the finding that non-steroidal anti-inflammatory drugs (NSAIDs), which inhibit COX2, can reduce the risk of breast cancer.[33] Despite the diverse mechanisms of action of these inflammatory molecules in breast cancer, one potential point of convergence is that all of these pathways can modulate the production of aromatase.[34] Aromatase is vital in the production of estrogens, a critical factor in the etiology and progression of breast cancer.

Estrogens and Oxidative Stress in Breast Cancer

For over 100 years, a link between breast cancer and estrogen has been acknowledged, with current data strongly supporting this idea.[35] Estrogens and estrogenic quinone-metabolites, as shown in Figure 1.2, can act as ROS themselves and alkylate or damage DNA and proteins or bind to estrogen receptors (ER) and activate estrogen response elements (EREs), which in turn can increase levels of ROS. This is supported by the observation that estrogen-responsive cells like MCF-7 show increased mitochondrial membrane potential, increased ROS production, and resultant compensatory changes in antioxidants.[36] Further, the higher the ERα/ERβ ratio in breast cancer cells, the higher the levels of ROS generated by estrogen.[37]

Lifestyle, diet, and environment influence oxidative stress levels and are linked to breast cancer initiation and disease progression. Of particular interest is the correlation between obesity and increased breast cancer risk in postmenopausal women. One potential cause of this risk factor is attributed to higher circulating estrogens caused by increased levels of adipose tissue, which can then drive the growth of ER-dependent tumors in postmenopausal women.[38] Further, aromatase expression was found to be higher in obese women with breast cancer and this correlated with markers of inflammation, such as COX-2 and PGE2.[38] Thus, a link between oxidative stress, obesity, and breast cancer can be drawn, with estrogen as the point of convergence between these complex processes as shown in Figure 1.2.

Age-Related Changes and Oxidative Stress in Breast Cancer

Telomeres are chromosomal end-caps that are shortened throughout the cellular aging process, and when telomeres shorten enough, cellular senescence occurs. Indeed, aging is associated with increased oxidative stress[39] and higher levels of oxidative stress have been observed to increase the rate of telomere shortening.[40] While telomerase may slow the effects of cellular aging, uncontrolled telomerase activity allows cells to replicate indefinitely, leading to the survival of cells with potentially carcinogenic mutations[41] and providing a mechanism by which cancer cells survive. This is also supported by a recent finding wherein increased ROS was identified to induce transport of the telomerase reverse transcriptase protein (TERT) from the nucleus to the mitochondria, which then prevents nuclear DNA damage and cell death, a mechanism of cancer cell survival and resistance.[42] Together, these data reveal an interesting interplay between telomerase, natural aging-related oxidative stress in cells, and implications in cancer therapy as a potential factor in age-related differences in therapeutic outcomes in breast cancer.[43]

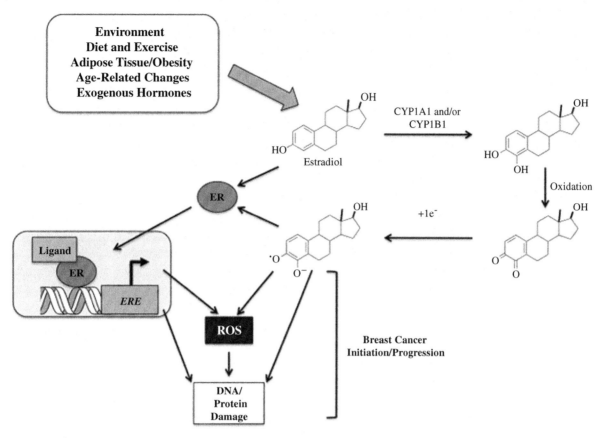

FIGURE 1.2 **Estrogens and Breast Cancer Initiation and Progression.** A variety of factors can impact the levels of estrogens in women.[35] Estrogens can then go on to bind estrogen receptor to promote expression of genes with estrogen response elements, which can induce increased levels of ROS and subsequently DNA damage and carcinogenesis.[73] Further, estrogens can form radical species through cytochrome p450 metabolism[73] that can either bind estrogen receptor or induce DNA damage directly.[74]

It has also been seen that increased age correlates with an increase in the number of mitochondrial DNA (mtDNA) mutations, which leads to age-related mitochondrial dysfunction due to increased oxidative stress in older individuals.[44] It has been observed that depletion of mtDNA-encoded genes can cause carcinogenesis in breast epithelial cells, mtDNA site mutations are associated with increased breast cancer risk, and there is a higher frequency of mtDNA mutations observed in breast cancer tissue.[44] Interestingly, levels of oxidative stress have not been found to increase as a result of increased mtDNA mutations *in vivo*, indicating that oxidative stress may not mediate the carcinogenesis found with mtDNA mutations.[45] However, the higher frequency of mtDNA mutations as a result of aging may be one reason for a greater than eight-fold likelihood of being diagnosed with cancer at age 70 than age 30.[38]

Epigenetics and Oxidative Stress in Breast Cancer

While oxidative stress is known to cause genetic changes through direct DNA damage, it can also cause epigenetic changes leading to cancer initiation. ROS can cause an increase in DNA methyltransferase (DNMT) levels directly, leading to increased DNA methylation, or ROS-dependent DNMT increases can be due to other redox-sensitive factors, such as Ras or PI3K/Akt activation (Figure 1.3 [A–C]). ROS can also directly cause DNA methylation on redox-sensitive promoters, such as the one found on the gene encoding for p16, a tumor suppressor gene (Figure 1.3 [D]). Additionally, it has been found that DNA damage caused by ROS can lead to DNA lesions, such as 8-hydroxyguanine, O6-methylguanine and single-stranded DNA breaks, which in turn induce global hypomethylation through a lower interaction with DNMTs.[46] Oxidative stress can cause genomic instability and tumor formation through hypomethylation of both satellites and interspersed repeat sequences.[46] Recruitment of DNMTs to promoter regions has also been associated with recruitment of HDACs, which cause epigenetic modifications through remodeling of the chromatin structure of DNA. This modulates the accessibility of chromosomal loci, affecting the level of translation of certain genes.[46]

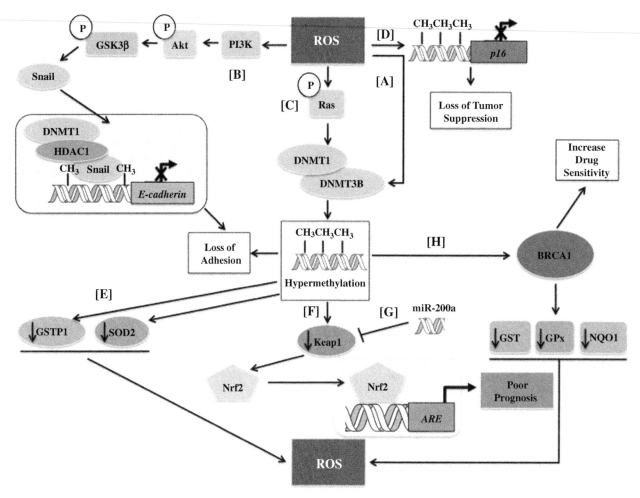

FIGURE 1.3   Epigenetics, ROS, and Breast Cancer. [A] ROS can directly induce epigenetic changes through an increase in DNA methyltransferase 3b (dnmt3b) levels.[46] **[B]** Activation of the PI3K/Akt pathway by ROS can lead to increased levels of Snail, inducing DNA methylation of the E-cadherin promoter, decreasing E-cadherin levels and leading to tumorigenic loss of cellular adhesion.[46] **[C]** Activation of the Ras pathway by ROS leads to increased dnmt expression, causing a global increase in DNA methylation.[46] **[D]** ROS can lead to inactivation of the tumor suppressor gene *p16*, promoting tumorigenesis.[75] **[E]** Hypermethylation of the promoters of antioxidants glutathione-*S*-transferase P1 (GSTP1)[76] and superoxide dismutase 2 (SOD2)[77] has been observed in a number of breast cancer patients, which can lead to an increase in cellular reactive oxygen species (ROS). **[F]** Dysregulated methylation of Keap1 leads to breast tumor formation and increased methylation of Keap1 is associated with poor prognosis in triple-negative breast cancer.[78] Keap1 is a negative regulator of Nrf2, which can bind to antioxidant response elements in the promoter region of antioxidant genes and promote their expression. With increased Keap1 methylation, Nrf2 can activate antioxidant response elements, promoting antioxidant expression and leading to poor outcomes in triple-negative breast cancer.[78] **[G]** Lower levels of miR-200a, which can degrade Keap1, are found in breast cancer, leading to higher levels of Keap1, blockade of Nrf2 and a decreased expression of antioxidants.[79] **[H]** BRCA1, a breast cancer susceptibility gene, is known to increase levels of antioxidants GST, GPx, and NQO1.[80] Hypermethylation of the BRCA1 promoter can lead to increased sensitivity to platinum-based chemotherapy,[81] most likely through a decrease in antioxidants leading to higher levels of therapy-derived ROS.

Breast Cancer Specific Epigenetic Modifications of Oxidative Stress Genes

Lower expression of antioxidant proteins known to be epigenetically regulated, such as glutathione-*S*-transferase *P1* (*GSTP1*) and SOD2, is seen in breast cancer tissue, and increased promoter methylation has been identified as an important mechanism of this lower expression (Figure 1.3 [E]). Decreased levels of these proteins can cause ROS upregulation, leading to further activation of the downstream epigenetic-ROS signaling cascade.

Another oxidative stress factor that is under epigenetic control in breast cancer is Keap1. Keap1 is a known sensor of oxidative stress and negative-regulator of Nrf2. In breast cancer, it was found that aberrant methylation of Keap1 can lead to breast carcinogenesis and that increased methylation in triple-negative breast cancer is correlated with a worse prognosis, potentially through promotion of survival signaling (Figure 1.3 [F]). Interestingly, Keap1 methylation in ER-positive tumors contributes to a better prognosis. It is posited that the induction of Nrf2 that occurs with the epigenetic silencing of

Keap1 leads to an increase in NQO1, which can prevent estrogen-mediated generation of ROS. Additionally, miR-200a, a micro RNA that is found to be epigenetically repressed in breast cancer, can negatively regulate Keap1, leading to Nrf2 activation (Figure 1.3 [G]).

The well-known breast cancer-susceptibility gene *BRCA1*, a marker for cancer development as well as chemosensitivity, is susceptible to both germline mutations and methylation. In fact, hypermethylation of this gene has been correlated with enhanced sensitivity to chemotherapy, better survival, and longer time to relapse (Figure 1.3 [H]). Further, BRCA1 expression can upregulate multiple genes related to ROS homeostasis, which can have profound effects on therapeutic outcomes in breast cancer patients discussed in the next section.

OXIDATIVE STRESS RESPONSE AND ADAPTATION MECHANISMS IN BREAST CANCER

In order to compensate for increased oxidative stress, cancer cells have been identified to garner redox adaptive mechanisms that enhance their ability to detoxify ROS; exposure to constant oxidative stress selects for cells that can adapt to these conditions through a number of mechanisms. The strongest and most clear evidence for redox adaptation in breast cancer is the concurrent elevation of tissue markers of oxidative stress and increased expression and activity of antioxidants in breast cancer tissue samples relative to their normal counterparts.

The glutathione system (GSH- reduced, GSSG-oxidized) is the most abundant redox buffer within the cell,[47] and any changes in the GSH to GSSG ratio will directly or indirectly affect various redox-sensitive cellular components. In a recent study, high GSH expression was associated with metastasis in breast cancer patients receiving chemotherapy.[48] Perry et al. noted that not only were GSH levels increased by two-fold in breast cancer tissue relative to normal breast specimens, but that tissue from lymph node metastases showed a four-fold increase in GSH over normal tissue. Similarly, breast cancer brain metastases, which rely heavily on oxidative phosphorylation for energy generation and thus produce high levels of ROS, showed significant upregulation of glutathione-associated enzymes including glutathione reductase (GSR) and GSTP1, which help them maintain a reduced cellular environment.[49] These observations indicate that redox adaptation is crucial for metastatic breast cancer, as metabolic pressures such as nutrient deprivation that are associated with a foreign environment can promote oxidative stress.[50] Interestingly, BRCA1 upregulates the expression of GSTs and promotes a reduced state within the cell by promoting an increase in the GSH:GSSG ratio.[51] Thus, while these functions protect nontransformed cells against oxidative stress

as a mechanism of cancer prevention, BRCA1 can promote a redox adaptive state in cells that have undergone carcinogenic transformation.

Another mechanism of redox adaptation is increased levels of SODs in breast cancer. *In vitro* studies have shown that overexpression of either SOD1 or SOD2 can inhibit breast cancer cell growth.[52] *In vivo*, infection of animals with an adenovirus expressing SOD1 or SOD2 decreased xenograft growth compared to controls.[52] Kattan et al. showed that there are differences in SOD2 expression between estrogen-dependent and estrogen-independent cancer cell lines, and that this expression regulates not only tumor cell growth and colony formation, but also doubling time, providing a link between SOD and the cell cycle.[53] A 2004 survey of breast cancer patients in Taiwan showed that SOD2, but not SOD1, expression was higher in cancer tissue than in malignancy-free tissue.[54] In studies conducted in our laboratory in two cellular models of aggressive breast cancer, redox-adapted populations of cells exposed to constitutive oxidative stress mediated by a small molecule dual kinase (ErbB1/2) inhibitor, lapatinib, exhibited enhanced expression of SOD1/2, overexpression of antiapoptotic XIAP, and increased GSH content relative to parental, lapatinib-sensitive cells.[55] These lapatinib-resistant cell lines were also cross-resistant to classical ROS-inducing treatments such as hydrogen peroxide and paraquat,[55] as well as other commonly used anticancer drugs including sunitinib, gefitinib, bleomycin, capecitabine,[56] and TRAIL.[57]

Indeed, identification of redox sensitive markers has been used by several groups to investigate the relationship between antioxidants and oxidative stress in breast cancer patient tissue. One such widely used marker is malondialdehyde (MDA), which has been observed to positively correlate with increases in SOD1/2 and GPx expression and activity in comparison to healthy breast tissue controls.[58–61] A similar study also observed increased SOD and GPx activity in breast cancer tissue relative to normal tissue, but they determined that MDA levels were slightly lower in the cancerous tissue, though serum levels of MDA were found to be elevated in breast cancer patients.[62] These studies also observed different results when measuring catalase activity, with some reporting an increase[60,61] while others found a reduction[58,59,62] in tumor tissue compared to healthy controls. Another study found that MDA, LOOH (lipid hydroperoxides), and CD (conjugated dienes) were elevated in breast cancer tissue, and that this correlated with increased expression of SOD, catalase, GSH, and GPx relative to uninvolved adjacent tissue.[63] Interestingly, the degree of increase in oxidative stress and antioxidant expression correlated with advanced disease; greater increases in both were observed in Stage III patient samples than in Stage I and II samples. Small sample size and the inherent heterogeneity of the disease likely played a role in these incongruous findings, but the common theme of enhanced antioxidant capacity in the presence of elevated oxidative stress remains consistent. Additionally, tumor stage may play a role in the finding that low antioxidant expression promotes oncogenesis, but tumor progression and increased ROS ultimately results in the need for upregulation of antioxidants to cope with increased oxidative stress.

Oxidative stress serves as a selective pressure that promotes the survival of cells with increased antioxidant capacity, activation of survival signaling, induction of antiapoptotic proteins, and alterations in drug metabolism, all of which contribute toward drug resistance mechanisms. Thus, redox adaption is not only involved in cancer progression and metastasis, but also in the development of drug resistance. Many common breast cancer therapies work through the generation of ROS and thus may be rendered ineffective in cell populations that have adapted to cope with oxidative stress. Chemotherapies widely used in breast cancer include anthracyclines (doxorubicin), taxanes (paclitaxel, docetaxel), alkylating agents and platinum compounds (cisplatin, carboplatin), as well as radiation therapy, and all of these agents rely heavily on the induction of oxidative stress-induced apoptosis for their antitumor activity.[16,24,64] We have summarized the studies in the past 10 years that highlight the importance of redox adaptive mechanisms in breast cancer progression and therapeutic resistance (Tables 1.1–1.3).

INTERESTING FINDINGS, LIMITATIONS, AND FUTURE DIRECTIONS

- The levels of antioxidants found in breast cancer patients vary, with low levels of antioxidants hypothesized during progression (as this increases reactive oxygen species and leads to mutations) and high levels in late stages (compensatory and promotes proliferation/carcinogenesis). Peroxiredoxin is found in higher levels in malignant tumors over normal tissue but no link between these increased levels and clinical features have been found.[65] It is unknown if activation of receptor tyrosine kinase/G-protein coupled receptor signaling by H_2O_2 is due to inhibition of protein tyrosine phosphatases, or if there is direct activation.[14] These studies are some examples to highlight the fact that oxidative stress pathways are complex and measurements are challenging. This is compounded by the instability of compounds, variability in assays, and a need for further development of newer assays that can allow for strong and reproducible

TABLE 1.1   Antioxidant Expression and Function as Redox-Adaptive Mechanisms in Breast Cancer

Sample Type	Description
Biochemical	GSH binds/inactivates metabolic intermediates of alkylating agents and platinum compounds, preventing cell damage [a]
MCF-7	GPx overexpression correlates with resistance to doxorubicin [b]
	SOD expression linked to doxorubicin- and radio-resistance [c]
SUM149, SUM190	Resistant cells had higher levels of SOD1/2 and GSH, relative to parental cells [d,e,f]
BCM2 Xenograft	Increased intracellular GSH associated with resistance to ROS-based drugs; brain metastatic cells 60-fold less sensitive to bortezomib due to treatment-mediated upregulation of GSH [g]
Patient (n = 63)	Increased Trx associated with low docetaxel response [h]
Patient (n = 44)	Trx, GST, Prx gene expression pattern may predict taxane response [i]
Patient (n = 63)	Increased GSH and GPx in hormone-negative tumors and expression predicted risk of metastasis [j]

[a] Manda G, Nechifor MT, Neagu T. Reactive Oxygen Species, Cancer and Anti-Cancer Therapies. Current Chemical Biology. 2009; 3: 342-66.

[b] Kalinina EV, Chernov NN, Saprin AN, Kotova YN, Andreev YA, Solomka VS, Scherbak NP. Changes in expression of genes encoding antioxidant enzymes, heme oxygenase-1, Bcl-2, and Bcl-xl and in level of reactive oxygen species in tumor cells resistant to doxorubicin. Biochemistry (Mosc). 2006; **71**(11): 1200-6.

[c] Guo G, Yan-Sanders Y, Lyn-Cook BD, Wang T, Tamae D, Ogi J, Khaletskiy A, Li Z, Weydert C, Longmate JA, Huang TT, Spitz DR, Oberley LW, Li JJ. Manganese superoxide dismutase-mediated gene expression in radiation-induced adaptive responses. Molecular and cellular biology. 2003; **23**(7): 2362-78.

[d] Aird KM, Allensworth JL, Batinic-Haberle I, Lyerly HK, Dewhirst MW, Devi GR. ErbB1/2 tyrosine kinase inhibitor mediates oxidative stress-induced apoptosis in inflammatory breast cancer cells. Breast cancer research and treatment. 2012; **132**(1): 109-19.

[e] Allensworth JL, Aird KM, Aldrich AJ, Batinic-Haberle I, Devi GR. XIAP Inhibition and Generation of Reactive Oxygen Species Enhances TRAIL Sensitivity in Inflammatory Breast Cancer Cells. Molecular cancer therapeutics. 2012.

[f] Williams KP, Allensworth JL, Ingram SM, Smith GR, Aldrich AJ, Sexton JZ, Devi GR. Quantitative high-throughput efficacy profiling of approved oncology drugs in inflammatory breast cancer models of acquired drug resistance and re-sensitization. Cancer letters. 2013; **17**(13): 00386-8.

[g] Chen EI, Hewel J, Krueger JS, Tiraby C, Weber MR, Kralli A, Becker K, Yates JR, 3rd, Felding-Habermann B. Adaptation of energy metabolism in breast cancer brain metastases. Cancer Res. 2007; **67**(4): 1472-86.

[h] Kim SJ, Miyoshi Y, Taguchi T, Tamaki Y, Nakamura H, Yodoi J, Kato K, Noguchi S. High thioredoxin expression is associated with resistance to docetaxel in primary breast cancer. Clinical cancer research: An official journal of the American Association for Cancer Research. 2005; **11**(23): 8425-30.

[i] Iwao-Koizumi K, Matoba R, Ueno N, Kim SJ, Ando A, Miyoshi Y, Maeda E, Noguchi S, Kato K. Prediction of docetaxel response in human breast cancer by gene expression profiling. J Clin Oncol. 2005; **23**(3): 422-31.

[j] Jardim BV, Moschetta MG, Leonel C, Gelaleti GB, Regiani VR, Ferreira LC, Lopes JR, de Campos Zuccari DA. Glutathione and glutathione peroxidase expression in breast cancer: An immunohistochemical and molecular study. Oncology reports. 2013.

TABLE 1.2 Modulation of Nrf2 and NF-κB Transcription Factors as Redox-Adaptive Mechanisms in Breast Cancer

Target	Sample Type	Description
Nrf2	Panel of cell lines	Increased expression led to higher GSH levels, increasing resistance to electrophilic drugs;[a,b] activation led to high GSH levels, increased NF-κB activity, and decreased ROS-mediated apoptosis;[c] overexpression causes chemoresistance;[d] Nrf2 inhibitor buthionine sulfoximine sensitized cells to paclitaxel;[e] siRNA knockdown of Nrf2 increased doxorubicin sensitivity [b]
	MCF-7	Silencing Nrf2 ubiquitin ligase, Cul3, resulted in resistance to oxidative stress by H_2O_2, paclitaxel, and doxorubicin[f]
		Activation upregulated γ-GCL, HO-1, Trx, and Prx and decreased ROS production [g]
NF-κB	MCF-7	Constitutive activation correlated with taxane- and radio-resistance; NF-κB inhibition reversed resistance [h,i]
	BT-474	NF-κB pathway activated by doxorubicin; NF-κB siRNA or inhibitory peptide reversed resistance [j]
	Cell lines, Patient (n = 439)	Activation associated with resistance to tamoxifen/aromatase inhibitors and decreased time to metastatic relapse despite adjuvant tamoxifen therapy [k]
	BT-474, Patient (n = 35)	Lapatinib induced NF-κB subunit RelA cytoprotective stress response; tumor biopsies show inverse correlation of p-RelA levels to lapatinib response [l]

[a] *Syed Alwi SS, Cavell BE, Donlevy A, Packham G. Differential induction of apoptosis in human breast cancer cell lines by phenethyl isothiocyanate, a glutathione depleting agent. Cell Stress Chaperones. 2012; **17**(5): 529-38. http://dx.doi.org/10.1007/s12192-012-0329-3. Epub 2012 Feb 17.*

[b] *Zhong Y, Zhang F, Sun Z, Zhou W, Li ZY, You QD, Guo QL, Hu R. Drug resistance associates with activation of Nrf2 in MCF-7/DOX cells, and wogonin reverses it by down-regulating Nrf2-mediated cellular defense response. Molecular carcinogenesis. 2012; **16**(10): 21921.*

[c] *Bellezza I, Mierla AL, Minelli A. Nrf2 and NF-κB and Their Concerted Modulation in Cancer Pathogenesis and Progression. Cancers. 2010; **2**(2): 483-97.*

[d] *Wang XJ, Sun Z, Villeneuve NF, Zhang S, Zhao F, Li Y, Chen W, Yi X, Zheng W, Wondrak GT, Wong PK, Zhang DD. Nrf2 enhances resistance of cancer cells to chemotherapeutic drugs, the dark side of Nrf2. Carcinogenesis. 2008; **29**(6): 1235-43.*

[e] *Ramanathan B, Jan KY, Chen CH, Hour TC, Yu HJ, Pu YS. Resistance to paclitaxel is proportional to cellular total antioxidant capacity. Cancer research. 2005; **65**(18): 8455-60.*

[f] *Loignon M, Miao W, Hu L, Bier A, Bismar TA, Scrivens PJ, Mann K, Basik M, Bouchard A, Fiset PO, Batist Z, Batist G. Cul3 overexpression depletes Nrf2 in breast cancer and is associated with sensitivity to carcinogens, to oxidative stress, and to chemotherapy. Mol Cancer Ther. 2009; **8**(8): 2432-40. http://dx.doi.org/10.1158/535-7163.MCT-08-1186. Epub 2009 Jul 28.*

[g] *Kim SK, Yang JW, Kim MR, Roh SH, Kim HG, Lee KY, Jeong HG, Kang KW. Increased expression of Nrf2/ARE-dependent anti-oxidant proteins in tamoxifen-resistant breast cancer cells. Free Radic Biol Med. 2008; **45**(4): 537-46. http://dx.doi.org/10.1016/j.freeradbiomed.2008.05.011. Epub May 24.*

[h] *Sprowl JA, Reed K, Armstrong SR, Lanner C, Guo B, Kalatskaya I, Stein L, Hembruff SL, Tam A, Parissenti AM. Alterations in tumor necrosis factor signaling pathways are associated with cytotoxicity and resistance to taxanes: A study in isogenic resistant tumor cells. Breast Cancer Res. 2012; **14**(1): R2.*

[i] *Guo G, Yan-Sanders Y, Lyn-Cook BD, Wang T, Tamae D, Ogi J, Khaletskiy A, Li Z, Weydert C, Longmate JA, Huang TT, Spitz DR, Oberley LW, Li JJ. Manganese superoxide dismutase-mediated gene expression in radiation-induced adaptive responses. Molecular and cellular biology. 2003; **23**(7): 2362-78.*

[j] *Tapia MA, Gonzalez-Navarrete I, Dalmases A, Bosch M, Rodriguez-Fanjul V, Rolfe M, Ross JS, Mezquita J, Mezquita C, Bachs O, Gascon P, Rojo F, Perona R, Rovira A, Albanell J. Inhibition of the canonical IKK/NF kappa B pathway sensitizes human cancer cells to doxorubicin. Cell Cycle. 2007; **6**(18): 2284-92. Epub 007 Jul 10.*

[k] *Zhou Y, Yau C, Gray JW, Chew K, Dairkee SH, Moore DH, Eppenberger U, Eppenberger-Castori S, Benz CC. Enhanced NF kappa B and AP-1 transcriptional activity associated with antiestrogen resistant breast cancer. BMC Cancer. 2007; **7**: 59.*

[l] *Xia W, Bacus S, Husain I, Liu L, Zhao S, Liu Z, Moseley MA, 3rd, Thompson JW, Chen FL, Koch KM, Spector NL. Resistance to ErbB2 tyrosine kinase inhibitors in breast cancer is mediated by calcium-dependent activation of RelA. Molecular cancer therapeutics. 2010; **9**(2): 292-9.*

correlations between oxidative stress and breast cancer clinical studies.[17]

- Role of estrogen in oxidative stress: Conflicting results show an inverse correlation between glutathione peroxidase expression and estrogen receptor (ER) status but little to no correlation when a larger number of cell lines are used.[20] Further, despite lower levels of circulating estrogen, aging women have a higher incidence of ERα–negative tumors than younger women, and have a better prognosis if they have triple-negative breast cancer.[43]

- Inflammatory signals: TNF-α has been shown to act as both antitumorigenic (inflammation can induce apoptosis and inhibit tumorigenesis) and pro-tumorigenic (damage of DNA, inhibition of DNA repair, autocrine production of growth/survival factors, matrix metalloproteinase-induced remodeling, and stimulation of NF-κB in resistant cell lines to increase resistance).[66] Further, although inflammatory molecules have varied mechanisms of action in breast cancer, modulation of aromatase seems to be a common point of convergence.[31]

CONCLUDING REMARKS

It is clear that there is an intricate cross-talk between signaling pathways that regulate antioxidant capacity, redox-sensitive transcription factors, cell survival/death signaling, and antiapoptotic proteins in oxidative stress

TABLE 1.3 Modulation of PI3K/AKT and ERK Signaling as Redox-Adaptive Mechanisms in Breast Cancer

Sample Type	Description
MCF-7	Tamoxifen resistant cells had increased p-ERK [a]
	Bcl-2 overexpression increased resistance to GSH modulatory agent neocarzinostatin [b]
	Introduction of constitutively active Ras or Akt conferred radioresistance to cells; inhibition of PI3K reversed Ras-mediated but not Akt-mediated radioresistance [c]
	Ectopic constitutively active Akt renders cells resistant to tamoxifen through increased p-NF-κB [d]
BT-474	Endogenous constitutive activation of PI3K/Akt pathway associated with radiation resistance [e]
SKBR-3, BT-474 cells, xenograft, patient samples (n = 84)	PTEN null cells resistant to trastuzumab *in vitro* and *in vivo* [f]
Patient samples (n = 252)	Increased p-Akt correlated with lower response to aromatase inhibitors or selective estrogen receptor modulators and was associated with worse disease-free survival in patients receiving hormonal therapy [g]
Patient samples (n = 109)	High ERK positivity correlated with anthracycline resistance and poor survival following relapse [h]
Patient samples (n = 886)	ER-positive breast cancer patients with >1% p-ERK1/2 did not respond to tamoxifen, but p-ERK negative patients did respond [i]

[a] *Li Z, Wang N, Fang J, Huang J, Tian F, Li C, Xie F. Role of PKC-ERK signaling in tamoxifen-induced apoptosis and tamoxifen resistance in human breast cancer cells. Oncology reports. 2012; 27(6): 1879-86.*
[b] *Schor NF, Kagan VE, Liang Y, Yan C, Tyurina Y, Tyurin V, Nylander KD. Exploiting oxidative stress and signaling in chemotherapy of resistant neoplasms. Biochemistry (Mosc). 2004; 69(1): 38-44.*
[c] *Liang K, Jin W, Knuefermann C, Schmidt M, Mills GB, Ang KK, Milas L, Fan Z. Targeting the phosphatidylinositol 3-kinase/Akt pathway for enhancing breast cancer cells to radiotherapy. Mol Cancer Ther. 2003; 2(4): 353-60.*
[d] *DeGraffenried LA, Chandrasekar B, Friedrichs WE, Donzis E, Silva J, Hidalgo M, Freeman JW, Weiss GR. NF-kappa B inhibition markedly enhances sensitivity of resistant breast cancer tumor cells to tamoxifen. Ann Oncol. 2004; 15(6): 885-90.*
[e] *Soderlund K, Perez-Tenorio G, Stal O. Activation of the phosphatidylinositol 3-kinase/Akt pathway prevents radiation-induced apoptosis in breast cancer cells. Int J Oncol. 2005; 26(1): 25-32.*
[f] *Nagata Y, Lan KH, Zhou X, Tan M, Esteva FJ, Sahin AA, Klos KS, Li P, Monia BP, Nguyen NT, Hortobagyi GN, Hung MC, Yu D. PTEN activation contributes to tumor inhibition by trastuzumab, and loss of PTEN predicts trastuzumab resistance in patients. Cancer Cell. 2004; 6(2): 117-27.*
[g] *Tokunaga E, Kimura Y, Oki E, Ueda N, Futatsugi M, Mashino K, Yamamoto M, Ikebe M, Kakeji Y, Baba H, Maehara Y. Akt is frequently activated in HER2/neu-positive breast cancers and associated with poor prognosis among hormone-treated patients. Int J Cancer. 2006; 118(2): 284-9.*
[h] *Eralp Y, Derin D, Ozluk Y, Yavuz E, Guney N, Saip P, Muslumanoglu M, Igci A, Kucucuk S, Dincer M, Aydiner A, Topuz E. MAPK overexpression is associated with anthracycline resistance and increased risk for recurrence in patients with triple-negative breast cancer. Annals of oncology: Official journal of the European Society for Medical Oncology / ESMO. 2008; 19(4): 669-74.*
[i] *Svensson S, Jirstrom K, Ryden L, Roos G, Emdin S, Ostrowski MC, Landberg G. ERK phosphorylation is linked to VEGFR2 expression and Ets-2 phosphorylation in breast cancer and is associated with tamoxifen treatment resistance and small tumours with good prognosis. Oncogene. 2005; 24(27): 4370-9.*

response and redox adaptation in cancer. Further, breast cancer is a highly heterogeneous disease and genetics, lifestyle, epigenetics, age, and hormonal status, as discussed in this chapter, are in a dynamic relationship with redox status in cancer cells and its microenvironment. These redox adaptive mechanisms work in concert to regulate one another and participate in feedback loops. If the ultimate outcome is a decrease in oxidative stress sensitivity, there is a potential for development of acquired resistance to many anticancer therapies whose mechanism of action involve the generation of oxidative stress and/or selection of cancer cells highly refractory to therapeutic intervention.

SUMMARY POINTS

- Activation of key survival signaling pathways (Figure 1.1), such as PI3K/Akt and ERK1/2, is mediated via hydrogen peroxide in many subtypes of breast cancer.
- Survival pathways have the ability to activate redox-sensitive transcription factors such as Nrf2 and NF-κB; alternatively, these transcription factors can be activated by oxidative modification of their inhibitors or active site cysteines.
- Nrf2 and NF-κB can themselves regulate antioxidants (GSTs, NQOs, GPxs, catalase, SOD1/2, Trxs, metallothionein, HO-1, and γ-GCS).
- Transcription factors and downstream redox-sensitive antiapoptotic proteins (Bcl-2, Bcl-xL, cIAP1/2, XIAP, survivin, c-FLIP, and TRAF1/2),[67-69] regulate each other during oxidative stress response in breast cancer cells.

Acknowledgements

This work was partly supported by the Duke Cancer Institute, Cancer and Environment Initiative P3917733 sub-award (GRD); NIH Training Grant Number T32CA009111 (SJS); American Cancer Society Research Scholar Grant RSG-08-290-01-CCE (GRD).

References

1. Droge W. Free radicals in the physiological control of cell function. *Physiol Rev* 2002;**82**(1):47–95.
2. Han D, Williams E, Cadenas E. Mitochondrial respiratory chain-dependent generation of superoxide anion and its release into the intermembrane space. *Biochem J* 2001;**353**(Pt 2):411–6.
3. Winterbourn CC. Reconciling the chemistry and biology of reactive oxygen species. *Nat Chem Biol* 2008;**4**(5):278–86.
4. Hirst DG, Robson T. Nitrosative stress in cancer therapy. *Front Biosci* 2007;**12**:3406–18.
5. Nitta M, Kozono D, Kennedy R, Stommel J, Ng K, Zinn PO, et al. Targeting EGFR induced oxidative stress by PARP1 inhibition in glioblastoma therapy. *PloS one* 2010;**5**(5):e10767.

6. Ziech D, Franco R, Georgakilas AG, Georgakila S, Malamou-Mitsi V, Schoneveld O, et al. The role of reactive oxygen species and oxidative stress in environmental carcinogenesis and biomarker development. *Chem Biol Interact* 2010;**188**(2):334–9.

7. McCord JM, Fridovich I. Superoxide dismutase. An enzymic function for erythrocuprein (hemocuprein). *J Biol Chem* 1969;**244**(22):6049–55.

8. Samoylenko A, Hossain JA, Mennerich D, Kellokumpu S, Hiltunen JK, Kietzmann T. Nutritional Countermeasures Targeting Reactive Oxygen Species in Cancer: From Mechanisms to Biomarkers and Clinical Evidence. *Antioxid Redox Signal* 2013.

9. Wells PG, McCallum GP, Chen CS, Henderson JT, Lee CJ, Perstin J, et al. Oxidative stress in developmental origins of disease: Teratogenesis, neurodevelopmental deficits, and cancer. *Toxicol Soc* 2009;**108**(1):4–18.

10. Indran IR, Hande MP, Pervaiz S. hTERT overexpression alleviates intracellular ROS production, improves mitochondrial function, and inhibits ROS-mediated apoptosis in cancer cells. *Cancer Res* 2011;**71**(1):266–76.

11. Matoba S, Kang JG, Patino WD, Wragg A, Boehm M, Gavrilova O, et al. p53 regulates mitochondrial respiration. *Science (New York, NY)* 2006;**312**(5780):1650–3.

12. Yagoda N, von Rechenberg M, Zaganjor E, Bauer AJ, Yang WS, Fridman DJ, et al. RAS-RAF-MEK-dependent oxidative cell death involving voltage-dependent anion channels. *Nature* 2007;**447**(7146):864–8.

13. Ambrosone CB. Oxidants and antioxidants in breast cancer. *Antioxid Redox Signal* 2000;**2**(4):903–17.

14. Bartosz G. Reactive oxygen species: Destroyers or messengers?. *Biochem Pharmacol* 2009;**77**(8):1303–15 doi: 10.016/j.bcp.2008.11.009. Epub Nov 24.

15. Forman HJ, Maiorino M, Ursini F. Signaling functions of reactive oxygen species. *Biochemistry* 2010;**49**(5):835–42. http://dx.doi.org/10.1021/bi9020378.

16. Trachootham D, Alexandre J, Huang P. Targeting cancer cells by ROS-mediated mechanisms: A radical therapeutic approach? *Nat Rev Drug Discov* 2009;**8**(7):579–91.

17. Zhao Y, Robbins D. Manganese superoxide dismutase in cancer prevention. *Antioxid Redox Signal* 2013.

18. Robinson MF, Godfrey PJ, Thomson CD, Rea HM, van Rij AM. Blood selenium and glutathione peroxidase activity in normal subjects and in surgical patients with and without cancer in New Zealand. *Am J Clin Nutr* 1979;**32**(7):1477–85.

19. Maitra A, Wistuba II, Washington C, Virmani AK, Ashfaq R, Milchgrub S, et al. High-resolution chromosome 3p allelotyping of breast carcinomas and precursor lesions demonstrates frequent loss of heterozygosity and a discontinuous pattern of allele loss. *Am J Pathol* 2001;**159**(1):119–30.

20. Esworthy RS, Baker MA, Chu FF. Expression of selenium-dependent glutathione peroxidase in human breast tumor cell lines. *Cancer Res* 1995;**55**(4):957–62.

21. Williams KJ, Cowen RL, Stratford IJ. Hypoxia and oxidative stress. Tumour hypoxia—Therapeutic considerations. *Breast Cancer Res* 2001;**3**(5):328–31.

22. Generali D, Berruti A, Brizzi MP, Campo L, Bonardi S, Wigfield S, et al. Hypoxia-inducible factor-1alpha expression predicts a poor response to primary chemoendocrine therapy and disease-free survival in primary human breast cancer. *Clin Cancer Res* 2006;**12**(15):4562–8.

23. Zhong H, De Marzo AM, Laughner E, Lim M, Hilton DA, Zagzag D, et al. Overexpression of hypoxia-inducible factor 1alpha in common human cancers and their metastases. *Cancer Res* 1999;**59**(22):5830–5.

24. Brown NS, Bicknell R. Hypoxia and oxidative stress in breast cancer. Oxidative stress: its effects on the growth, metastatic potential and response to therapy of breast cancer. *Breast Cancer Res* 2001;**3**(5):323–7.

25. Wincewicz A, Koda M, Sulkowska M, Kanczuga-Koda L, Wincewicz D, Sulkowski S. STAT3 and hypoxia induced proteins—HIF-1alpha, EPO and EPOR in relation with Bax and Bcl-xL in nodal metastases of ductal breast cancers. Folia histochemica et cytobiologica / Polish Academy of Sciences. *Folia Histochem Cytobiol* 2009;**47**(3):425–30.

26. Whelan KA, Schwab LP, Karakashev SV, Franchetti L, Johannes GJ, Seagroves TN, et al. The Oncogene HER2/neu (ERBB2) Requires the Hypoxia-inducible Factor HIF-1 for Mammary Tumor Growth and Anoikis Resistance. *J Biol Chem* 2013;**288**(22):15865–77.

27. Chen CL, Chu JS, Su WC, Huang SC, Lee WY. Hypoxia and metabolic phenotypes during breast carcinogenesis: expression of HIF-1alpha, GLUT1, and CAIX. *Virchows Arch* 2010;**457**(1):53–61.

28. Blancher C, Moore JW, Talks KL, Houlbrook S, Harris AL. Relationship of hypoxia-inducible factor (HIF)-1alpha and HIF-2alpha expression to vascular endothelial growth factor induction and hypoxia survival in human breast cancer cell lines. *Cancer Res* 2000;**60**(24):7106–13.

29. Lal A, Peters H, St Croix B, Haroon ZA, Dewhirst MW, Strausberg RL, et al. Transcriptional response to hypoxia in human tumors. *J Natl Cancer Inst* 2001;**93**(17):1337–43.

30. Lithgow D, Covington C. Chronic inflammation and breast pathology: A theoretical model. *Biol Res Nurs* 2005;**7**(2):118–29.

31. Simpson ER, Brown KA. Minireview: Obesity and breast cancer: A tale of inflammation and dysregulated metabolism. *Mol Endocrinol* 2013;**27**(5):715–25.

32. Ristimaki A, Sivula A, Lundin J, Lundin M, Salminen T, Haglund C, et al. Prognostic significance of elevated cyclooxygenase-2 expression in breast cancer. *Cancer Res* 2002;**62**(3):632–5.

33. Khuder SA, Mutgi AB. Breast cancer and NSAID use: A meta-analysis. *Br J Cancer* 2001;**84**(9):1188–92.

34. Simpson ER, Clyne C, Rubin G, Boon WC, Robertson K, Britt K, et al. Aromatase—A brief overview. *Annu Rev Physiol* 2002;**64**:93–127.

35. Clemons M, Goss P. Estrogen and the risk of breast cancer. *N Engl J Med* 2001;**344**(4):276–85.

36. Sastre-Serra J, Valle A, Company MM, Garau I, Oliver J, Roca P. Estrogen down-regulates uncoupling proteins and increases oxidative stress in breast cancer. *Free Radic Biol Med* 2010;**48**(4):506–12.

37. Nadal-Serrano M, Sastre-Serra J, Pons DG, Miro AM, Oliver J, Roca P. The ERalpha/ERbeta ratio determines oxidative stress in breast cancer cell lines in response to 17beta-estradiol. *J Cell Biochem* 2012;**113**(10):3178–85.

38. Howlander N, Noone AM, Krapcho M, Garshell J, Neyman N, Altedruse SF, Kosary CL, Yu M, Ruhl J, Tatalovich Z, Cho H, Mariotto A, D.R. L, Chen HS, Feuer EJ, Cronin KA (eds). SEER Cancer Statistics Review, 1975-2010. National Cancer Institute Bethesda, MD, http://seercancergov/csr/1975_2010/, based on November 2012 SEER data submission. (based on November 2012 SEER data submission, posted to the SEER web site, April 2013).

39. Mates JM, Segura JA, Alonso FJ, Marquez J. Intracellular redox status and oxidative stress: Implications for cell proliferation, apoptosis, and carcinogenesis. *Arch Toxicol* 2008;**82**(5):273–99.

40. Hou Z, Falcone DJ, Subbaramaiah K, Dannenberg AJ. Macrophages induce COX-2 expression in breast cancer cells: Role of IL-1beta autoamplification. *Carcinogenesis* 2011;**32**(5):695–702.

41. Donate LE, Blasco MA. Telomeres in cancer and ageing. Philosophical transactions of the Royal Society of London Series B. *Biol Sci* 2011;**366**(1561):76–84.

42. Singhapol C, Pal D, Czapiewski R, Porika M, Nelson G, Saretzki GC. Mitochondrial telomerase protects cancer cells from nuclear DNA damage and apoptosis. *PLoS one* 2013;**8**(1):e52989.

43. Aapro M, Wildiers H. Triple-negative breast cancer in the older population. *Ann Oncol* 2012;**23**(Suppl. 6) vi52-5.

44. Cui H, Kong Y, Zhang H. Oxidative stress, mitochondrial dysfunction, and aging. *J Signal Transduct* 2012;**2012**:646354.

45. Rohan TE, Wong LJ, Wang T, Haines J, Kabat GC. Do alterations in mitochondrial DNA play a role in breast carcinogenesis? *J Oncol* 2010;**2010**:604304.

46. Ziech D, Franco R, Pappa A, Panayiotidis MI. Reactive oxygen species (ROS)—induced genetic and epigenetic alterations in human carcinogenesis. *Mutat Res* 2011;**711**(1-2):167–73.

47. Schafer FQ, Buettner GR. Redox environment of the cell as viewed through the redox state of the glutathione disulfide/glutathione couple. *Free Radic Biol Med* 2001;**30**(11):1191–212.

48. Jardim BV, Moschetta MG, Leonel C, Gelaleti GB, Regiani VR, Ferreira LC, et al. Glutathione and glutathione peroxidase expression in breast cancer: An immunohistochemical and molecular study. *Oncol Rep* 2013.

49. Chen EI, Hewel J, Krueger JS, Tiraby C, Weber MR, Kralli A, et al. Adaptation of energy metabolism in breast cancer brain metastases. *Cancer Res* 2007;**67**(4):1472–86.

50. Singh B, Tai K, Madan S, Raythatha MR, Cady AM, Braunlin M, et al. Selection of metastatic breast cancer cells based on adaptability of their metabolic state. *PLoS One* 2012;**7**(5):e36510. http://dx.doi.org/10.1371/journal.pone.0036510 Epub 2012 May 3.

51. Bae I, Fan S, Meng Q, Rih JK, Kim HJ, Kang HJ, et al. BRCA1 induces antioxidant gene expression and resistance to oxidative stress. *Cancer Res* 2004;**64**(21):7893–909.

52. Weydert CJ, Waugh TA, Ritchie JM, Iyer KS, Smith JL, Li L, et al. Overexpression of manganese or copper-zinc superoxide dismutase inhibits breast cancer growth. *Free Radic Biol Med* 2006;**41**(2):226–37.

53. Kattan Z, Minig V, Leroy P, Dauca M, Becuwe P. Role of manganese superoxide dismutase on growth and invasive properties of human estrogen-independent breast cancer cells. *Breast Cancer Res Treat* 2008;**108**(2):203–15.

54. Er TK, Hou MF, Tsa EM, Lee JN, Tsai LY. Differential expression of manganese containing superoxide dismutase in patients with breast cancer in Taiwan. *Ann Clin Lab Sci* 2004;**34**(2):159–64.

55. Aird KM, Allensworth JL, Batinic-Haberle I, Lyerly HK, Dewhirst MW, Devi GR. ErbB1/2 tyrosine kinase inhibitor mediates oxidative stress-induced apoptosis in inflammatory breast cancer cells. *Breast Cancer Res Treat* 2012;**132**(1):109–19.

56. Williams KP, Allensworth JL, Ingram SM, Smith GR, Aldrich AJ, Sexton JZ, et al. Quantitative high-throughput efficacy profiling of approved oncology drugs in inflammatory breast cancer models of acquired drug resistance and re-sensitization. *Cancer Lett* 2013;**17**(13):00386–8.

57. Allensworth JL, Aird KM, Aldrich AJ, Batinic-Haberle I, Devi GR. XIAP inhibition and generation of reactive oxygen species enhances TRAIL sensitivity in inflammatory breast cancer cells. *Mol Cancer Ther* 2012;**11**(7):1518–27.

58. Tas F, Hansel H, Belce A, Ilvan S, Argon A, Camlica H, et al. Oxidative stress in breast cancer. *Med Oncol* 2005;**22**(1):11–5.

59. Ray G, Batra S, Shukla NK, Deo S, Raina V, Ashok S, et al. Lipid peroxidation, free radical production and antioxidant status in breast cancer. *Breast Cancer Res Treat* 2000;**59**(2):163–70.

60. Portakal O, Ozkaya O, Erden Inal M, Bozan B, Kosan M, Sayek I. Coenzyme Q10 concentrations and antioxidant status in tissues of breast cancer patients. *Clin Biochem* 2000;**33**(4):279–84.

61. Kumaraguruparan R, Subapriya R, Viswanathan P, Nagini S. Tissue lipid peroxidation and antioxidant status in patients with adenocarcinoma of the breast. *Clin Chim Acta* 2002;**325**(1-2):165–70.

62. Punnonen K, Ahotupa M, Asaishi K, Hyoty M, Kudo R, Punnonen R. Antioxidant enzyme activities and oxidative stress in human breast cancer. *J Cancer Res Clin Oncol* 1994;**120**(6):374–7.

63. Kumaraguruparan R, Kabalimoorthy J, Nagini S. Correlation of tissue lipid peroxidation and antioxidants with clinical stage and menopausal status in patients with adenocarcinoma of the breast. *Clin Biochem* 2005;**38**(2):154–8.

64. Manda G, Nechifor MT, Neagu T. Reactive Oxygen Species, Cancer and Anti-Cancer Therapies. *Curr Chem Biol* 2009;**3**:342–66.

65. Noh DY, Ahn SJ, Lee RA, Kim SW, Park IA, Chae HZ. Overexpression of peroxiredoxin in human breast cancer. *Anticancer Res* 2001;**21**(3B):2085–90.

66. Balkwill F. Tumor necrosis factor or tumor promoting factor? *Cytokine Growth Factor Rev* 2002;**13**(2):135–41.

67. Tian H, Zhang B, Di J, Jiang G, Chen F, Li H, et al. Keap1: One stone kills three birds Nrf2, IKKbeta and Bcl-2/Bcl-xL. *Cancer Lett.* 2012;**325**(1):26–34. http://dx.doi.org/10.1016/j.canlet.2012.06.007 Epub Jun 26.

68. Bharti AC, Aggarwal BB. Nuclear factor-kappa B and cancer: Its role in prevention and therapy. *Biochem Pharmacol* 2002;**64**(5-6):883–8.

69. Morgan MJ, Liu ZG. Crosstalk of reactive oxygen species and NF-kappaB signaling. *Cell Res* 2011;**21**(1):103–15.

70. Yook JI, Li XY, Ota I, Hu C, Kim HS, Kim NH, et al. A Wnt-Axin2-GSK3beta cascade regulates Snail1 activity in breast cancer cells. *Nat Cell Biol* 2006;**8**(12):1398–406.

71. Lu Z, Xu S. ERK1/2 MAP kinases in cell survival and apoptosis. *IUBMB Life* 2006;**58**(11):621–31.

72. Truong TH, Carroll KS. Redox regulation of epidermal growth factor receptor signaling through cysteine oxidation. *Biochemistry* 2012;**51**(50):9954–65.

73. Badawi AF, Cavalieri EL, Rogan EG. Role of human cytochrome P450 1A1, 1A2, 1B1, and 3A4 in the 2–, 4–, and 16alpha-hydroxylation of 17beta-estradiol. *Metabolism* 2001;**50**(9):1001–3.

74. Bolton JL, Thatcher GR. Potential mechanisms of estrogen quinone carcinogenesis. *Chem Res Toxicol* 2008;**21**(1):93–101.

75. Tanaka T, Iwasa Y, Kondo S, Hiai H, Toyokuni S. High incidence of allelic loss on chromosome 5 and inactivation of p15INK4B and p16INK4A tumor suppressor genes in oxystress-induced renal cell carcinoma of rats. *Oncogene* 1999;**18**(25):3793–7.

76. Browne EP, Punska EC, Lenington S, Otis CN, Anderton DL, Arcaro KF. Increased promoter methylation in exfoliated breast epithelial cells in women with a previous breast biopsy. *Epigenetics* 2011;**6**(12):1425–35.

77. Hitchler MJ, Wikainapakul K, Yu L, Powers K, Attatippaholkun W, Domann FE. Epigenetic regulation of manganese superoxide dismutase expression in human breast cancer cells. *Epigenetics* 2006;**1**(4):163–71.

78. Barbano R, Muscarella LA, Pasculli B, Valori VM, Fontana A, Coco M, et al. Aberrant Keap1 methylation in breast cancer and association with clinicopathological features. *Epigenetics* 2013;**8**(1):105–12.

79. Eades G, Yang M, Yao Y, Zhang Y, Zhou Q. miR-200a regulates Nrf2 activation by targeting Keap1 mRNA in breast cancer cells. *J Biol Chem* 2011;**286**(47):40725–33.

80. Acharya A, Das I, Chandhok D, Saha T. Redox regulation in cancer: A double-edged sword with therapeutic potential. *Oxidative medicine and cellular longevity* 2010;**3**(1):23–34.

81. Stefansson OA, Villanueva A, Vidal A, Marti L, Esteller M. BRCA1 epigenetic inactivation predicts sensitivity to platinum-based chemotherapy in breast and ovarian cancer. *Epigenetics* 2012;**7**(11):1225–9.

2

Oxidative Stress and Prostate Cancer

Masaki Shiota, Akira Yokomizo, Seiji Naito

Department of Urology, Graduate School of Medical Sciences, Kyushu University, Fukuoka, Japan

List of Abbreviations

AP-1 activator protein 1
AR androgen receptor
COX cyclooxygenase
HIF-1α hypoxia-inducible factor-1α
GSTP1 glutathione S-transferase π
mtDNA mitochondrial DNA
PI3K phosphoinositide-3 kinase
MAPK mitogen-activated protein kinase
Nox NAPDH oxidase
Nrf2 NF-E2-related factor 2
NFκB nuclear factor-κ B
NSAID nonsteroidal anti-inflammatory drug
PIA proliferative inflammatory atrophy
PIN prostatic intraepithelial neoplasia
ROS reactive oxygen species
SNP single nucleotide polymorphism
SOD superoxide dismutase
TRAMP transgenic adenocarcinoma of the mouse model of prostate cancer
4-HNE 4-hydroxy-2-nonenal
8-OHdG 8-hydroxydeoxyguanosine

INTRODUCTION

Prostate cancer is a major cause of death among men in developed countries, affecting mainly older men. Under both physiological and pathological conditions, metabolic processes generate hydroxyl radicals, hydrogen peroxide, and superoxides, defined as reactive oxygen species (ROS). While increased ROS have traditionally been thought to cause tissue injury and/or damage to intracellular components, they also participate in a wide range of crucial physiological as well as pathological processes, including cell proliferation, cell-cycle progression, antiapoptosis, invasion, metastasis, and angiogenesis, which contribute to prostate carcinogenesis and prostate cancer progression.[1] Several impairments of prooxidant and antioxidant systems resulting in increased oxidative stress and oxidation damage have been identified in prostate cancer. Oxidative stress has thus been proposed to contribute to the pathogenesis of prostate cancer. Definitive risk factors for the development of prostate cancer include age, race, and family history, while androgens, inflammation, diet, and lifestyle have been suggested as potential risk factors. These risk factors may be linked to oxidative stress, and close relationships between oxidative stress and prostate cancer risk factors have been suggested. Here, we summarize the findings regarding functional links between oxidative stress and prostate cancer.

CAUSES OF OXIDATIVE STRESS IN PROSTATE CANCER

ROS are generated endogenously from mitochondrial bioenergetics, as well as the NADPH oxidase (Nox) complex, cyclooxygenase (COX), and exogenously from the cellular microenvironment under conditions such as hypoxia.[2] To protect against elevated ROS and oxidative stress, cells are equipped with antioxidant defense systems,[3] including low-molecular-weight antioxidant compounds and reducing buffer systems, as well as antioxidant enzymes such as superoxide dismutases (SODs), catalase, glutathione peroxidase, and peroxiredoxins.[3] Transcription factors regulating the reduction-oxidation (redox) state in response to oxidative stress, such as NF-E2-related factor 2 (Nrf2), nuclear factor-κB (NFκB), and activator protein 1 (AP-1), are also included as important regulators of the antioxidant defense system. These proteins maintain the intracellular homeostasis of the redox states and the balance of reducing/oxidizing equivalents.

However, impairments of the redox systems can result in a redox imbalance as a result of increased ROS production and/or decreased antioxidant defenses, as shown in Figure 2.1. This section describes the deregulation of ROS production and antioxidant systems in prostate cancer.

© 2014 Elsevier Inc. All rights reserved.

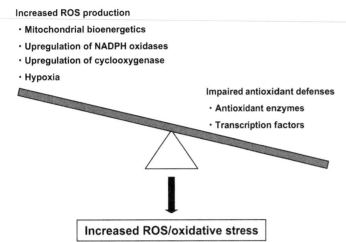

FIGURE 2.1 Increased ROS production and impaired antioxidant defenses contribute to increased oxidative stress.

Increased ROS Production

Altered Mitochondrial Bioenergetics

Altered mitochondrial bioenergetics is thought to be one of the major sources of increased ROS generation. Metabolic transformation is a common feature of tumors, including prostate cancer. Mutation rates in mitochondrial DNA (mtDNA) are high, compared with nuclear DNA, presumably because of the high concentrations of ROS around the mtDNA, the lack of histone protection, and inadequate DNA repair systems. Accelerated mtDNA mutations are thought to increase ROS production and oxidative stress. Petros et al. demonstrated the role of mtDNA-mutation-induced oxidative stress in prostate tumorigenesis using a mtDNA mutation (Thr8993Gly) known to cause increased mitochondrial ROS production.[4] Indeed, frequent mtDNA mutations have been identified in clinically relevant human prostate cancer.[5] In addition to mtDNA mutations, alterations of mitochondrial enzymes have also been thought to contribute to prostate carcinogenesis, as suggested by the fact that ROS generation was elevated by increased expression of mitochondrial glycerophosphate dehydrogenase in prostate cancer progression.[6]

Upregulation of Nox Enzymes

Upregulation of membrane-bound Nox is a potential source of increased intracellular ROS production. The Nox enzyme family consists of Nox1–5 and Duox, and catalyzes the production of superoxides utilizing oxygen as a substrate and NADPH as a cofactor.[7]

Nox1 overexpression has been reported to increase superoxide production and cause malignant transformation.[7] Nox1, Nox2, and Nox4 were expressed and further upregulated in rat prostate after castration, indicating a possible link between androgens and oxidative stress.[8] Similarly, Nox1 and Nox2 are also expressed in human

prostate, and increased Nox1 was correlated with elevated hydrogen peroxide levels.[7] In addition, Nox2, Nox4, and Nox5 levels were elevated in human prostate cancer cells, while conversely, the Nox inhibitor diphenyliodonium and down-regulation of Nox5 suppressed ROS production, leading to decreased cell proliferation and increased apoptosis.[9] Taken together, these lines of evidence indicate that upregulation of the Nox ROS producers contribute to prostate cancer growth.

Upregulation of COX Enzymes

COXs comprise two isoforms, COX-1 and COX-2, which are rate-limiting enzymes in prostaglandin biosynthesis, and which produce ROS during their processes. Although COX-1 is constitutively expressed ubiquitously, COX-2 is only induced by stimulation. COX-2 expression was shown to be increased in high-grade prostatic intraepithelial neoplasia (PIN) lesions compared with normal prostate tissue in a transgenic adenocarcinoma mouse model of prostate cancer (TRAMP).[10] In addition, COX-2 expression was significantly higher in cancerous prostate tissues compared with benign prostate tissues, and higher in high-grade tumors.[11] Based on these findings, nonsteroidal anti-inflammatory drugs (NSAIDs), which inhibit COX activity, may represent attractive potential chemopreventive agents. Several preclinical studies have demonstrated growth-suppressive and apoptosis-inducing effects of NSAIDs in prostate cancer.[12]

Hypoxia

Tumor hypoxia is an inevitable characteristic of advanced prostate cancer, and increases intracellular ROS as an exogenous source.[2] Hypoxia-induced ROS can activate numerous signaling components and stabilize hypoxia-inducible factor-1α (HIF-1α) through activation of proto-oncogenic signal-transduction pathways, including the phosphoinositide-3 kinase (PI3K)/Akt and mitogen-activated protein kinase (MAPK) pathway, which was inhibited by the ROS scavenger N-acetyl-cysteine.[2] HIF-1α regulates its target genes, which are pivotal in regulating cellular metabolism, cell-cycle progression, anti-apoptosis, metastasis, and angiogenesis.[2] Under hypoxic conditions, human prostate cancer cells showed increased levels of ROS and accumulation of HIF-1α. Hypoxia thus acts as a regulator of oxidative stress as an exogenous source in prostate cancer, resulting in activation of proto-oncogenic pathways by HIF-1α.

Impaired Antioxidant Defenses

Altered Antioxidant Enzymes

Dismutases catalyzing the conversion of superoxides to water include the intracellular copper-zinc SOD (SOD1) in the nucleus/cytoplasm, manganese SOD

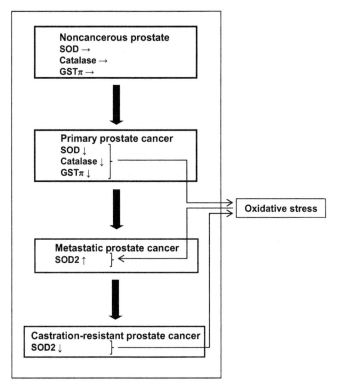

FIGURE 2.2 Changes in antioxidant enzymes in prostate carcinogenesis, prostate cancer progression, and castration resistance.

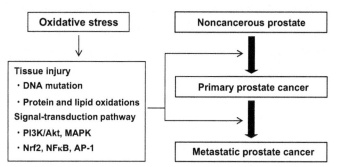

FIGURE 2.3 Mechanisms of prostate carcinogenesis and prostate cancer progression by oxidative stress.

(SOD2) in mitochondria, and extracellular SOD. Catalase catalyzes the conversion of hydrogen peroxide to water and is located within peroxisomes and the cytoplasm. Several lines of evidence have shown a close relationship between antioxidant defense systems and prostate cancer. SOD1, SOD2, and catalase were suppressed in cancerous regions of human prostate cancer tissues, compared with noncancer regions.[13] Similarly, Bostwick et al. found lower expression of SOD1, SOD2, and catalase in human high-grade PIN and prostate cancer, compared with benign prostate epithelium.[14] Oberley et al., however, found lower levels of SOD2 in human primary prostate cancer tissues and higher levels in metastatic tissues, compared with normal tissues.[3] Taken together, these results indicate that SOD and catalase expression levels are decreased in tumor tissues compared with nontumor tissues, although SOD2 was elevated in metastatic sites compared with primary regions, as depicted in Figure 2.2.

Glutathione S-transferase π (GSTP1) plays an important role in the detoxification of electrophilic compounds, such as carcinogens and cytotoxic drugs, by glutathione conjugation. Inactivation of GSTP1 can impair cellular antioxidant systems and augment intracellular oxidative stress. Hypermethylation of the promoter region in the *GSTP1* gene is a common feature identified during prostate carcinogenesis, and *GSTP1* gene methylation has been detected in 50 to 70% of PIN and 70 to 95% of prostate cancers.[15]

Thus, as shown in Figure 2.3, a decrease in antioxidant enzymes including SOD, catalase, and GSTP1 may lead to elevated levels of oxidative stress in cancers, while augmented oxidative stress may induce a compensating increase in antioxidant enzymes. However, SOD2 levels were reduced in castration-resistant prostate cancer,[16] indicating that the alterations in antioxidant enzyme activities are complex.

Altered Transcriptional Factors Related to Redox Balance

Transcription factors represent another player in the antioxidant system. These include Nrf2, NFκB, and AP-1, which are modulated by redox status and inversely control the redox balance through regulation of their downstream genes.[17]

The transcription factor Nrf2 regulates the expression of phase II key protective enzymes through the antioxidant-response element.[18] Recent studies indicated that Nrf2 and several of its target genes were significantly down-regulated in human prostate cancer, resulting in continuous exposure to increased oxidative stress.[19] The transcription factors NFκB and AP-1 have been shown to be induced by oxidative stress, and regulate the expression of genes involved in several processes, including cell proliferation, antiapoptosis, invasion, metastasis, and angiogenesis.[20] Both NFκB and AP-1 were thought to be constitutively activated and contribute to malignant transformation and progression, as well as castration resistance, in prostate cancer.[20] Conversely, suppression of NFκB and AP-1 was shown to reduce the invasive and metastatic properties of prostate cancer.[21,22] These results suggest that increased oxidative stress resulting from Nrf2 suppression may contribute to an upregulation of the proto-oncogenic transcription factors NFκB and AP-1.

ROLE OF OXIDATIVE STRESS IN PATHOGENESIS OF PROSTATE CANCER

Although above-physiological levels of ROS are toxic, physiological levels of ROS regulate many cellular functions such as proliferation, cell-cycle progression, invasion, metastasis, and angiogenesis, possibly resulting in prostate carcinogenesis and prostate cancer progression through oxidative modifications of cellular components. Signaling induced by oxidation can activate signal transduction molecules such as PI3K/Akt and MAPK, and elevate the transcriptional activity of Nrf2, NFκB, and AP-1 via phosphorylation of Jun, and dissociation of Nrf2 and NFκB from inhibitory protein complexes, respectively.[17] Low doses of hydrogen peroxide have been demonstrated to stimulate prostate cancer cell growth and migration.[23] In contrast, oxidative stress has been shown to be involved in the inactivation of several key proteins, including those involved in DNA repair, apoptosis, cell signaling, and essential enzymatic pathways.[24] Thus, oxidative stress is thought to contribute to the pathogenesis of prostate cancer through oncogenic and antiapoptotic pathways.

Oxidative Stress in Carcinogenesis and Cancer Progression

Increased oxidative damage has been shown to correlate with tumor progression in mouse models of prostate tumorigenesis.[25,26] DNA and proteins were increasingly damaged by oxidation during tumor progression in Nkx3.1/Pten knockout mice.[26] Similarly, Tam et al. demonstrated increased ROS damage in the prostate gland of TRAMP mice during tumorigenesis.[25]

Also, in humans, numerous studies have shown a close relationship between oxidative stress in prostate carcinogenesis and prostate cancer development. Clearly, Kumar et al. demonstrated more production of ROS in prostate cancer cell lines compared with benign ones.[9] In addition, oxidative stress, represented by urinary F2-isoprostane levels, was increased in high-grade PIN as well as in prostate cancer.[27] Conversely, serum levels of the antioxidant α-tocopherol were decreased in prostate cancer compared with a control group; intriguingly, most patients with metastatic prostate cancer received androgen-deprivation therapy, suggesting that androgen-deprivation therapy may affect serum antioxidant and oxidative stress levels.[28] Similarly, oxidative stress, represented by the ratio of 8-hydroxydeoxyguanosine (8-OHdG) to creatinine in the urine, was increased in prostate cancer compared with age-matched healthy controls.[29] Additionally, peroxide levels and the total equivalent antioxidant capacity in the serum were increased and decreased in prostate cancer and nonprostate disease patients, respectively.[30]

Furthermore, changes in the redox state have been demonstrated with progression of human prostate cancer. High levels of 8-OHdG were detected in primary prostate cancer compared with normal prostate epithelium, whereas metastatic prostate cancer showed higher levels of ROS-damage products (4-hydroxy-2-nonenal-modified proteins and 8-OHdG) than either primary cancer or normal

prostate epithelium.[31] In addition, higher oxidative stress levels have been detected in aggressive prostate cancer cells compared with normal prostatic epithelium-derived cells and localized prostate cancer cells.[32] Matrix metalloproteinases, known as invasion/metastasis promoters, were shown to be activated by ROS and attenuated by ROS-production blockade through Nox inhibition and extracellular SOD overexpression in prostate cancer cells, indicating that matrix metalloproteinases may mediate prostate cancer progression through increased oxidative stress.[9] Oxidative stress is also known to promote angiogenesis in cancer cells.[2] The ROS inducer Nox consistently induced angiogenesis in prostate cancer,[33] whereas heme oxygenase 1, which counteracts oxidative stress, suppressed angiogenesis,[34] indicating that oxidative stress regulates cancer progression through induction of angiogenesis.

Oxidative Stress in Castration Resistance

Despite the initial success of androgen-deprivation therapy in suppressing advanced prostate cancer, castration-resistant cancer eventually evolves, with critical consequences. Oxidative stress is known to be implicated in castration resistance of prostate cancer.[35] Oxidative stress aberrantly activates androgen receptor (AR) signaling under low androgen levels by a variety of pathways, including AR overexpression, AR cofactors, and signal-transduction pathways, as indicated in Figure 2.4,[35] thus promoting castration resistance. Hydrogen peroxide-resistant derivatives

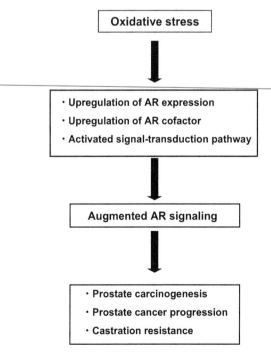

FIGURE 2.4　Mechanism of acquired castration resistance by oxidative stress through augmented AR signaling.

of androgen-dependent prostate cancer cells express high levels of ARs and exhibit a castration-resistant phenotype.[36] In addition, several lines of evidence have shown that oxidative stress is increased in castration-resistant cells, as indicated by higher intracellular ROS levels in castration-resistant cells compared with androgen-dependent cells,[37] and higher antioxidant protein levels and an increased ability to scavenge ROS in castration-resistant cells and tumors.[35,38] AR activation by oxidative stress is thus thought to render prostate cancer cells resistant to castration.

PROSTATE CANCER RISK FACTORS AND THEIR LINKS TO OXIDATIVE STRESS

Several definitive prostate cancer risk factors have been identified, including age, race, and family history, as well as other possible risk factors including androgens, inflammation, diet, and lifestyle. These risk factors may result in increased oxidative stress, as shown in Figure 2.5. In order to understand the relationship between oxidative stress and prostate cancer, this section reviews the possible links between risk factors affecting prostate cancer and oxidative stress.

Aging

The incidence of prostate cancer increases with age; the incidence of prostate cancer is low in men younger than 40, but it becomes common in men older than 80. Several studies have suggested that 42 to 80% of men will develop prostate cancer in their eighth decade.[39] Exposure to oxidative stress, generated endogenously by the by-products of normal metabolic processes and exogenously by environmental exposure to toxic substances, has been thought to increase with aging. Indeed,

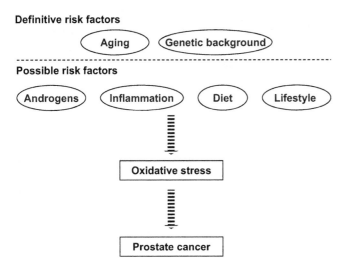

FIGURE 2.5 Relationships among risk factors affecting prostate cancer and oxidative stress.

the progressive accumulation of DNA strand breaks, DNA adducts, and oxidative modification of enzymes for DNA repair have been documented to increase while antioxidant defenses including ROS detoxification enzyme activities decrease with age. Thus, increased oxidative stress associated with aging is thought to contribute to the pathogenesis of prostate cancer.

Genetic Background (Race and Family History)

African-American and Caucasian men are at high risk for clinically relevant prostate cancer compared with Asian people. For example, Japanese men living or born in the United States have a lower risk of prostate cancer than African and Caucasian men in the United States,[40] indicating a racial difference in prostate cancer development. Furthermore, the response to androgen-deprivation therapy is also better in Japanese men compared with American men,[41] suggesting that the racial difference affects not only prostate carcinogenesis, but also castration resistance. Family history of prostate cancer is also a well-known risk factor.[42]

Genetic background also contributes to the differential incidence rates of prostate cancer among races and families, even after excluding the effects of exogenous factors. Several genetic variations have been reported to influence prostate cancer incidence. Associations between prostate cancer and the single nucleotide polymorphism (SNP) Ala16Val SNP in the *SOD2* gene have been reported. A meta-analysis of 34 studies that examined the association between the *SOD2* Ala16Val polymorphism and prostate cancer found a significant association between Ala/Ala and prostate cancer, compared with Val/Val.[43] In addition, SNPs in the gene for the high-density-lipoprotein-associated enzyme paraoxonase 1, which acts to decrease oxidative stress, have been reportedly associated with an increased risk of aggressive prostate cancer, suggesting that increased oxidative stress by altered paraoxonase 1 activity may promote aggressiveness of prostate cancer.[44] In addition, an SNP in the catalase gene was correlated with increased risk of prostate cancer diagnosed under the age of 65.[45] Similarly, correlations between prostate cancer incidence and SNPs in xenobiotic metabolic enzyme genes including glutamate cysteine ligase, thioredoxin reductase 2, microsomal epoxide hydrolase 1, and myeloperoxidase have been observed.[46]

Androgens

The importance of androgens in prostate carcinogenesis has been suggested by classical observations of young castrated men and men deficient in 5α-reductase, which converts testosterone to the more potent dihydrotestosterone. Further recent prevention studies (PCPT and

FIGURE 2.6 Relationship between androgen level and oxidative stress level.

REDUCE trials using the 5α-reductase inhibitors finasteride and dutasteride, respectively) also showed that 5α-reductase inhibitors reduced the total prostate cancer detection rate among men with high prostate-specific antigen levels.[47,48] Furthermore, the growth and progression of most prostate cancers are androgen-dependent, as indicated by initial remarkable therapeutic effect of androgen-deprivation therapy.[49]

Androgens have been implicated in oxidative stress. ROS levels can be increased by androgens via the transcription factor JunD and the mitochondrial redox regulator p66Shc.[50,51] On the other hand, castration clearly induced an oxidative state in the rat prostate, while testosterone replacement in castrated rats reduced oxidative stress levels by down-regulating Nox expression and recovering antioxidants.[8] Similarly, androgen deprivation induced oxidative stress in prostate cancer cells and in human prostate cancer tissues.[30,52] Intriguingly, ROS levels were decreased at physiological androgen levels that stimulated cell growth, while oxidative stress levels dramatically increased at higher nonphysiological doses that inhibited cell growth.[53] Taken together, these reports suggest that a nonphysiological androgen milieu involving either androgen deprivation or overload induces oxidative stress in the prostate, as shown in Figure 2.6.

Inflammation

ROS production by activated inflammatory cells and/or secreted inflammatory cytokines is thought to cause cellular injury and accumulation of genomic damage, leading to prostate carcinogenesis, prostate cancer progression, and castration resistance.[24,54] This is supported by evidence that several inflammatory conditions in humans are accompanied by increased levels of oxidative DNA damage.[55]

Chronic inflammation and proliferative inflammatory atrophy (PIA) are thought to be a first step in prostate carcinogenesis, as suggested by the pathological observations that chronic inflammation and PIA accompany high-grade PIN.[54] Epidemiologically, a prior history of certain sexually transmitted infections or prostatitis increased the relative risk of prostate cancer,[56] whereas the use of NSAIDs reduced prostate cancer risk, as described earlier. Thus, several lines of evidence suggest that inflammation may be closely implicated in prostate cancer by promoting oxidative damage.

Diet

Japanese men who emigrated to Hawaii had a higher rate of prostate cancer than those living in Japan.[40] This observation was supported by another report showing that Japanese men born in the United States had higher incidence rates of prostate cancer than Japanese immigrants born in Japan.[57] In addition, epidemiological data regarding the ethnic and geographic variations in prostate cancer incidence and mortality suggest that dietary factors may influence prostate carcinogenesis and prostate cancer progression. The idea that high-calorie and high-fat diets may affect prostate carcinogenesis is supported by the finding that these diets, which are thought to increase ROS production through metabolic processes, promote prostate cancer proliferation in mice. In contrast, vegetables that are rich in antioxidants are thought to protect against prostate cancer, and several antioxidants including green tea, isoflavones, lycopene, vitamin E, and resveratrol are all considered to help prevent prostate carcinogenesis.[58]

Lifestyle

Similar to dietary factors, lifestyle may also affect the incidence of prostate cancer. Smoking is a well-known risk factor for lung cancer, head and neck cancer, and bladder cancer. Although the contribution of smoking to prostate cancer is obscure, the smoking rate was reportedly higher in men with advanced and critical prostate cancer,[59] suggesting that it may contribute to prostate cancer progression. In addition, a meta-analysis of 24 cohort studies showed an increased incidence of prostate cancer among heavy smokers.[60] Physiological activity and obesity, which can be linked to oxidative stress, have also been suggested to influence prostate cancer incidence.[61]

CONCLUSIONS

Increased production of ROS by altered mitochondrial bioenergetics, upregulation of Nox and COX, hypoxia, and/or impairment of antioxidant defenses result in redox imbalance and oxidative stress. Physiological levels of ROS promote broad cellular processes such as cell proliferation, invasion, metastasis, and angiogenesis through the accumulation of genetic and epigenetic damage, as well as by activating proto-oncogenic signaling (PI3K/Akt and MAPK) and proto-oncogenic transcription factors (NFκB, AP-1, and AR). Some factors including age, race, family history, androgens, inflammation, diet, and lifestyle are associated with oxidative stress and are involved in prostate cancer incidence and progression, as well as castration resistance. Oxidative stress is thus implicated in prostate carcinogenesis, prostate cancer progression, and castration resistance. However, current levels of knowledge are insufficient to ascertain the precise role of ROS/oxidative stress in the pathogenesis of prostate cancer, and further studies are thus needed to elucidate the implications of oxidative stress in prostate cancer.

SUMMARY POINTS

- ROS production by altered mitochondrial bioenergetics, upregulation of NADPH oxidase, cyclooxygenase, and hypoxia is increased in prostate cancer.
- Antioxidant enzymes and their regulating transcription factors are deregulated in prostate cancer.
- Oxidative stress caused by redox imbalances between prooxidants and antioxidants contributes to prostate carcinogenesis and prostate cancer progression.
- Oxidative stress promotes castration resistance through aberrant activation of AR.
- Definitive and possible risk factors affecting prostate cancer are linked to oxidative stress, implicating these factors in the pathogenesis of prostate cancer through oxidative stress.

References

1. Sauer H, Wartenberg M, Hescheler J. Reactive oxygen species as intracellular messengers during cell growth and differentiation. *Cell Physiol Biochem* 2001;**11**:173–86.
2. Galanis A, Pappa A, Giannakakis A, Lanitis E, Dangaj D, Sandaltzopoulos R. Reactive oxygen species and HIF-1 signalling in cancer. *Cancer Lett* 2008;**266**:12–20.
3. Oberley TD. Oxidative damage and cancer. *Am J Pathol* 2002;**160**:403–8.
4. Petros JA, Baumann AK, Ruiz-Pesini E, Amin MB, Sun CQ, Hall J, et al. mtDNA mutations increase tumorigenicity in prostate cancer. *Proc Natl Acad Sci U S A* 2005;**102**:719–24.
5. Yu JJ, Yan T. Effect of mtDNA mutation on tumor malignant degree in patients with prostate cancer. *Aging Male* 2010;**13**:159–65.
6. Chowdhury SK, Raha S, Tarnopolsky MA, Singh G. Increased expression of mitochondrial glycerophosphate dehydrogenase and antioxidant enzymes in prostate cancer cell lines/cancer. *Free Radic Res* 2007;**41**:1116–24.
7. Bedard K, Krause KH. The NOX family of ROS-generating NADPH oxidases: Physiology and pathophysiology. *Physiol Rev* 2007;**87**:245–313.
8. Tam NN, Gao Y, Leung YK, Ho SM. Androgenic regulation of oxidative stress in the rat prostate: Involvement of NAD(P)H oxidases and antioxidant defense machinery during prostatic involution and regrowth. *Am J Pathol* 2003;**163**:2513–22.
9. Kumar B, Koul S, Khandrika L, Meacham RB, Koul HK. Oxidative stress is inherent in prostate cancer cells and is required for aggressive phenotype. *Cancer Res* 2008;**68**:1777–85.
10. Shappell SB, Olson SJ, Hannah SE, Manning S, Roberts RL, Masumori N, et al. Elevated expression of 12/15-lipoxygenase and cyclooxygenase-2 in a transgenic mouse model of prostate carcinoma. *Cancer Res* 2003;**63**:2256–67.
11. Kirschenbaum A, Liu X, Yao S, Levine AC. The role of cyclooxygenase-2 in prostate cancer. *Urology* 2001;**58**:127–31.
12. Harris RE. Cyclooxygenase-2 (cox-2) blockade in the chemoprevention of cancers of the colon, breast, prostate, and lung. *Inflammopharmacology* 2009;**17**:55–67.
13. Baker AM, Oberley LW, Cohen MB. Expression of antioxidant enzymes in human prostatic adenocarcinoma. *Prostate* 1997;**32**:229–33.
14. Bostwick DG, Alexander EE, Singh R, Shan A, Qian J, Santella RM, et al. Antioxidant enzyme expression and reactive oxygen species damage in prostatic intraepithelial neoplasia and cancer. *Cancer* 2000;**89**:123–34.
15. Li LC. Epigenetics of prostate cancer. *Front Biosci* 2007;**12**:3377–97.
16. Sharifi N, Hurt EM, Thomas SB, Farrar WL. Effects of manganese superoxide dismutase silencing on androgen receptor function and gene regulation: Implications for castration-resistant prostate cancer. *Clin Cancer Res* 2008;**14**:6073–80.
17. Reuter S, Gupta SC, Chaturvedi MM, Aggarwal BB. Oxidative stress, inflammation, and cancer: How are they linked? *Free Radic Biol Med* 2010;**49**:1603–16.
18. Taguchi K, Motohashi H, Yamamoto M. Molecular mechanisms of the Keap1–Nrf2 pathway in stress response and cancer evolution. *Genes Cells* 2011;**16**:123–40.
19. Frohlich DA, McCabe MT, Arnold RS, Day ML. The role of Nrf2 in increased reactive oxygen species and DNA damage in prostate tumorigenesis. *Oncogene* 2008;**27**:4353–62.
20. Karin M, Takahashi T, Kapahi P, Delhase M, Chen Y, Makris C, et al. Oxidative stress and gene expression: The AP-1 and NF-κB connections. *Biofactors* 2001;**15**:87–9.
21. Huang S, Pettaway CA, Uehara H, Bucana CD, Fidler IJ. Blockade of NF-κB activity in human prostate cancer cells is associated with suppression of angiogenesis, invasion, and metastasis. *Oncogene* 2001;**20**:4188–97.
22. Ozanne BW, McGarry L, Spence HJ, Johnston I, Winnie J, Meagher L, et al. Transcriptional regulation of cell invasion: AP-1 regulation of a multigenic invasion programme. *Eur J Cancer* 2000;**36**:1640–8.
23. Polytarchou C, Hatziapostolou M, Papadimitriou E. Hydrogen peroxide stimulates proliferation and migration of human prostate cancer cells through activation of activator protein-1 and upregulation of the heparin affin regulatory peptide gene. *J Biol Chem* 2005;**280**:40428–35.
24. Palapattu GS, Sutcliffe S, Bastian PJ, Platz EA, De Marzo AM, Isaacs WB, et al. Prostate carcinogenesis and inflammation: Emerging insights. *Carcinogenesis* 2005;**26**:1170–81.
25. Tam NN, Nyska A, Maronpot RR, Kissling G, Lomnitski L, Suttie A, et al. Differential attenuation of oxidative/nitrosative injuries in early prostatic neoplastic lesions in TRAMP mice by dietary antioxidants. *Prostate* 2006;**66**:57–69.
26. Ouyang X, DeWeese TL, Nelson WG, Abate-Shen C. Loss-of-function of Nkx3.1 promotes increased oxidative damage in prostate carcinogenesis. *Cancer Res* 2005;**65**:6773–9.
27. Barocas DA, Motley S, Cookson MS, Chang SS, Penson DF, Dai Q, et al. Oxidative stress measured by urine F2-isoprostane level is associated with prostate cancer. *J Urol* 2011;**185**:2102–7.
28. Yossepowitch O, Pinchuk I, Gur U, Neumann A, Lichtenberg D, Baniel J. Advanced but not localized prostate cancer is associated with increased oxidative stress. *J Urol* 2007;**178**:1238–43.
29. Miyake H, Hara I, Kamidono S, Eto H. Oxidative DNA damage in patients with prostate cancer and its response to treatment. *J Urol* 2004;**171**:1533–6.

30. Pace G, Di Massimo C, De Amicis D, Corbacelli C, Di Renzo L, Vicentini C, et al. Oxidative stress in benign prostatic hyperplasia and prostate cancer. *Urol Int* 2010;**85**:328–33.

31. Oberley TD, Zhong W, Szweda LI, Oberley LW. Localization of antioxidant enzymes and oxidative damage products in normal and malignant prostate epithelium. *Prostate* 2000;**44**:144–55.

32. Freitas M, Baldeiras I, Proença T, Alves V, Mota-Pinto A, Sarmento-Ribeiro A. Oxidative stress adaptation in aggressive prostate cancer may be counteracted by the reduction of glutathione reductase. *FEBS Open Bio* 2012;**2**:119–28.

33. Kim J, Koyanagi T, Mochly-Rosen D. PKCδ activation mediates angiogenesis via NADPH oxidase activity in PC-3 prostate cancer cells. *Prostate* 2011;**71**:946–54.

34. Ferrando M, Gueron G, Elguero B, Giudice J, Salles A, Leskow FC, et al. Heme oxygenase 1 (HO-1) challenges the angiogenic switch in prostate cancer. *Angiogenesis* 2011;**14**:467–79.

35. Shiota M, Yokomizo A, Naito S. Oxidative stress and androgen receptor signaling in the development and progression of castration-resistant prostate cancer. *Free Radic Biol Med* 2011;**51**:1320–8.

36. Shiota M, Yokomizo A, Tada Y, Inokuchi J, Kashiwagi E, Masubuchi D, et al. Castration resistance of prostate cancer cells caused by castration-induced oxidative stress through Twist1 and androgen receptor overexpression. *Oncogene* 2010;**29**:237–50.

37. Shigemura K, Sung SY, Kubo H, Arnold RS, Fujisawa M, Gotoh A, et al. Reactive oxygen species mediate androgen receptor- and serum starvation-elicited downstream signaling of ADAM9 expression in human prostate cancer cells. *Prostate* 2007;**67**:722–31.

38. Kuruma H, Egawa S, Oh-Ishi M, Kodera Y, Satoh M, Chen W, et al. High molecular mass proteome of androgen-independent prostate cancer. *Proteomics* 2005;**5**:1097–112.

39. Minelli A, Bellezza I, Conte C. Culig, Z. Oxidative stress-related aging: A role for prostate cancer? *Biochim Biophys Acta* 2009;**1795**:83–91.

40. Hirayama T. Comparative epidemiology of cancer in the U.S. and Japan. Morbidity. *Japan Society for the Promotion of Science* 1978:43–62.

41. Fukagai T, Namiki TS, Carlile RG, Yoshida H, Namiki M. Comparison of the clinical outcome after hormonal therapy for prostate cancer between Japanese and Caucasian men. *BJU Int* 2006;**97**:1190–3.

42. Madersbacher S, Alcaraz A, Emberton M, Hammerer P, Ponholzer A, Schröder FH, et al. The influence of family history on prostate cancer risk: Implications for clinical management. *BJU Int* 2011;**107**:716–21.

43. Wang S, Wang F, Shi X, Dai J, Peng Y, Guo X, et al. Association between manganese superoxide dismutase (MnSOD) Val-9Ala polymorphism and cancer risk—A meta-analysis. *Eur J Cancer* 2009;**45**:2874–81.

44. Stevens VL, Rodriguez C, Talbot JT, Pavluck AL, Thun MJ, Calle EE. Paraoxonase 1 (PON1) polymorphisms and prostate cancer in the CPS-II Nutrition Cohort. *Prostate* 2008;**68**:1336–40.

45. Choi JY, Neuhouser ML, Barnett M, Hudson M, Kristal AR, Thornquist M, et al. Polymorphisms in oxidative stress-related genes are not associated with prostate cancer risk in heavy smokers. *Cancer Epidemiol Biomarkers Prev* 2007;**16**:1115–20.

46. Koutros S, Andreotti G, Berndt SI, Hughes Barry K, Lubin JH, Hoppin JA, et al. Xenobiotic-metabolizing gene variants, pesticide use, and the risk of prostate cancer. *Pharmacogenet Genomics* 2011;**21**:615–23.

47. Thompson IM, Goodman PJ, Tangen CM, Lucia MS, Miller GJ, Ford LG, et al. The influence of finasteride on the development of prostate cancer. *N Engl J Med* 2003;**349**:215–24.

48. Andriole GL, Bostwick DG, Brawley OW, Gomella LG, Marberger M, Montorsi F, et al. Effect of dutasteride on the risk of prostate cancer. *N Engl J Med* 2010;**362**:1192–202.

49. Miyamoto H, Messing EM, Chang C. Androgen deprivation therapy for prostate cancer: Current status and future prospects. *Prostate* 2004;**61**:332–53.

50. Mehraein-Ghomi F, Lee E, Church DR, Thompson TA, Basu HS, Wilding G. JunD mediates androgen-induce oxidative stress in androgen dependent LNCaP human prostate cancer cells. *Prostate* 2008;**68**:924–34.

51. Veeramani S, Yuan TC, Lin FF, Lin MF. Mitochondrial redox signaling by p66Shc is involved in regulating androgenic growth stimulation of human prostate cancer cells. *Oncogene* 2008;**27**:5057–68.

52. Shiota M, Song Y, Takeuchi A, Yokomizo A, Kashiwagi E, Kuroiwa K, et al. Antioxidant therapy alleviates oxidative stress by androgen deprivation and prevents conversion from androgen dependent to castration resistant prostate cancer. *J Urol* 2012;**187**:707–14.

53. Ripple MO, Henry WF, Rago RP, Wilding G. Prooxidant-antioxidant shift induced by androgen treatment of human prostate carcinoma cells. *J Natl Cancer Inst* 1997;**89**:40–8.

54. De Marzo AM, Platz EA, Sutcliffe S, Xu J, Grönberg H, Drake CG, et al. Inflammation in prostate carcinogenesis. *Nat Rev Cancer* 2007;**7**:256–69.

55. Shen Z, Wu W, Hazen SL. Activated leukocytes oxidatively damage DNA, RNA, and the nucleotide pool through halide-dependent formation of hydroxyl radical. *Biochemistry* 2000;**39**: 5474–82.

56. Sutcliffe S. Sexually transmitted infections and risk of prostate cancer: Review of historical and emerging hypotheses. *Future Oncol* 2010;**6**:1289–311.

57. Cook LS, Goldoft M, Schwartz SM, Weiss NS. Incidence of adenocarcinoma of the prostate in Asian immigrants to the United States and their descendants. *J Urol* 1999;**161**:152–5.

58. Hori S, Butler E, McLoughlin J. Prostate cancer and diet: Food for thought? *BJU Int* 2011;**107**:1348–59.

59. Plaskon LA, Penson DF, Vaughan TL, Stanford JL. Cigarette smoking and risk of prostate cancer in middle-aged men. *Cancer Epidemiol Biomarkers Prev* 2003;**12**:604–9.

60. Huncharek M, Haddock KS, Reid R, Kupelnick B. Smoking as a risk factor for prostate cancer: A meta-analysis of 24 prospective cohort studies. *Am J Public Health* 2010;**100**:693–701.

61. Stein CJ, Colditz GA. Modifiable risk factors for cancer. *Br J Cancer* 2004;**90**:299–303.

Oxidative Stress in Lung Cancer

Wei Sheng Joshua Loke*, Mann Ying Lim*

Inflammation and Infection Research Centre, Faculty of Medicine, University of New South Wales and Department of
Respiratory Medicine, Prince of Wales Hospital, Randwick, Sydney NSW 2031, Australia

Craig R. Lewis

Department of Medical Oncology, Prince of Wales Hospital, Randwick, Sydney NSW 2031, Australia

Paul S. Thomas

Inflammation and Infection Research Centre, Faculty of Medicine, University of New South Wales and Department of
Respiratory Medicine, Prince of Wales Hospital, Randwick, Sydney NSW 2031, Australia

List of Abbreviations

AP-1 activator protein-1
CDKN2A cyclin-dependent kinase inhibitor 2A
H$_2$O$_2$ hydrogen peroxide
IARC International Agency for Research on Cancer
IL-6 interleukin-6
IL-8 interleukin-8
KRAS kirsten rat sarcoma
NF-κB nuclear factor kappa B
NO nitric oxide
NO$_2$ nitrogen dioxide
O$_2^{\bullet-}$ superoxide
O$_3$ ozone
OH$^-$ hydroxyl radicals
ONOO$^-$ peroxynitrite
PAH polycyclic aromatic hydrocarbons
Q$^{\bullet-}$ semiquinone
R$^{\bullet}$ carbon-centered radials
RNS reactive nitrogen species
RO$^{\bullet}$ alkoxyl radicals
ROO$^{\bullet}$ peroxyl radicals
ROS reactive oxygen species
SO$_2$ sulphur dioxide
STK11 serine/threonine kinase 11
TB tuberculosis
TNF-α tumor necrosis factor α
TP53 tumor protein 53
UVA ultraviolet-A

*These authors have made an equal contribution to this work.

INTRODUCTION

The lung is a highly specialized human organ. It is richly vascularized and facilitates gaseous exchange through coordinated interactions between the chest and diaphragmatic musculature, the central nervous and the cardiovascular systems. Upon maturity, the lung has a surface area of approximately 140 m^2. With the exception of the skin, the lung is the organ with the highest exposure to the ambient air.[1,2]

The average adult inhales about 10,000 L of air daily, which can be contaminated with cigarette smoke, diesel soot, vehicle exhaust, organic and mineral dusts, gases such as sulphur dioxide (SO$_2$), ozone (O$_3$), nitrogen dioxide, viruses, and microbial pathogens. Inhalation of these substances can result in the production of reactive oxygen and nitrogen species that are able to, through oxidation and nitrosylation, initiate a cascade of signaling events that induce in the production of pro-inflammatory chemokines and cytokines that injure the lung. Persistent inhalation results in enhanced production of reactive oxygen and nitrogen species, which may lead to chronic inflammation that precipitates pulmonary carcinogenesis as discussed in the following.[1,2]

Apart from exogenous sources, oxygen and nitrogen species can be produced endogenously by reducing molecular oxygen to water in the mitochondrial electron

Cancer
http://dx.doi.org/10.1016/B978-0-12-405205-5.00003-9

© 2014 Elsevier Inc. All rights reserved.

transport during respiration, by cellular enzymes (e.g., xanthine oxidase and P450 oxidase) and via the cyclooxygenase pathway.[3] Furthermore, nitrogen species can be produced in response to hypoxic conditions.[4] In order to mitigate these insults, the epithelium of the lung is covered by a thin layer of lining fluid (respiratory tract lining fluid) vested with several antioxidants,[5] which can be categorized into enzymatic antioxidants (i.e., superoxide dismutase, catalase, glutathione peroxidase, and glutathione-S-transferase) and nonenzymatic antioxidants (i.e., glutathione, cysteine, thioredoxin, vitamins C and E, beta-carotene, and uric acid).[6]

This review outlines the etiology of lung cancer and the role of reactive oxygen and nitrogen species in pulmonary carcinogenesis.

LUNG CANCER

Lung cancer is the most common type of cancer in the world. In 2008, there were an estimated 1.2 million new cases (12.7% of new cancers) and 1.3 million cancer deaths (18.2% of cancer deaths). Lung cancer is uncommon in young adults, with the average age of occurrence being more than 60. The prognosis of lung cancer remains poor, especially in patients presenting with advanced disease. The expected survival at five years for all patients is only 5 to 10%. Although recent studies have suggested a role for low-dose CT screening in high risk individuals based on smoking history, this has yet to be implemented into standard clinical practice.[7]

Lung cancer can be broadly categorized into two main histopathological subtypes, including small cell and non-small cell lung cancer (NSCLC). Adenocarcinoma, squamous cell carcinoma, and large cell carcinoma form the main histological types of NSCLC and account for about 85% of all lung cancers.[8]

ETIOLOGY OF LUNG CANCER

The majority of lung cancers are associated with smoking, but it may occur in up to 10% of individuals who have no smoking history. Epidemiological studies have identified other factors associated with the risk of having lung cancer, including infection and pollution (environmental, occupational, and domestic).[9]

Tobacco Smoke

Tobacco smoking remains the major risk factor for lung cancer. In most populations, tobacco smoking

accounts for 90% and 70 to 80% of lung cancer cases in men and women, respectively, and because it is such a prominent risk factor, the geographical and temporal patterns of lung cancer largely reflect population-level changes in duration, type, and dose of tobacco used and in smoking behavior.[8]

The smoke from cigarettes and other tobacco products contains numerous carcinogens and agents that cause inflammation. The relative risk of lung cancer in smokers compared to never-smokers is 8 to 15 in men and 3 to 10 in women. Although smoking increases the risk of all histological subtypes of lung cancer, the risk is strongest for squamous cell carcinoma, and thereafter small cell carcinoma and adenocarcinoma. Nonetheless, in the last few decades, the percentage of NSCLC as squamous cell carcinoma, which had been the predominant histological type of lung cancer, has been decreasing while the incidence of adenocarcinomas has been increasing. This has been attributed to changes in the smoking behavior (e.g., using filtered cigarettes, reducing inhalation) and the composition of tobacco products.[8]

Exposure to Environmental Smoke

There exists an association between lung cancer and second-hand smoke. Lung cancer risk increases with both the number of cigarettes smoked by the smoking partner and the duration of exposure to second-hand smoke. When confounding factors such as active smoking and diet have been taken into account, the increased risk of lung cancer is in the order of 20 to 25%. This is supported by biological data. For instance, tobacco metabolites have been found in 90% of urine samples from children whose parents smoke.[10] Moreover, it has been observed that lung cancer rates are higher in cities than in rural settings. Although this might be confounded by occupational exposure and tobacco smoking, the combined evidence suggests that urban air pollution is a possible risk factor for lung cancer. Nonetheless, the excess risk remains more than 20% in most urban areas.[8]

Infections

The Human Papillomavirus (types 16 and 18) are well-known carcinogens and are commonly present in lung tumor tissue.[11] In addition, individuals who have been infected with *Mycobacterium tuberculosis* (TB) are shown to be at higher risk of lung cancer, which is independent of their smoking status. Patients with old TB lesions are predisposed to having epidermal growth factor receptor (EGFR) mutations in these neoplasms, particularly exon 19 deletions.[12]

Occupational Exposure

Respiratory cancers are the most frequently acquired occupational malignancy. Asbestos, crystalline silica, polycyclic aromatic hydrocarbons, and heavy metals (e.g., cadmium) are recognized by the International Agency for Research on Cancer (IARC) as pulmonary carcinogens.[9] In addition, radon, associated with uranium mines, has been established as a risk factor for lung cancer. The risk of lung cancer in never-smokers who are exposed to 0, 100, and 400 Bq/m^3 of radon is 0.4%, 0.5%, and 0.7%, respectively. These risks are approximately 25 times higher in smokers.[13]

Domestic Exposure

Indoor air pollution elevates one's risk of having lung cancer. The strongest evidence is for fumes from high-temperature cooking using unrefined vegetable oil (e.g., rapeseed oil) and from the burning of coal and other solid fuels.[14]

TOBACCO SMOKING AND OXIDANTS

Tobacco smoke is a complex amalgamation of more than 4,700 chemical compounds, which are dispersed in the tar and gas phases. The radicals present in these phases differ.[15]

Tobacco smoke in the tar phase consists of extremely high concentrations of radicals (1017 per gram). These radicals are stable and are largely organic. Semiquinone ($Q^{\bullet-}$), for example, is held in the tar matrix. It reacts with oxygen to form superoxide ($O_2^{\bullet-}$; Equation 3.1), which consequently dismutates to form hydrogen peroxide (H_2O_2; Equation 3.2).

$$Q^{\bullet-} + O_2 \rightarrow Q + O_2^{\bullet-} \tag{3.1}$$

$$2\,O_2^{\bullet-} + 2H^+ \rightarrow O_2 + H_2O_2 \tag{3.2}$$

Furthermore, the tar contains metal ions (e.g., iron) that, through the Fenton reaction, generate highly oxidizing hydroxyl radicals (OH^-) from hydrogen peroxide (Equation 3.3).

$$H_2O_2 + Fe\,(II) \rightarrow HO^{\bullet} + HO^- + Fe\,(III) \tag{3.3}$$

Hydroxyl radicals can also be formed as a result of the decomposition of peroxynitrite ($ONOO^-$), which is a product of a reaction between nitric oxide and superoxide (Equation 3.4).[16,17]

$$O2^{\bullet-} + NO \rightarrow ONOO^- \rightarrow HO^{\bullet} + NO_2 \tag{3.4}$$

Tobacco smoke in the gas phase contains more than 10^{15} radicals per puff. Contrary to stable radicals found in the tar phase, the organic radicals in the gas phase are transient (i.e., lifetimes of less than 1 second) reactive oxygen- and carbon-centered radicals. They are quickly quenched by the respiratory tract lining fluid. It is a paradox that, in spite of their short lifetimes, high concentrations of radicals can be maintained and even increased in the gas phase for more than 10 minutes. This is because gas phase radicals exist in a steady state where they are continuously made and destroyed.[17] It has been postulated that this steady state involves the slow oxidation of nitric oxide (NO) to nitrogen dioxide (NO_2; Equation 3.5).

$$NO + \frac{1}{2}O_2 \rightarrow NO_2 \tag{3.5}$$

Nitrogen dioxide then reacts with isoprene present in tobacco smoke to form carbon-centered radicals ($R^{\bullet}$; Equation 3.6). Carbon radicals then react with oxygen, forming peroxyl radicals ($ROO^{\bullet}$; Equation 3.7). These react with nitric oxide to form alkoxyl radicals ($RO^{\bullet}$) and more nitrogen dioxide (Equation 3.8).[16]

$$NO_2 + \text{(isoprene)} \rightarrow \ \ NO_2 \ (R^{\bullet}) \tag{3.6}$$

$$R^{\bullet} + O_2 \rightarrow ROO^{\bullet} \tag{3.7}$$

$$ROO^{\bullet} + NO \rightarrow RO^{\bullet} + NO_2 \tag{3.8}$$

The oxidant burden placed on the lung by the previously mentioned exogenously-derived oxidants is further intensified in smokers who have higher numbers of alveolar macrophages (by two- to four-fold) and leukocytes (by 10-fold). Moreover, compared to nonsmokers, alveolar macrophages and leukocytes from tobacco smokers spontaneously release increased amounts of superoxide and hydrogen peroxide, thereby exacerbating the oxidative burden in the lung (Figure 4).[18] In addition, hydrogen peroxide has been detected in increased amounts in the exhaled breath condensate in those with lung cancer.[19]

SILICA AND OXIDANTS

Silica can be inhaled during hard-rock mining, sand-blasting, or grinding. There are different forms of silica: vitreous, crystalline, synthetic/mineral, amorphous/natural, and biogenic. Silica exposure results in severe alveolar inflammation sustained by

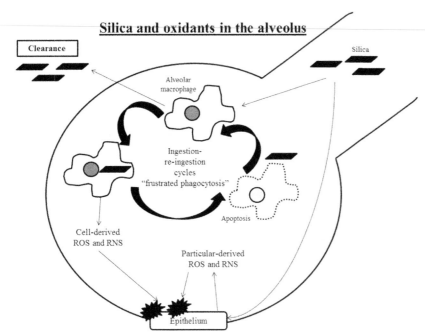

Silica and oxidants in the alveolus

FIGURE 3.1 **Silica and Oxidants in the Alveolus.** The inhaled silica reaches the alveolus and is ingested by alveolar macrophages. Alveolar macrophage ingested silica is either cleared or persists in the macrophage, resulting in a process called "frustrated phagocytosis," which releases ROS and RNS (cell-derived). The alveolar macrophages then undergo apoptosis and release the silica, perpetuating the aforementioned process. Inhaled silica can also cause the alveolar epithelium to release ROS and RNS (particular-derived). The ROS and RNS produced via these two sources increase the oxidative burden on the alveolar epithelium.

oxidants present in the alveolar space. Upon inhalation, silica reaches the alveolar space where it is phagocytosed by alveolar macrophages. Depending on the surface characteristics, the silica particles are either cleared from the lungs by the macrophages or activate macrophages at the molecular and cellular levels. This results in the release of ROS and RNS. Eventually the macrophages undergo apoptosis and release the silica particles. Subsequent ingestion–reingestion cycles result in the release of more cell-derived ROS and RNS. Additionally, silica particles can react directly with alveolar and bronchiolar epithelium to form particle-derived ROS and RNS (Figure 3.1). These damage the alveolar and bronchiolar epithelium and may react with cell-derived ROS and RNS to yield peroxynitrite (ONOO⁻) from superoxide ($O_2^{\bullet-}$) and nitric oxide (NO).[20]

Particle-Generated Oxidants

Oxidants can be either bound to the silica surface (surface radical) or formed when silica is placed in aqueous suspensions. The former are formed when silica is fractured or ground. When this occurs, the silicon–oxygen bonds are cleaved.[21] Molecular oxygen then reacts at the sites of cleavage and produces several "surface-bound ROS" – $SiO_3^{\bullet}$, $SiO_2^{\bullet}$, $Si^+O_2^{\bullet-}$.[22]

Oxidants can also be generated when silica is suspended in aqueous solution. Iron ions that are located in the redox and coordinative positions on the silica

surface and silicon-based surface radicals serve as two active centers for oxidant production. The iron centers yield $HO^{\bullet}$ radicals via the Fenton reaction (Equation 3.9) or the Haber-Weiss cycle (Equations 3.10–3.13) when a reductant and trace amounts of iron are present.

$$Fe^{2+} \rightarrow H_2O_2 \rightarrow Fe^{3+} + OH^- + HO^{\bullet} \qquad (3.9)$$

$$Fe^{3+} \rightarrow reductant \rightarrow Fe^{2+} + reductant \qquad (3.10)$$

$$Fe^{2+} \rightarrow O_2 \rightarrow Fe^{3+} + O2^{-\bullet} \qquad (3.11)$$

$$O_2^{-\bullet} + H_2O \rightarrow HO_2^{\bullet} + OH^- \text{ or } O_2^{\bullet} + 2H^+ + e^- \rightarrow H_2O_2 \qquad (3.12)$$

$$2HO_2^{\bullet} \rightarrow H_2O_2 + O_2O_2^{-\bullet} + H_2O_2 \rightarrow HO^{\bullet} + OH^- + O_2 \qquad (3.13)$$

Moreover, hydroxyl radicals can also be formed when surface radicals ($SiO^{\bullet}$, $SiO_2^{\bullet}$, $SiO_3^{\bullet}$, Si^+-$O_2^{\bullet-}$) come into contact with water (Equation 3.14) or hydrogen peroxide (Equations 3.15 and 3.16).[23]

$$\text{-SiO}^{\bullet} + H_2O \rightarrow \text{-SiOH} + HO^{\bullet} \qquad (3.14)$$

$$\text{-SiOO}^{\bullet} + H_2O_2 \rightarrow \text{-SiOH} + HO^{\bullet} + O_2 \qquad (3.15)$$

$$\text{-Si}^+O_2^{\bullet} + H_2O_2 \rightarrow \text{-SiOH} + HO^{\bullet} + O_2 \qquad (3.16)$$

ASBESTOS AND OXIDANTS

Asbestos fibers may be inhaled during the mining, extraction, processing, and use of this fiber. It is now most commonly a problem for those in the construction industry, but previously was used widely in ship-building, boiler-making, plumbing, roofing, as well as insulation for heat and electricity. It is still a common problem and continues to be used in countries such as Russia, India, and those in Southeast Asia. The inhalation of asbestos results in the accumulation of macrophages in the alveolar space.[24] The mechanisms underlying ROS production following asbestos inhalation are similar to those of silica (see earlier). Alveolar macrophages engulf the asbestos fibers and undergo a process of "frustrated phagocytosis."[25] Post inhalation, the asbestos fiber acquires a redox-active iron on its surface, which encourages the development of extremely reactive hydroxyl radicals from hydrogen peroxide through Fenton-catalyzed Haber-Weiss reactions (Equation 3.17). The iron can also catalyze the production of alkoxyl radical from organic hydroperoxides (Equation 3.18).[25]

$$O_2^- + H_2O_2 \rightarrow HO^- + HO^\bullet + O_2 \qquad (3.17)$$

$$Fe^{2+} + ROOH \rightarrow Fe^{3+} + RO^\bullet + HO^- \qquad (3.18)$$

Alkoxyl and hydroxyl radical production leads to ROS-induced perturbation of DNA structure and function, which leads to pulmonary carcinogenesis.[26]

DOMESTIC COOKING AND OXIDANTS

When food is grilled, fried, or stir-fried at high temperatures with cooking oil, the combustion of cooking oils release polycyclic aromatic hydrocarbons (PAH) into the environment. Moreover high temperature cooking also causes sugar and fat degradation and amino acid and protein pyrolysis. This adds to the concentration of PAH in the air.[27]

Upon ultraviolet-A (UVA) photo-irradiation, PAHs oxidize to a myriad of hydroxylated products, namely oxygenated PAH, PAH quinones, nitro-PAH, and halo-PAH.[28] UVA photo-irradiation of these PAH-derived products can absorb light energy and form various photo-excited substrates.[29] They transfer energy via electron transfer to molecular oxygen to produce singlet oxygen ($^1O_2$) and superoxide radicals. These ROS participate in lipid peroxidation and can cause DNA damage (Figure 3.2).[30]

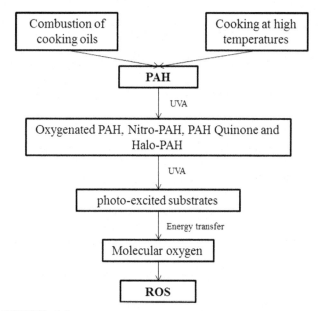

FIGURE 3.2 **Domestic Cooking and Oxidants.** Combusting of cooking oils and cooking at high temperature results in the production of polycyclic aromatic hydrocarbons (PAH). These, on exposure to ultraviolet radiation (UVA), become photo-excited and can produce reactive oxygen species (ROS) by transferring their energy to molecular oxygen.

ROLE OF OXIDANTS IN NORMAL PHYSIOLOGY

Oxidants have several roles in normal physiology. One example is the regulation of normal cell growth through the activation of transcription factors such as nuclear factor kappa B (NF-κB) and activator protein-1 (AP-1).[31] Apart from cell growth, activation of transcriptional factors NF-κB and AP-1 also initiates inflammation involved in the host defense of the lung.[15] Moreover, oxidants are also important in inducing angiogenesis, apoptosis of mutated/damaged cells, and senescence, which are crucial in the conversion of normal cells to neoplastic ones.[19]

INFLAMMATION AND OXIDATIVE STRESS

Upon exposure of the lung epithelium to oxidants, a protective mechanism, inflammation, is triggered in the lungs in an attempt to eliminate the toxins.[32] As will be discussed next, the inflammatory process can induce genetic mutations through oxidative stress, resulting in lung cancer.

Oxidants initiate inflammation by activating transcription factor NF-κB and AP-1 in the airway epithelial cells and macrophages.[6,33] Transcription factors NF-κB and AP-1 are responsible for the gene transcription of downstream inflammatory cytokines such as interleukin-8 (IL-8), tumor necrosis factor α (TNF-α), and interleukin-6 (IL-6), which attract more inflammatory cells such as alveolar macrophages, neutrophils, and eosinophils to generate an inflammatory cascade.[6,34,35] The recruited leukocytes in turn eliminate pathogens by producing oxidants as mentioned, including superoxide, nitric oxide, hydrogen peroxide, hydroxyl radical, peroxynitrite, and hydrochlorous acid. The oxidants can inactivate pathogens via halogenation or protein or lipid peroxidation. Following the destruction and removal of these foreign pathogens, inflammation settles.[36] As such, in the attempt to eliminate toxins, neutrophils and macrophages produce oxidants that further augment the inflammation response.

OXIDATIVE STRESS LEADS TO DNA MUTATIONS AND LUNG CANCER

Oxidants are highly reactive oxidizing agents due to the unpaired electrons, which readily attack DNA, proteins, and lipids (such as those in the cell membrane).[37]

Oxidants readily react with DNA bases to form DNA adducts, which are complexes formed from the covalent binding of DNA to molecules including carcinogens.[38] DNA adducts can cause miscoding during DNA replication when an incorrect base is paired, resulting in permanent mutation following replication.[39] Other changes to DNA include base alteration, base insertion, deletion, chromosomal translocation, single- or double-strand breaks, microsatellite instability, and the activation of oncogenes, which are directly associated with lung cancer (Figure 3.3).[6,36,40–44] DNA changes are also seen in mitochondrial DNA, which can be detected in the breath condensate.[45]

When sufficient damage from oxidative stress has accumulated, irreversible changes to DNA may confer the cells a survival advantage. This constitutes the "initiation step" in the three-step progression of cancer, namely initiation, promotion, and progression.[42] This is especially true when mutations occur in critical coding regions such as those of oncogene or tumor suppressor genes, which will result in a loss of normal growth regulation, followed by uncontrolled cell proliferation.[39,46,47]

Chronic pulmonary obstructive disease (COPD secondary to smoking) is a well-established disease of chronic inflammation that is also triggered by oxidative stress from tobacco smoking.[46] Genetic mutation is similarly conferred in COPD patients and COPD itself is known to accrue a 4.5-fold increase in risk of lung cancer[48] and is thus viewed as a stepping stone toward lung cancer progression.

Genetic mutation in lung cancer can occur in oncogenes such as K-ras, jun, and myc, or tumor suppressor genes such as TP53, CDKN2A, and STK11 although they most commonly occur in oncogene K-ras and tumor suppressor gene TP53.[39,49-51]

The most frequently observed mutations on TP53 in lung cancer are guanine→ thymine transversions followed by guanine→ adenine transitions. They mostly occur at codons 157, 158, 245, 248, 249, and 273. Mutations on K-ras, on the other hand, are mainly guanine→ thymine transversions with smaller numbers of guanine→ adenine transitions, and most commonly occur in codon 12.[39]

Proofreading mechanisms of DNA replication may attempt to repair or remove the damaged DNA via direct repair, double-strand break repair, cross-link repair, nucleotide excision, or base excision.[52] When

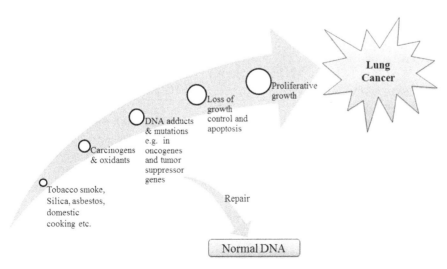

FIGURE 3.3 **Stepwise Progression toward Lung Carcinogenesis.** Tobacco smoke and other sources of oxidants cause DNA adduct formation and mutation. These lead to mutations and loss of cellular growth controls, resulting in unencumbered proliferative growth and, finally, lung cancer. This process can be mitigated or halted when DNA is repaired.

damaged beyond repair, the cell usually undergoes apoptosis.[46] However, if any of the steps of reparation fails, or if damage to DNA is too extensive beyond the control of reparation, permanent mutations may occur in the DNA, resulting in oncogenesis.

LIPID AND PROTEIN PEROXIDATION

Apart from DNA, lipids and proteins can also be attacked by oxidants, further increasing risk of mutations and malignancy. Lipid peroxidation is the breakdown of polyunsaturated phospholipids into more reactive lipid peroxides; for instance hydroperoxides, lipoperoxides, and toxic aldehydes (e.g., malondialdehyde).[6,43,53] Not only does lipid peroxidation damage cells by impairing membrane function,[53] products of lipid peroxidation also behave in a similar manner as free radicals with the ability to cause oxidative stress and react with DNA.[43] While lipid peroxidation produces by-products that react with DNA, protein peroxidation may interfere with enzymes involved in the DNA reparative system such as DNA polymerase, thereby promoting genetic mutations.[36]

Simply put, oxidants promote proliferative cell growth via DNA toxicity as well as lipid and protein damage. Oxidation of DNA, lipids, and proteins may also lead to increased reactive species production, contributing to the vicious cycle of oxidative stress.

OXIDANT/ANTIOXIDANT DISEQUILIBRIUM

Under normal physiology, oxidants are counterbalanced by endogenous antioxidants.

In response to elevated levels of oxidants and oxidant-producing inflammatory cells, the local capacity of antioxidants can increase. Superoxide dismutase, catalase, glutathione associated enzymes, and manganese superoxide dismutase are among the antioxidants that have been demonstrated to be raised in smokers, suggesting that a counterbalance to the continuing insult is attempted.[54]

The balance between oxidants and antioxidants is delicate. In the event where the endogenous system is unable to completely ward off the pathogens or when the insults are persistent, chronic inflammation occurs. During chronic inflammation, there is simultaneous tissue injury (from inflammation) and repair. The production of oxidants is persistently increased as a result, exhausting the buffering capacity of antioxidants. This oxidant/antioxidant disequilibrium, favoring the former, results in oxidative stress.

ANTIOXIDANTS AND LUNG CANCER CHEMOPREVENTION

There has been significant interest in the role of antioxidants as potential chemoprevention of lung cancer. Several large randomized clinical trials have examined this question. In patients with previously treated early stage lung cancer, placebo controlled studies randomized patients to receive antioxidants or retinoids with primary study outcomes including prevention of second primary tumors, recurrence, and survival.[55,56] Neither of these trials showed any benefit in favor of the antioxidant treatment arm. Several studies and a subsequent meta-analysis[57–60] have failed to show any impact on lung cancer incidence with antioxidant agents in patient populations at risk of lung cancer (based on smoking history) and two of these studies[58,60] identified possible increased risk with antioxidant therapy. Based on results of these large trials there is no substantive evidence to support the role of antioxidants as chemopreventative agents for lung cancer.

CONCLUSION

In summary, exogenous and endogenous sources of oxidants can result in oxidative stress. The body reacts by orchestrating an acute inflammatory response to alleviate this insult. While acute inflammation results in restoration of normal physiological functions in the majority of situations by replacing injured tissue with scar tissue, the persistence of oxidants in the alveolar microvasculature leads to chronic inflammation, which exacerbates the production of ROS/NOS. Elevated levels of ROS/RNS result in preneoplastic DNA mutations and growth factor activation, which eventually lead to malignant transformations. Therefore, lung diseases and events associated with chronic inflammation are the major risk factors of lung cancer. Cigarette smoking is not only in itself a major source of exogenous oxidants, but the chronic inflammation it triggers also leads to elevated levels of ROS and RNS, both of which contribute to oxidative stress in lung tissues.

SUMMARY POINTS

- There exist numerous endogenous and exogenous sources of oxidants to which the lung is exposed.
- This surplus of oxidants depletes natural alveolar antioxidants, resulting in oxidative stress.
- Oxidative stress leads to the release of pro-inflammatory cytokines and damage at the cellular level.

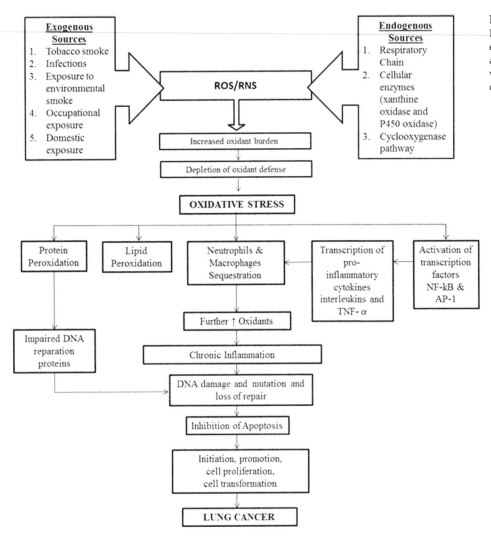

FIGURE 3.4 **Oxidative Stress and Lung Carcinogenesis.** Endogenous and exogenous sources of oxidants result in an oxidant/antioxidant disequilibrium, which causes cell damage. Prolonged oxidative injury can lead to lung cancer.

- Prolonged inflammation results in DNA damage and mutation and an inhibition of cellular repair mechanisms, which predispose the cell to neoplastic transformation.
- Based on results from several large trials, there exists no evidence to support the role of antioxidants as chemopreventative agents for lung cancer.
- The mechanism underpinning oxidative stress and lung cancer is summarized in Figure 3.4.

References

1. D'Amato G, Liccardi G, D'Amato M, Cazzola M. Respiratory allergic diseases induced by outdoor air pollution in urban areas. *Monaldi Archives for Chest Disease* 2002;**57**:161–3.
2. Schwela D. Air pollution and health in urban areas. *Rev Environ Health* 2000;**15**:13–42.
3. Li X, Fang P, Mai J, Choi ET, Wang H, Yang XF. Targeting mitochondrial reactive oxygen species as novel therapy for inflammatory diseases and cancers. *J Hematol Oncol* 2013;**19**.
4. Araneda OF, Tuesta M. Lung oxidative damage by hypoxia. *Oxid Med Cell Longevity* 2012.
5. Gould NS, Min E, Gauthier S, Chu HW, Martin R, Day BJ. Aging adversely affects the cigarette smoke-induced glutathione adaptive response in the lung. *Am J Respir Crit Care Med* 2010;**182**:1114–22.
6. Federico A, Morgillo F, Tuccillo C, Ciardiello F, Loguercio C. Chronic inflammation and oxidative stress in human carcinogenesis. *Int J Cancer* 2007;**121**:2381–6.
7. National Lung Screening Trial Research Team Aberle DR, Adams AM, Berg CD, Black WC, Clapp JD, Fagerstrom RM, et al. Reduced lung-cancer mortality with low-dose computed tomographic screening. *N Engl J Med* 2011;**365**:395–409.
8. Parkin M, Tyczynski JE, Boffetta P, Samet J, Shields P, Caporaso N. Tumours of the lung. In: Travis WDBE, Müller-Hermelink HK, Harris CC, editors. Pathology and genetics of tumours of the lung. WHO Publications Center; 2004. p. 12–5.
9. Couraud S, Zalcman G, Milleron B, Morin F, Souquet PJ. Lung cancer in never smokers - A review. *Eur J Cancer* 2012;**48**:1299–311.
10. Thomas JL, Guo H, Carmella SG, Balbo S, Han S, Davis A, et al. Metabolites of a tobacco-specific lung carcinogen in children exposed to secondhand or thirdhand tobacco smoke in their homes. *Cancer Epidemiol Biomark Prev* 2011;**20**:1213–21.
11. Koshiol J, Rotunno M, Gillison ML, Van Doorn LJ, Chaturvedi AK, Tarantini L, et al. Assessment of human papillomavirus in lung tumor tissue. *J Natl Cancer Inst* 2011;**103**:501–7.

12. Shiels MS, Albanes D, Virtamo J, Engels EA. Increased risk of lung cancer in men with tuberculosis in the alpha-tocopherol, beta-carotene cancer prevention study. *Cancer Epidemiol Biomark Prev* 2011;**20**:672–8.

13. Darby S, Hill D, Auvinen A, Barros-Dios JM, Baysson H, Bochicchio F, et al. Radon in homes and risk of lung cancer: Collaborative analysis of individual data from 13 European case-control studies. *Br Med J* 2005;**330**:223–6.

14. Sun S, Schiller JH, Gazdar AF. Lung cancer in never smokers—A different disease. *Nat Rev Cancer* 2007;**7**:778–90.

15. Rahman I, Biswas SK, Kode A. Oxidant and antioxidant balance in the airways and airway diseases. *Eur J Pharmacol* 2006;**533**:222–39.

16. Pryor WA, Stone K, Cross CE, Machlin L, Packer L. Oxidants in cigarette smoke: Radicals, hydrogen peroxide, peroxynitrate, and peroxynitrite. *Ann N Y Acad Sci* 1993;**686**:12–28.

17. Valavanidis A, Vlachogianni T, Fiotakis K. Tobacco smoke: Involvement of reactive oxygen species and stable free radicals in mechanisms of oxidative damage, carcinogenesis and synergistic effects with other respirable particles. *Int J Environ Res Public Health* 2009;**6**:445–62.

18. Sarir H, Mortaz E, Janse WT, Givi ME, Nijkamp FP, Folkerts G. IL-8 production by macrophages is synergistically enhanced when cigarette smoke is combined with TNF-α. *Biochem Pharmacol* 2010;**79**(5):698–705.

19. Chan HP, Tran V, Lewis C, Thomas PS. Elevated levels of oxidative stress markers in exhaled breath condensate. *J Thorac Oncol* 2009;**4**:172–8.

20. Fubini B, Hubbard A. Reactive oxygen species (ROS) and reactive nitrogen species (RNS) generation by silica in inflammation and fibrosis. *Free Radical Biol Med* 2003;**34**:1507–16.

21. Fubini B, Giamello E, Volante M, Bolis V. Chemical functionalities at the silica surface determining its reactivity when inhaled. Formation and reactivity of surface radicals. *Toxicol Ind Health* 1990;**6**:571–98.

22. Fubini B, Bolis V, Giamello E. The surface chemistry of crushed quartz dust in relation to its pathogenicity. *Inorg Chim Acta* 1987;**138**:193–7.

23. Fubini B, Fenoglio I, Elias Z, Poirot O. Variability of biological responses to silicas: Effect of origin, crystallinity, and state of surface on generation of reactive oxygen species and morphological transformation of mammalian cells. *J Environ Pathol, Toxicol Oncol* 2001;**20**:95–108.

24. Guidotti TL, Miller A, Christiani D, Harbut MR, Hillerdal G, Balmes JR, et al. Diagnosis and initial management of nonmalignant diseases related to asbestos. *Am J Respir Crit Care Med* 2004;**170**:691–715.

25. Kamp DW, Graceffa P, Pryor WA, Weitzman SA. The role of free radicals in asbestos-induced diseases. *Free Radical Biol Med* 1992;**12**:293–315.

26. Liu G, Beri R, Mueller A, Kamp DW. Molecular mechanisms of asbestos-induced lung epithelial cell apoptosis. *Chem-Biol Interact* 2010;**188**:309–18.

27. Shuguang L, Dinhua P, Guoxiong W. Analysis of polycyclic aromatic hydrocarbons in cooking oil fumes. *Arch Environ Health* 1994;**49**:119–22.

28. Fu PP, Xia Q, Sun X, Yu H. Phototoxity and environmental transformation of polycyclic aromatic hydrocarbons (PAHs)-light-induced reactive oxygen species, lipid peroxidation, and DNA damage. *J Environ Sci Health, Part C: Environ Carcinog Ecotoxicol Rev* 2012;**30**:1–41.

29. Yu H. Environmental carcinogenic polycyclic aromatic hydrocarbons: Photochemistry and phototoxicity. *J Environ Sci Health, Part C: Environ Carcinog Ecotoxicol Rev* 2002;**20**:149–83.

30. Foote CS. Definition of type I and type II photosensitized oxidation. *Photochem Photobiol* 1991;**54**:659.

31. Sauer H, Wartenberg M, Hescheler J. Reactive oxygen species as intracellular messengers during cell growth and differentiation. *Cell Physiol Biochem* 2001;**11**:173–86.

32. Emmendoerffer A, Hecht M, Boeker T, Mueller M, Heinrich U. Role of inflammation in chemical-induced lung cancer. *Toxicol Lett* 2000;**112-113**:185–91.

33. Brody JS, Spira A. Chronic obstructive pulmonary disease, inflammation, and lung cancer. *Proc Am Thorac Soc* 2006;**3**:535–7.

34. Barnes PJ. Immunology of asthma and chronic obstructive pulmonary disease. *Nat Rev Immunol* 2008;**8**:183–92.

35. Drost EM, Skwarski KM, Sauleda J, Roca J, Agusti A, MacNee W. Oxidative stress and airway inflammation in severe exacerbations of COPD. *Thorax* 2005;**60**:293–300.

36. Azad N, Rojanasakul Y, Vallyathan V. Inflammation and lung cancer: Roles of reactive oxygen/nitrogen species. *J Toxicol Environ Health, Part B* 2008;**11**:1–15.

37. Zieba M, Suwalski M, Kwiatkowska S, Piasecka G, Grzelewskarzymowska I, Stolarek R, et al. Comparison of hydrogen peroxide generation and the content of lipid peroxidation products in lung cancer tissue and pulmonary parenchyma. *Respir Med* 2000;**94**:800–5.

38. Hecht SS. Lung carcinogenesis by tobacco smoke. *Int J Cancer* 2012;**131**:2724–32.

39. Hecht SS. Cigarette smoking and lung cancer: Chemical mechanisms and approaches to prevention. *Lancet Oncol* 2002;**3**:461–9.

40. Adcock IM, Caramori G, Barnes PJ. Chronic Obstructive pulmonary disease and lung cancer: New molecular insights. *Respir Int Rev Thoracic Dis* 2011;**81**:265–84.

41. Feig DI, Reid TM, Loeb LA. Reactive oxygen species in tumorigenesis. *Cancer Res* 1994;**54** 1890s-4s.

42. Klaunig JE, Kamendulis LM. The Role of Oxidative Stress in Carcinogenesis. *Annu Rev Pharmacol Toxicol* 2004;**44**:239–67.

43. Marnett LJ. Oxyradicals and DNA damage. *Carcinogenesis* 2000;**21**:361–70.

44. Szabó C, Ohshima H. DNA damage induced by peroxynitrite: Subsequent biological effects. *Nitric Oxide* 1997;**1**:373–85.

45. Ai SSY, Hsu K, Cheng Z, Hunt J, Lewis CR, Thomas PS. Mitochondrial DNA mutations in exhaled breath condensate of patients with lung cancer. *Respir Med* 2013;**18**.

46. Dela Cruz CS, Tanoue LT, Matthay RA. Lung Cancer: Epidemiology, Etiology, and Prevention. *Clin Chest Med* 2011;**32**:605–44.

47. Dubey S, Powell CA. Update in lung cancer 2007. *Am J Respir Crit Care Med* 2008;**177**:941–6.

48. Punturieri A, Szabo E, Croxton TL, Shapiro SD, Dubinett SM. Lung cancer and chronic obstructive pulmonary disease: Needs and opportunities for integrated research. *J Natl Cancer Inst* 2009;**101**:554–9.

49. Belinsky SA. Role of the cytosine DNA-methyltransferase and p16(INK4a) genes in the development of mouse lung tumors. *Exp Lung Res* 1998;**24**:463–79.

50. Bennett WP, Colby TV, Travis WD, Borkowski A, Jones RT, Lane DP, et al. p53 Protein accumulates frequently in early bronchial neoplasia. *Cancer Res* 1993;**53**:4817–22.

51. Volm M, Van Kaick G, Mattern J. Analysis of c-fos, c-jun, c-erbB1, c-erbB2 and c-myc in primary lung carcinomas and their lymph node metastases. *Clin Exp Metastasis* 1994;**12**:329–34.

52. Sancar A, Lindsey-Boltz LA, Ünsal-Kaçmaz K, Linn S. Molecular mechanisms of mammalian DNA repair and the DNA damage checkpoints. *Annu Rev Biochem* 2004;**73**:39.

53. Nagorni-Obradović L, Pesut D, Skodrić-Trifunović V, Adzić T. Influence of tobacco smoke on the appearance of oxidative stress in patients with lung cancer and chronic obstructive pulmonary diseases. *Vojnosanit Pregl* 2006;**63**:893–5.

54. Kinnula VL. Focus on antioxidant enzymes and antioxidant strategies in smoking related airway diseases. *Thorax* 2005;**60**:693–700.

1. OXIDATIVE STRESS AND CANCER

55. Lippman SM, Lee JJ, Karp DD, Vokes EE, Benner SE, Goodman GE, et al. Randomized phase III intergroup trial of isotretinoin to prevent second primary tumors in stage I non-small-cell lung cancer. *J Natl Cancer Inst* 2001;**93**:605–17.

56. Van Zandwijk N, Dalesio O, Pastorino U, De Vries N, Van Tinteren H. EUROSCAN, a randomized trial of vitamin A and N-acetylcysteine in patients with head and neck cancer or lung cancer. *J Natl Cancer Inst* 2000;**92**:977–86.

57. Cortés-Jofré M, Rueda JR, Corsini-Muñoz G, Fonseca-Cortés C, Caraballoso M, Bonfill Cosp X. Drugs for preventing lung cancer in healthy people. *Cochrane database of systematic reviews (Online)* 2012;**10**.

58. Group TA- TBCCPS. The effect of vitamin E and beta carotene on the incidence of lung cancer and other cancers in male smokers. *N Engl J Med* 1994;**330**:1029–35.

59. Kelly K, Kittelson J, Franklin WA, Kennedy TC, Klein CE, Keith RL, et al. A randomized phase II chemoprevention trial of 13-cis retinoic acid with or without α tocopherol or observation in subjects at high risk for lung cancer. *Cancer Prev Res* 2009;**2**:440–9.

60. Omenn GS, Goodman GE, Thornquist MD, Balmes J, Cullen MR, Glass A, et al. Effects of a combination of beta carotene and vitamin A on lung cancer and cardiovascular disease. *N Engl J Med* 1996;**334**:1150–5.

4

Oxidative Stress and Stomach Cancer

Hidekazu Suzuki

Division of Gastroenterology and Hepatology, Department of Internal Medicine, Keio University School of Medicine, Tokyo, Japan

Toshihiro Nishizawa

Division of Gastroenterology, National Hospital Organization Tokyo Medical Center, Tokyo, Japan

List of Abbreviations

8-hydroxy-dGTP 8-hydroxy-deoxyguanosine triphosphate
8-OHdG 8-hydroxy-2-deoxyguanosine
8-oxo-Gua 8-oxo-7,8-dihydroguanine
95% CI 95% confidence interval
CagA cytotoxin-associated gene A product
DNMT1 DNA methyltransferase 1
GGT γ-glutamyltranspeptidase
GPx glutathione peroxidase
HR hazard ratio
HDAC1 histone deacetylase 1
iNOS inducible nitric oxide synthase
MPO myeloperoxidase,
NAC *N*-acetyl cysteine
Nrf2 NF-E2 p45-related factor-2
NO nitric oxide
NSAIDs nonsteroidal anti-inflammatory drugs
NER nucleotide excision repair
RNS reactive nitrogen species
ROS reactive oxygen species
RUNX3 Runt domain transcription factor 3
SeCys selenocysteine
SePs selenoproteins
TrxR thioredoxin reductase
SePP selenoprotein P
VacA vacuolating cytotoxin A

INTRODUCTION

Oxidative stress is the disturbance of oxidant–antioxidant homeostasis, leading to potential cellular damage. Inflammatory cells, which are activated during inflammation, trigger the generation of oxidant-generating enzymes such as inducible nitric oxide synthase (iNOS), NADPH oxidase, and myeloperoxidase (MPO) to produce high concentrations of free radicals, including reactive nitrogen species (RNS) and reactive oxygen species (ROS). Although ROS are crucial regulators of cellular signal transduction and energy transmission, disturbance in the balance between the generating and scavenging capability of ROS may lead to cellular damage. ROS and RNS can react with cellular proteins or lipids, transforming them into oxidized forms, or bind with nucleic acids, turning them into mutated forms that can play a role in the multistage carcinogenic process.

Helicobacter pylori (*H. pylori*) is the main cause of chronic gastritis and peptic ulcers, and is a potential risk factor for gastric adenocarcinoma. The molecular mechanisms of *H. pylori*-induced production of ROS/ RNS are wide ranging and include the activation of neutrophils by *H. pylori* itself. It is currently thought that gastric cancer arises from multiple "hits" including oxidative stress and environmental toxins, which increase DNA mutation rates. In addition, many other factors might be involved in the pathogenesis of gastric cancer, such as host factors (e.g., interleukin-8 genetic polymorphism, cyclooxygenase expression, and heat shock protein 70 expression), bacterial factors (e.g., cytotoxin-associated gene A product [CagA] and vacuolating cytotoxin A [VacA]; Table 4.1). As a result, it is difficult to understand the key mechanisms underlying *H. pylori* contributions to gastric carcinogenesis.

This chapter summarizes oxidative stress in gastric mucosa, the possible mechanisms of gastric carcinogenesis, and the role of antioxidants in gastric cancer (Table 4.2).

© 2014 Elsevier Inc. All rights reserved.

TABLE 4.1 Pathogenic Factors of *H. pylori*

Pathogenic Factors	Function
Cytotoxin-associated gene A product (CagA)	Interference with multiple host cell signaling cascades, cell proliferation
Type-IV secretion system	Injection of bacterial toxins
Vacuolating cytotoxin A (VacA)	Vacuolation of gastric epithelial cells
Urease	Neutralization of gastric acid
Flagella	Motility
OipA, BabA, AlpA/B, SabA	Bacterial adhesion to gastric mucosa
Duodena ulcer promoting gene (DupA)	Association with an increased risk of duodenal ulcer

TABLE 4.2 Stomach Cancer Risk Factors

Evidence Level	Risk Factor	Inhibitory Factor
Sufficient	*H. pylori* infection	*H. pylori* eradication
Fair	High salt, smoking	Fruits, vegetables, green tea
Insufficient	Alcohol consumption	Selenium, β-carotene, sulforaphane

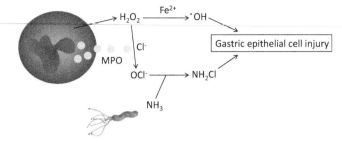

FIGURE 4.1 *Helicobacter pylori*-associated gastric oxidative injury: monochloramine pathway. MPO: myeloperoxidase. (See the color plate.)

H. PYLORI AND OXIDATIVE STRESS

Oxygen radicals or ROS released from activated neutrophils are considered to be potential toxic factors involved in *H. pylori*-induced gastric mucosal injury. *H. pylori* exhibits chemotactic activity for neutrophils (polymorphonuclear cells), and neutrophil infiltration of the gastric mucosa leads to the development of the initial lesions of *H. pylori*-associated active gastritis. Neutrophils express an enzyme, MPO, which produces the hypochlorous anion (OCl^-), a potent oxidant, generated from H_2O_2 in the presence of Cl^-. This hypochlorous anion reacts with ammonia, produced from urea by *H. pylori*-derived urease, to yield a lipophilic cytotoxic oxidant, monochloramine (NH_2Cl), which freely penetrates biological membranes to oxidize intracellular components[1,2,3] (Figure 4.1).

In addition, *H. pylori* produces γ-glutamyltranspeptidase (EC 2.3.2.2; GGT) in the periplasm.[4] *H. pylori* GGT catalyzes the transpeptidation and hydrolysis of the γ-glutamyl groups of glutamine and glutathione. Interestingly, these findings indicate that *H. pylori* GGT performs two functions. First, GGT plays a role in the physiological functioning of *H. pylori*, which is unable to take up extracellular glutamine and glutathione directly, by hydrolyzing these substances to glutamate. Glutamate is then transported into *H. pylori* cells via a Na^+-dependent reaction and is mainly incorporated into the TCA cycle.[5] Second, GGT acts as a virulence factor by disrupting the

antioxidant ability of host cells. Specifically, glutathione has antioxidant potential in host cells, and *H. pylori* GGT reduces extracellular glutathione levels. As a result, *H. pylori* GGT lowers the resistance of the host cells to ROS, thereby inducing their apoptosis or necrosis.[4,6]

Excess ROS are produced in *H. pylori*-colonized human stomachs; both the gastric mucosa and *H. pylori* are exposed to oxidative stress. However, because *H. pylori* has the deft capability of detoxifying ROS using a variety of enzymes to establish long-term colonization,[7,8] excess ROS production leads solely to host cell damage.

We previously reported greater enhancement of neutrophil-derived gastric mucosal ROS levels, detected using luminol-dependent chemiluminescence, in *cagA*-positive patients than in *cagA*-negative patients.[9] *CagA*-positive patients are characterized by increased oxidative DNA damage, both overall and at younger ages, in the presence of multifocal mucosal atrophy.

Infection of gastric epithelial cells with *H. pylori* increases the accumulation of intracellular ROS. Increased intracellular oxidative stress may play a role in the induction of programmed cell death.[10] Increased mitochondrial ROS production is detected in *H. pylori*-infected gastric epithelial cells. This finding suggests that *H. pylori* infection alters mitochondrial function. In addition, the antioxidant *N*-acetyl cysteine (NAC) inhibits the induction of autophagy and induces the accumulation of intracellular CagA, suggesting that intracellular oxidative stress is involved in the induction of autophagy required for CagA degradation.[11]

Runt domain transcription factor 3 (RUNX3) is a tumor suppressor that is silenced in cancer via hypermethylation of its promoter region. RUNX3 mRNA and protein expressions are downregulated in response to hydrogen peroxide (H_2O_2) in the human colorectal cancer cell line SNU-407. This downregulation is abolished with NAC pretreatment. Moreover, methylation-specific PCR data revealed that H_2O_2 treatment increases RUNX3 promoter methylation, whereas NAC and the cytosine methylation inhibitor 5-aza-2-deoxycytidine decrease it, suggesting that an epigenetic regulatory mechanism involving ROS-induced methylation may mediate RUNX3 silencing. H_2O_2 treatment increases DNA methyltransferase 1 (DNMT1) and histone deacetylase 1

(HDAC1) expression and activity, the binding of DNMT1 to HDAC1, and DNMT1 binding to the RUNX3 promoter. In addition, 5-aza-2-deoxycytidine treatment prevents the decrease in RUNX3 mRNA and protein levels caused by H_2O_2 treatment. Additionally, H_2O_2 inhibits the nuclear localization and expression of RUNX3, which is abolished by NAC. Furthermore, downregulation of RUNX3 expression by H_2O_2 also influences cell proliferation.[12] These data provide further support for the proposition that ROS silence the tumor suppressor RUNX3 via epigenetic regulation. The eradication of *H. pylori* also increases RUNX3 expression in glandular epithelial cells in enlarged fold gastritis.

Recently, we reported that RUNX3 is expressed in gastric epithelial cells and that *H. pylori* eradication significantly increases RUNX3 expression in the glandular epithelium of the corpus; however, no changes were observed in the antrum.[13] The mucosal levels of ROS, measured using chemiluminescence, are fourfold higher in the corpus than in the antrum, whereas *H. pylori* eradication significantly decreases the mucosal chemiluminescence values in both portions of the stomach to nearly undetectable levels. These data suggest that the glandular epithelium is exposed to high levels of carcinogenic oxidative stress, which results in reduced expression of the tumor-suppressing molecule RUNX3. The expression of RUNX3 was restored after the eradication of the bacterium, suggesting that a high risk of carcinogenesis is associated with *H. pylori*-induced enlarged fold gastritis especially in the corpus mucosa.

ETHANOL

In the gut, ROS can also be generated by ethanol, nonsteroidal anti-inflammatory drugs (NSAIDs), and ischemia/reperfusion. The effect of ethanol on gastric mucosa is a complicated and multifaceted process. It may be associated with disturbance of the balance between gastric mucosal defense and offensive factors. Gastric mucosa contains gastric acid, pepsin, stimulant, and such, whereas the defensive factors contain gastric slime layer, mucosal blood flow, HCO^{3-}, prostaglandins, epidermal growth factor, and epithelial cell renewing. Ethanol induces vascular endothelium injury of gastric mucosa, disorder of microcirculation, and ischemia as a result of more production of oxygen free radicals. Pan et al. reported the role of mitochondrial energy disorder in the pathogenesis of ethanol-induced gastric mucosal injury.[14] The gastric mucosal lesion index was correlated with the malondialdehyde content in gastric mucosa. As the concentration of ethanol was elevated and the exposure time to ethanol was extended, the content of malondialdehyde in gastric mucosa increased and the extent of damage aggravated. The ultrastructure of mitochondria was positively related to the ethanol concentration and exposure time. The expression of mitochondrial DNA ATPase subunits 6 and 8 mRNA declined with the increasing malondialdehyde content in gastric mucosa after gavage with ethanol. Ethanol-induced gastric mucosa injury is related to oxidative stress, which disturbs energy metabolism of mitochondria and plays a critical role in the pathogenesis of ethanol-induced gastric mucosa injury.

Duell et al. evaluated the association between alcohol consumption and gastric cancer risk.[15] They conducted a prospective analysis in the European Prospective Investigation into Cancer and Nutrition cohort, which included 444 cases of first primary gastric adenocarcinoma. Heavy (compared with very light) alcohol consumption (≥ 60 compared with 0.1–4.9 g/d) at baseline was positively associated with gastric cancer risk (Hazard ratio (HR): 1.65; 95% confidence interval (95% CI): 1.06–2.58), whereas lower consumption amounts (< 60 g/d) were not. When gastric cancer risk was analyzed by type of alcoholic beverage, there was a positive association for beer (≥ 30 g/d; HR: 1.75; 95% CI: 1.13–2.73) but not for wine or liquor. Associations were primarily observed at the highest amounts of drinking in men and limited to noncardia subsite and intestinal histology. Heavy (but not light or moderate) consumption of alcohol (mainly from beer) is associated with intestinal-type noncardia gastric cancer risk in men.

NONSTEROIDAL ANTI-INFLAMMATORY DRUGS AND ASPIRIN

In addition to inhibiting cyclooxygenase and decreasing prostaglandin production, NSAIDs induce mucosal damage through ROS produced by recruited leukocytes. ROS-mediated mitochondrial damage, oxidation of lipids, proteins, and DNA leads to cellular apoptosis and mucosal injury. Proton pump inhibitor therapy is thought to primarily protect gastric mucosa by inhibiting gastric acid secretion. Nevertheless, Maity et al. recently demonstrated that the lansoprazole also inhibits NSAID-induced gastropathy by inhibiting mitochondrial and Fas-mediated apoptosis pathway.[16] Lansoprazole antiapoptotic activity appears to be mediated by preventing NSAID-induced reductions in the antiapoptotic genes Bcl and Bcl-2, while inhibiting increases in Fas and Fas ligand and the proapoptotic genes Bax and Bak.

Aspirin increases the permeability of cultured gastric epithelial cell monolayers. The disruption in barrier integrity was mediated by p38 mitogen-activated protein kinase and involves down-regulation of claudin-7, a component protein of tight junctions.[17] This differs from the effects of other NSAIDs, which increase epithelial permeability coupled to cyclooxygenase-1 inhibition and are restored by the administration of prostaglandin E_2.

ISCHEMIA/REPERFUSION INJURY

Ischemia/reperfusion damages the gastric mucosa by inducing oxidative stress. Specifically, ROS such as superoxide and hydrogen peroxide induce inflammatory responses and tissue damage by fragmenting cellular DNA. NADPH oxidase found on phagocytic cells, vascular muscular cells, endothelial cells, fibroblasts, and adipocytes converts oxygen into superoxide anions. Nakagiri et al. recently reported that NADPH oxidase activity is increased in ischemia and ischemia/reperfusion and is involved in the resulting gastric mucosal damage.[18] The increased NADPH oxidase activity may also induce up-regulation of cyclooxygenase-2.

Peskar et al. reported that during ischemia/reperfusion, inhibitors of the cyclooxygenase and lipoxygenase pathways increased gastric mucosal damage in a dose-dependent manner.[19] Synergism observed with the combination of cyclooxygenase and lipoxygenase pathway inhibitors suggests that both pathways are important in gastric mucosal defense during ischemia/reperfusion. Prostaglandin E_2 antagonized the effects of cyclooxygenase and lipoxygenase pathway inhibitors. Similarly, lipoxin A_4, a lipoxygenase-derived product of arachidonate metabolism, also antagonized the effects of cyclooxygenase and lipoxygenase pathway inhibitors, and could replace prostaglandins E_2 in the prevention of gastric mucosal damage caused by cyclooxygenase inhibitor during ischemia/reperfusion.

We previously reported altered gastric motility after gastric ischemia/reperfusion.[20] Ghrelin is a gastrointestinal peptide that stimulates food intake and gastrointestinal motility. Wistar rats were exposed to 80-min gastric ischemia, followed by 12-h or 48-h reperfusion. Food intake, the plasma ghrelin levels, the count of ghrelin-immunoreactive cells corrected by the percentage areas of the remaining mucosa, and the expression levels of preproghrelin mRNA in the stomach were significantly reduced at 12 h and 48 h after ischemia/reperfusion compared with the levels in the sham-operated rats. Intraperitoneal administration of ghrelin significantly reversed the decrease of food intake after ischemia/reperfusion.[21] These data show that gastric ischemia and reperfusion evoked anorexia with decreased plasma ghrelin levels and ghrelin production, which appears to be attributable to the ischemia/reperfusion-induced gastric mucosal injuries.

REACTIVE OXYGEN SPECIES

When DNA is attacked by ROS, ultraviolet light, or genotoxic agents, guanine is easily oxidized into 8-oxo-7,8-dihydroguanine (8-oxo-Gua). The existence of this oxidized guanine in genomic DNA can cause transversional mutations such as G → T or G → A, accumulation of which has detrimental consequences.[22] Fortunately, mammalian cells have multiple repair systems such as base excision repair enzymes and nucleotide excision repair (NER) enzymes, which counteract the hazardous effects of 8-oxo-Gua. Consequently, 8-hydroxy-2-deoxyguanosine (8-OHdG), a nucleoside form of 8-oxo-Gua, is generated either from damaged oligomers containing 8-oxo-Gua by NER enzymes or from cytoplasmic oxidized nucleotides such as 8-hydroxy-deoxyguanosine triphosphate (8-hydroxy-dGTP; Figure 4.2). Exogenously administered 8-OHdG cannot be reincorporated into genomic DNA because the activity of deoxynucleotide kinase, which converts 8-OHdG into 8-hydroxy-dGTP, is very low, although wild deoxyguanosine can be readily converted to deoxyguanosine triphosphate, which can be used as a substrate for DNA polymerase.

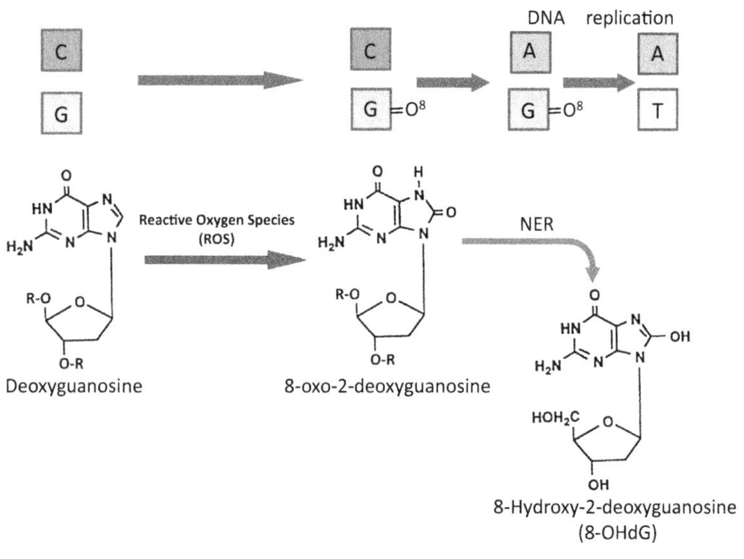

FIGURE 4.2　Point mutation induced by 8-oxo-2-deoxyguanosine via induction of the G:C → T:A transversion. NER: nucleotide excision repair enzyme. (See the color plate.)

Unlike any other molecule that contains oxidized guanine, 8-OHdG can pass through the cellular membrane, and is usually detected in the urine or sera of patients with diseases associated with oxidative stress. Thus, this metabolite has been used widely in many studies as a biomarker for the measurement of endogenous oxidative DNA damage and as a risk factor for many diseases including cancers. Urinary 8-OHdG has proven to be a good biomarker for the risk assessment of various cancers and other degenerative diseases.

Patients with *cagA*-positive strains of *H. pylori* have higher 8-OHdG levels than do *cagA*-negative or *H. pylori*-negative patients. 8-OHdG staining in the tissues of chronic atrophic gastritis and gastric cancer patients is stronger than in control subjects.[23]

REACTIVE NITROGEN SPECIES

Nitric oxide (NO), a primary initiator of RNS, is generated specifically during inflammation via iNOS in inflammatory and epithelial cells. Overproduction of NO induces the generation of peroxynitrite (ONOO⁻), which can lead to the formation of 8-nitroguanine, an indicator of nitrative DNA damage. 8-Nitroguanine undergoes spontaneous depurination in DNA, resulting in the formation of an apurinic site. Incorporated adenine can form a pair with apurinic sites during DNA replication, leading to a G → T transversion. Therefore, 8-nitroguanine is a potential mutagenic DNA lesion involved in inflammation-mediated carcinogenesis (Figure 4.3).[24]

In patients with *H. pylori*-positive gastritis or gastric ulcers, iNOS is expressed in the infiltrating inflammatory cells. The expression of iNOS mRNA and protein in the gastric epithelial cells is significantly increased in patients with *H. pylori*-positive gastritis compared to that

in patients who are *H. pylori* negative. Ma et al. showed that the levels of 8-nitroguanine and 8-OHdG in gastric gland epithelium are significantly higher in gastritis patients with an *H. pylori* infection than in those without such an infection.[25] Therefore, 8-nitroguanine may not only be a promising biomarker for inflammation but also a useful indicator of the risk of developing gastric cancer in response to chronic *H. pylori* infection.

ANTIOXIDANTS AND STOMACH CANCER

The association of the dietary intake of fruits, vegetables, and antioxidants as well as the baseline serum levels of antioxidants with the subsequent incidence of gastric cardia cancer and gastric noncardia cancer was determined in a prospective cohort study.[26] Participants of the study were 29,133 male smokers recruited into the α-Tocopherol, β-Carotene Cancer Prevention study between 1985 and 1988. At baseline, a self-administered food use questionnaire with 276 food items was used to assess dietary intake. Baseline serum samples were stored at –70 °C. During a median follow-up of 12 years, 243 incidents of gastric adenocarcinomas (64 gastric cardia cancer and 179 gastric noncardia cancer) were diagnosed in this cohort. For gastric cardia cancer, high dietary intake of retinol was protective (HR 0.46; 95% confidence interval [95% CI], 0.27–0.78), but high intake of α-tocopherol (HR, 2.06; 95% CI, 1.20–3.54) or γ-tocopherol (HR, 1.94; 95% CI, 1.13–3.34) increased risk. For gastric noncardia cancer, higher intake of fruits (HR, 0.51; 95% CI, 0.37–0.71), vitamin C (HR, 0.60; 95% CI, 0.41–0.86), α-tocopherol (HR, 0.78; 95% CI, 0.55–1.10), γ-tocopherol (HR, 0.69; 95% CI, 0.49–0.96), or lycopene (HR, 0.67; 95% CI, 0.47–0.95) were protective. The results

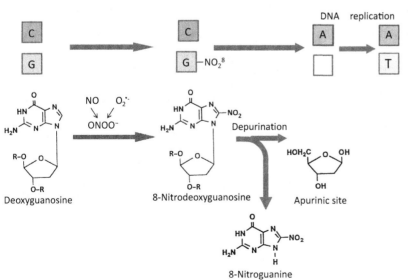

FIGURE 4.3 Point mutation induced by 8-nitrodeoxyguanosine via induction of the G:C → T:A transversion. (See the color plate.)

TABLE 4.3 Stomach Cancer Prevention with Antioxidants

Antioxidants	Study Design	Efficacy
Fruits, vegetables	Prospective cohort study	Effective
β-Carotene, vitamins A, C	Meta-analyses	Not effective
Vitamin E	Meta-analyses Randomized trial	Not effective Increased risk (gastric cardia cancer)
Green tea	Prospective cohort study	Effective
Selenium	Meta-analyses Four randomized trials	Not effective Effective
Sulforaphane	Animal study	Effective

suggest a difference in the effect of some of these exposures on gastric cardia cancer and gastric noncardia cancer. Tocopherols were associated with a higher risk of gastric cardia cancer, whereas dietary intake of fruits, vitamin C, tocopherols, and lycopene seemed protective for gastric noncardia cancer.

Epidemiologic evidence indicates that diets high in fruits and vegetables are associated with a reduced risk of several cancers (Table 4.3). In China, Blot et al. sought to determine if dietary supplementation with specific vitamins and minerals can lower mortality from or incidence of cancer as well as mortality from other diseases.[27] Mortality and cancer incidence between March 1986 and May 1991 were ascertained for 29,584 adults who received daily vitamin and mineral supplementation throughout this period. The subjects were randomly assigned to four intervention groups: (A) retinol and zinc; (B) riboflavin and niacin; (C) vitamin C and molybdenum; and (D) β-carotene, vitamin E, and selenium. A total of 2,127 deaths occurred among trial participants during the intervention period. Cancer was the leading cause of death, with 32% of all deaths caused by esophageal or stomach cancer, followed by cerebrovascular disease (25%). Significantly lower total mortality (relative risk, 0.91; 95% CI, 0.84–0.99) was observed among those receiving supplementation with β-carotene, vitamin E, and selenium (P = 0.03). The reduction was mainly due to lower cancer rates (relative risk, 0.87; 95% CI, 0.75–1.00), especially stomach cancer (relative risk, 0.79; 95% CI, 0.64–0.99), with the reduced risk beginning to arise approximately 1 to 2 years after the start of supplementation with these vitamins and minerals. No significant effects on mortality rates from all causes were found for supplementation with retinol and zinc, riboflavin and niacin, or vitamin C and molybdenum. Patterns of cancer incidence, on the basis of 1,298 cases, generally resembled those for cancer mortality, indicating that vitamin and mineral supplementation, particularly the combination of β-carotene, vitamin E, and selenium, may cause a reduction in cancer risk in this population.

Bjelakovic et al. reviewed all randomized trials comparing antioxidant supplements with placebo for the prevention of gastrointestinal cancers.[28] Fourteen randomized trials (n = 170,525) were identified and trial quality was generally high. Neither the fixed-effect (relative risk, 0.96; 95% CI, 0.88–1.04) nor random-effects (relative risk, 0.90, 95% CI, 0.77–1.05) meta-analyses showed significant effects of supplementation with β-carotene; vitamins A, C, or E; or selenium (alone or in combination) versus placebo on esophageal, gastric, colorectal, pancreatic, or liver cancer incidences. In seven high-quality trials (n = 131,727), the fixed-effect model showed that antioxidants significantly increased mortality (relative risk, 1.06; 95% CI, 1.02–1.10), whereas the random-effects meta-analysis did not (relative risk, 1.06; 95% CI, 0.98–1.15). The authors could not find any evidence that antioxidant supplements can prevent gastrointestinal cancers. On the contrary, they appear to increase overall mortality. However, in four trials, selenium showed significant beneficial effect on the incidence of gastrointestinal cancer.

Selenium (Se) exists in the human body in the form of selenocysteine (SeCys), a component of selenoproteins (SePs), and appears to have important functions associated with antioxidant activity. Twenty-five SePs have been identified to date, out of a possible 50 thought to exist. Human defense mechanisms against ROS are amplified by Se. In particular, SeCys reduces the levels of H_2O_2 and may act as an antitumor agent. Glutathione peroxidase (GPx), thioredoxin reductase (TrxR), and selenoprotein P (SePP) are the main SePs containing molecular Se within their active center. SePs regulate the cellular antioxidant defense system, DNA damage, and protein function. They also control cell-mediated immunity and B-cell function. The antiproliferative action of Se in the G1-phase of the cell cycle has been well documented in both normal and neoplastic cells. In addition, Se impairs the expression of *c-fos* and *c-myc* oncogenes.[29] The potential chemopreventive effects of Se should be adequately studied in randomized trials.

Glucoraphanin, the inert glucosinolate precursor of sulforaphane (the biologically active isothiocyanate) in cruciferous plants, is hydrolyzed by the enzyme myrosinase, which is present in fresh (uncooked) broccoli and broccoli sprouts.[30] Sulforaphane stimulates transcription factor NF-E2 p45-related factor-2 (Nrf2)-dependent antioxidant enzyme activity, thereby protecting cells from oxidative injury. Recent studies have clearly shown that the genes encoding Nrf2 play an important role in the induction of antioxidant enzymes against oxidative stress. Interestingly, tissue uptake of sulforaphane is greatest in the stomach.[31] These findings may provide a basis for further studies on the chemopreventive activity of sulforaphane in gastric cancer.

Previous experimental studies have suggested many possible anticancer mechanisms for green tea, but epidemiological evidence for the effect of green tea consumption on gastric cancer risk is contradictory. Inoue et al. examined the association between green tea consumption and gastric cancer. They analyzed original data from six cohort studies that measured green tea consumption using validated questionnaires at baseline. During 2,285,968 person-years of follow-up for a total of 219, 080 subjects, 3,577 cases of gastric cancer were identified. Compared with those drinking less than 1 cup/day, no significant risk reduction for gastric cancer was observed with increased green tea consumption in men, even in stratified analyses by smoking status. However, in women, a significantly decreased risk was observed for those with a consumption of 5 or more cups/day (multivariate-adjusted pooled HR, 0.79; 95% CI, 0.65–0.96). This decrease was also significant for the distal subsite (HR, 0.70; 95% CI, 0.50–0.96). In contrast, a lack of association for proximal gastric cancer was consistently seen in both men and women.[32]

Sasazuki et al. recently evaluated the association between green tea consumption and risk for gastric cancer risk among the Japanese population based on a systematic review of epidemiologic evidence. Eight cohort studies and three case-control studies were identified. Overall, no preventive effect was observed for green tea intake in cohort studies. However, a small, consistent risk reduction limited to women was observed, which was confirmed by pooling data of six cohort studies (HR, 0.79; 95% CI, 0.65–0.96 with ≥ 5 cups/day of green tea intake). Thus, epidemiologic evidence indicates that green tea may decrease the risk of gastric cancer in women.[33]

Because *H. pylori* induces oxidative stress in the gastric mucosa, drugs with antioxidant properties are expected to reduce *H. pylori*-induced gastric diseases. Some gastroprotective drugs have been reported to have antioxidant activity. For example, rebamipide, 2-(4-chlorobenzoylamino)-3-[2(1H)-quinolinone-4-yl]-propionic acid, is used for mucosal protection, healing of gastroduodenal ulcers, and treatment of gastritis.[34] This drug exerts its effects by scavenging oxygen radicals such as O_2^-, as shown *in vitro* experiments.[35] Because rebamipide also suppresses iNOS expression in *H. pylori*-infected gastric mucosa, rebamipide is also expected to have a suppressive effect on carcinogenesis.[36] Another gastroprotective drug, polaprezinc, has been reported to suppress NH_2Cl-induced gastric injury (DNA fragmentation of gastric epithelial cells) in *H. pylori*-infected Mongolian gerbils via an O_2-scavenging mechanism.[37,38]

CONCLUSION

As discussed in this chapter, the mechanism by which *H. pylori* induces oxidative stress in the stomach and its relationship with gastric carcinogenesis has been extensively explored. Recent therapeutic options such as gastroprotective agents with antioxidant properties can modulate the level of oxidative stress and enhance anti-inflammatory or antioxidant capacity.[34] However, further research is needed to elucidate the precise mechanism by which *H. pylori* induces gastric carcinogenesis and to provide useful strategies to combat carcinogenic *H. pylori*.

SUMMARY POINTS

- NH_3, generated from urea by *H. pylori*-associated urease, reacts with OCl^-, produced by activated neutrophils, to form highly toxic monochloramine (NH_2Cl) in the stomach.
- Mediators of oxidative stress, such as reactive oxygen species, attack DNA. Guanine is readily oxidized into 8-oxo-2-dihydroguanine, which can cause transversional mutations in genomic DNA, such as $G \rightarrow T$ or $G \rightarrow A$.
- 8-Hydroxy-2-deoxyguanosine (8-OHdG) has been used widely as a biomarker for the measurement of endogenous oxidative DNA damage.
- Peroxynitrite ($ONOO^-$) can lead to the formation of 8-nitroguanine, which undergoes spontaneous depurination in DNA, resulting in the formation of an apurinic site.
- Incorporated adenine can pair with apurinic sites during DNA replication, leading to $G \rightarrow T$ transversions.
- 8-Nitroguanine is an indicator for the measurement of nitrative DNA damage.
- Epidemiologic evidence indicates that diets high in fruits and vegetables are associated with a reduced risk of several cancers, including gastric cancer.

References

1. Suzuki M, Miura S, Suematsu M, Fukumura D, Kurose I, Suzuki H, et al. *Helicobacter pylori*-associated ammonia production enhances neutrophil-dependent gastric mucosal cell injury. *Am J Physiol* 1992;**263**(5 Pt 1):G719–25.
2. Suzuki H, Mori M, Suzuki M, Sakurai K, Miura S, Ishii H. Extensive DNA damage induced by monochloramine in gastric cells. *Cancer Lett* 1997;**115**(2):243–8.
3. Suzuki H, Seto K, Mori M, Suzuki M, Miura S, Ishii H. Monochloramine induced DNA fragmentation in gastric cell line MKN45. *Am J Physiol* 1998;**275**(4 Pt 1):G712–6.
4. Shibayama K, Kamachi K, Nagata N, Yagi T, Nada T, Doi Y, et al. A novel apoptosis-inducing protein from *Helicobacter pylori*. *Mol Microbiol* 2003;**47**(2):443–51.

5. Shibayama K, Wachino J, Arakawa Y, Saidijam M, Rutherford NG, Henderson PJ. Metabolism of glutamine and glutathione via gamma-glutamyltranspeptidase and glutamate transport in *Helicobacter pylori*: Possible significance in the pathophysiology of the organism. *Mol Microbiol* 2007;**64**(2):396–406.

6. Flahou B, Haesebrouck F, Chiers K, Van Deun K, De Smet L, Devreese B, et al. Gastric epithelial cell death caused by *Helicobacter suis* and *Helicobacter pylori* gamma-glutamyl transpeptidase is mainly glutathione degradation-dependent. *Cell Microbiol* 2011. Epub ahead of print.

7. Allen LA. Phagocytosis and persistence of *Helicobacter pylori*. *Cell Microbiol* 2007;**9**(4):817–28.

8. Wang G, Alamuri P, Maier RJ. The diverse antioxidant systems of *Helicobacter pylori*. *Mol Microbiol* 2006;**61**(4):847–60.

9. Suzuki H, Suzuki M, Mori M, Kitahora T, Yokoyama H, Miura S, et al. Augmented levels of gastric mucosal leucocyte activation by infection with cagA gene-positive *Helicobacter pylori*. *J Gastroenterol Hepatol* 1998;**13**(3):294–300.

10. Ding SZ, Minohara Y, Fan XJ, Wang J, Reyes VE, Patel J, et al. *Helicobacter pylori* infection induces oxidative stress and programmed cell death in human gastric epithelial cells. *Infect Immun* 2007;**75**(8):4030–9.

11. Tsugawa H, Suzuki H, Saya H, Hatakeyama M, Hirayama T, Hirata K, et al. *Helicobacter pylori* CagA degradation by reactive oxygen species-mediated autophagy is escaped in CD44 variant-expressing cancer stem-like cells. *Cell Host & Microbe* 2013.

12. Kang KA, Zhang R, Kim GY, Bae SC, Hyun JW. Epigenetic changes induced by oxidative stress in colorectal cancer cells: methylation of tumor suppressor RUNX3. *Tumour Biol* 2012;**33**(2):403–12.

13. Suzuki M, Suzuki H, Minegishi Y, Ito K, Nishizawa T, Hibi T. *H. pylori*-Eradication Therapy Increases RUNX3 Expression in the Glandular Epithelial Cells in Enlarged-Fold Gastritis. *J Clin Biochem Nutr* 2010;**46**(3):259–64.

14. Pan JS, He SZ, Xu HZ, Zhan XJ, Yang XN, Xiao HM, et al. Oxidative stress disturbs energy metabolism of mitochondria in ethanol-induced gastric mucosa injury. *World J Gastroenterol* 2008;**14**(38):5857–67.

15. Duell EJ, Travier N, Lujan-Barroso L, Clavel-Chapelon F, Boutron-Ruault MC, Morois S, et al. Alcohol consumption and gastric cancer risk in the European Prospective Investigation into Cancer and Nutrition (EPIC) cohort. *Am J Clin Nutr* 2011;**94**(5):1266–75.

16. Maity P, Bindu S, Choubey V, Alam A, Mitra K, Goyal M, et al. Lansoprazole protects and heals gastric mucosa from non-steroidal anti-inflammatory drug (NSAID)-induced gastropathy by inhibiting mitochondrial as well as Fas-mediated death pathways with concurrent induction of mucosal cell renewal. *J Biol Chem* 2008;**283**(21):14391–401.

17. Oshima T, Miwa H, Joh T. Aspirin induces gastric epithelial barrier dysfunction by activating p38 MAPK via claudin-7. *Am J Physiol Cell Physiol* 2008;**295**(3):C800–6.

18. Nakagiri A, Murakami M. Roles of NADPH oxidase in occurrence of gastric damage and expression of cyclooxygenase-2 during ischemia/reperfusion in rat stomachs. *J Pharmacol Sci* 2009;**111**(4):352–60.

19. Peskar BM, Ehrlich K, Schuligoi R, Peskar BA. Role of lipoxygenases and lipoxin A(4)/annexin-1 receptor in gastric protection induced by 20% ethanol or sodium salicylate in rats. *Pharmacology* 2009;**84**(5):310–3.

20. Suzuki S, Suzuki H, Horiguchi K, Tsugawa H, Matsuzaki J, Takagi T. et al. Delayed gastric emptying and disruption of the interstitial cells of Cajal network after gastric ischaemia and reperfusion. *Neurogastroenterol Motil* 2010;**22**(5):585–93.

21. Mogami S, Suzuki H, Fukuhara S, Matsuzaki J, Kangawa K, Hibi T. Reduced ghrelin production induced anorexia after rat gastric ischemia and reperfusion. *Am J Physiol Gastrointest Liver Physiol* 2012;**302**(3):G359–64.

22. Suzuki H, Nishizawa T, Tsugawa H, Mogami S, Hibi T. Roles of oxidative stress in stomach disorders. *J Clin Biochem Nutr* 2012;**50**(1):35–9.

23. Ni J, Mei M, Sun L. Oxidative DNA damage and repair in chronic atrophic gastritis and gastric cancer. *Hepatogastroenterology* 2012;**59**(115):671–5.

24. Handa O, Naito Y, Yoshikawa T. *Helicobacter pylori*: A ROS-inducing bacterial species in the stomach. *Inflamm Res* 2010;**59**(12):997–1003.

25. Ma N, Adachi Y, Hiraku Y, Horiki N, Horiike S, Imoto I, et al. Accumulation of 8-nitroguanine in human gastric epithelium induced by *Helicobacter pylori* infection. *Biochem Biophys Res Commun* 2004;**319**(2):506–10.

26. Nouraie M, Pietinen P, Kamangar F, Dawsey SM, Abnet CC, Albanes D, et al. Fruits, vegetables, and antioxidants and risk of gastric cancer among male smokers. *Cancer Epidemiol Biomarkers Prev* 2005;**14**(9):2087–92.

27. Blot WJ, Li JY, Taylor PR, Guo W, Dawsey S, Wang GQ, et al. Nutrition intervention trials in Linxian, China: Supplementation with specific vitamin/mineral combinations, cancer incidence, and disease-specific mortality in the general population. *J Natl Cancer Inst* 1993;**85**(18):1483–92.

28. Bjelakovic G, Nikolova D, Simonetti RG, Gluud C. Antioxidant supplements for preventing gastrointestinal cancers. *Cochrane Database Syst Rev* 2008;(3):CD004183.

29. Charalabopoulos K, Kotsalos A, Batistatou A, Charalabopoulos A, Peschos D, Vezyraki P, et al. Serum and tissue selenium levels in gastric cancer patients and correlation with CEA. *Anticancer Res* 2009;**29**(8):3465–7.

30. Yanaka A. Sulforaphane enhances protection and repair of gastric mucosa against oxidative stress in vitro, and demonstrates anti-inflammatory effects on *Helicobacter pylori*-infected gastric mucosae in mice and human subjects. *Curr Pharm Des* 2011;**17**(16):1532–40.

31. Veeranki OL, Bhattacharya A, Marshall JR, Zhang Y. Organ-specific exposure and response to sulforaphane, a key chemopreventive ingredient in broccoli: Implications for cancer prevention. *Br J Nutr* 2012:1–8.

32. Inoue M, Sasazuki S, Wakai K, Suzuki T, Matsuo K, Shimazu T, et al. Green tea consumption and gastric cancer in Japanese: A pooled analysis of six cohort studies. *Gut* 2009;**58**(10):1323–32.

33. Sasazuki S, Tamakoshi A, Matsuo K, Ito H, Wakai K, Nagata C, et al. Green tea consumption and gastric cancer risk: an evaluation based on a systematic review of epidemiologic evidence among the Japanese population. *Jpn J Clin Oncol* 2012;**42**(4):335–46.

34. Nishizawa T, Suzuki H, Nakagawa I, Minegishi Y, Masaoka T, Iwasaki E, et al. Rebamipide-promoted restoration of gastric mucosal sonic hedgehog expression after early *Helicobacter pylori* eradication. *Digestion* 2009;**79**(4):259–62.

35. Naito Y, Yoshikawa T, Tanigawa T, Sakurai K, Yamasaki K, Uchida M, et al. Hydroxyl radical scavenging by rebamipide and related compounds: Electron paramagnetic resonance study. *Free Radic Biol Med* 1995;**18**(1):117–23.

36. Haruma K, Ito M, Kido S, Manabe N, Kitadai Y, Sumii M, et al. Long-term rebamipide therapy improves *Helicobacter pylori*-associated chronic gastritis. *Dig Dis Sci* 2002;**47**(4):862–7.

37. Suzuki H, Mori M, Seto K, Miyazawa M, Kai A, Suematsu M, et al. Polaprezinc attenuates the *Helicobacter pylori*-induced gastric mucosal leucocyte activation in Mongolian gerbils—A study using intravital videomicroscopy. *Aliment Pharmacol Ther* 2001;**15**(5):715–25.

38. Suzuki H, Mori M, Seto K, Nagahashi S, Kawaguchi C, Morita H, et al. Polaprezinc, a gastroprotective agent: attenuation of monochloramine-evoked gastric DNA fragmentation. *J Gastroenterol* 1999;**34**(Suppl. 11):43–6.

The Role of Oxidative Stress in Ovarian Cancer: Implications for the Treatment of Patients

Matthew White, Joshua Cohen, Charles Hummel, Robert Burky, Ana Cruz, Robin Farias-Eisner

UCLA David Geffen School of Medicine, Department of Obstetrics and Gynecology, Los Angeles, CA, USA

List of Abbreviations

OSE ovarian surface epithelium
OCP oral contraceptive pill
OS oxidative stress
ROS reactive oxygen species
SOD superoxide dismutase
LH luteinizing hormone
FSH follicle stimulating hormone
NO nitric oxide
8-oxodG 8-oxodeoxyguanosine
GC guanine-cytosine
TA thymine-adenine
Gpx glutathione peroxidase
Trx Thioredoxin

INTRODUCTION

As of 2008, ovarian cancer is the eighth most common cancer in the world and the seventh leading cause of cancer death in women worldwide.[1] Every year 220,000 new cases of ovarian cancer are diagnosed and 140,000 women die as a result of the disease. Ovarian cancer is the most deadly gynecologic cancer in the developed world, and in the United States deaths due to ovarian cancer exceed those due to all other gynecological cancers. Unfortunately, progression of the disease is not well understood. It is often undetected until disease has reached an advanced stage with the majority of patients found to have Stage III or IV disease. Those patients with Stage III disease who undergo standard of care treatment with tumor debulking and adjuvant chemotherapy will likely have disease recurrence with an overall survival of approximately 50% five years after diagnosis. Stage IV

disease presentation carries an even more dismal prognosis, demonstrating the importance to better understand the etiology of this aggressive malignancy.

Although the definitive cause of ovarian cancer remains unknown, significant risk factors for the disease have been elucidated. Research has largely focused on the ovarian surface epithelium (OSE), where ovarian cancers may originate. Over 40 years ago Fathalla proposed the incessant ovulation hypothesis, which maintains that the repeated destruction and repair of the OSE at the site of ovulation during the ovulatory cycle provides a mechanism for the pathogenesis of ovarian cancer.[2] Indeed, the protective effects of lactation, multiparity, and oral contraceptive pills (OCPs), which reduce the lifetime number of ovulatory cycles, also reduce the risk of ovarian cancer by as much as 50%.[3,4]

Current research indicates that oxidative stress (OS) plays a major role in many instances of ovarian cancer. The act of ovulation at the ovarian surface involves a series of coordinated biochemical steps, and ovarian cancers are thought to frequently arise from inclusion cysts that may be related to this process.[5] Most notably, inflammatory mediators and reactive oxygen species (ROS) are generated during ovulation, especially during the follicular rupture stage. These ROS can damage nucleic acids, proteins, and lipids and result in DNA strand breaks. When the reparative capacity of epithelial cells is overcome by mutations, apoptosis fails to occur in response to these alterations, and genetic mutations begin to accumulate across generations of cells (see Figure 5.1). Because of the more limited evolutionary pressure on ovarian epithelial cells, which replicate far less often

© 2014 Elsevier Inc. All rights reserved.

FIGURE 5.1 Oxidative Stress Can Damage Critical Ovarian Cell Components. ROS are created endogenously through normal reactions within the cell. Antioxidant enzymes within the cell normally convert these ROS into more harmless substances in the cell. However, when these antioxidant mechanisms are overwhelmed, superoxide radicals are capable of breaking down the outer mitochondrial membrane, initiating a cascade that results in apoptosis. Free radicals within the cytosol can also enter the nucleus and react with DNA. If apoptosis does not occur, mutations within the DNA can be carried on to future generations of cells, potentially deactivating crucial tumor suppressor genes. (See the color plate.)

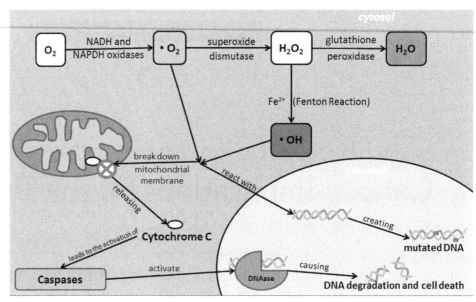

than skin and gastrointestinal mucosal cells, it is possible that the DNA repair mechanism is less vigorous in the surface epithelial cells of the ovary.[6] Consistent with the "multihit hypothesis" of cancer, a decreased repair capacity would accelerate further genetic transformation in which tumor suppressor genes become mutated and cells become malignant. These malignant cells subsequently clonally divide and tumor growth proliferates.[6]

Generally ROS are generated by the reduction of oxygen to water. This occurs via the transformation of oxygen to superoxide, followed by conversion to hydrogen peroxide, the hydroxide radical, and ultimately water. The Fenton reaction, electron transfer, and oxygenase reactions are major contributors of ROS, and their harmful products—sulfoxidation of disulfide bonds, modification of DNA bases, conversion of lipids to lipid peroxides and aldehydes—must be reduced by the cellular redox system.[7] This system is generally in balance via well-coordinated enzymatic processes that control these systematic reactions as well as the detrimental effects of their ROS byproducts. Such enzymes include the superoxide dismutases (SODs), peroxidases, and catalases, as well as enzymes that control the iron concentration of the cell.[8] Despite these systems, up to 2% of oxygen reduction does not go to completion.[8] This leaves excess superoxide and hydrogen peroxide within the cell, which can participate in the Haber-Weiss reaction to form hydroxide radicals, which have the ability to damage all cellular components.[8] Normally antioxidants within the cell, obtained directly from the organism's diet or indirectly through intracellular enzymatic mechanisms, limit the OS induced by these hydroxide radicals by eliminating them upon their creation.[9] However, if the concentration of ROS exceeds the scavenging ability of the cell's antioxidant mechanisms, the cell

is left in a state of OS in which many of the cells' vital regulatory elements can become incapacitated. This provides a mechanism for DNA damage and the malignant transformation of previously normal cells.

The link between OS, cancer pathogenesis, and a possible role for the dietary and pharmacologic prevention of epithelial ovarian cancer remains an area of intense research interest. Given the potentially significant role of OS in the pathophysiology of ovarian cancer, dietary antioxidant consumption may have an impact on the treatment and prevention of ovarian cancer. Here we review the relevant literature and discuss the effects of OS on the development of ovarian cancer.

OVARIAN BIOLOGY

The ovulatory cycle is critical to ovarian function, but this process has also been linked to the development of ovarian cancer. Ovulation along with other events that take place within the ovary produce ROS. These ROS have been implicated both physiologically in various signaling processes and pathologically as causes of lipid damage, inhibition of protein synthesis, and depletion of ATP.[10,11] ROS are generated at all stages in the ovulatory cycle and continue to be produced after ovulation has ceased due to the loss of estrogens, which exert an antioxidant effect against low-density lipoproteins.[10] Therefore, it is important to take a closer look at this process and the potential for cellular damage caused by OS in the development of ovarian cancer.

The incessant ovulation hypothesis maintains that repeated cycles of apoptotic cell death and repair at the OSE during the ovulatory cycle lead to accumulation of genetic damage and predispose females to an increased risk of ovarian cancer.[12] Numerous studies have confirmed women with a history of oral contraceptive pill (OCP) use and/or multiparity have a reduced risk of ovarian cancer.[2] Further support is garnered by studies of animals—which do not ovulate with nearly the same frequency as humans—that show the disease to be rare in almost all instances. One important exception is the domestic hen, which ovulates frequently and shows a marked increase in ovarian cancer incidence.[13]

Pregnancy, OCP use, and lactation prevent many of the physiologic changes that take place during the ovulatory cycle. Pregnancy, which prevents ovulation, is accompanied by a decrease in the levels of luteinizing hormone (LH) and follicle stimulating hormone (FSH), and an increase in the levels of estrogen and progesterone. OCPs maintain high estrogen and progesterone levels in the blood through the consumption of synthetic estrogen and progesterone. As a result, the pituitary releases less LH and FSH resulting in ovulation suppression. Breast-feeding has some of the same effects, leading to low estrogen and LH levels, which also leads to suppression of ovulation.[14,15] All of these events are shown to be protective against OSE tumorogenesis, most likely because they limit the number of lifetime ovulations and the accompanying stress that this process places on the ovary.

Follicular Phase

During the preovulatory or follicular phase, many hormonal changes take place in preparation for ovulation. There is a rise in FSH, which causes the secretion of estrogen and inhibin. LH also rises at this time and generates a positive feedback loop with estrogen, leading to a spike in LH levels. An increase in cellular ROS and the depletion of ascorbic acid accompany this spike, followed ultimately by ovulation. It is thought that estrogen plays a role in the antioxidant defense mechanism and the response to increased OS before ovulation.[11,14,16]

In addition to the increase in cellular ROS, LH also induces the expression of nitric oxide synthase III (NOS III), one of three genes that codes for nitric oxide synthase, the enzyme that catalyzes the formation of NO.[7] The increase in NO caused by the expression of NOS III stimulates the ovulatory process and inhibits estradiol secretion through the inhibition of aromatase.[7] It appears that this local NO generation is required for follicular growth and ovulation since NOS III knockout mice oocytes do not meitotically mature.[7]

During this stage, numerous SODs are present to combat the OS induced by NO and other molecules. Mn-SOD is present throughout, whereas Cu/Zn-SOD is present during the antral phase of folliculogenesis, and these SODs serve to protect the follicle from ROS during steroidogenesis.[17]

Ovulation

Following the spike of LH and FSH, LH causes a signal transduction cascade. This leads to the production of lysosomal bodies in surface epithelial cells that migrate to the basal region of the cells and release proteolytic enzymes. These enzymes aid in follicular rupture and breakdown of the tunica albuginea, leaving a wound at the stigma for ovum release.[2,5] The stigma is the area along the ovarian surface where the follicle will break through at the time of ovulation. At follicular rupture, tumor necrosis factor alpha (TNF-alpha), leukocyte, macrophage, neutrophil, IL-1, and IL-6 concentrations are all elevated, which leads to a proliferation of free radicals.[10,13,14,16] Additionally, maximal levels of NO in follicular fluid have been measured at the time of ovulation.[18] Repeated ovulation increases the articulation of ovarian surface epithelial cells to OS, inflammation, and cytokines. This inflammatory response, which is modulated by OS, could ultimately promote the development of cancer.[10]

The wound at the stigma is then repaired by a proliferation of surface epithelial cells, the cells likely responsible for the vast majority of ovarian carcinomas.[5] This proliferation is unique because rather than utilizing stem cells, each cell yields two daughter cells with equal potential for subsequent growth. This allows mutations to be passed on exponentially to future generations as they continue to accumulate.[5] During this repair process it is possible for epithelial cells to become trapped in the stroma and form inclusion cysts. Steroid hormones and gonadotropins are present at high levels and can cause these transformations. Inclusion cysts such as these are frequently found in the normal ovary, and their frequency increases with age. Incessant ovulation would increase the opportunity for these cysts to develop with the repeated rupture and remodeling that accompany this process.[2] At the time of rupture, the likelihood of mutations from spontaneous errors in DNA synthesis is increased. While trapped, these epithelial cells can differentiate, proliferate, and undergo malignant transformation by direct gonadotropin or estrogen stimulation via gonadotropin-induced steroidogenesis.[14] This can occur as a result of the close relationship of the ovary and the Mullerian system, whose differentiation depends on surface epithelium sensitivity to estrogen, progesterone, and androsenedione. Receptors for estrogen and progesterone have been discovered on ovarian tumor cells and have been shown to play a role in cell growth and function.[14,15] DNA damage as a result of OS seems to be greatest during this proliferation stage.[19]

Luteal Phase

Following ovulation, the ruptured ovarian follicle enters the luteal phase, where the remaining LH and

FSH cause the follicle to fold in on itself and form the corpus luteum. The corpus luteum continues to grow and produce significant amounts of progesterone and moderate amounts of estrogen. As the estrogen levels fall, there is an increase in TNF-α, which leads to an increase in Mn-SOD mRNA transcription.[8] This causes the suppression of LH and FSH production and the death of the corpus luteum through the activity of PGF-2α, which is associated with the generation of superoxide, hydrogen peroxide, and lipid peroxides. Hydrogen peroxide and the lipid peroxides act to mediate the antigonadotropic and antisteroidogenic effects of superoxide by decreasing gonadotropin action and progesterone secretion in granulosa and luteal cells.[11] They also act to interrupt LH action by inhibiting cAMP-dependent steroidogenesis. These ROS, however, also have the effect of depleting ascorbic acid, which is required to regenerate the active state of peroxidases. In addition, leukocyte infiltration takes place during luteal regression, causing further superoxide and hydrogen peroxide to be generated. Estrogen and progesterone levels then fall as a result, leading to an increase in FSH levels and the recruitment of follicles for the next cycle.

During luteal regression, ROS levels have been shown to be increased in both rats and humans.[8,17] To combat this, vitamin E is elevated in the corpus luteum along with carotenoids, which act as highly effective oxygen scavengers.[11] CuZn-SOD activity increases during the early to mid-luteal phase and then decreases as the corpus luteum regresses.[10] Despite potential damaging effects, ROS do play a critical role in signaling the initiation of apoptosis in luteal cells. Once ROS reach a prohibitively high level, Cox-2, p53, and Bax mRNA transcripts are upregulated, triggering downstream apoptosis signals.[20]

With the effects of OS and ROS that potentially alter DNA, the incessant ovulation hypothesis explains why certain characteristics are protective against ovarian cancer and others place women at higher risk (see Figure 5.2). Still though, the number of ovulatory cycles does not appear to be the sole driver of ovarian cancer risk. The incessant menstruation hypothesis identifies OS as a major cause of ovarian carcinoma, but unlike the incessant ovulation hypothesis, it identifies a different source of ROS (see Figure 5.3).[21] Iron, located in heme, is associated with the generation of ROS via the Fenton reaction. When antioxidants are unable to remove ROS, OS results in DNA mutations. Retrograde menstruation leads to pooling of blood in the Pouch of Douglas and release of endometrial cells in the peritoneal cavity. The fimbrae of fallopian tubes in the Pouch of Douglas are exposed to chronic OS due to lysis of red blood cells by macrophages.[21] Endometriomas develop from endometrial cells with high concentrations of iron, leading to chronic OS as ROS are generated.[21]

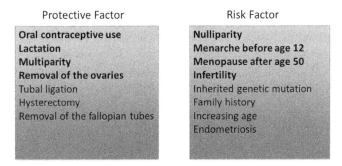

FIGURE 5.2 **Modifying Factors in the Development of Ovarian Cancer.** There are modifying risk factors associated with ovarian cancer. Those in bold are consistent with the theory of incessant ovulation because they impact the number of lifetime ovulatory cycles.

FIGURE 5.3 **Oxidative Stress in Ovarian Cancer.** OSE, ROS. In both the incessant ovulation model and the incessant menstruation model oxidative stress results in DNA mutations and malignant transformation.[2,21] Although neither model explains all aspects of ovarian cancer pathogenesis, they indicate the key role that oxidative stress likely plays in development of this malignancy.

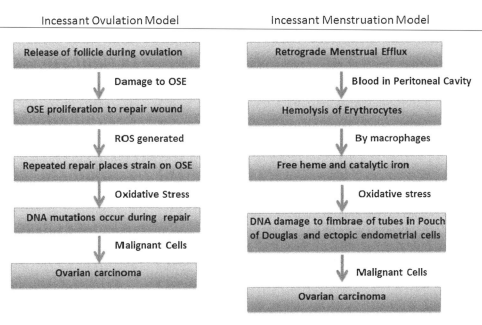

As a result, the theory of incessant menstruation may help explain why tubal cancers develop in the fimbrae and why ovarian endometrioid or clear cell cancers exist.

Menopause

Beyond the extreme oxidative environment created during the ovulatory cycle, normal aging also provides some additional challenges for the ovary in its response to OS. With aging, OS begins to play a more important role in cancer development than exogenous factors, with human DNA taking 10,000 oxidative hits per day, and mitochondrial DNA taking about 10 times more.[22] Some of these factors are global, such as the accumulation of mitochondrial DNA mutations that reduce ROS detection, scavenging, and removal throughout the body.[23] DNA repair mechanisms are also overwhelmed and impaired by cumulative genetic and cellular damage. In the ovary this manifests in diminished antioxidant gene expression, increased OS, and damage to DNA, protein, and lipids, as observed in a recent study of aging mouse ovaries.[24] The antioxidant levels in blood are also shown to decrease with age, which decreases the ability to combat OS.[23] Oocytes also show more DNA fragmentation as women age and they become less viable, possibly as a result of ROS damage.[10,25] This contributes to both age-related declines in ovulatory function and fertility, as well as the pathogenesis of cancer.

Menopause causes significant hormonal changes in and around the ovary. During menopause there is a loss of estrogens, which have a protective effect as antioxidants on LDLs.[10] FSH, LH, human chorionic gonadotropin, and prolactin are all known to have growth stimulating properties, and they remain elevated for a long period following menopause. FSH levels have been found to increase by up to 15 times and LH levels by as much as 5 times.[2] Receptors for FSH and LH are frequently found in ovarian cancer specimens and have been shown to be elevated for an extended time following menopause.[5] Early menopause and premature menopause failure—resulting from smoking, irradiation, chemicals, and viruses—have also been shown to be associated with higher gonadotropin levels, and ovarian cancer incidence rates for this group follow a similar time delay from menopause to other groups.[15]

Endometriosis

Aside from the normal physiology of the ovary, a common gynecologic condition is endometriosis, which is estimated to affect up to 10% of women. When adjusting for oral contraceptive use, age, parity, family history of ovarian cancer, and tubal ligation, women with endometriosis appear to be at higher risk of ovarian cancer compared to those without endometriosis, adjust OR 1.32, 95% CI 1.06–1.65.[26] OS conditions found within endometriotic lesions have been linked to the subsequent development of clear cell carcinoma of the ovary, an aggressive subtype of epithelial ovarian carcinoma.[27] Aberrant inflammation promotes the growth and invasion of ectopic endometrium and leads to the development of endometriosis. One potential cause of endometriosis could be an estrogen rich and progesterone poor environment. The follicular fluid and serum of patients with endometriosis have been shown to have increased markers of OS over controls.[28,29] This is similar to results from the serum of patients with ovarian cancer, although the results may be nonspecific, as cervical cancer patients also show these increases.[30,31]

COMMON OVARIAN CANCER MUTATIONS LINKED TO ROS

Common mutations in ovarian cancer can be linked to OS (see Table 5.1). Many of these mutations are found in genes involved in the metabolism of oxygen products.[10] ROS involvement is relevant to various stages of cancer development, initiation, promotion, and progression; damage to DNA caused by the hydroxide radical is involved at each of these points. ROS are capable of reacting with components of DNA, and often result in the mutation of the deoxyguanine base to 8-oxodeoxyguanosine (8-oxodG), which leads to guanine-cytosine (GC) to thymine-adenine (TA) transversions.[32] Proofreading is blocked in this type of mutation, which impairs the ability of the cell to repair the mismatch.[33] 8-oxodGTP can also form and cause AT to CG mutations.[34] In addition OS may alter methylation of DNA and impact the function of methyltransferase.[35] For example, methyltransferase at cytosine C-5 is important for organizing the genome, and abnormal patterns of methylation are consistently found in human tumors.[35]

The *p53* gene is a critical regulatory control gene, and its errant function has been implicated in 50 to 80% of ovarian cancers.[13,36,37,38] Normally, *p53* is involved in triggering apoptosis to prevent the proliferation of cells with unregulated G1-S control.[36] It also gives cells an opportunity to fix oxidative disturbances to DNA prior to driving them toward the apoptotic pathway. In a case-case and case-control study, women whose ovarian tumors overexpressed mutant *p53* protein were found to have a higher number of ovulatory cycles, indicating ovulation number may be associated with increased amounts of proliferation associated DNA damage.[19]

Ras mutations are also commonly observed in ovarian cancer, and cells with these mutations appear to be resistant to OS induced apoptosis.[38] *Ras* mutations lead to the upregulation of thioredoxin peroxidase, peroxiredoxin 3,

TABLE 5.1 Common OSE Gene Mutations Associated with Oxidative Stress

Mutated Gene	Functional Result
Tumor protein 53 (p53)	Implicated in 50–80% of ovarian cancers. Normally prevents proliferation of cells with unregulated G1-S control and gives opportunity for DNA oxidative disturbances to be repaired.
Retinoblastoma (Rb)	Implicated in about 30% of ovarian cancers Speculated that errant methylations cause many of the destructive mutations
Rat sarcoma viral oncogene homolog (Ras)	Causes upregulation of thioredoxin peroxidase, peroxiredoxin 3, selenophosphate synthase, NADH dehydrogenase, ubiquinone, and Fe/S protein. Leads to resistance against oxidative stress-induced apoptosis.
8-oxoguanine DNA glycosylase (OGG1)	Repairs oxidized guanine to prevent the C to T transitions that take place as a result of mutation of deoxyguanine to 8-oxodeoxyguanosine.
mutS homolog 2 (hMSH2)	Involved in DNA mismatch repair.

selenophosphate synthase, NADH dehydrogenase, ubiquinone, and Fe/S protein. These proteins protect a cell from OS induced apoptosis and are vital to the normal functioning of the cell. When *Ras*-signaling pathways are upregulated in malignancy the antioxidant capacity of tumor cells is enhanced, protecting these altered cells from undergoing apoptosis that would normally be induced by ROS.[38]

Despite the many negative actions that take place with high levels of OS and resulting ROS, a certain level of OS is needed to initiate the destruction of mutated cells through apoptosis. Mutations that arise from the action of ROS and the damage they inflict on DNA, especially the conversion of guanine to 8-oxodG, likely contribute to the development of ovarian cancer. Such mutations can lead to the destruction of a variety of caretaker genes, including *p53*. Ironically, these mutations allow cancer cells to overcome the OS that would normally initiate their apoptosis. Oxidative damage and associated mutations can occur at any stage in the progression of the cancer, highlighting the importance of the delicate balance of ROS required by cells (see Figure 5.4).

ANTIOXIDANT MECHANISMS IN THE OVARY

A strong oxidative defense system, composed of both enzymatic and nonenzymatic antioxidants, is necessary to combat the oxidative damage that persists in the ovary.

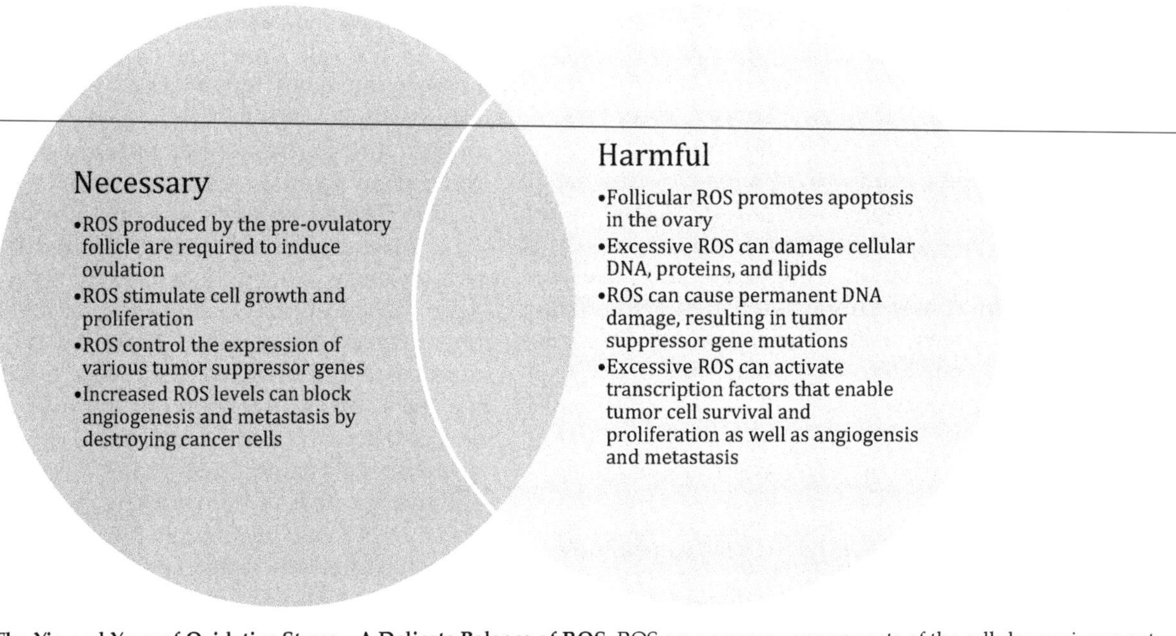

Necessary

- ROS produced by the pre-ovulatory follicle are required to induce ovulation
- ROS stimulate cell growth and proliferation
- ROS control the expression of various tumor suppressor genes
- Increased ROS levels can block angiogenesis and metastasis by destroying cancer cells

Harmful

- Follicular ROS promotes apoptosis in the ovary
- Excessive ROS can damage cellular DNA, proteins, and lipids
- ROS can cause permanent DNA damage, resulting in tumor suppressor gene mutations
- Excessive ROS can activate transcription factors that enable tumor cell survival and proliferation as well as angiogensis and metastasis

FIGURE 5.4 The Yin and Yang of Oxidative Stress—A Delicate Balance of ROS. ROS are necessary components of the cellular environment that are responsible for key regulatory functions of the cell. However, in excess, ROS can become harmful to the cell and initiate processes that often result in tumorogenesis. (See the color plate.)

These antioxidants, derived endogenously in the cell as well as through exogenous consumption of fruits, vegetables, and supplements, have been studied extensively both *in vivo* and *in vitro* in numerous cancers. Detailed studies of water buffalo ovaries provide evidence that the follicular fluid is a milieu rich with antioxidants.[39] The lipid peroxide concentration of this fluid is also low, providing further evidence of high antioxidant activity.[17] The enzymatic antioxidants include SOD, catalase, glutathione peroxidase (Gpx), and glutathione reductase.[2] Important nonenzymatic antioxidants include vitamins C and E, selenium, zinc, taurine, hypotaurine, glutathione, beta carotene, and carotene.[10] These different mechanisms often work in coordination to provide protection to the ovary.

The backbone of the cellular antioxidant defense system is SOD, catalase, and Gpx.[13] A decrease in the activity of these enzymes or an unbalanced increase in any one of them can increase the vulnerability of the cell to oxygen free radicals.[13] SOD is involved in the reaction of superoxide to hydrogen peroxide, which is then degraded by catalase, Gpx, and other peroxidases.[11] Overexpression of Mn-SOD however leads to cell cycle arrest, intracellular hydrogen peroxide accumulation, and induction of mRNA for metalloprotease-1, which is involved in carcinogenesis.[23] Cu,Zn-SOD prevents the accumulation of superoxide, and when its levels are decreased, lipid peroxide concentrations increase. Thioredoxin (Trx) serves as an important electron donor for ribonucleotide reductase and regulates enzymes and transactivating factors of genes.[34] Trx is critical to cell growth, differentiation, and death. It also acts as a protein disulfide isomerase, correcting erroneous disulfide bridges and donating electrons to peroxiredoxin.[34]

Ascorbic acid is particularly abundant in and around the ovary, and its levels fluctuate significantly during the course of the ovulatory cycle, suggesting that the burden of OS changes dramatically. The ovary even has transport mechanisms to regulate ascorbic acid levels.

Vitamin E is found in cellular membranes where it acts to stabilize the phospholipid membrane and protect against OS. Vitamin E has been shown to prevent the increase in 8-oxoguanine caused by OS in OSE cells.[13] Myeloperoxidase and eosinophil peroxidase effectively scavenge hydrogen peroxide in the presence of ascorbic acid or thiocyanate lactoperoxidase.[11] Reduced glutathione, which acts as a substrate for Gpx, can directly repair oxidative DNA lesions, and can be used as an electron donor to scavenge free radicals including superoxide, hydroxide radical, and lipid hyperoxides.[13] Selenocysteine is also important to the function of Gpx.[9] Selenium, a widely studied metal cofactor, is part of the active catalytic site of SOD. It has been shown to help reduce ROS burden within the cells.[40] Overall levels of antioxidants have been found to be lower in cancer patients, although this could be the result of numerous other confounders.[41]

ANTIOXIDANT THERAPIES, OVARIAN CANCER

Since ROS appear to play such an active role in the development of ovarian cancer, there has been a large amount of work carried out to investigate the role of dietary antioxidants in the prevention of the disease. Studies have focused on the effects of increasing the intake of vitamins, carotenoids, and minerals in the diet. A critical concern of dietary antioxidants is the bioavailability of the compounds following their administration and how this contributes to their overall effectiveness. Of the antioxidants reviewed, few have demonstrated evidence of significant improvement compared to control, while most have led to no difference in ovarian cancer occurrence based on dietary antioxidant intake (see Table 5.2).

Vitamins A, C, E, and folate have antioxidant properties and have been investigated to determine if they have protective roles against the development of ovarian cancer. Increasing vitamin A intake demonstrated a modest protective effect in three studies,[42,43,44] while showing no evidence of prevention in four other studies.[45,46,47,48] A high vitamin C diet showed no protective benefit against ovarian cancer in all studies reviewed.[42,45,46,47,49,50] Studies investigating vitamin E reveal mixed results. Two studies reviewed demonstrated a slight preventative benefit from increased vitamin E,[42,51] while three others revealed no benefit.[45,46,50] Additionally, no benefit was demonstrated by increased Vitamin D or folate intake.[48] Overall, the dietary or supplemental intake of vitamins does not appear to strongly impact the development of ovarian cancer.

Carotenoids are another class of antioxidants that have been extensively studied for a potential dietary role in inhibiting the development of ovarian cancer. Unfortunately, much like vitamins, this class of compounds does not show a consistent beneficial effect. The carotenoids reviewed included α-carotene, β-carotene, lycopene, lutein, zeaxanthin, and cryptoxanthin. Total carotenoids were measured in three of the studies reviewed and one showed a small decrease in odds ratio for developing ovarian cancer of 0.64 (95% CI 0.45–0.98) between the highest and lowest quartiles of intake,[42] while the two other studies showed no significant effect.[44,46] Of the three studies evaluating α-carotene, there was no significant difference in cancer risk between different groups.[42,44,46] The results for β-carotene intake were mixed, with four studies showing a reduced risk for ovarian cancer development[42,43,47,51] and four showing no benefit.[44,45,46,50] Lycopene demonstrated no benefit in four studies[42,44,46,50] and only a minute benefit in one study.[43] Lutein and zeaxanthin showed a protective effect in the highest intake groups in two studies,[43,51] but four other studies demonstrated no effect.[42,44,46,50] Cryptoxanthin intake was also shown to have no effect.[42,44,46] Overall, dietary carotenoids do not seem to provide significant if any protection against ovarian cancer development,

TABLE 5.2 Dietary Antioxidant Studies for the Prevention of Ovarian Cancer

Antioxidant	Study (by Reference #)	Conclusion (OR = Odds Ratio) (RR = Relative Risk)
Vitamin A	McCann et al.[42]	OR = 0.66 95% CI: 0.45–0.98
	Jeong et al.[43]	OR = 0.45 95% CI: 0.21–0.98
	Tung et al.[44]	OR = 0.61 95% CI: 0.42–0.89
	Fairfield et al.[45] Thomson et al.[46] Slattery et al.[47] Chang et al.[48]	No significant difference between control and test group
Vitamin C	McCann et al.[42] Byers et al.[49] Fairfield et al.[45] Thomson et al.[46] Slattery et al.[47] Gifkins et al.[50]	No significant difference between control and test group
Vitamin E	McCann et al.[42]	OR = 0.58 95% CI: 0.38–0.88
	Bidoli et al.[51]	OR = 0.60 95% CI: 0.5–0.8
	Fairfield et al.[45] Thomson et al.[46] Gifkins et al.[50]	No significant difference between control and test group
Vitamin D	Chang et al.[48]	No significant difference between control and test group
Folate	Chang et al.[48]	No significant difference between control and test group
Alpha-carotene	McCann et al.[42] Tung et al.[44] Thomson et al.[46]	No significant difference between control and test group
Beta-carotene	Jeong et al.[43]	OR = 0.12 95% CI: 0.04–0.37
	McCann et al.[42]	OR = 0.68 95% CI: 0.46–0.98
	Slattery et al.[47]	OR = 0.5 95% CI: 0.3–1.0
	Bidoli et al.[51]	OR = 0.8 95% CI: 0.6–1.0
Lycopene	Jeong et al.[43]	OR = 0.09 95% CI: 0.03–0.32
	McCann et al.[42] Tung et al.[44] Thomson et al.[46] Gifkins et al.[50]	No significant difference between control and test group

(Continued)

TABLE 5.2 Dietary Antioxidant Studies for the Prevention of Ovarian Cancer   cont'd

Antioxidant	Study (by Reference #)	Conclusion (OR = Odds Ratio) (RR = Relative Risk)
Lutein and Zeaxanthin	Jeong et al.[43]	OR = 0.21 95% CI: 0.09–0.52
	Bidoli et al.[51]	OR = 0.6 95% CI: 0.5–0.8
	McCann et al.[42] Tung et al.[44] Thomson et al.[46] Gifkins et al.[50]	No significant difference between control and test group
Cryptoxanthin	McCann et al.[42] Tung et al.[44] Thomson et al.[46]	No significant difference between control and test group
Calcium	Bidoli et al.[51]	OR = 0.7 95% CI: 0.6–1.0
	Chang et al.[48]	No significant difference between control and test group
Selenium	Gifkins et al.[50]	OR = 0.41 95% CI: 0.20–0.85
	Thomson et al.[46]	No significant difference between control and test group
Phosphorous	Chang et al.[48]	No significant difference between control and test group

and many of the modestly successful results are based on serum tests of patients who have already been diagnosed with ovarian cancer, a potential confounding factor.

Dietary minerals such as calcium, selenium, and phosphorus make up a third class of antioxidants, but they too appear to have a limited benefit for the prevention of ovarian cancer. A study by Bidoli et al. showed a small benefit with an odds ratio of 0.7 (95% CI 0.6–1.0) between the highest and lowest groups for calcium intake,[51] while a study by Chang et al. showed no effect.[48] Selenium also showed mixed results, with one study showing a mild protective effect with an odds ratio of 0.41 (95% CI 0.20–0.85) and another showing no difference.[50,46] Only one study looked at dietary phosphorus and it found no effect against ovarian cancer incidence.[48]

A small number of studies have evaluated broad dietary categories such as total intake of calories, fruits, vegetables, fiber, fats, carbohydrates, protein, and cholesterol. The results are inconsistent with regard to prevention of ovarian cancer.[47,52,53,54] A study by McCann et al. demonstrated an odds ratio of 0.62 (95% CI 0.42–0.92) for combined fruit and vegetable intake between the highest and lowest quartile groups and an odds ratio of 0.57 (95% CI 0.38–0.87) for fiber alone.[42] Childhood diet may be a more important factor, with one study demonstrating that high fruit and vegetable intake early in life was associated with a decreased risk of ovarian cancer development. Fairfield et al. showed that greater than 2.5 servings of fruits and vegetables a day during adolescence decreased the relative risk of ovarian cancer, 0.54 (95% CI 0.29–1.03).[45] Nonetheless, further investigation will be needed to provide evidence either for or against the use of such antioxidants as a worthwhile preventative measure against the development of ovarian cancer.

CONCLUSION

OS appears to play a major role in the development and progression of ovarian cancer, as various events linked to a high ROS environment are significant risk factors for the development of this devastating disease. Although one unifying hypothesis does not exist to explain the origin of ovarian cancer, existing theories identify OS as an important source of malignant transformation. Many changes take place in and around the ovaries during ovulation, and it is clear that OS develops during these processes and is capable of causing mutations that lead to ovarian cancer.

Multiparity, lactation, and OCP use all offer a modicum of protection against ovarian cancer development, and the common thread between these factors is a reduction in the number of lifetime ovulations. With regard to prevention, decreasing the number of lifetime ovulations currently appears to be the best method for risk reduction. Despite the clear association between OS and ovarian cancer, it does not appear that dietary antioxidant intake plays a significant role in prevention. Most studies show that various classes of antioxidants including vitamins, carotenoids, and minerals do not provide any protection against developing the disease. Though a small number of studies demonstrated a modest effect at best, the majority of results remain inconclusive. This lack of effect could be a result of the limited bioavailability of these substances when consumed in the diet or simply because the normal dietary intake of these compounds is already sufficient.

Despite these mixed results, research investigating mechanisms to reverse the damaging effects of OS as it relates to ovarian cancer may still be beneficial in prevention and/or treatment of this malignancy. Greater success may arise by increasing the prevalence of endogenous enzymatic antioxidants or by finding additional methods that can decrease the amount of ROS present in the ovarian environment.

SUMMARY POINTS

- As of 2008, ovarian cancer is the eighth most common cancer in the world and the seventh leading cause of cancer death in women worldwide.

- It is often undetected until disease has reached advanced stage with the majority of patients found to have Stage III or IV disease on presentation.
- Repeated destruction and repair of the ovarian surface epithelium (OSE) at the site of ovulation provides a mechanism for the pathogenesis of ovarian cancer through oxidative stress (OS).
- Inflammatory mediators and reactive oxygen species (ROS) are generated during ovulation, especially during the follicular rupture stage. These ROS can damage nucleic acids, proteins, and lipids and result in DNA strand breaks.
- Lactation, multiparity, and oral contraceptive pills (OCPs), which reduce the lifetime number of ovulatory cycles, also reduce the risk of ovarian cancer by as much as 50%.
- Mutations that arise from the action of ROS and the damage they inflict on DNA, especially the conversion of guanine to 8-oxodG, likely contribute to the development of ovarian cancer.
- The *p53* gene is a critical regulatory control gene, and its errant function has been implicated in 50 to 80% of ovarian cancers.
- The enzymatic antioxidants of the ovary include SOD, catalase, glutathione peroxidase (Gpx), and glutathione reductase.[2] Important nonenzymatic antioxidants include vitamins C and E, selenium, zinc, taurine, hypotaurine, glutathione, beta carotene, and carotene.
- Overall, the increased intake of dietary or supplemental vitamins, carotenoids, and minerals does not appear to impact the risk of developing of ovarian cancer.

References

1. Ferlay J, Shin H, Bray F, Forman D, Mathers C, Parkin DM. Estimates of worldwide burden of cancer in 2008: GLOBOCAN 2008. *Int J Cancer* 2010;**127**(12):2893–917.
2. Fathalla MF. Incessant ovulation—A factor in ovarian neoplasia?. *Lancet (London, England)* 1971;**2**(7716) 163–163.
3. Beral V, Doll R, Hermon C, et al. Ovarian cancer and oral contraceptives: collaborative reanalysis of data from 45 epidemiological studies including 23,257 women with ovarian cancer and 87,303 controls. *Lancet (London, England)* 2008;**371**(9609):303–14.
4. McCann SE, Freudenheim JL, Marshall JR, et al. Risk of human ovarian cancer is related to dietary intake of selected nutrients, phytochemicals and food groups. *J Nutr* 2003;**133**(6):1937–42.
5. Godwin AK, Testa JR, Hamilton TC. The biology of ovarian cancer development. *Cancer* 1993;**71**(2 Suppl):530–6.
6. Murdoch WJ, McDonnel AC. Roles of the ovarian surface epithelium in ovulation and carcinogenesis. *Reproduction* 2002;**123**(6): 743–50.
7. Fujii J, Iuchi Y, Okada F. Fundamental roles of reactive oxygen species and protective mechanisms in the female reproductive system. *Reprod Biol Endocrinol* 2005;**3** 43–43.
8. Rosen GM, Pou S, Ramos CL, et al. Free radicals and phagocytic cells. *FASEB J* 1995;**9**(2):200–9.

9. Ray SD, Lam TS, Rotollo JA, et al. Oxidative stress is the master operator of drug and chemically-induced programmed and unprogrammed cell death: Implications of natural antioxidants in vivo. *BioFactors* 2004;**21**(1-4):223–32.

10. Agarwal A, Gupta S, Sharma RK. Role of oxidative stress in female reproduction. *Reprod Biol Endocrinol* 2005;**3** 28–28.

11. Behrman HR, Kodaman PH, Preston SL, et al. Oxidative stress and the ovary. *J Soc Gynecol Invest* 2001;**8**(1 Suppl Proceedings):S40–2.

12. Ho S. Estrogen, progesterone and epithelial ovarian cancer. *Reprod Biol Endocrinol* 2003;**1** 73–73.

13. Murdoch WJ, Martinchick JF. Oxidative damage to DNA of ovarian surface epithelial cells affected by ovulation: carcinogenic implication and chemoprevention. *Exp Biol Med* 2004;**229**(6):546–52.

14. Ness RB, Cottreau C. Possible role of ovarian epithelial inflammation in ovarian cancer. *J National Cancer Inst* 1999;**91**(17):1459–67.

15. Cramer DW, Welch WR, Hutchison GB, et al. Dietary animal fat in relation to ovarian cancer risk. *Obstet Gynecol* 1984;**63**(6):833–8.

16. Jozwik M, Wolczynski S, Szamatowicz M. Oxidative stress markers in preovulatory follicular fluid in humans. *MHR: Basic Science of Reproductive Medicine* 1999;**5**(5):409–13.

17. Suzuki T, Sugino N, Fukaya T, et al. Superoxide dismutase in normal cycling human ovaries: immunohistochemical localization and characterization. *Fertil Steril* 1999;**72**(4):720–6.

18. Chung HT, Pae HO, Choi BM, et al. Nitric oxide as a bioregulator of apoptosis. *Biochem Biophys Res Commun* 2001;**282**(5):1075–9.

19. Schildkraut JM, Bastos E, Berchuck A. Relationship between lifetime ovulatory cycles and overexpression of mutant p53 in epithelial ovarian cancer. *J National Cancer Inst* 1997;**89**(13):932–8.

20. Nakamura T, Sakamoto K. Reactive oxygen species up-regulates cyclooxygenase-2, p53, and Bax mRNA expression in bovine luteal cells. *Biochem Biophys Res Commun* 2001;**284**(1):203–10.

21. Vercellini P, Crosignani P, Somigliana E, et al. The incessant menstruation hypothesis: a mechanistic ovarian cancer model with implications for prevention. *Human Reprod* 2011;**26**(9):2262–73.

22. Ames BN, Gold LS, Willett WC. The causes and prevention of cancer. *Proc Natl Acad Sci USA* 1995;**92**(12):5258–65.

23. Wei Y, Lee H. Oxidative stress, mitochondrial DNA mutation, and impairment of antioxidant enzymes in aging. *Exp Biol Med* 2002;**227**(9):671–82.

24. Lim J, Luderer U. Oxidative damage increases and antioxidant gene expression decreases with aging in the mouse ovary. *Biol Reprod* 2011;**84**(4):775–82.

25. Fujino Y, Ozaki K, Yamamasu S, et al. DNA fragmentation of oocytes in aged mice. *Human Reprod* 1996;**11**(7):1480–3.

26. Modugno F, Ness RB, Allen GO, et al. Oral contraceptive use, reproductive history, and risk of epithelial ovarian cancer in women with and without endometriosis. *Am J Obstet Gynecol* 2004;**191**(3):733–40.

27. Penson RT, Dizon DS, Birrer MJ. Clear cell cancer of the ovary. *Curr Opin Oncol* 2013;**25**(5):553–7.

28. Prieto L, Quesada JF, Cambero O, et al. Analysis of follicular fluid and serum markers of oxidative stress in women with infertility related to endometriosis. *Fertil Steril* 2012;**98**(1):126–30.

29. Lambrinoudaki IV, Augoulea A, Christodoulakos GE, et al. Measurable serum markers of oxidative stress response in women with endometriosis. *Fertil Steril* 2009;**91**(1):46–50.

30. Senthil K, Aranganathan S, Nalini N. Evidence of oxidative stress in the circulation of ovarian cancer patients. *Clin Chim Acta* 2004;**339**(1-2):27–32.

31. Manju V, Kalaivani Sailaja J, Nalini N. Circulating lipid peroxidation and antioxidant status in cervical cancer patients: a case-control study. *Clin Biochem* 2002;**35**(8):621–5.

32. Grollman AP, Moriya M. Mutagenesis by 8-oxoguanine: an enemy within. *Trends Genet* 1993;**9**(7):246–9.

33. Cunningham RP. DNA repair: Caretakers of the genome? *Curr Biol* 1997;**7**(9):R576–9.

34. Dreher D, Junod AF. Role of oxygen free radicals in cancer development. *Eur J Cancer* 1996;**32A**(1):30–8.

35. Cerda S, Weitzman SA. Influence of oxygen radical injury on DNA methylation. Mutation research. *Fundam Mol Mech Mutagen* 1997;**386**(2):141–52.

36. Hashiguchi Y, Tsuda H, Yamamoto K, et al. Combined analysis of p53 and RB pathways in epithelial ovarian cancer. *Human Pathol* 2001;**32**(9):988–96.

37. Okamoto A, Sameshima Y, Yokoyama S, et al. Frequent allelic losses and mutations of the p53 gene in human ovarian cancer. *Cancer Res* 1991;**51**(19):5171–6.

38. Young TW, Mei FC, Yang G, et al. Activation of antioxidant pathways in ras-mediated oncogenic transformation of human surface ovarian epithelial cells revealed by functional proteomics and mass spectrometry. *Cancer Res* 2004;**64**(13):4577–84.

39. Cassano E, Tosto L, Balestrieri M, et al. Antioxidant defense in the follicular fluid of water buffalo. *Cell Physiol Biochem* 1999;**9**(2):106–16.

40. Hu Y, Rosen DG, Zhou Y, et al. Mitochondrial manganese-superoxide dismutase expression in ovarian cancer: role in cell proliferation and response to oxidative stress. *J Biol Chem* 2005;**280**(47):39485–92.

41. Sanchez M, Torres JV, Tormos C, et al. Impairment of antioxidant enzymes, lipid peroxidation and 8-oxo-2′-deoxyguanosine in advanced epithelial ovarian carcinoma of a Spanish community. *Cancer Lett* 2006;**233**(1):28–35.

42. McCann SE, Moysich KB, Mettlin C. Intakes of selected nutrients and food groups and risk of ovarian cancer. *Nutr Cancer* 2001;**39**(1):19–28.

43. Jeong N, Song E, Lee J, et al. Plasma carotenoids, retinol and tocopherol levels and the risk of ovarian cancer. *Acta Obstet Gynecol Scand* 2009;**88**(4):457–62.

44. Tung K, Wilkens LR, Wu AH, et al. Association of dietary vitamin A, carotenoids, and other antioxidants with the risk of ovarian cancer. *Cancer Epidemiol Biomark Prev* 2005;**14**(3):669–76.

45. Fairfield KM, Hankinson SE, Rosner BA, et al. Risk of ovarian carcinoma and consumption of vitamins A, C, and E and specific carotenoids: a prospective analysis. *Cancer* 2001;**92**(9):2318–26.

46. Thomson CA, Neuhouser ML, Shikany JM, et al. The role of antioxidants and vitamin A in ovarian cancer: Results from the Women's Health Initiative. *Nutr Cancer* 2008;**60**(6):710–9.

47. Slattery ML, Schuman KL, West DW, et al. Nutrient intake and ovarian cancer. *Am J Epidemiol* 1989;**130**(3):497–502.

48. Chang ET, Lee VS, Canchola AJ, et al. Diet and risk of ovarian cancer in the California Teachers Study cohort. *Am J Epidemiol* 2007;**165**(7):802–13.

49. Byers T, Marshall J, Graham S, et al. A case-control study of dietary and nondietary factors in ovarian cancer. *J National Cancer Inst* 1983;**71**(4):681–6.

50. Gifkins D, Olson SH, Paddock L, King M, Demissie K, Lu S-E, et al. Total and individual antioxidant intake and risk of epithelial ovarian cancer. BMC Cancer 2012;**12**:211.

51. Bidoli E, La Vecchia C, Talamini R, et al. Micronutrients and ovarian cancer: an Italian case-control study. *IARC Sci Publ* 2002;**156**:357–60.

52. Mommers M, Schouten LJ, Goldbohm RA, et al. Consumption of vegetables and fruits and risk of ovarian carcinoma. *Cancer* 2005;**104**(7):1512–9.

53. Koushik A, Hunter DJ, Spiegelman D, et al. Fruits and vegetables and ovarian cancer risk in a pooled analysis of 12 cohort studies. *Cancer Epidemiol Biomark Prev* 2005;**14**(9):2160–7.

54. Chandran U, Bandera EV, Williams King MG, et al. Healthy eating index and ovarian cancer risk. CCC. *Cancer Causes Control* 2011;**22**(4):563–71.

Role of Oxidative Stress in Human Papillomavirus-Driven Cervical Carcinogenesis

Cesira Foppoli

CNR Institute of Molecular Biology and Pathology, Sapienza University of Rome, Rome, Italy

Raffaella Coccia, Marzia Perluigi

Department of Biochemical Sciences, Sapienza University of Rome, Rome, Italy

List of Abbreviations

AP-1 Activator protein-1
CAT catalase
CDK cyclin-dependent kinase
CIN cervical intraepithelial neoplasia
COX-2 cycloxygenase2
δ ALA-D δ-aminolevulinate dehydratase
ERp57 endoplasmic reticulum protein 57
hTERT human telomerase reverse transcriptase
GAPDH gliceraldehyde 3-phosphate dehydrogenase
GPx glutathione peroxidase
GST glutathione S-transferase
HPV human papillomavirus
HR-HPV high-risk HPV
HSIL high-grade squamous intraepithelial lesion
iNOS inducible nitric oxide synthase
LCR long control region
LPO lipid peroxidation
LR-HPV low-risk HPV
LSIL low-grade squamous intraepithelial lesion
MAGUK membrane associated guanylate kinase
MDA malondialdehyde
NHEK normal human epithelial keratinocytes
pRB retinoblastoma protein
RNS reactive nitrogen species
ROS reactive oxygen species
SOD superoxide dismutase
TBARS thiobarbituric acid reactive substances
URR up-stream regulatory region

INTRODUCTION

Cervical cancer is the second most common cancer among women worldwide. At the beginning of the 1970s Harold zur Hausen[1] hypothesized a correlation between cervical neoplasia and human papillomavirus (HPV) infection. After this first assumption, a plethora of laboratory and epidemiologic data indicated HPV as the major etiological factor in cervical carcinogenesis.

HPVs are a group of DNA viruses highly species-specific and exclusively tissue tropic, undergoing a complete infectious cycle only in fully differentiating squamous epithelium.

More than 100 HPV types are described,[2] clinically classified as "low-risk" (LR-HPV) and "high-risk" (HR-HPV) based on the relative propensity to cause benign or cancerous lesions. LR-HPVs generate mild dysplasia and genital warts, while HR-HPVs are associated with high grade lesions and invasive cancer. Virtually all cervical cancers (over 99%) contain the genes of HR-HPV types, especially HPV 16, HPV 18, HPV 31, and HPV 33. HPV 16 is by far the most prevalent type, found in more than 50% of all cases.

The oncogenic potential of HPV infection mostly depends on the activity of the viral oncogenes E6, E7, and to a lesser extent, E5. E6 and E7 can modulate cellular proliferation and apoptosis by interfering with the function of two tumor suppressor proteins, p53 and Rb, respectively, and their expression is required for cell transformation and for maintenance of the transformed state. In addition, E6 and E7 oncoproteins play key roles throughout the whole disease process by manipulating the function of a variety of host regulatory proteins. However, viral oncogenes expression, although necessary, is not *per se* sufficient to induce cervical cancer, and other factors are needed to drive the neoplastic progression.

Cancer
http://dx.doi.org/10.1016/B978-0-12-405205-5.00006-4

© 2014 Elsevier Inc. All rights reserved.

Despite the large number of studies on viral, host, and environmental putative cofactors, results have been largely disappointing and our present understanding of cervical cancer progression remains largely unsatisfactory.

Oxidative stress represents an interesting and underexplored candidate as a promoting factor in HPV-driven carcinogenesis. The role of reactive oxygen species (ROS) in tumor progression is well established,[3] as well as the notion that oxidative stress perturbs the cellular redox status, thus leading to alteration of gene expression responses and triggering a signaling cascade that affects both cell growth and cell death.

In this chapter, we report results from several studies that support the role of oxidative stress in cervical cancer.

CERVICAL CANCER

Cervical cancer is a slow-evolution disease, arising from dysplastic lesions after long persistent infection. It is in fact a multistep process, in which progressive histologic and cytologic changes occur, that can be divided into early lesions, currently indicated as cervical intraepithelial neoplasia 1 (CIN 1) or low grade squamous intraepithelial lesion (LSIL); and high grade lesions, known as cervical intraepithelial neoplasia 2/3 (CIN 2/CIN 3) or high grade squamous intraepithelial lesion (HSIL).[4] The majority of HPV infections (80%) are subclinical and transitory and are successfully cleared by an efficient cell-mediated immune response exerted by the host. The importance of the immune system in counteracting HPV infection is also demonstrated by the fact that regressing warts are infiltrated by T lymphocytes and macrophages and that in immunosuppressed individuals a higher prevalence of HPV-induced lesions and HPV-related tumors are observed.[5]

However, in case the infection is not properly cleared in those individuals with "less efficient" immune response, it progresses to mild dysplasia, CIN 1. Also CIN 1 lesions typically regress spontaneously, but a few lesions progress to moderate and then severe dysplasia (CIN 2/CIN 3) and eventually to carcinoma in situ and invasive carcinoma.[6] The rates of regression, persistence, or progression vary according to the different CIN grades and these data are shown in Figure 6.1.

HPV STRUCTURE

The viral genome is constituted by a circular double-strand DNA containing 6000 to 8000 base pairs and comprises eight gene sequences, codifying for proteins with

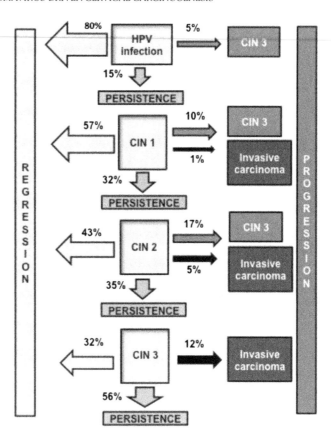

FIGURE 6.1   Rates of Regression, Persistence, or Progression of Different Grades of Cervical Intraepithelial Neoplasia (CIN). Only a minor part of cervical HPV infections tend to progress to cancer.

early (E) or late (L) functions. L region encodes the two capsid proteins L1 and L2. E region contains six genes expressed in the initial phase of replicative cycle, codifying for the E1, E2, E4, E5, E6, and E7 proteins. In addition, viral genome contains a region, named Long Control Region (LCR) or Up-stream Regulatory Region (URR), where the site for plasmide replication origin, several sequences implied in viral replication, and numerous binding sites for cellular transcription factors have been identified. This specific region is crucial in determining the host range and tissue tropism of each HPV type and also regulates viral gene expression upon infection.

E6, E7, and in minor grade E5 are the oncoproteins, well-recognized as responsible for cell transformation events. Molecular mechanisms of oncogenic action of E6 and E7 proteins from HR-HPVs are summarized in Figure 6.2.

E6 Oncoprotein

The most characterized role of HR-HPV E6 is the induction of tumor suppressor protein p53 degradation. p53 represents a major defense to viral replication since,

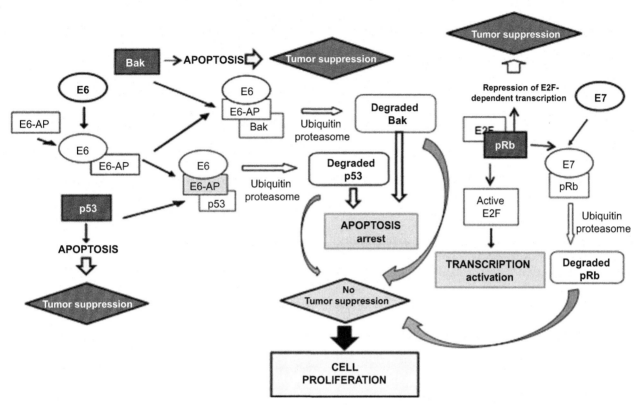

FIGURE 6.2 **Oncogenic Activities of E6 and E7 Proteins of High-Risk Human Papillomavirus.** E6 oncogenic power is due to its ability to bind the tumor suppressor protein p53 and to induce its degradation through the ubiquitin pathway. Therefore, apoptosis is inhibited. Apoptosis arrest can also be provoked by E6 through the inactivation of Bak. E7 binds to the retinoblastoma (Rb) family of tumor suppressor proteins and induces its degradation through the ubiquitin pathway. In this way, the Rb/E2F complexes are disrupted, E2F-dependent transcription is no longer repressed and cell division is driven.

once activated, it can promote cell cycle arrest or apoptosis of the infected cell. p53 is considered the "guardian of the genome" by maintaining its integrity through the inhibition of cell division if DNA damage occurs or eventually reprogramming it once the damage has been eliminated. If the replication is already active, even in the presence of damaged DNA, p53 can modulate the cell cycle by forcing it into the apoptotic pathway. By doing this, it prevents the selection of potentially transformed progeny.

A major strategy employed by the HR-HPV E6 protein to abrogate the oncosuppressive function of p53 is to induce its degradation through the ubiquitin-proteasome pathway. HPV-16 E6 binds to E6-associated protein (E6AP), cellular protein with ubiquitin protein ligase function. The E6/E6AP complex then binds p53, which is targeted for degradation. E6 from HR-HPV types can also interact with other additional ubiquitin ligases that may be involved in HPV-mediated p53 degradation.[7] Since p53 plays a key role in preserving genome integrity, HPV infected cells tend to accumulate chromosomal abnormalities, thus increasing the probability of a malignant evolution. Exclusively the E6 proteins from the HR-HPV types have the ability to induce degradation of

p53, whereas no p53 degradation was observed in cells expressing E6 from LR-HPVs.[8]

HPV E6 can modulate the apoptotic pathways independently of p53, through the pro-apoptotic effector Bak that displays E6 binding affinity. E6 inhibits Bak-induced apoptosis and this is mediated by an interaction between the E6 and Bak proteins, resulting in degradation of the Bak protein.[9] In addition, HPV E6 upregulates the expression of the inhibitors of apoptosis proteins, as c-IAP2 and survivin, through a mechanism that involves NF-jB.

Another characteristic feature of E6 from HR-HPV types is the ability to bind members of the membrane associated guanylate kinase (MAGUK) family, large proteins localized in the cytoplasmic membrane involved in cell-to-cell contact. These kinase family members contain multiple protein/protein interaction domains, including also PDZ motifs, which are those interacting with E6, at its four amino acid PDZ-binding motif.[10] The E6/MAGUK association is detrimental because it causes the loss of cell-to-cell contact and cell polarity.

HPV16 E6 also has the ability to activate the transcription of the hTERT (human telomerase reverse transcriptase) gene encoding for one of the catalytic subunits of the telomerase complex. In somatic cells telomerase activity is usually very low or completely lacking; conversely, HPV 16-infected cells show high levels of telomerase activity, which allow telomere length maintenance and indefinite proliferation. The induction of hTERT activation by E6 is accomplished through the degradation of transcriptional repressors, such as NFX1-91, or the activation of hTERT promoter, such as Myc.[11] In contrast to E6, E7 alone has no effect on hTERT expression; however, the concomitant expression of E7 and E6 significantly enhances E6-induced h-TERT transcription.[12]

E7 Oncoprotein

E7, one of the most efficient cell cycle deregulators, is a small nuclear phosphoprotein known to bind to the Retinoblastoma protein (pRb), promoting its degradation via the ubiquitin/proteasome route.[13] pRb is an inhibitory protein that negatively regulates, via direct association, the activity of several transcription factors, including members of the E2F family, stimulators of the expression of multiple genes involved in the progression of cell cycle and DNA synthesis.

In quiescent cells, pRb is present in a hypophosphorylated form and associates with E2F molecules, thereby inhibiting their transcriptional activity. When quiescent cells are exposed to mitogenic signals, pRb undergoes phosphorylation in mid-G1 phase, through cyclin-dependent kinase (CDK) activity. Phosphorylation of pRb leads to the disruption of pRb/E2F complexes, with consequent activation of E2F, which in turn activates the transcription of a group of genes encoding proteins essential for cell cycle progression, such as cyclin E and cyclin A. pRB remains phosphorylated until late M when it again becomes hypophosphorylated through the action of a specific phosphatase. E7 binding to pRb mimics its phosphorylation. Thus, E7 expressing cells can enter S phase in the absence of mitogenic signals.

E7 has the ability to alter the cell cycle via several mechanisms. First, HPV 16 E7 can associate with the CDK inhibitors p21WAF1/CIP1 and p27KIP1, thus neutralizing their inhibitory effects on the cell cycle, or it can also directly and/or indirectly interact with the cyclin A/CDK2 complex.[14,15] Second, it has been suggested that E7 may modulate kinase complexes, thus favoring phosphorylation of a different set of substrates involved in progression/completion of the viral life cycle.

The correlation between pRb-binding efficiency and E7 transforming activity varies consistently among HPV types. For example, E7s from certain noncarcinogenic HPV types, such as the cutaneous HPV 1 and the mucosal HPV 32, strongly associate with pRb but do not display any *in vitro* transforming activities.[16]

Although HPV 1, 32, and 16 E7s associate with pRb with similar affinity, the consequences of these interactions can be different. Exclusively in the case of HPV 16 E7, the association is responsible for pRb degradation via an ubiquitin/proteasome-dependent mechanism,[17] whereas HPV 1 and 32 E7 proteins do not. Thus, only the HR-HPV types are able to promote pRb degradation, with all its downstream effects for cell replication and survival.

Cancer cells are characterized by accumulation of genetic damage, and also by alteration of intracellular pH (i.e., alkalinization). It has been demonstrated that HPV 16 E7 has the ability to induce Na+/H+ exchanger NHE-1 in NIH3T3 cells and primary human keratinocytes.[18] This effect is blocked by treatment with specific NHE-1 inhibitors, thus inhibiting the development of the transformed phenotype, supporting the importance of this event in HPV-mediated carcinogenesis.

HR-HPV oncoproteins also cause mitotic abnormalities, which result in genomic instability and contribute to malignant progression. Several evidences showed that cervical cancer cells accumulated a plethora of numerical and structural chromosomal aberrations.[19] In detail, HR-HPV but not LR-HPV infections are responsible for genomic instability that can already be evidenced in premalignant lesions prior to the integration of HPV genomes into host chromosomes.[20] These data led to hypothesizing a causal role for HR-HPV in subversion of genomic stability. Accordingly, several cytogenetic abnormalities have been detected in HPV immortalized keratinocytes, which confirm that HR-HPV oncogene expression may promote genomic destabilization.

E5 Oncoprotein

Compared with E6 and E7, the oncotic role of E5 protein has not been fully elucidated. During HR-HPV infection steps, the HPV E5 protein is expressed in precancerous stages but not after viral integration. It is reasonable that E5 plays a role mostly in the early stage of cervical carcinogenesis. The finding supports this hypothesis that the E5 gene frequently is deleted when the HPV genome is integrated during the progression from low-grade to malignant disease. Further, it is likely that E5 may also contribute to cell transformation by modulating cellular signaling pathways in addition to potentiating the immortalizing ability of E6 and E7.

Recent evidences have shown that HPV 16 E5 may contribute to cervical carcinogenesis in part via stimulation of IFN-β. This effect is mediated by the induction of interferon regulatory factor 1 (IRF-1), which, in turn, induces transcriptional activation of IRF-1-targeted interferon-stimulated genes (ISGs).[21]

HPV Oncoproteins and Transcription Factors

A number of transcription factors have been linked to the development of human tumors. E6 and E7 oncoproteins stimulate proliferation by manipulating the function of a variety of host regulatory proteins. In turn, the binding of transcription factors to URR of HPV genome facilitates the expression of viral oncoproteins; the carcinogenic power is also dependent on the availability of a defined set of transcription factors derived from the infected host cell.

Intracellular signaling cascade occurs through a sequence of molecular events initiated by the specific interaction of an extracellular ligand with its receptor. This complex series of signals are further amplified by the action of second messengers that are able to modulate a number of target proteins ultimately affecting several cellular functions. Within this frame, ROS/RNS have been proposed as second messengers able to activate several signaling pathways leading to mitogenesis or apoptosis. It is well established that gene expression is significantly regulated in response to oxidative stress and requires the activation of specific redox-sensitive transcription factors, such as Nrf2, NF-κB, AP-1, and MAPKs, necessary to ensure cell survival. In response to changes of the activities of these transcription factors, most likely through oxidation of critical cysteine residues found on these proteins, the cellular redox status is modified.[22]

Activator Protein-1

Activator protein-1 (AP-1) is closely linked with proliferation and transformation of tumor cells. This factor, normally consisting of a heterodimer between c-Fos and c-Jun, seems to play a central role in transcriptional regulation of viral oncogene expression. In fact, while AP-1 binding was very low or absent in normal as well as in premalignant lesions, AP-1 transcription and binding activity were found to be very high in malignant tissues.[23] It was also shown that antioxidant-induced changes of the AP-1 transcription complex were paralleled by a selective suppression of HPV transcription.[24] The AP-1 is a direct target for the E7 oncogene, which post-translationally interacts with c-Jun in elevating its transactivation activity.[25]

AP-1 in response to low levels of oxidants binds to specific DNA sequence, thus resulting in activation of gene expression. Once activated, AP-1 in turn induces JNK activity, which phosphorylates the c-Jun transactivation domain. Conversely, when ROS are produced at high concentrations, AP-1 and AP-1-induced gene expression is inhibited. ROS cause oxidation of specific cysteine residues in c-Jun's DNA binding region, thus blocking AP-1/DNA interactions.

NF-κB

Similarly, NF-κB contains a redox-sensitive critical cysteine residue that is involved in DNA binding.[22] NF-κB is a protein complex that controls the transcription of DNA and is involved in cellular responses to stimuli such as stress, cytokines, free radicals, ultraviolet irradiation, oxidized LDL, and bacterial or viral antigens. NF-κB plays a key role in regulating the immune response to infection and its incorrect regulation has been linked to several diseases, including cancer. This transcription factor is a dimeric complex composed of different members of the Rel/NFkB family of polypeptides. NF-κB is normally located in the cytosol complexed with the inhibitory protein IκBα, but under oxidative conditions, IκBα sequentially undergoes phosphorylation by IκB kinase (IKK), ubiquitination, and subsequent degradation by the proteosome. Once it is dissociated by IκBα, the activated NF-κB is then translocated into the nucleus and binds to specific sequences of DNA adjacent to the genes that it regulates. Though ROS production is necessary to promote the initial events leading to the dissociation of the NF-κB/IκB complex, excessive ROS levels result in the oxidation of cysteine residues without affecting its translocation to the nucleus, but rather its DNA binding.

Activated NFkB suppresses apoptosis in a wide variety of tumor cells[26] and several proliferative, proinflammatory, and proangiogenic factors associated with aggressive tumor growth are regulated by NFkB. Among these is cyclooxygenase-2 (COX-2), an enzyme regulating prostaglandins production, that has been implicated in participating in both carcinogenesis and cancer progression in various malignancies, including cervical cancer.[27]

COX-2 transcription was shown to be stimulated by HPV 16 E6 and E7[28] and also by E5[29] through NFkB and AP-1. COX-2 expression was found to increase from LSIL to HSIL, the highest score being noted in HSIL corresponding to CIN 3 lesions, suggesting that this enzyme may have a role in the development and progression of cervical squamous intraepithelial lesions.[30] Results from different groups demonstrated that COX-2 expression promotes the progression of cervical cancer by increasing lymphonode metastasis and resistance to radiation therapy.

It is worth mentioning that several naturally occurring compounds are used against cervical cancer for their antioxidant and anti-inflammatory properties. In fact, their protective effects rely on the ability to modulate the activation/repression of multiple redox-sensitive transcription factors.[31,32]

OXIDATIVE STRESS AND CERVICAL CANCER

Despite the evidence that HPV is strongly implicated as the causative agent of cervical cancer, oncogenic HPV infection alone is not sufficient for tumor development. Other factors have to be involved in the progression of infected cells to the full neoplastic phenotype.

Oxidative stress appears to be a good candidate as a cancer promoting factor. It is known that cancer cells are characterized by enhanced oxidative stress. Elevated rates of ROS have been detected in almost all cancers, where they act as second messengers in intracellular signaling cascades, promoting many aspects of tumor development and progression.

It is well documented that some risk factors known to be implicated in cervical cancer development, such as tobacco smoking and chronic inflammation, determine oxidative stress increase[33–35] and several evidences depose for the occurrence of oxidative stress in cervical cancer.

Oxidative/Nitrosative Stress Markers

Studies performed on blood from patients with lesions diagnosed as CIN or cervical cancer evidenced significant changes in the levels of oxidative and nitrosative stress indicators (Table 6.1).

TABLE 6.1 Changes in Oxidative Stress Marker Levels in Samples from Patients with Lesions at Different Grades or Cervical Carcinoma with Respect to Samples from Control Healthy Subjects

Oxidative Stress Marker	Human Sample	Lesion	Pathological Sample vs Control	References
MDA [a]	Plasma	CC,[d] CIN [e]	Increase	[36–38]
MDA [a]	Erythrocyte	CC [d]	Increase	[36]
TBARS [b]	Plasma	CC [b]	Increase	[39]
TBARS [b]	Erythrocyte	LSIL,[f] HSIL,[g] CC [d]	Increase	[40]
NO products	Plasma	CC [d]	Increase	[36]
δ ALA-D [c]	Blood	LSIL,[d] HSIL,[e] CC [d]	Increase	[40]
Protein carbonyls	Cervical tissue	CIN,[e] CC [d]	Increase	[47]

[a] MDA = malondialdehyde
[b] TBARS = thiobarbituric acid reactive substances
[c] δ ALA-D = δ-amino levulinate dehydratase
[d] CC = cervical carcinoma
[e] CIN = cervical intraepithelial neoplasia
[f] LSIL = low grade squamous intraepithelial lesion
[g] HSIL = high grade squamous intraepithelial lesion

Lipid Peroxidation Products

An important role in the control of cell division is played by lipid peroxidation (LPO). Oxidative stress and LPO associated with infections and chronic inflammation may induce several human cancers. LPO products are reactive to nucleic acids, proteins, and cellular thiols and have profound mutagenic potential. Modifications of proteins with LPO products may regulate cellular processes like apoptosis, cell signaling, and senescence.

A significant increase of plasma levels of malondialdehyde (MDA), a product of lipid peroxidation initiated by ROS, was found in women with invasive cervical cancer[36,37] and CIN[37,38] compared to those of healthy subjects. Also erythrocyte MDA level was found significantly more elevated in patients with cervical cancer with respect to controls.[36]

An estimation of LPO, as indexed by the measurement of thiobarbituric acid reactive substances (TBARS), indicated increased TBARS levels in plasma from patients with cervical carcinoma.[39] Higher TBARS than control subjects were also demonstrated in the erythrocytes of LSIL, HSIL, and carcinoma patients,[40] indicating that oxidative stress occurs precociously with premalignant states.

δ-Aminolevulinate Dehydratase

As an index of overproduction of free radicals, the reactivation index of δ-aminolevulinate dehydratase (δ ALA-D) in the blood of premalignant and malignant stages of cervical cancer was also determined.[40] This enzyme plays a fundamental role in most aerobic organisms by participating in heme biosynthesis pathway and its inhibition can lead to δ-ALA accumulation, which in turn can enhance the generation of free radicals, aggravating oxidative damage to cellular components. The reactivation index of δ ALA-D activity is a good tool to evaluate oxidative stress due to the high sensitivity of the enzyme to oxidation of its SH groups. Results obtained by Gonçalves et al.[40] indicated that the enzyme is more oxidized in LSIL, HSIL, and carcinoma patients compared to the control group.

Nitric Oxide

Nitrosative stress was also evidenced in patients with cervical carcinoma, as shown by increased plasmatic nitric oxide (NO) levels.[36] NO is a crucial factor in the regulation of many homeostatic mechanisms. However, beside its regulatory effects, it can react with superoxide anion, forming peroxynitrite, a reactive and oxidant compound able to induce nitration and nitrosation of proteins, so altering their functions and therefore causing severe oxidative damage to the cell structures. NO and its derivatives, owing to their cytotoxicity and capacity to cause direct or indirect genetic damage, have been

TABLE 6.2 Changes in Antioxidant Levels in Samples from Patients with Lesions at Different Grades or Cervical Carcinoma with Respect to Samples from Control Healthy Subjects

Enzymatic Antioxidants	Human Sample	Lesion	Pathological Sample vs Control	References
SOD [a]	Plasma	CC [e]	Decrease	[36,39,41]
CAT [b]	Plasma	CC [e]	Decrease	[36,41]
GPX [c]	Plasma	CC,[e] CIN [f]	Decrease	[36,37,39]

Nonenzymatic Antioxidants	Human Sample	Lesion	Pathological Sample vs Control	References
Vitamin C	Plasma	LSIL,[g] HSIL,[h] CC [e]	Decrease	[38–40]
Vitamin E	Plasma	CIN,[f] CC [e]	Decrease	[37–39,42]
Vitamin A, lutein, lycopene, zeaxanthin	Plasma	CIN,[f] CC [e]	Decrease	[37]
GSH [d]	Plasma	CC [e]	Decrease	[39]
Coenzyme Q10	Plasma	CIN,[f] CC [e]	Decrease	[43]

[a] SOD = superoxide dismutase
[b] CAT = catalase
[c] GPX = glutathione peroxidase
[d] GSH = reduced glutathione
[e] CC = cervical carcinoma
[f] CIN = cervical intraepithelial neoplasia
[g] LSIL = low grade squamous intraepithelial lesion
[h] HSIL = high grade squamous intraepithelial lesion

suggested to play a role in carcinogenetic processes. NO can stimulate ROS-induced LPO; in fact, a direct correlation between NO levels and erythrocyte MDA in cervical cancer was reported by Beevi et al.[36]

Antioxidant Systems

The cells counteract the oxidative stress by an array of many different defense mechanisms, including free radical scavengers and antioxidant enzymes to protect themselves from the potentially deleterious effects of ROS/RNS. However, in pathological conditions, despite this wide set of protecting mechanisms, oxidative damage eventually accumulates, contributing to a number of degenerative diseases including cancer. Oxidative burden attenuates the antioxidant defense systems and reduction in antioxidant defense mechanisms correlates with the emergence of malignant phenotype.

In cervical cancer a reduction of both enzymatic and nonenzymatic antioxidant defense has been shown to occur (Table 6.2).

A significant depletion in the activities of primary antioxidant enzymes, superoxide dismutase (SOD),

TABLE 6.3 Changes in Stress Response Marker Levels in Samples from Patients with Lesions at Different Grades or Cervical Carcinoma with Respect to Samples from Control Healthy Subjects

Stress Response Markers	Human Sample	Lesion	Pathological Sample vs Control	References
GST [a]	Erythrocyte	CC [d]	Increase	[36]
GST [a]	Cervical tissue	CIN,[e] CC [d]	Increase	[47]
iNOS [b]	Cervical tissue	CIN,[e] CC [d]	Decrease	[47]
ERp57 [c]	Cervical tissue	CIN,[e] CC [d]	Increase	[47]

[a] GST = Glutathione S-transferase
[b] iNOS = inducible Nitric Oxide Synthase
[c] ERp57 = Endoplasmic Reticulum protein 57
[d] CC = cervical carcinoma
[e] CIN = cervical intraepithelial lesion

glutathione peroxidase (GPx), and catalase (CAT) was observed in patients with cervical cancer[36,39,41] and CIN[37] as compared to healthy subjects. Either an increased utilization or a direct ROS/RNS-mediated suppression could be the cause of the impairment in the antioxidant enzymes.

Also plasma levels of antioxidant vitamins and compounds such as α-tocopherol,[37–39,42] vitamin C,[38–40] lutein, beta-carotene, lycopene, and zeaxanthin[37] were all significantly lower in women with cervical CIN or invasive cervical cancer compared to those of the normal control group.

A decrease in the level of two other potent antioxidants, reduced glutathione and coenzyme Q10, was also found in plasma of patients with various grades of CIN and cervical cancer compared to controls.[39,43]

Stress Response Markers

An increase of the activity of glutathione S-transferase (GST) was observed in erythrocytes of cervical cancer patients[36] (Table 6.3). GST catalyzes the nucleophilic addition of the thiol of GSH to a variety of electrophiles and plays a prominent role in the protection of cells against many cytotoxic and carcinogenic chemicals. Increased expression of various GST isoenzymes has been found in human tumor cell lines of different histological origin, and increased GST activity has been observed in most of the human cancers[44] arguably as an adaptive response to oxidative/nitrosative stress.

Modulation of stress response in the different types of lesions was also detected in studies on patients' tissues (see later).

Protein Oxidation

Proteins represent important targets of oxidative stress. Several studies demonstrated that protein oxidation is frequently associated with a decrease or complete loss of protein catalytic functions and triggers the formation of high molecular, potentially cytotoxic aggregates.

To elucidate the molecular events and mechanisms associated with cervical carcinogenesis and to identify prognostic or predictive markers, alterations of protein profiles were investigated. For this purpose, a proteomics approach was utilized because this technique gives qualitative and quantitative information not only about the native proteins but also about their post-translation modifications. Among these, the formation of carbonyl groups is the most widely studied index of protein oxidation. In particular, redox proteomics allows investigation of protein oxidative modifications and is revealed to be a useful method to identify those proteins specifically altered in pathological conditions.

Protein oxidation in cervical cancer has been recently investigated by studies on cell cultures and patients' tissues.

Studies on Cell Cultures

A convenient experimental method to study the molecular mechanisms of tumor progression is represented by cell lines containing HPV genome, models of preneoplastic and neoplastic stages of cervical cancer.

Perluigi et al.[45,46] applied a proteomic approach to study the effects of oxidative stress induced by UVB irradiation on HK-168, a keratinocyte cell line transfected with the whole genome of the high risk type HPV-16, in comparison with normal human epithelial keratinocytes (NHEK).

Significant changes in expression profile and specific protein oxidation were evidenced in both cell types. Proteins involved in folding (glucose-regulated protein 78, HSPs, glucosidase II), protein synthesis (elongation factors), maintenance of cytoskeleton integrity (cytokeratins, actin-related protein 3), and cell–cell interaction (protein disulfide isomerase) became more oxidized upon UVB irradiation.

On the other hand, proteins implicated in defense systems were not efficiently activated in HK-168, thus making the cells more susceptible to accumulate damaged and toxic proteins, so driving the virus-transformed cell toward a neoplastic phenotype. Conversely, even if UVB irradiation caused significant protein alterations also in NHEK, these cells maintained their ability to counteract oxidative stress conditions by modulating the expression levels of protective proteins, such as tumor suppressors, heat-shock proteins, and proteasome components, which participate in cell defense and repair.

Studies on Patients' Tissues

A recent study has been performed on tissues from high grade dysplastic HPV-16 lesions (CIN 2/3) and invasive squamous cervical carcinoma tissues.[47] It was shown that protein carbonyls were significantly increased in histological samples of dysplastic tissues, while levels detected in neoplastic tissues were not significantly different from controls (normal cervical tissue specimens). This unexpected trend was also paralleled by the extent of oxidative DNA damage, which was clearly increased in dysplastic tissues with respect to both cancer and controls.

By the redox proteomic approach five proteins having increased carbonyls levels in dysplastic samples with respect to controls were identified: cytokeratin 6, actin, cornulin, retinal dehydrogenase, and gliceraldehyde 3-phosphate dehydrogenase (GAPDH). Comparison of cancer with dysplastic tissues evidenced that in carcinoma samples five proteins (peptidyl-prolyl cis-trans isomerase A, ERp57, serpin B3, annexin 2, and GAPDH) showed lower carbonylation than in dysplastic tissues (Table 6.4).

Interestingly, GAPDH was found more oxidized in dysplastic lesion versus either controls or neoplastic tissue. GAPDH-increased oxidation was paralleled by decreased enzymatic activity in dysplastic tissue with respect to control and, conversely, a reduced oxidation level in cancer tissue was accompanied by a considerable retrieval of enzymatic activity. Besides the

TABLE 6.4 List of Proteins Identified as More Oxidized in Dysplastic Lesions with Respect to Controls (A) or in Cervical Carcinoma with Respect to Dysplasia (B)

Dysplasia vs Control	
Protein	**Function**
Cytokeratin 6	Cytoskeleton integrity
Gliceraldehyde 3-phosphate dehydrogenase	Energy metabolism, etc.
Cornulin	Cell differentiation
Retinal dehydrogenase	Cell differentiation
Actin	Cytoskeleton integrity
Carcinoma vs Dysplasia	
Protein	**Function**
Gliceraldehyde 3-phosphate dehydrogenase	Energy metabolism, etc.
Peptidyl-proly l cis-trans isomerase A	Cell signaling/division
Endoplasmic reticulum protein 57	Protein folding
Serpin B2	Tumor proliferation
Annexin A2	Membrane trafficking

(Main functions of listed protein are also reported)
Data are from the redox proteomic analysis of De Marco et al. [47] on histological samples from high grade dysplastic lesions (CIN 2/3) and invasive squamous cervical carcinoma tissues.

well-known function in aerobic metabolism of glucose, GAPDH has been implicated in several cell pathways and its role in cell death regulation and carcinogenesis has been suggested.[48]

These redox proteomics results indicate the occurrence—in dysplastic lesions—of a selective oxidation of specific proteins involved in cell signaling/division, morphogenesis, and differentiation, leading to cytoskeleton derangement and suppression of terminal differentiation.

In addition, the oxidative modification of protein implied in the synthesis of compounds able to operate a suppressive modulation of E6/E7 oncogenes (i.e., increased carbonylation of retinal dehydrogenase in dysplasia) can lead to a reduced control on viral oncogenes activity.

Conversely, cancer tissue seems to attain an improved control on oxidative damage, as shown by lower oxidation of key detoxifying/prosurvival proteins with respect to dysplasia. This compensation to oxidative conditions could be explained by the modulation of some stress response markers in the different types of lesions (Table 6.3).

Expression of both endoplasmic reticulum protein 57 (ERp57), a protein assisting the maturation and transport of unfolded secretory proteins by facilitating disulphide bond formation, and GST, able to offer protection against many cytotoxic oxidative/nitrosative stress-related metabolites, were significantly higher in neoplastic tissue compared with both dysplastic and controls tissues. On the contrary, a progressive down-regulation of the inducible form of nitric oxide synthase (iNOS) expression during the different stages of cervical carcinogenesis—from the control to

dysplastic and to neoplastic samples—was evidenced by different authors.[47,49] This trend, which is divergent from that observed in most cancer types, suggests that low iNOS expression may address toward a malignant phenotype, promoting tumor progression rather than an antitumor response. However, the role of this enzyme in carcinogenesis is far from being completely clear.

It is likely that activated detoxifying systems (ERp57, GST) and reduced iNOS in carcinoma cells might be part of a complex adaptive metabolic profile allowing cell survival in an increasingly oxidant environment.

On the whole, these findings support the view that dysplastic state is highly vulnerable to oxidative damage, a major factor of genetic instability, providing the conditions for the neoplastic evolution of transformed cells. Thus, oxidative injury is a prominent factor even in the early steps of carcinogenesis. Conversely, tumor cells adapt their metabolism in order to support their growth and survival, seemingly creating a paradox of high ROS production in the presence of high antioxidant levels, seeming to fit well with stress conditions (Figure 6.3).

CONCLUDING REMARKS

Elevated rates of ROS have been detected in almost all cancers, where they act as second messengers in intracellular signaling cascades, promoting many aspects of tumor development and progression. Indeed, it is well established that some risk factors implicated in cervical cancer development, such as tobacco smoking and chronic inflammation, are responsible for increased oxidative stress conditions. However, the molecular events associated with oxidative damage still remain unclear. The present poor knowledge about the mechanisms involved in neoplastic progression has dramatic consequences on the clinical side. Because it is not possible to predict the clinical outcome of precancerous lesions, even though just a very minor part of them tend to progress, all of them have to be regarded as potentially progressive. Further research should be directed to evaluate if the impairment of protein function, consequent to oxidative modifications, may correlate with a specific state of neoplastic transformation/progression. The identification of specific proteins that are irreversibly modified by oxidative damage could shed light on the molecular mechanism targeted by oxidative stress in cell transformation and tumor development. Future studies should explore alternative therapeutic strategies, which can effectively target oxidative stress-related pathways during various phases of viral infection and its oncogenic expression.

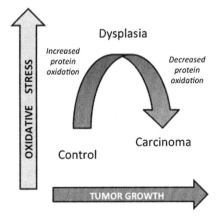

FIGURE 6.3 **Interplay between Tumor Growth and Oxidative Stress.** In dysplastic lesions a selective oxidation of specific proteins occurs. Conversely, cancer tissue seems to attain an improved control on oxidative damage, as shown by a lower protein oxidation. It is a likely hypothesis that tumor cells have the ability to withstand high levels of ROS by protecting cellular components from oxidative insult.

SUMMARY POINTS

- Cervical cancer is the second most common type of cancer in women worldwide.
- High-risk Human Papillomavirus (HR-HPV) are recognized as the main etio-pathogenetic factors of cervical cancer.
- Viral oncogenes expression is not sufficient per se to induce cervical cancer; a number of risk factors have to play a role.
- Oxidative stress is an underexplored candidate risk factor in cervical carcinogenesis.
- The oncogenic proteins modulate redox-sensitive transcription factors affecting both cell growth and cell death.
- Oxidative stress markers have been detected in human tissues and body fluids from patients with cervical cancer.
- In cervical cancer a reduction of both enzymatic and nonenzymatic antioxidant defense has been demonstrated.
- Increased carbonyls levels of specific proteins in precancerous and carcinoma lesions have been shown.
- Novel therapeutic strategies for cervical cancer should include the use of antioxidant compounds.

Acknowledgments

We would like to thank Dr Fabio Di Domenico and Dr Federico De Marco for their critical review of the manuscript.

References

1. zur Hausen H, Meinhof W, Scheiber W, Bornkamm GW. Attempts to detect virus-specific DNA in human tumors. I. Nucleic acid hybridizations with complementary RNA of human wart virus. *Int J Cancer* 1974;**13**:650–6.
2. De Villiers EM, Fauquet C, Broker TR, Bernard HU, zur Hausen H. Classification of papillomaviruses. *Virology* 2004;**324**:17–27.
3. Storz P. Reactive oxygen species in tumor progression. *Front Biosci* 2005;**10**:1881–96.
4. Unger ER, Steinau M, Rajeevan MS, Swan D, Lee DR, Vernon SD. Molecular markers for early detection of cervical neoplasia. *Dis Markers* 2004;**20**:103–16.
5. Shamanin V, Glover M, Rausch C, Proby C, Leigh IM, zur Hausen H, et al. Specific types of human papillomavirus found in benign proliferations and carcinomas of the skin in immunosuppressed patients. *Cancer Res* 1994;**54**:4610–3.
6. Baak JP, Kruse AJ, Robboy SJ, Janssen EA, van Diermen B, Skaland I. Dynamic behavioural interpretation of cervical intraepithelial neoplasia with molecular biomarkers. *J Clin Pathol* 2006;**59**:1017–28.
7. Massimi P, Shai A, Lambert P, Banks L. HPV E6 degradation of p53 and PDZ containing substrates in an E6AP null background. *Oncogene* 2008;**27**:1800–4.
8. Hiller T, Poppelreuther S, Stubenrauch F, Iftner T. Comparative analysis of 19 genital human papillomavirus types with regard to p53 degradation, immortalization, phylogeny, and epidemiologic risk classification. *Cancer Epidemiol Biomarkers Prev* 2006;**15**:1262–7.
9. Thomas M, Banks L. Inhibition of Bak-induced apoptosis by HPV-18 E6. *Oncogene* 1998;**17**:2943–54.
10. Watson RA, Thomas M, Banks L, Roberts S. Activity of the human papillomavirus E6 PDZ-binding motif correlates with an enhanced morphological transformation of immortalized human keratinocytes. *J Cell Sci* 2003;**116**:4925–34.
11. Veldman T, Liu X, Yuan H, Schlegel R. Human papillomavirus E6 and Myc proteins associate in vivo and bind to and cooperatively activate the telomerase reverse transcriptase promoter. *Proc Natl Acad Sci USA* 2003;**100**:8211–6.
12. Liu X, Roberts J, Dakic A, Zhang Y, Schlegel R. HPV E7 contributes to the telomerase activity of immortalized and tumorigenic cells and augments E6-induced hTERT promoter function. *Virology* 2008;**375**:611–23.
13. Münger K, Basile JR, Duensing S, Eichten A, Gonzalez SL, Grace M, et al. Biological activities and molecular targets of the human papillomavirus E7 oncoprotein. *Oncogene* 2001;**20**:7888–98.
14. Zerfass-Thome K, Zwerschke W, Mannhardt B, Tindle R, Botz JW, Jansen-Dürr P. Inactivation of the cdk inhibitor p27KIP1 by the human papillomavirus type 16 E7 oncoprotein. *Oncogene* 1996;**13**:2323–30.
15. He W, Staples D, Smith C, Fisher C. Direct activation of cyclin-dependent kinase 2 by human papillomavirus E7. *J Virol* 2003;**77**:10566–74.
16. Caldeira S, Dong W, Tomakidi P, Paradiso A, Tommasino M. Human papillomavirus type 32 does not display in vitro transforming properties. *Virology* 2002;**301**:157–64.
17. Jones DL, Münger K. Analysis of the p53-mediated G1 growth arrest pathway in cells expressing the human papillomavirus type 16 E7 oncoprotein. *J Virol* 1997;**71**:2905–12.
18. Reshkin SJ, Bellizzi A, Caldeira S, Albarani V, Malanchi I, Poignee M, et al. Na$^+$/H$^+$ exchanger-dependent intracellular alkalinization is an early event in malignant transformation and plays an essential role in the development of subsequent transformation-associated phenotypes. *FASEB J* 2000;**14**:2185–97.
19. Mitelman F, Johansson B, Mertens F. Mitelman database of chromosome aberrations in cancer. http://cgap.nci.nih.gov/Chromosomes/Mitelman; 2007.
20. Bulten J, Poddighe PJ, Robben JC, Gemmink JH, de Wilde PC, Hanselaar AG. Interphase cytogenetic analysis of cervical intraepithelial neoplasia. *Am J Pathol* 1998;**152**:495–503.
21. Muto V, Stellacci E, Lamberti AG, Perrotti E, Carrabba A, Matera G, et al. Human papillomavirus type 16 E5 protein induces expression of beta interferon through interferon regulatory factor 1 in human keratinocytes. *J Virol* 2011;**85**:5070–80.
22. Matthews JR, Kaszubska W, Turcatti G, Wells TN, Hay RT. Role of cysteine62 in DNA recognition by the P50 subunit of NF-kappa B. *Nucleic Acids Res* 1993;**21**:1727–34.
23. Prusty BK, Das BC. Constitutive activation of transcription factor AP-1 in cervical cancer and suppression of human papillomavirus (HPV) transcription and AP-1 activity in HeLa cells by curcumin. *Int J Cancer* 2005;**113**:951–60.
24. Rösl F, Das BC, Lengert M, Geletneky K, zur Hausen H. Antioxidant-induced changes of the AP-1 transcription complex are paralleled by a selective suppression of human papillomavirus transcription. *J Virol* 1997;**71**:362–70.
25. Antinore MJ, Birrer MJ, Patel D, Nader L, McCance DJ. The human papillomavirus type 16 E7 gene product interacts with and transactivates the AP1 family of transcription factors. *EMBO J* 1996;**15**:1950–60.
26. Wang CY, Mayo MW, Kornelul RG, Goeddel DV, Baldwin Jr AS. NF-kappaB antiapoptosis: Induction of TRAF1 and TRAF2 and c-IAP1 and c-IAP2 to suppress caspase-8 activation. *Science* 1998;**281**:1680–3.
27. Kim MH, Seo SS, Song YS, Kang DH, Park IA, Kang SB, et al. Expression of cyclooxygenase-1 and -2 associated with expression of VEGF in primary cervical cancer and at metastatic lymphnodes. *Gynecol Oncol* 2003;**90**:83–90.

28. Subbaramaiah K, Dannenberg AJ. Cyclooxygenase-2 transcription is regulated by human papillomavirus 16 E6 and E7 oncoproteins: Evidence of a corepressor/coactivator exchange. *Cancer Res* 2007;**67**:3976–85.

29. Kim SH, Oh JM, No JH, Bang YJ, Juhnn YS, Song YS. Involvement of NF-kB and AP-1 in COX-2 upregulation by human papillomavirus 16 E5 oncoprotein. *Carcinogenesis* 2009;**30**:753–7.

30. Balan R, Amălinei C, Giuşcă SE, Ditescu D, Gheorghiţă V, Crauciuc E, et al. Immunohistochemical evaluation of COX-2 expression in HPV positive cervical squamous intraepithelial lesions. *Rom J Morphol Embryol* 2011;**52**:39–43.

31. Prusty BK, Bhudev CD. Constitutive activation of transcription factor AP-1 in cervical cancer and suppression of human papillomavirus (HPV) transcription and AP-1 activity in HeLa cells by curcumin. *Int J Cancer* 2004;**113**:951–60.

32. Di Domenico F, Foppoli C, Coccia R, Perluigi M. Antioxidants in cervical cancer: Chemopreventive and chemotherapeutic effects of polyphenols. *Biochim Biophys Acta* 2012;**1822**:737–47.

33. Moktar A, Singh R, Vadhanam MV, Ravoori S, Lillard JW, Gairola CG, et al. Cigarette smoke condensate-induced oxidative DNA damage and its removal in human cervical cancer cells. *Int J Oncol* 2011;**39**:941–7.

34. Boccardo E, Lepique AP, Villa LL. The role of inflammation in HPV carcinogenesis. *Carcinogenesis* 2010;**31**:1905–12.

35. Williams VM, Filippova M, Soto U, Duerksen-Hughes PJ. HPV-DNA integration and carcinogenesis: Putative roles for inflammation and oxidative stress. *Future Virol* 2011;**6**:45–57.

36. Beevi SS, Rasheed MH, Geetha A. Evidence of oxidative and nitrosative stress in patients with cervical squamous cell carcinoma. *Clin Chim Acta* 2007;**375**:119–23.

37. Kim YT, Kim JW, Choi JS, Kim SH, Choi EK, Cho NH. Relation between deranged antioxidant system and cervical neoplasia. *Int J Gynecol Cancer* 2004;**14**:889–95.

38. Lee GJ, Chung HW, Lee KH, Ahn HS. Antioxidant vitamins and lipid peroxidation in patients with cervical intraepithelial neoplasia. *J Korean Med Sci* 2005;**20**:267–72.

39. Manju V, Kalaivani Sailaja J, Nalini N. Circulating lipid peroxidation and antioxidant status in cervical cancer patients: A case-control study. *Clin Biochem* 2002;**35**:621–5.

40. Gonçalves TL, Erthal F, Corte CL, Müller LG, Piovezan CM, Nogueira CW, et al. Involvement of oxidative stress in the premalignant and malignant states of cervical cancer in women. *Clin Biochem* 2005;**38**:1071–5.

41. Manju V, Balasubramanian V, Nalini N. Oxidative stress and tumor markers in cervical cancer patients. *J Biochem Mol Biol Biophys* 2002;**6**:387–90.

42. Palan PR, Woodall AL, Anderson PS, Mikhail MS. Alpha-tocopherol and alpha-tocopheryl quinone levels in cervical intraepithelial neoplasia and cervical cancer. *Am J Obstet Gynecol* 2004;**190**:1407–10.

43. Palan PR, Mikhail MS, Shaban DW, Romney SL. Plasma concentrations of coenzyme Q10 and tocopherols in cervical intraepithelial neoplasia and cervical cancer. *Eur J Cancer Prev* 2003;**12**:321–6.

44. Hayes JD, Pulford DJ. The glutathione S-transferase supergene family: regulation of GST and the contribution of the isoenzymes to cancer chemoprotection and drug resistance. *Crit Rev Biochem Mol Biol* 1995;**30**:445–600.

45. Perluigi M, Giorgi A, Blarzino C, De Marco F, Foppoli C, Di Domenico F, et al. Proteomics analysis of protein expression and specific protein oxidation in human papillomavirus transformed keratinocytes upon UVB irradiation. *J Cell Mol Med* 2009;**13**:1809–22.

46. Perluigi M, Di Domenico F, Blarzino C, Foppoli C, Cini C, Giorgi A, et al. Effects of UVB-induced oxidative stress on protein expression and specific protein oxidation in normal human epithelial keratinocytes: A proteomic approach. *Proteome Sci* 2010;**8**:8–13.

47. De Marco F, Bucaj E, Foppoli C, Fiorini A, Blarzino C, Filipi K, et al. Oxidative stress in HPV-driven viral carcinogenesis: Redox proteomics analysis of HPV-16 dysplastic and neoplastic tissues. *PLoS One* 2012;**7**:e34366.

48. Colell A, Green DR, Ricci JE. Novel roles for GAPDH in cell death and carcinogenesis. *Cell Death Differ* 2009;**16**:1573–81.

49. Mazibrada J, Ritta M, Mondini M, De Andrea M, Azzimonti B, Borgogna C, et al. Interaction between inflammation and angiogenesis during different stages of cervical carcinogenesis. *Gynecol Oncol* 2008;**108**:112–20.

Inflammation and Oxidative DNA Damage: A Dangerous Synergistic Pathway to Cancer

Olga A. Martin

Department of Radiation Oncology & Laboratory of Molecular Radiation Biology, Peter MacCallum Cancer Centre, and The Sir Peter MacCallum Department of Oncology, University of Melbourne, Melbourne, VIC, Australia

Somaira Nowsheen

Medical Scientist Training Program, Mayo Graduate School, Mayo Clinic, College of Medicine, Rochester, MN, USA

Shankar Siva

Department of Radiation Oncology, Peter MacCallum Cancer Centre and The Sir Peter MacCallum Department of Oncology, University of Melbourne, VIC, Australia

Khaled Aziz

Medical Scientist Training Program, Mayo Graduate School, Mayo Clinic, College of Medicine, Rochester, MN, USA

Vasiliki I. Hatzi

Laboratory of Health Physics & Environmental Health, Institute of Nuclear Technology & Radiation Protection, National Center for Scientific Research "Demokritos," 153 10 Aghia Paraskevi, Athens, Greece

Alexandros G. Georgakilas

Physics Department, School of Applied Mathematics and Physical Sciences, National Technical University of Athens (NTUA), Zografou 15780, Athens, Greece

List of Abbreviations

AG aminoguanidine
AID activation-induced cytidine deaminase
BER base excision repair
COX-2 cyclooxygenase-2
c-PTIO 2-(4-carboxyphenyl)-4,4,5,5-tetramethylimidazoline-1-oxyl-3-oxide
CRT calreticulin
CSFS colony-stimulating factors
DC dendritic cell
DDR DNA Damage Response
DMSO dimethyl sulfoxide
DSB double-strand break
EGFR epidermal growth factor receptor

HIF hypoxia-inducible factor
HMGB1 high-mobility group-protein B1
IBD inflammatory bowel diseases
ICAM intercellular adhesion molecule
IL interleukin
JNK c-Jun N-terminal kinase
L-NNA N(omega)-nitro-L-arginine
MDSC myeloid-derived suppressor cells
MHC major histocompatibility complex
MIF 1 macrophage migration inhibitory factor
NADP nicotinamide adenine dinucleotide phosphate
NER nucleotide excision repair
NFκB nuclear factor kappa beta
NO nitric oxide
NSCLC non-small cell lung cancer

Cancer
http://dx.doi.org/10.1016/B978-0-12-405205-5.00007-6

© 2014 Elsevier Inc. All rights reserved.

OCDL oxidatively induced clustered DNA lesions
PTC papillary thyroid carcinoma
PTEN phosphatase and tensin homologue
RNS reactive nitrogen species
ROS reactive oxygen species
SABR stereotactic ablative body radiotherapy
SASP senescence-associated secretory phenotype
STAT signal transducer and activator of transcription
TAM tumor associated macrophages
TGFβ transforming growth factor-β
TLR4 toll-like receptor 4
TNF tumor necrosis factor
VHL von Hippel Lindau

INTRODUCTION

Cancer is a very real threat to people of all ages and despite decades of research we have failed to conquer the disease. Oncogenic transformation is due to accumulation of various mutations, whether acquired or inherited, and caused by endogenous and/or exogenous agents. These bestow prosurvival capacities to the transformed cells and often allow them to evade and modify the immune response. Multiple genes in the human body precisely control cell growth. Errors in these genes lead to further alterations or mutations. Accumulation of many mutations over time usually leads to a malignant state manifested by high chromosomal instability.

The human body is under continuous attack from both external and internal insults which results in numerous DNA lesions per cell per day (10,000–100,000).[1] These lesions can block DNA replication and transcription leading to mutations and possibly transformation and carcinogenesis. Just one unrepaired double-strand break (DSB) can be lethal to the cell or highly mutagenic. Failure to repair any DNA damage leads to apoptotic or necrotic cell death. The DNA damage response (DDR) network detects DNA lesions, signals their presence, and promotes DNA repair. Defects in this pathway are often seen in cancer.

DNA damage can be induced by oxidation and this may eventually progress to carcinogenesis. In addition, cancer is considered a pro-inflammatory disease and a number of current therapies target this pro-inflammatory state within the tumor microenvironment. Thus, in this chapter we discuss the role(s) of oxidatively induced DNA damage and inflammation in cancer. Overall, a better understanding of the synergy between oxidative DNA damage, inflammation, and cancer (i.e., a "lethal triptych") will provide the center for future therapies.

OXIDATIVE DNA DAMAGE

Oxidative DNA damage is an inevitable consequence of endogenous and exogenous events, such as cellular metabolism and toxic insults such as exposure to chemicals or ionizing radiation. Oxidative stress has been associated with various serious diseases including cancer, Alzheimer's, arteriosclerosis, and diabetes. Oxidative damage occurs when the body is exposed to excessive amounts of electrically charged, aggressive oxygen and nitrogen compounds: reactive oxygen and nitrogen species (ROS, RNS). The purine and pyrimidine bases and sugar moieties can be affected by oxidation. Oxidatively induced DNA lesions, multiple DNA lesions in close proximity (clusters or OCDLs), play a critical role in carcinogenesis.[2] DNA-protein crosslinks can also result from oxidation. Though oxidative modifications occur in proteins, lipids, and DNA, since proteins and lipids are readily degraded and resynthesized, the most significant consequences of the oxidative stress are the modifications to the DNA, which can cause mutations and lead to genomic instability (Figure 7.1).

Mechanisms of Induction

Oxidation is a critical component of energy production by mitochondria, the inflammatory response and, in general, by the cellular defense system. Acute inflammatory response recruits activated leukocytes that can cause extensive DNA damage by secreting various chemical mediators. Some of oxygen-derived products include hydroxyl radical and superoxide radical. The hydroxyl radical reacts with biological molecules such as DNA, causing damage to the heterocyclic DNA bases and the sugar moiety by a variety of mechanisms. Hydroxyl radical reacts with purines and pyrimidines of DNA by addition to double bonds, and by abstraction of an H from the methyl group of thymine and from each of the C–H bonds of 2'-deoxyribose leading to modifications.[3] This oxidative stress can also lead to DSBs.

Pathways of Repair

Oxidative DNA damage is repaired by multiple, overlapping DNA repair pathways. Two major mechanisms exist to repair oxidatively induced DNA lesions: base-excision repair (BER) and nucleotide-excision repair (NER). In BER-mediated repair, DNA glycosylase usually detects the damaged base and mediates base removal prior to nuclease, polymerase, and ligase proteins bridging the gap and completing the repair process. On the other hand, NER-mediated repair recognizes base lesions that distort the helical structure. The damaged base is excised as a 22–30 base oligonucleotide resulting in single-stranded DNA that is repaired by proteins such as DNA polymerase before proceeding to ligation. There are two pathways that differ in the mechanism of helix recognition: transcription-coupled NER specifically targets lesions of that transcription, while global-genome NER covers the other lesions. Other repair pathways include mismatch repair, nonhomologous end joining, and homologous recombination, all of which repair DSBs.[4]

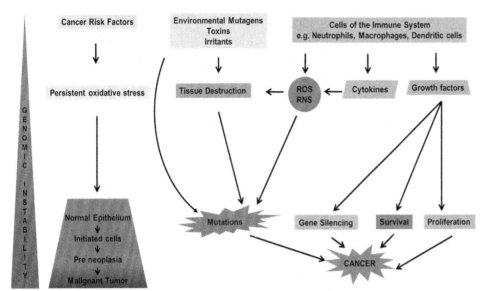

FIGURE 7.1 **Persistent Oxidative Stress Leads to Neoplastic Transformation.** The key steps are the production of reactive oxygen and nitrogen species (ROS/RNS), which can lead to accumulation of mutations and therefore preneoplastic states. In this first initiation step, components of the immune system seem to play a pivotal role. The occurrence of a tissue malignancy seems always to coincide with genomic and chromosomal instability.

ROLE OF INFLAMMATION IN INDUCTION OF OXIDATIVE STRESS AND DNA DAMAGE LEADING TO CANCER

Inflammation is a key component of the tumor microenvironment and a recognized hallmark of cancer.[5,6] The causal linkage between inflammation and cancer was initially suggested in the nineteenth century following the observation that tumors often developed in settings of chronic inflammation and that pro-inflammatory cells were present in biopsied tumor specimens.[7] Accumulating evidence shows that chronic inflammation, in fact, is associated with increased risk of cancer development. Moreover, chronic inflammation is linked to 15 to 20% of worldwide cancer deaths.[6]

The interlink between inflammation and cancer involves two major interconnected pathways: an extrinsic mechanism, where a constant inflammatory state (chronic inflammation) contributes to increased cancer risk; and an intrinsic mechanism, where genetic events (e.g., oncogenes) induce neoplastic transformation triggering the inflammatory cascade.[6] The relationship between cancer and inflammation is discussed in detail next and is summarized in Figures 7.1 through 7.4.

Extrinsic Pathway of Carcinogenesis

Inflammatory or infectious conditions can increase cancer risk via the extrinsic pathway. Leukocytes producing inflammatory mediators are primarily responsible for triggering inflammation. Chronic inflammation can be induced by, among other sources, chronic infections, exposure to noxious agents that trigger inflammation (e.g., gastric acid reflux, tobacco, asbestos, and other chemicals), and auto-immune conditions.[6] Pathogenic infections such as those due to Hepatitis B and C viruses or *Helicobacter pylori* results in chronic inflammation that favors initiation and progression of tumors.[8] In all cases the production of DNA damage and accumulation of mutations and epigenetic changes is considered critical. Autoimmune diseases such as inflammatory bowel disease for colon cancer and prostatitis for prostate cancer, mechanical, radiation, and chemical insults can also induce inflammation associated with human malignancy (Figure 7.4).

The role of the tumor microenvironment (stroma) is being increasingly appreciated as being a critical part of carcinogenesis. Inflammatory cells are an important component of the stroma and milieu fosters proliferation, survival, and migration.[9] Figures 7.2, 7.4 and 7.5 illustrate how chronic inflammation may contribute to carcinogenesis.

Chronic inflammation promotes the development of blood vessels and the remodeling of the extracellular matrix, fostering the perfect environment in which a mutation bearing normal cells can turn potentially malignant. In addition, immune cells like neutrophils and macrophages produce ROS via a plasma membrane bound nicotinamide adenine dinucleotide phosphate (NADP). Based on *in vitro* and *in vivo* data, ROS and RNS that play a vital role in normal cellular metabolism are inflammation-generated mediators of DNA damage. An increase in oxidative stress leads to a spike in ROS/RNS formation.[10] These highly reactive species can easily bind to proteins, lipids, and DNA. Since proteins and lipids are usually turned over, damage to these macromolecules is usually not detrimental to the cell. However, damage to the DNA can lead to cancer.[11] Interestingly, tumor promoters are able to recruit inflammatory cells and stimulate them to generate ROS/RNS, which, in turn, generate DNA lesions and lead to mutations. Numerous reports suggest that tumor growth *in vivo*, inflammation, and OCDLs are interconnected.

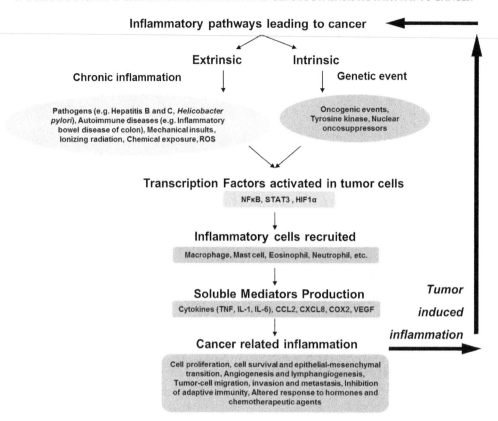

FIGURE 7.2   **Inflammatory Pathways Lead to Cancer.** As explained in the text the major events and sources contributing to persistent inflammation can be of an extrinsic or intrinsic nature. The two pathways can be identified as major contributors to the inflammatory milieu: the intrinsic pathway where genetic events (e.g., mutations in oncogenes) induce neoplastic transformation triggering the inflammatory cascade and the extrinsic pathway where chronic inflammation (e.g., infections, low doses of ionizing radiation) significantly increases the risk for different types of cancer. The two pathways converge, resulting in the activation of transcription factors (e.g., NFκB, STAT3, HIF) that coordinate the production of inflammatory mediators and the activation of various leukocytes generating an inflammatory microenvironment that nurtures cancer progression. The resulting activation of several transcription factors, inflammatory cells, and chemical mediators like cytokines has been closely bonded to the creation of uncontrolled cell proliferation, mitigation of apoptosis, abnormal angiogenesis, and other molecular changes leading to cancer. The production of DNA damage is considered the critical step. In addition, a high inflammatory response has been related with tumor cell migration and metastatic ability.

Moreover, tumors can induce their DNA damaging effects in distant tissues and organs.[12] Since clustered DNA lesions (both DSBs and OCDLs) are highly mutagenic, these results are biologically relevant.[13] Reports also suggest that OCDLs can be induced by a cytokine CCL2-based mechanism.[14] Researchers are actively pursuing these avenues of inflammation-induced carcinogenesis.

Several recent reports support the involvement of ROS in cancer-related processes. For instance, ROS production has been demonstrated to be required for mediating K-ras-induced lung cancer in mice.[15] Moreover, ROS released by damaged cells can induce inflammation and trigger production of pro-inflammatory cytokines by functioning as signaling molecules. New molecular pathways involving mitochondrial damage and ROS production are being actively investigated. These not only play a significant role in DNA damage and activation of oncogenes, but also in different aspects of inflammation. This suggests that ROS play an important role in the promotion of inflammation and tumorigenesis by modulating cancer-related signaling pathways.

Clinical data indicate that chronic inflammation promotes carcinogenesis. For instance, patients with inflammatory bowel diseases (IBD, ulcerative colitis, and Crohn's disease) have a five- to seven-fold increased risk of developing colorectal cancer. Alarmingly, 43% of patients with ulcerative colitis develop colorectal cancer after 25 to 35 years.[16] Chronic airway inflammatory conditions such as asbestosis, silicosis, exposure to airborne particulate matter, idiopathic pulmonary fibrosis, and tuberculosis have been reported to trigger nonsmoking related cancer development.[17] Another form of lung disease is mesothelioma, which is caused by exposure to asbestos and asbestos-induced chronic inflammation, and subsequent production of ROS and DNA damage.

Chronic inflammation also leads to gastric cancer. Aberrant expression of activation-induced cytidine deaminase (AID), a member of the cytidine-deaminase family that acts as a DNA- and RNA-editing enzyme, is induced by *Helicobacter pylori* and is observed in this malignancy.[18] *H. pylori*-mediated upregulation of AID results in accumulation of nucleotide alterations in gastric cells, which ultimately

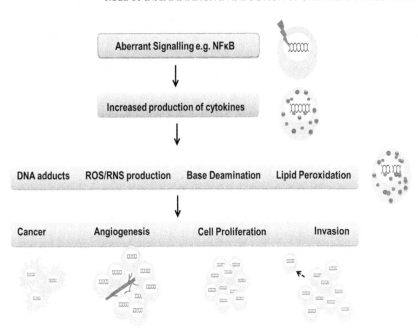

FIGURE 7.3 Signaling of the Inflammatory Pathways Leading to Tissue Abnormal Changes. Aberrant signaling can lead to increased angiogenesis, cell proliferation, and invasion, which, in turn, can lead to abnormal growth premalignant and finally malignant states.

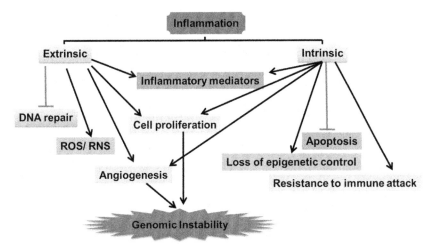

FIGURE 7.4 Overview of Pathways Leading to Cancer. As described in the text, the association of inflammation with the generation of ROS/RNS and DNA damage induces several relating pathways of DNA damage response (DDR) like signaling and induction of DNA repair proteins, cell cycle arrest, and proliferation changes. In every step, there are two major safeguarding mechanisms: DNA repair and apoptosis, assuming they work properly. A premalignant state will be characterized as loss of all these control mechanisms due to accumulation of mutations, epigenetic changes, and finally genomic and chromosomal instability.

leads to the development of gastric cancer.[18] Moreover, tumor necrosis factor (TNF) stimulation in human bile-duct cells induces ectopic AID production, which results in chronic biliary inflammation and the development of cholangiocarcinoma. Therefore, AID may be the link between chronic inflammation and DNA damage in these tumors.

Oxidative DNA damage and inflammation is also implicated in schistosomiasis,[19] lung, liver, and breast cancers. Elevated levels of DNA damage is seen upon urine analysis of patients with schistosomiasis.[19] The cells are more prone to DNA damage induced by the ROS/RNS produced by activated inflammatory cells. This, in turn, leads to an increased risk for bladder cancer in adults.[19] Chronic infection with hepatitis B or C viruses or ingestion of aflatoxin that causes ROS and subsequent DNA damage production leads to hepatocellular carcinoma and is considered a significant cause of cancer-associated mortality in Asia and Africa.[20]

Oxidative DNA damage may be involved in the development of breast cancer as well. Increased steady-state levels of DNA damage and ROS have been reported in invasive ductal carcinoma.[20] Whether the changes are due to decreased DNA repair and/or increased oxidative DNA damage remains to be confirmed.

Intrinsic Pathway

The intrinsic pathway is induced by genetic events such as activation of various types of oncogenes by mutation, chromosomal rearrangement, or amplification, and the inactivation of tumor-suppressor genes. These cells produce inflammatory mediators, which create an inflammatory microenvironment in tumors without prior underlying inflammatory conditions.[21]

Results reported in the literature show that various oncogenic mechanisms are involved in cancer-related

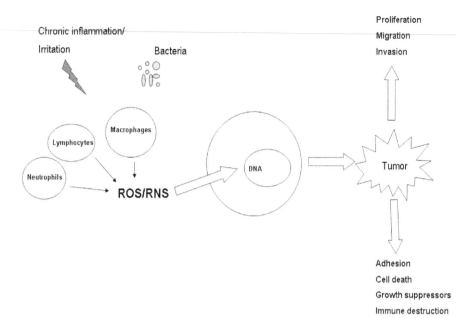

FIGURE 7.5 **An Overview of the Suggested Interplay between Inflammation and Associated Oxidative Stress.** This interplay can affect key aspects of tumorigenesis, angiogenesis, and metastasis. In this model, the generation of oxidative stress and potentially DNA damage through the inflammatory responses involving macrophages and different cytokines like the MCP-1/IL-6 are considered critical.[9]

inflammation pathways[6] (Figures 7.2 and 7.3). For example, in human papillary thyroid carcinoma (PTC), a tumor characterized by the presence of chemokine-guided macrophage and dendritic cell infiltration, rearrangements in the protein tyrosine kinase RET play a key role in the pathogenesis.[22] One of the signaling molecules in PTC colony-stimulating factors (CSFS) promotes leukocyte recruitment and survival. Interleukin 1β (IL-1β) is also secreted and is one of the main inflammatory cytokines.

Cyclo-oxygenase-2 (COX-2) is frequently expressed in cancer and is involved in the synthesis of prostaglandin E(2), which can promote tumor growth by binding its receptors and activating signaling pathways that regulate cell proliferation, migration, apoptosis, and angiogenesis. Chemokines attracting monocytes and dendritic cells such as CCL2 and CCL20 are also secreted and, as expected, these molecules have been reported to be pro-tumorigenic.

Angiogenic chemokines such as CXCL8 coordinate induction and inhibition of matrix-degrading enzymes, which promote tumor progression and survival. Upregulation of L-selectin and expression of the chemokine receptor CXCR4 that promote metastasis are also observed in these cells.[6,22] Thus, an early, causative, and sufficient genetic event promotes an inflammatory microenvironment, which, in turn, leads to tumor formation. The inflammatory cascade and tumor progression can be triggered by activation of oncogenes or inactivation of tumor suppressors. For example in the *ras* family oncogenes activation induces expression and production of inflammatory mediators. Expression of *ras* in a cervical carcinoma cell line induces the production of CXCL8, which promotes angiogenesis and tumor progression.[6] Moreover, mild chronic pancreatitis

and K-ras mutation induces pancreatic intra-epithelial neoplasia and invasive ductal carcinoma.[23] Similarly, *Braf*, which is frequently activated in malignant melanoma, induces cytokines that create a pro-tumorigenic microenvironment.[24]

The *myc* oncogene encodes a transcription factor that is overexpressed in many human tumors. Deregulation of *myc* is important in the initiation and maintenance of key aspects of the tumor phenotype. In association with inflammatory cells and mediators, *myc* promotes cell proliferation and remodeling of the extracellular milieu. *Myc*-mediated alterations include secretion of chemokines, which recruit mast cells and help sustain the formation of new vessels and tumor growth.[6] The epidermal growth factor receptor (EGFR) family plays an important role in cancer. EGFR activation in glioma induces COX-2 expression, which is involved in the synthesis of prostaglandin E(2). This can promote tumor growth by binding its receptors and activating signaling pathways that regulate cell proliferation, migration, apoptosis, and angiogenesis. COX-2 expression is an independent prognostic factor in glioma.

Production of inflammatory mediators can also be regulated by tumor suppressor proteins such as von Hippel Lindau/hypoxia-inducible factor (VHL/HIF), transforming growth factor-β (TGF-β), and phosphatase and tensin homologue (PTEN). The chemokine receptor CXCR4 is frequently expressed on malignant cells and has been implicated in cell survival and metastasis.CXCR4 and TNF-α lie downstream of the VHL/HIF axis in human renal-cell carcinoma.[6] Mutation of PTEN in non-small cell lung cancer (NSCLC) results in upregulation of HIF-1 activity and subsequent HIF-1-dependent transcription of the CXCR4 gene. CXCR4 regulates migration of lung cells through activation of Rac1 and matrix metalloproteinases. CXCR4 also modulates the action of ERK, IKK, NFκB, and integrins, which promote metastasis of the lung cancer.[25] Data from breast carcinoma suggests that inactivation of the gene encoding the type II TGF-β receptor stimulates the production of CXCL5 and CXCL12, which draws myeloid-derived suppressor cells (MDSC). CXCL5 induces Raf/MEK/ERK activation, Elk-1 phosphorylation and Snail upregulation. Activation of Elk-1 facilitates recruitment of phosphorylated mitogen- and stress-activated protein kinase 1, which in turn enhances histone H3 acetylation and phosphorylation of Snail promoter, resulting in Snail enhancement and E-cadherin down-regulation. This facilitates metastasis of breast cancer.[6,26] Thus, oncogenes and tumor suppressor genes can induce inflammation.

The Link between Extrinsic and Intrinsic Pathways

Some of the molecules involved in both the intrinsic and extrinsic pathways include transcription factors, such as nuclear factor-κB (NF-κB), signal transducer and activator of transcription 3 (STAT3), and hypoxia-inducible factor 1α (HIF1α).[27] These transcription factors modulate the inflammatory response and promote tumorigenesis via soluble mediators including cytokines (e.g., IL-1, IL-6, and IL-23), chemokines (e.g., CCL2 and CXCL8), and other cellular components (e.g., tumor-associated macrophages).[6] These factors recruit and activate various leukocytes, mainly of myelomonocytic lineage. The cytokines also activate key transcription factors in various cell types such as inflammatory, stromal, and tumor cells. This results in a cascade where even more inflammatory mediators are generated and a cancer-related inflammatory milieu is created.[21]

NFκB is a key transcription factor that potentially is a link between tumor cells and inflammatory cells. NFκB has a variety of roles including facilitating proliferation and survival of malignant cells by activating genes that regulate cell cycle progression (e.g., cyclin D and c-myc) and apoptosis (e.g., cIAPs, A1/BFL1, Bcl2, c-Flip), promoting angiogenesis and metastasis, disrupting adaptive immunity, and altering responses to hormones and chemotherapeutic agents.[6] In this respect, NFκB induces the expression of inflammatory cytokines such as key enzymes in the prostaglandin synthase pathway (COX-2), adhesion molecules, nitric oxide (NO) synthase, and angiogenic factors that promote inflammation as well as tumorigenesis. Hepatocarcinogenesis is substantially reliant on NFκB activation in both parenchymal (hepatocytes) and nonparenchymal liver cells.[6]

STAT3 is also implicated in both extrinsic and intrinsic pathways.[27] Constitutively activated STAT3 increases tumor cell proliferation, survival, and invasion and subdues antitumor immunity. Persistent activation of STAT3 leads to inflammation, which promotes tumor formation. This dual role of STAT3 in tumor inflammation and immunity involves upregulation of pro-oncogenic inflammatory pathways, including NFκB and IL-6-gp130-JAK pathways, and down-regulation of STAT1 and NFκB-mediated Th1 antitumor immune responses.

Soluble Mediators and Cellular Components

Inflammation is sustained by molecules such as TNF-α. Tumor-derived TNF-α supports the growth and development of skin, pancreatic, liver, and bowel tumors.[28] Constitutively produced TNF-α is associated with increased release of chemokines such as CCL2, CXCL12, CXCL8, CXCL1, CXCL13, CCL5, CCL17, and CCL22, IL-1, IL-6, VEGF, and macrophage migration inhibitory factor (MIF-1).[6]

Tumor-associated macrophages (TAM) represent the major inflammatory component of the stroma of many tumors and affect different aspects of the tumor.[6] TAM accumulation has been reported to promote angiogenesis via production VEGF and platelet-derived endothelial cell growth factor.[29] Moreover, myeloid cells in the tumor milieu also play a role as an angiogenic switch at different levels. Mast cells, eosinophils, neutrophils, and effectors of the adaptive immunity response are capable of tolerating the inflammatory reactions that lead to cancer.

Tissue Injury

There are significant overlaps between key features of wound healing and tumor development. These include stem cell and myofibroblast activation, enhanced cell proliferation, inflammation, and neo-angiogenesis. Chronic injury results in an aberrant healing and regenerative response that ultimately stimulates growth and development of initiated cells. Indeed, in the initial phase, the body interprets tumors as wounds and similar to healing tissues, activated platelets are present in tumors. This phase of tumor growth is governed by the actions of the stroma, which is similar to physiologic tissue repair.[11] However, during late tumor growth, the tumor becomes independent of stromal signaling for progression and survival. So far we have focused on the interaction of extrinsic and intrinsic mechanisms of inflammation and their role in induction of carcinogenesis. It should be noted though that the inflammatory response is also critical in other aspects of tumor progression as well, such as tissue invasion and metastasis. Angiogenesis significantly augments vascular invasion of migrating cells. Matrix metalloproteases and their inhibitors are essential for angiogenesis and remodeling of extracellular matrix. Similar to cancer, cell proliferation is enhanced in a wound, which results in tissue regeneration. However, unlike cancer, cell proliferation and inflammation subside after the foreign particle is removed or the repair is complete.[11]

Chronic tissue damage and inflammation can indeed promote the growth and progression of cancer. For instance, *v-Src* oncogene cannot induce cancer unless supplemented by tissue injury and ensuing tissue renewal.[30] Similarly, pancreatic insult is required to unravel the oncogenic potential of activated *K-Ras*. Finally, tissue injury inflicted by tobacco smoke influences lung cancer development, and suppression of cell death related pathways (c-Jun N-terminal kinase, JNK) or death-induced pro-inflammatory cytokines (TNF and IL-6) reduces tumor development. Taken together, these results support the notion that a substantial fraction of all cancer cases are likely to be initiated and promoted by chronic tissue injury. Given that persistent inflammation promotes genetic instability,[6] targeting cancer-related inflammation is a possible treatment strategy that can minimize normal tissue injury. The global inflammation that prevails in cancer can be targeted to restore normal tissue homeostasis and, perhaps, can be used in cancer prevention.

NONTARGETED EFFECTS, INFLAMMATION, OXIDATIVE STRESS AND DNA DAMAGE

Bystander and Abscopal Effects

The term "bystander" effect was first used in radiation biology to explain the results obtained in cell cultures irradiated with α-particles (energetic helium nuclei with the short range of absorption that can be produced by cyclotrons or synchrotrons). Although only a few cells were traversed by α-particles, many more exhibited sister chromatid exchanges, indicating that nontargeted cells also sustained damage.[31] Subsequently the term has been used in various scenarios to describe the ability of cells affected by an agent to convey manifestations of damage to other cells not directly targeted by the agent or necessarily susceptible to it themselves.[32] These indirectly affected, bystander cells exhibited various types of genomic destabilization such as altered clonogenic survival, changed frequency of gene mutations, induction of apoptosis and micronuclei, altered expression of stress-related genes, elevated frequencies of malignant transformation of mammalian cells *in vitro*, altered DNA damage and repair and senescence arrest, and various epigenetic changes.[33] (See review.[32,34]). All these indirect consequences in bystander cells are delayed effects, and though similar to the direct effects of radiation, can follow different kinetics. For instance, elevated levels of phosphorylated histone H2AX (γ-H2AX) have been found in both irradiated and bystander cultures indicating the presence of DNA DSBs.[35,36]

Not all cell types and not every cell in a bystander culture are equally responsive.[37] For example, rapidly proliferating cancer cells in culture and, generally, proliferating cells in S-phase appear to be highly susceptible.[38] However, other data suggest that this is not the only factor determining bystander vulnerability.[39] In such cells, ROS generated directly or indirectly as a result of cell damage interact with bystander cell DNA, producing lesions ranging from base or sugar modifications to abasic sites and single-stranded breaks. ROS-induced DNA damage can interfere with both replication and transcription in proliferating cells and transcription in nonproliferating cells, leading to DSB formation. Bystander effects have been noted in response to a number of cellular stresses including UV exposure and nonradiation sources of cellular damage such as media from tumor and aging cells.[39] Therefore, bystander effects can be generalized as an overall cell population response to the presence of cells undergoing stresses of various types.[40]

Of particular interest in regards to human health are the nontargeted effects that have been reported by radiation oncologists for decades—reactions in normal unirradiated tissues after radiation therapy of a particular part of the body. These out-of-field, or abscopal, effects have been described as an action at a distance from the irradiated volume but within the same organism.[41] Abscopal effects have been also shown to be a general phenomenon; they result from a number of other localized stimuli, for example surgery, hyperthermia, and laser immunotherapy.[34] The discovery of the radiation-induced bystander effect (RIBE) has prompted the description of abscopal effects as distant *in vivo* bystander effects, and their further investigation. Several studies reported *in vivo* radiation-induced bystander effects in animal models. Using strategies that involve partial head or body 1Gy X-ray irradiation, profound genetic and epigenetic changes were identified in shielded organs, such as skin and spleen.[42,43] The results of the *in vivo* RIBE can be transmitted to future generations. The radiation-induced bystander effect is a potential contributor to the well-documented clinical phenomenon of secondary cancers, a major concern in cancer RT, affecting more than 1% of patients.[34]

Bystander Signaling *In Vitro*

Two major mechanisms have been identified as playing a role in the transmission of bystander responses in cellular models: physical contact, via gap junction mediation, and soluble factors, such as ROS, NO, and cytokines (see review.[34]). In the absence of gap junctions (for example, in the medium transfer experiments), several studies found suppression of bystander responses when various inhibitors and ROS and NO scavengers were added to the media of donor cells and recipient cells. The use of 2-(4-carboxyphenyl)-4,4,5,5-tetramethylimidazoline-1-oxyl-3-oxide (c-PTIO) as a NO-specific scavenger, reduced the incidence of bystander micronuclei yields, which indicated that NO contributes to the bystander effect.[44]

Various inflammation-related cytokines have been found at elevated levels in media conditioned by irradiated cells. Notably, stress events other than ionizing radiation, such as UVC, UVA, and unirradiated cancer cells, similar to irradiated cultures, release cytokines in the culture medium.[39] They can target bystander cells directly; cytokine TGF-β, when added to cell cultures, induces elevated levels of DSBs similar to those induced by the conditioned medium, and addition of the blocking anti-TGF-β antibody reversed the effect.[39,45] Indirectly, through activation of cytokine receptors-mediated pathways, bystander cells also start expression and production of IL-8, IL-6, IL-33, RGE2, and other factors.[46] When the gene expression profile was compared with unirradiated and bystander normal human fibroblasts, transcription level of COX-2 was found consistently upregulated in bystander cells by more than three-fold. Addition of COX-2 inhibitor NS-398 suppressed COX-2 activity and decreased bystander mutagenesis.[47]

Recently several publications addressed the role of the stimulatory neurotransmitter 5-hydroxytryptamine (serotonin), which is a serum component in a cell culture medium, on the radiation-induced bystander effect. Some publications report a trend for increasing bystander response with increasing serotonin concentration, while others found no effect.[48-50]

Role of Cytokines for Bystander Signaling

The cell-cell communication *in vivo* is mediated by the immune system, and is more complex. Recently, the abscopal DNA damage response in normal tissues has been described, which was influenced by early-stage tumors growing in mice.[51] The presence of a tumor has been shown to induce inflammatory and DNA damage responses in the immediate tumor microenvironment, possibly due to the production of ROS and cytokines, similar to the radiation-induced bystander effect signaling. Results obtained in cell culture indicate that tumors could influence normal cell cultures; normal cells sustain elevated levels of DNA damage when incubated with medium previously conditioned on tumor cells, and similar cytokines released into medium of unirradiated tumor and irradiated normal cells.[39] Syngeneic tumors (B16 melanoma, reticulum cell sarcoma, and colon adenocarcinoma) were implanted subcutaneously into mice, and two weeks later, the levels of two types of DNA damage in different tissues were measured. Elevated levels of DSBs, as marked by γ-H2AX foci, and

OCDLs, were present in several distant tissues, such as duodenum, colon, stomach, rectum, and skin.[12] Both DSBs and OCDLs are potentially serious lesions and lead to genome instability if not fully repaired.[13] Ovary and lung did not exhibit elevated γ-H2AX foci, but had elevated OCDL levels. This wider incidence of elevated OCDLs versus γ-H2AX foci may be attributable to the mechanisms of lesion formation. γ-H2AX foci form in tissues with larger fractions of proliferating cells, such as those in the gastrointestinal tract, in which replication forks may participate in DSB formation, while OCDLs form equally well in every tissue.

Out of 59 cytokines measured in mouse serum, CCL2/MCP-1, CCL7/MCP-3, and CXCL1/IP-10 were over three-fold elevated in tumor-bearing mice. Elevated numbers of activated macrophages were found in gastrointestinal tract organs and skin. This suggested that macrophages in these distant tissues secrete ROS that induce DNA damage in the cells of the host organs. This also substantiated an association between inflammation and bystander DNA damage responses in these tumor-bearing mice. Interestingly, this systemic oxidative DNA damage in normal tissues both neighboring to and distant from injected tumors can be abrogated by feeding mice with the antioxidant tempol-rich food, suggesting that the endogenous antioxidant systems could be efficiently boosted by a well-designed antioxidant therapy to suppress the oxidative load in the organism (unpublished results).

The role of CCL2 in cancer development has been controversial, with evidence of pro- and antitumorigenic effects. Recent studies suggest that CCL2 contributes to cancer growth. There is mounting evidence linking inflammation and cancer. ROS are secreted from activated immune cells and stressed epithelial cells, resulting in DNA damage and genomic instability that may contribute to carcinogenesis. The blockade of CCL2 can inhibit tumor growth of primary and metastatic disease in animal models of non-small cell lung and prostate cancer. CCL2 is also involved in elevated COX-2 production and TGF-β upregulation, both critical factors in bystander signaling and carcinogenesis.[34] To examine whether the association between CCL2 and tumor-induced bystander DNA damage was causal, the same tumors were implanted into CCL2 knock-out (KO) mice. Strikingly, there was no measurable increase in distant DNA damage in the tumor-bearing CCL2 KO mice, suggesting that CCL2 is essential in the tumor-induced genotoxic response *in vivo*. The proposed model states that the bystander DNA damage in the tumor-bearing mice is due to the presence of activated macrophages in the distant tissues.[51]

Another good example of systemic cytokine-mediated intercellular communication is the relationship between senescent cells and surrounding normal tissues in an organism. Cellular senescence is a part of normal aging, as well as a preventive strategy to stop proliferation of cells undergoing malignant transformation.[52] This antiproliferative response can be driven by oncogene activation or loss of tumor suppressor signaling. A direct connection between cellular senescence and inflammation was established recently, indicating a crucial role in oncogene-induced cellular senescence the senescence-associated secretory phenotype (SASP), a cross-talk between senescent cells and their environment by secretion of numerous cytokines, chemokines, growth factors, and proteases. For example, IL-6 and IL-8, two well-known pro-inflammatory cytokines, seem to play a central role in premature cellular senescence induction. CCL2 appears as the most upregulated factor and a critical component in the SASP from melanoma cells.[53] Moreover, the SASP from senescent melanoma cells or recombinant CCL2 induce DNA damage in naïve melanoma cells, another indication that CCL2 triggers bystander effects.[53] On the detrimental side of the SASP effects, the chronic presence of senescent cells secreting numerous proteins has been predicted to significantly alter normal tissue structure and functions, not only in the local milieu, but in the whole organism.

RADIATION-INDUCED INFLAMMATION

The concept that ionizing radiation as a stress factor interferes with both targeted and nontargeted tissues is supported by multiple sources of evidence of systemic response to radiation. Radiation induces cellular oxidative stress that results in damage of not only nuclear DNA, but also mitochondrial DNA, leading to a decrease in respiratory chain activity and loss of mitochondrial function. The outcome is persistent metabolic oxidative stress that could continue to cause further oxidative damage to critical biological structures after long radiation exposure.[54] This radiation-induced damage to mitochondrial DNA in directly targeted or bystander tissues could become heritable and contribute to radiation-induced genomic instability. Genomic instability in nonirradiated normal tissues has been reported to be mediated by late cytokine response, as in the case of long-lived COX-2 pathway cytokine-dependent DNA damage and apoptosis response in nonirradiated mouse bone marrow cells after bone marrow was retrospectively irradiated. Such mechanistic studies provide insight on the nature of signaling molecules participating in targeted and nontargeted effects that potentially can be manipulated to increase therapeutic gain in radiotherapy.

Exposure to ionizing radiation has long been known to modulate the immune capacity of irradiated subjects, with a recognized dose/effect relationship.[55] Radiation exposure directly damages hematopoietic stem cells and alters the capacity of bone marrow stromal elements to support and maintain hematopoiesis. Data from the atomic bomb victims suggest a threshold dose to the

acute radiation hematological syndrome characterized by severe immune-compromise and subsequent death. In solid tumors these forms of unscheduled cell death can lead to a pro-inflammatory environment and an increase in cell-to-cell signaling. In this scenario the innate immune system is important in mediating the antitumor effects of localized ionizing radiation. For example a preclinical murine study in as early as 1979 demonstrated that *in vivo* tumor control probability to radiation was profoundly influenced by the host immune-competence in a transplanted murine fibro-sarcoma model.[56] However, even in the presence of a competent immune system an established tumor system is usually adapted to avoid immune recognition in the absence of additional antitumor stimulus. This section focuses on the complex induced immune response of the tumor and host secondary to radiotherapy.

Local Tumor Environment and Radiation

Immune cells associated with the complex tumor microenvironment can function to promote or suppress the adaptive immune response. Tumor associated macrophages, which are consistently colocated within the tumor microenvironment, are pro-angiogenic and can assist in tumor growth. In established and advanced neoplasia, when persistent tumor cells have escaped the immune attack, M2-polarized macrophages predominate the tumor microenvironment and suppress adaptive immunity. The response to ionizing radiation can trigger inflammation; however, the interpretation of this process by the innate immune system appears to be dependent on a variety of factors. In tumor cells, doses of less than 0.5Gy (which are generally too low to directly induce cell death) result in the release of oxygen and nitrogen radicals that activate innate immune cells, such as macrophages, to release cytokines. Depending on the environment and genetic background, this process can result in chronic inflammation that causes genetic alterations and cell death as a secondary event. It is in this setting that the immune-modulating effects of radiation promote mostly a pro-tumorigenic role of the immune system. Conversely, at doses sufficient to directly provoke significant cell death, inflammatory cell signaling can result in an adaptive immune response. This inflammatory signaling cascade can promote antitumor immunity, for example through activation of M1-polarized macrophages. These M1 macrophages have antitumor activity, mediated directly by the ability to kill tumor cells, as well as indirectly by the activation of adaptive antitumor immunity.

Radiation Exposure and the Immunogenic Effect

Conventional radiotherapy is comprised of doses of 1.8–2Gy per fraction, delivered five days a week, for several weeks. *In vitro*, when mouse B16 melanoma cells were exposed to multiple daily doses of 2Gy to a total dose of 50Gy, mimicking clinical protocols, MHC-I expression was increased after the second week, when the total dose amounted to 20Gy.[57] This expression profile was stable for greater than 5 weeks after the last radiation fraction. Whereas immunogenic signaling may occur in a cumulative fashion during conventional radiotherapy, pro-inflammatory cytokines generally are produced by higher doses than are conventionally used in RT. This has particular importance as the role of stereotactic ablative body radiotherapy (SABR) has emerged.

Recent technological advances in precision radiotherapy delivery have allowed the safe clinical application of high-dose per fraction SABR.[58] Typical dose/fractionation schedules are in the ablative spectrum, and potentially augment the tumorocidal properties of radiation through some proportion of vascular damage, ceramide-induced endothelial cell damage, and increased apoptosis of tumor cells. Recent evidence also suggests that ablative doses of radiation evoke a particularly strong immune response. Lugade et al.[59] showed that cross-priming of T-cells against tumor antigens were induced by both 3Gy by 5 fractions and a single dose of 15Gy in the draining lymph nodes. Using the B16 mouse melanoma model, Lee et al.[60] showed that retardation of tumor was more pronounced with a single dose of 20Gy, and three fractions of 15Gy were comparable. This effect was markedly reduced by host CD8+ T-cell depletion, suggesting that both regimens can promote cross-priming of antitumor T-cells. In contrast, a nonablative dose of 5Gy by 4 fractions delivered over 2 weeks showed inferior tumor growth inhibition. These results suggest that RT induced adaptive immune response can be a dose-dependent phenomenon and may result in additional tumor cell kill beyond direct DNA damage.

CONCLUSION

Persistent stress is induced by self-perpetuating inflammatory processes resulting in buildup of DNA damage in target tissues. The resulting genetic changes act as a driving force in chronic inflammation-associated human disease pathogenesis. Therefore, steady-state levels of DNA damage due to pro-inflammatory molecules provide promising molecular signatures for predicting disease risk and may be potential targets and biomarkers for cancer.

SUMMARY POINTS

- DNA damage can be induced by oxidative stress and radiation and is resolved by DNA repair and activation of cell cycle checkpoints to arrest the cell to allow time for repair. If not properly repaired, it causes a threat to maintenance of genomic stability and may progress to carcinogenesis.

- Oxidative stress, DNA damage, and inflammatory responses are interconnected.
- Inflammatory responses are critical at different phases of tumor development including initiation, promotion, malignant conversion, invasion, and metastasis.
- Cell-signaling effects such as bystander and abscopal effects are mediated by inflammatory factors.
- Inflammation affects immune surveillance in response to cancer therapy. That results in antitumor immune activation and systemic effects, which are indicative of the potential efficacy of radiation as a cancer therapy that extends beyond classical direct DNA damage.

References

1. Kryston TB, Georgiev AB, Pissis P, Georgakilas AG. Role of oxidative stress and DNA damage in human carcinogenesis. *Mutat Res* 2011;**711**(1-2):193–201.
2. Nowsheen S, Aziz K, Kryston TB, Ferguson NF, Georgakilas A. The interplay between inflammation and oxidative stress in carcinogenesis. *Curr Mol Med* 2012;**12**(6):672–80.
3. Cadet J, Douki T, Ravanat JL. Oxidatively generated base damage to cellular DNA. *Free Radic Biol Med* 2010;**49**(1):9–21.
4. Aziz K, Nowsheen S, Pantelias G, Iliakis G, Gorgoulis VG, Georgakilas AG. Targeting DNA damage and repair: Embracing the pharmacological era for successful cancer therapy. *Pharmacol Ther* 2012;**133**(3):334–50.
5. Mantovani A, Garlanda C, Allavena P. Molecular pathways and targets in cancer-related inflammation. *Ann Med* 2010;**42**(3):161–70.
6. Del Prete A, Allavena P, Santoro G, Fumarulo R, Corsi MM, Mantovani A. Molecular pathways in cancer-related inflammation. *Biochem Med (Zagreb)* 2011;**21**(3):264–75.
7. Balkwill F, Mantovani A. Inflammation and cancer: Back to Virchow? *Lancet* 2001;**357**(9255):539–45.
8. Georgakilas AG, Mosley WG, Georgakila S, Ziech D, Panayiotidis MI. Viral-induced human carcinogenesis: An oxidative stress perspective. *Mol Biosyst* 2010;**6**(7):1162–72.
9. Nowsheen S, Aziz K, Panayiotidis MI, Georgakilas AG. Molecular markers for cancer prognosis and treatment: Have we struck gold? *Cancer Lett* 2012;**327**(1-2):142–52.
10. Martin OA, Redon CE, Nakamura AJ, Dickey JS, Georgakilas AG, Bonner WM. Systemic DNA damage related to cancer. *Cancer Res* 2011;**71**(10):3437–41.
11. Nowsheen S, Aziz K, Kryston TB, Ferguson NF, Georgakilas A. The interplay between inflammation and oxidative stress in carcinogenesis. *Curr Mol Med* 2011;**12**(6):672–80.
12. Redon CE, Dickey JS, Nakamura AJ, Kareva IG, Naf D, Nowsheen S, et al. Tumors induce complex DNA damage in distant proliferative tissues in vivo. *Proc Natl Acad Sci U S A* 2010;**107**(42):17992–7.
13. Georgakilas AG. Processing of DNA damage clusters in human cells: Current status of knowledge. *Mol Biosyst* 2008;**4**(1):30–5.
14. Bartek J, Mistrik M, Bartkova J. Long-distance inflammatory and genotoxic impact of cancer in vivo. *Proc Natl Acad Sci U S A* 2010;**107**(42):17861–2.
15. Weinberg F, Hamanaka R, Wheaton WW, Weinberg S, Joseph J, Lopez M, et al. Mitochondrial metabolism and ROS generation are essential for Kras-mediated tumorigenicity. *Proc Natl Acad Sci U S A* 2010;**107**(19):8788–93.
16. Ferrone C, Dranoff G. Dual roles for immunity in gastrointestinal cancers. *J Clin Oncol* 2010;**28**(26):4045–51.
17. Yao H, Rahman I. Current concepts on the role of inflammation in COPD and lung cancer. *Curr Opin Pharmacol* 2009;**9**(4):375–83.

18. Matsumoto Y, Marusawa H, Kinoshita K, Endo Y, Kou T, Morisawa T, et al. Helicobacter pylori infection triggers aberrant expression of activation-induced cytidine deaminase in gastric epithelium. Nat Med 2007;13(4):470–6.

19. Rosin MP, Anwar WA, Ward AJ. Inflammation, chromosomal instability, and cancer: The schistosomiasis model. Cancer Res 1994; 54(7 Suppl.):1929s–33s.

20. Wiseman H, Halliwell B. Damage to DNA by reactive oxygen and nitrogen species: Role in inflammatory disease and progression to cancer. Biochem J 1996;313(Pt 1):17–29.

21. Mantovani A, Allavena P, Sica A, Balkwill F. Cancer-related inflammation. Nature 2008;454(7203):436–44.

22. Borrello MG, Alberti L, Fischer A, Degl'innocenti D, Ferrario C, Gariboldi M, et al. Induction of a proinflammatory program in normal human thyrocytes by the RET/PTC1 oncogene. Proc Natl Acad Sci U S A 2005;102(41):14825–30.

23. Guerra C, Schuhmacher AJ, Canamero M, Grippo PJ, Verdaguer L, Perez-Gallego L, et al. Chronic pancreatitis is essential for induction of pancreatic ductal adenocarcinoma by K-Ras oncogenes in adult mice. Cancer Cell 2007;11(3):291–302.

24. Sumimoto H, Imabayashi F, Iwata T, Kawakami Y. The BRAF-MAPK signaling pathway is essential for cancer-immune evasion in human melanoma cells. J Exp Med 2006;203(7):1651–6.

25. Cavallaro S. CXCR4/CXCL12 in Non-Small-Cell Lung Cancer Metastasis to the Brain. Int J Mol Sci. 2013;14(1):1713–27.

26. Bierie B, Moses HL. TGF-beta and cancer. Cytokine Growth Factor Rev 2006;17(1-2):29–40.

27. Yu H, Pardoll D, Jove R. STATs in cancer inflammation and immunity: A leading role for STAT3. Nat Rev Cancer 2009;9(11):798–809.

28. Balkwill F. Tumour necrosis factor and cancer. Nat Rev Cancer 2009;9(5):361–71.

29. Stearman RS, Dwyer-Nield L, Grady MC, Malkinson AM, Geraci MW. A macrophage gene expression signature defines a field effect in the lung tumor microenvironment. Cancer Res 2008;68(1):34–43.

30. Kuraishy A, Karin M, Grivennikov SI. Tumor promotion via injury- and death-induced inflammation. Immunity 2012;35(4):467–77.

31. Nagasawa H, Little JB. Induction of sister chromatid exchanges by extremely low doses of alpha-particles. Cancer Res 1992;52(22): 6394–6.

32. Morgan WF. Non-targeted and delayed effects of exposure to ionizing radiation: I. Radiation-induced genomic instability and bystander effects in vitro. 2003. Radiat Res 2012;178(2):AV223–36.

33. Sedelnikova OA, Nakamura A, Kovalchuk O, Koturbash I, Mitchell SA, Marino SA, et al. DNA double-strand breaks form in bystander cells after microbeam irradiation of three-dimensional human tissue models. Cancer Res 2007;67(9):4295–302.

34. Prise KM, O'Sullivan JM. Radiation-induced bystander signalling in cancer therapy. Nat Rev Cancer 2009;9(5):351–60.

35. Bonner WM, Redon CE, Dickey JS, Nakamura AJ, Sedelnikova OA, Solier S, et al. GammaH2AX and cancer. Nat Rev Cancer 2008;8(12):957–67.

36. Sedelnikova OA, Pilch DR, Redon C, Bonner WM. Histone H2AX in DNA damage and repair. Cancer Biol Ther 2003;2(3):233–5.

37. Mothersill C, Rea D, Wright EG, Lorimore SA, Murphy D, Seymour CB, et al. Individual variation in the production of a 'bystander signal' following irradiation of primary cultures of normal human urothelium. Carcinogenesis 2001;22(9):1465–71.

38. Shao C, Folkard M, Prise KM. Role of TGF-beta1 and nitric oxide in the bystander response of irradiated glioma cells. Oncogene 2008;27(4):434–40.

39. Dickey JS, Baird BJ, Redon CE, Sokolov MV, Sedelnikova OA, Bonner WM. Intercellular communication of cellular stress monitored by gamma-H2AX induction. Carcinogenesis 2009;30(10):1686–95.

40. Sokolov MV, Dickey JS, Bonner WM, Sedelnikova OA. Gamma-H2AX in bystander cells: Not just a radiation-triggered event, a cellular response to stress mediated by intercellular communication. Cell Cycle 2007;6(18):2210–2.

41. Mole RH. Whole body irradiation; radiobiology or medicine? Br J Radiol 1953;26(305):234–41.

42. Koturbash I, Rugo RE, Hendricks CA, Loree J, Thibault B, Kutanzi K, et al. Irradiation induces DNA damage and modulates epigenetic effectors in distant bystander tissue in vivo. Oncogene 2006;25(31):4267–75.

43. Koturbash I, Loree J, Kutanzi K, Koganow C, Pogribny I, Kovalchuk O. In vivo bystander effect: Cranial X-irradiation leads to elevated DNA damage, altered cellular proliferation and apoptosis, and increased p53 levels in shielded spleen. Int J Radiat Oncol Biol Phys 2008;70(2):554–62.

44. Shao C, Stewart V, Folkard M, Michael BD, Prise KM. Nitric oxide-mediated signaling in the bystander response of individually targeted glioma cells. Cancer Res 2003;63(23):8437–42.

45. Dickey JS, Baird BJ, Redon CE, Avdoshina V, Palchik G, Wu J, et al. Susceptibility to bystander DNA damage is influenced by replication and transcriptional activity. Nucleic Acids Res 2012; 40(20):10274–86.

46. Hei TK, Zhou H, Chai Y, Ponnaiya B, Ivanov VN. Radiation induced non-targeted response: Mechanism and potential clinical implications. Curr Mol Pharmacol 2011;4(2):96–105.

47. Zhou H, Ivanov VN, Gillespie J, Geard CR, Amundson SA, Brenner DJ, et al. Mechanism of radiation-induced bystander effect: Role of the cyclooxygenase-2 signaling pathway. Proc Natl Acad Sci U S A 2005;102(41):14641–6.

48. Klammer H, Iliakis G. The impact of serotonin on the development of bystander damage assessed by gamma-H2AX foci analysis. Int J Radiat Biol 2012;88(10):777–80.

49. Fazzari J, Mersov A, Smith R, Seymour C, Mothersill C. Effect of 5-hydroxytryptamine (serotonin) receptor inhibitors on the radiation-induced bystander effect. Int J Radiat Biol 2012;88(10):786–90.

50. Chapman KL, Al-Mayah AH, Bowler DA, Irons SL, Kadhim MA. No influence of serotonin levels in foetal bovine sera on radiation-induced bystander effects and genomic instability. Int J Radiat Biol 2012;88(10):781–5.

51. Martin OA, Redon CE, Dickey JS, Nakamura AJ, Bonner WM. Para-inflammation mediates systemic DNA damage in response to tumor growth. Comm Integrat Biol 2011;4(1):78–81.

52. Kuilman T, Michaloglou C, Vredeveld LC, Douma S, van Doorn R, Desmet CJ, et al. Oncogene-induced senescence relayed by an interleukin-dependent inflammatory network. Cell 2008;133(6):1019–31.

53. Ohanna M, Giuliano S, Bonet C, Imbert V, Hofman V, Zangari J, et al. Senescent cells develop a PARP-1 and nuclear factor-{kappa}B-associated secretome (PNAS). Genes Dev 2011;25(12):1245–61.

54. Azzam EI, Jay-Gerin JP, Pain D. Ionizing radiation-induced metabolic oxidative stress and prolonged cell injury. Cancer Lett 2012; 327(1-2):48–60.

55. LeRoy GV. Hematology of atomic bomb casualties. Arch Intern Med 1950;86(5):691.

56. Slone HB, Peters LJ, Milas L. Effect of host immune capability on radiocurability and subsequent transplantability of a murine fibrosarcoma. J Natl Cancer Inst 1979;63(5):1229–35.

57. Hauser SH, Calorini L, Wazer DE, Gattoni-Celli S. Radiation-enhanced expression of major histocompatibility complex class I antigen H-2Db in B16 melanoma cells. Cancer Res 1993;53(8):1952–5.

58. Siva S, Chesson B, Aarons Y, Clements N, Kron T, MacManus M, et al. Implementation of a lung radiosurgery program: Technical considerations and quality assurance in an Australian institution. J Med Imaging Radiat Oncol 2012.

59. Lugade AA, Moran JP, Gerber SA, Rose RC, Frelinger JG, Lord EM. Local radiation therapy of B16 melanoma tumors increases the generation of tumor antigen-specific effector cells that traffic to the tumor. J Immunol 2005;174(12):7516–23.

60. Lee Y, Auh SL, Wang Y, Burnette B, Wang Y, Meng Y, et al. Therapeutic effects of ablative radiation on local tumor require CD8+ T cells: changing strategies for cancer treatment. Blood 2009; 114(3):589–95.

ANTIOXIDANTS AND CANCER

Molecular Approaches Toward Targeted Cancer Therapy with Some Food Plant Products: On the Role of Antioxidants

Anisur Rahman Khuda-Bukhsh, Santu Kumar Saha, Sreemanti Das

Cytogenetics and Molecular Biology Laboratory, Department of Zoology, University of Kalyani, Kalyani, India

List of Abbreviations

ROS Reactive oxygen species
WHO World Health Organization
PAH Polycyclic aromatic hydrocarbons
GSH Glutathione
GPX Glutathione peroxidase
PHGPX Phospholipid hydroperoxide glutathione peroxidase
MDA Malondialdehyde
COX2 Cyclooxygenase 2
MAPK Mitogen-activated protein kinase
ASK1 apoptosis signal-regulated kinase 1
PI3K Phosphoinositide 3-kinase
EGF Epidermal Growth Factor
PDGF Platelet-Derived Growth Factor
VEGF Vascular Endothelial Growth Factor
PTEN Phosphatase and Tensin Homology
Ref-1 Redox factor-1
Nrf2 Nuclear factor (erythroid-derived 2)-like 2
IRE Iron-responsive element
IRP Iron regulatory protein
ATM Ataxia–telangiectasia mutated
ATR Ataxia–telangiectasia and Rad3-related
EpRE/ARE Electrophile/antioxidant response element
NF-kB Nuclear factor-kB
RAR Retinoic acid receptor
RXR Retinoid X receptor
PPAR Peroxisome proliferator-activated receptor
IGF1 Insulin-like growth factor type 1
MMP Matrix metalloproteinase

INTRODUCTION

Cancer remains a major public health issue that profoundly affects human health as more than one million people are diagnosed each year with this disease.[1] Howlader's (2012)[1] report estimated that approx. 1,638,910 people (about 848,170 men and 790,740 women) are at risk of being diagnosed with cancer and about 577,190 of them are likely to die, and this is only in the United States. This surely is a situation of horror and needs careful attention as to how best we can manage this problem.

Because of modernization and explosive growth of various sectors of industry, including the automobile industry, pollutants are constantly being added to the environment with an increasing load of carcinogens. These environmental carcinogens from various sources contribute to an individual's total burden of oxidative stress and invade the antioxidant defense mechanisms of the human body and other biological systems. The resulting damage leads to genetic mutations with the outcome of carcinogenesis/malignancy[2] (Figure 8.1). The accumulation of oxidative damage has been implicated in both acute and chronic cell injuries including possible participation in the formation of cancer. Reactive oxygen species (ROS) appear to be involved in all stages of cancer development. Cellular oxidative stress can modify intercellular communication, protein kinase activity, membrane structure and function, and gene expression, and result in modulation of cell growth.

Many food plants are reported to contain antioxidants in adequate amounts that give them the ability to protect them from oxidative assaults/physiological stress generated by ROS, and this has made them the subject of further investigation to find whether they could also be utilized for reducing cancer risk.[3,4] Many epidemiological studies have consistently shown that adequate consumption of foods of plant origin, such as fruits, vegetables, whole grains, legumes, nuts, seeds, and tea, along with a change in related lifestyle, can decrease the risk of developing various cancers.[5] In fact, recent epigenetic research[6] has shown that DNA methylation

© 2014 Elsevier Inc. All rights reserved.

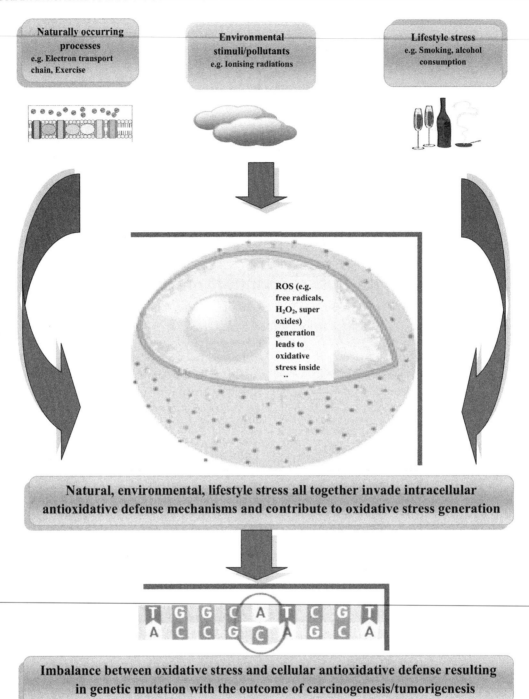

FIGURE 8.1   Oxidative stress leads to ROS generation with the ultimate outcome of carcinogenesis/tumorigenesis: natural, environmental, lifestyle stress all together contribute to oxidative stress generation, resulting in the imbalance of cellular antioxidant defense mechanisms that ultimately cause genetic mutation with the outcome of carcinogenesis. (See the color plate.)

occurs with the pre-intake of certain food plants, and the tumor suppressor effect becomes discernible in head and neck squamous cell carcinomas.

Chemopreventive potentials of naturally occurring dietary antioxidants and their bioactive phytochemicals are suggested to have effects on cellular signaling molecules as targets. These molecules trigger induction of cellular defense systems to detoxify the agents responsible for generating the oxidative stress by synthesizing the requisite amount of antioxidative enzymes. Further, as an additional measure of protection, inhibition of inflammatory and cell growth signaling pathways is initiated, which culminates in progression of the cells into the apoptotic pathway and/or brings the cell cycle to a halt.[7]

The principal aim of this chapter is to focus on the antioxidative properties of some food plants and their products, and to elucidate the molecular approaches (targeting mechanism of oxidative stress and ROS signaling events) of the dietary antioxidants that can be made for targeted cancer therapy.

OXIDATIVE STRESS AND REACTIVE OXYGEN SPECIES (ROS): THE "RUSTING DISORDER"

At an international conference of the World Health Organization (WHO) in the mid-1960s Professor Sanojki, a Russian toxicologist, described certain diseases, the origin of which could be attributed to oxygen toxicity; for an example, rheumatoid arthritis. He termed the phrase "the rusting diseases" for all such diseases. The diseases that have been directly related to oxidative stress for their origin include cancer, Parkinson's disease, Alzheimer's disease, http://en.wikipedia.org/wiki/Oxidative_stress - cite_note-1atherosclerosis, heart failure, myocardial infarction, schizophrenia, bipolar disorder, fragile X syndrome disorder, sickle cell anemia lichen planus, vitiligo, autism, chronic fatigue syndrome, and many others.[8]

OXIDATIVE STRESS: DEFINITION, SOURCES, AND CHEMISTRY IN GENERAL

Oxidative stress is literally defined as "an imbalance between the ROS generation and its detoxification by biological system leading to impairment of damage repair by cell/tissue".[9] ROS, which include superoxide anion radical ($O_2^-\cdot$), singlet oxygen ($1O_2$), hydrogen peroxide (H_2O_2), and the highly reactive hydroxyl radical ($\bullet OH$), can be generated due to naturally occurring processes such as mitochondrial electron transport, exercise, environmental stimuli and pollutants (e.g., ionizing radiation), change in atmospheric conditions (e.g., hypoxia), and lastly because of lifestyle stress (e.g., cigarette smoking, alcohol consumption, etc.).[7]

Cellular systems have evolved a range of metalloenzymes and regulatory systems that facilitate interaction between redox metals with O_2 using various catalytic pathways, the end products resulting in the generation of free radicals and ROS,[10] as shown:

$$Fe^{3+} + \bullet O_2^- \rightarrow Fe^{2+} + O_2 \text{ (step 1)}$$

$$Fe^{2+} + H_2O_2 \rightarrow Fe^{3+} + OH^- + \bullet OH \text{ (step 2) step 1} + \text{step 2}$$

$$\bullet O_2^- + H_2O_2 \rightarrow \bullet OH + HO^- + O_2 \text{ [Fenton reaction]}$$

Apart from the direct effect, ROS generation can also be achieved substantially by disruption of calcium channels with the help of metallo-enzymes.[11]

To counter the oxidant effects and to restore redox balance, cells must reset important homeostatic parameters.

CANCER: DEFINITION AND THE IMPACT OF FOOD AND FOOD HABITS

Cancer is a disease of uncontrolled proliferation and growth of cells at inappropriate times and locations in the body. When cells acquire mutations that abolish regulation of cell division, the cells multiply to form masses that we call tumors [12] and these cells consequently transform to malignant ones and invade other tissue as a result of metastasis.

There is a growing awareness that oxidative stress plays a role in various clinical conditions including malignant diseases.[13,14] The initial process of tumor promotion involves a nonlethal and inheritable mutation in cells by interaction of a chemical with DNA. This mutation confers a growth advantage to that cell. For the mutation to be set around, DNA synthesis must occur to lock in the mutation. The activation of the carcinogen to an electrophilic DNA-damaging moiety is a necessary step for this stage. ROS are believed to mediate the activation of such carcinogens through hydroperoxide-dependent oxidation that can be mediated by peroxyl radicals.[15] This occurs with aflatoxin B, aromatic amines, and polycyclic aromatic hydrocarbon dihydrodiols.[15] ROS or their byproduct of lipid peroxidation, MDA, can also directly react with DNA to form oxidative DNA adducts.[16] The presence of carcinogen-DNA adducts and oxidative DNA adducts generated by chemical carcinogens suggest an interactive role of ROS in cancer initiation. ROS, therefore, can have multiple effects in the initiation stage of carcinogenesis by mediating carcinogen activation, causing DNA damage, and interfering with the repair of the DNA damage.

The "2010 Dietary Guidelines for Americans"[17] has added some of the dietary and behavioral factors associated with developing cancer risk and also prepared food and lifestyle guidelines to lower the risk of getting cancer. The report suggests that intake of limited amount of fruits and vegetables, and excessive amounts of red meat, fat and alcohol consumption, smoking, overweight/obesity, and less physical activity all contribute to the elevation of cancer risk. The report also highlighted some food guidelines to help prevent cancers and other chronic diseases (Table 8.1).

Apart from the major dietary antioxidants just mentioned, some associated bioactive compounds/ phytochemicals that are used for cancer therapy are listed in Table 8.2 and their biochemical structures in Figure 8.2.

TABLE 8.1   Health Benefits/Harms of Some Dietary Components/Lifestyle

Diet/lifestyle	Health Benefits/Harms	Suggestive Doses
Fruits	Reduce the risk of getting cancer due to presence of dietary antioxidants.	1–2½ cups daily
Dark green and orange vegetables and legumes	Reduced the risk of getting cancer due to presence of dietary antioxidants.	1–4 cups daily
Red meat	The increased cancer risk may be because of the iron, fat, salt, and nitrates/nitrites in processed meat. Additionally, when meat is cooked at high temperatures, substances are formed that may be mutagenic or carcinogenic.	No more than 18 ounces a week
Monounsaturated and polyunsaturated fatty acids	Effects are still unclear; in fact more recent research on the effects of trans fatty acids also has yet to reach definitive conclusions.	Less than 10% of calories from saturated fatty acids and keeping trans fatty acid consumption as low as possible for general health and the prevention of chronic diseases, including cancer and heart disease
Drinking alcohol	Increases the risk of cancers of the mouth, esophagus, pharynx, larynx, and liver in men and women and of breast cancer in women.	A drink is defined as 12 ounces of regular beer, 5 ounces of wine, or 1.5 ounces of 80-proof liquor
Physical activity	Compelling evidence indicates that prevention of obesity reduces the risk for several types of cancer, such as colon, postmenopausal breast, uterine, esophageal, and renal cell cancers.	Physical activity at work or during leisure-time is linked to a 30 percent lower risk of getting colon cancer

Indicating some diet/lifestyles, their health benefits or harmful effects, if any, and suggestive doses.
Source: US Department of Agriculture and US Department of Health and Human Services (2010). Dietary Guidelines for Americans.

FINDING OUT THE CULPRIT: GENOME VERSUS OXIDATIVE STRESS

Dietary factors of cancer risk can be influenced by an individual's genetic susceptibility. Several of the more common variations in susceptibility among individuals result from polymorphisms in specific genes (including *CYP1A1*, *CPY1A2*, *CYP2D6*, *GSTM1*, and *NAT2*) that cause differences in metabolic or detoxification activities.[18] Increased risk of lung cancer, for example, has been correlated with variant forms of *CYP1A1* (catalyses the oxygenation of polyaromatic hydrocarbons (PAHs) and is induced by cigarette smoke), *CYP2D6* (acts on an unknown substrate, possibly tobacco-specific nitrosamines), and *GSTM1* (detoxifies reactive, electrophilic compounds such as PAHs).[18] The relative prevalence and/or distribution of such polymorphisms in populations targeted in epidemiological studies may affect individual responses to both risk and protective factors, including antioxidants and other dietary constituents.

AIMS OF TARGETED CANCER THERAPY

With regard to targeted cancer therapy, six essential alterations in cell physiology that are known to cumulatively lead to malignancy are: (1) self sufficiency in growth signals, (2) evasion of apoptosis, (3) limitless replicative potentials, (4) angiogenesis, (5) tissue invasion, and (6) metastasis.[12] Therapeutic approaches that can target more than one of these mechanisms gain preference and attention for drug formulation in cancer therapy.

CANCER THERAPEUTICS BY TARGETING ANTIOXIDATIVE MECHANISMS

Epidemiological studies have consistently shown that a high dietary intake of fruits and vegetables as well as whole grains is strongly associated with reduced risk of developing chronic diseases such as cancer.[19] Chemopreventive effects elicited by these natural dietary compounds at least in part are due to their antioxidant property, which is the main focus of this review.

Research in cancer therapeutic agents in the past few decades has demonstrated that the mechanisms by which dietary antioxidants exhibit their chemopreventive effects are multiple and complex. The therapeutic strategy (Figure 8.3) employed by any antioxidant against the generation of free radicals causing oxidative stress is either to directly combat and completely eliminate or to convert the ROS to a condition that is less

TABLE 8.2 Some Dietary Foods (Dietary Antioxidants) and Their Major Bioactive Phytochemicals/Compounds

Dietary Antioxidants	Active Compounds
Turmeric	Curcumin
Grapes	Resveratrol
Blueberries	Pterostilbene
Tea	Epigallocatechin-3-gallate(EGCG)
	Theaflavin
Chrysanthemum	Acacetin
Ginger	[6]-Gingerol
	[6]-Shogaol
Celery	Luteolin
Citrus peel	Tangeretin
	Nobiletin
	5-hydroxy-3,6,7,8,3',4'hexamethoxyflavone
Chili	Capsaicin
Cabbage	Indole-3-carbinol
Milk thistle	Silibinin
Broccoli	Sulforaphane
	Phenethyl isothiocyanate (PEITC)
Garlic	Allicin
	Diallyl sulphide
Soybeans	Genistein
Tomato	Lycopene
Spinach	Natural antioxidant mixture (NAO)
Cranberry	Quercetin
	Myricetin
	Cyanidine
	Peonidin
Aloe	Emodin
Cloves	Eugenol
Carrots	Beta carotene
Scutellaria	Apigenin
Cirisium	Apigenin
Rosemary	Carnosol
Brassica sp.	Sulforaphane
	Indole-3-carbinol
Pomegranate	Ellagic acid
Guggulu	Guggulsterone

A list of some dietary foods or dietary antioxidants and their major bioactive compounds useful for chemoprevention.

harmful and toxic. In general, the upstream mechanisms include biochemical steps preventing free radial generation, modulating metal-protein interaction, and promoting normal metal homeostasis, while the downstream mechanism involves processes like inflammation and free radical scavenging. To achieve therapeutic success, antioxidants may target different mechanisms and pathways. Among many, some of the probable mechanisms targeted by antioxidants are mentioned hereunder.

ROS Scavenging through Phase 2 Enzymes/Antioxidant Activity

Oxygen radical generation systems (e.g., superoxide anion generation by xanthine oxidase), inflammatory cells, tumor promotors, and free radical generating systems (e.g., benzoyl peroxide) produce a range of ROS. Consequently, scavenging ROS is a chemopreventive

mechanism; for example, dietary vitamin C, vitamin E, and retinoids (derived from citrus fruits, carrots, green vegetables, etc.) provide an integrated antioxidant system, with tissue GSH scavenging ROS and protecting tissues from ROS induced oxidative damage.[20] Also genistein, the active compound derived from soybeans and curcumin, derived from turmeric, which belongs to thiol groups such as N-acetylcysteine, is known to react with hydroxyl radicals and phenolic antioxidants.[21]

Reduction of Peroxides and Repair of Peroxidized Biological Membrane

The free radicals, peroxides, super oxides generated through oxidative stress, can be reduced to their non-toxic form to prevent damage of biological membranes; for example, the dietary mineral element selenium is essential for this aspect of antioxidant protection and disease prevention, as it is a vital component of the two peroxidase enzymes, GPX, which reduces soluble peroxides, and PHGPX, which removes lipid peroxides from biological membranes.[20]

Sequestration of Iron to Decrease ROS Formation

Regulation of metal (iron) formation and its management inside the cell is important in maintaining the balance of oxidative stress; for example, silicic acid, a ubiquitously distributed component of cereals and other foods, forms complexes with inorganic iron, which enables the safe sequestration of iron in tissues, decreasing its ability to generate ROS, initiates membrane lipid peroxidation, and mobilizes leukocytes.[22]

Utilization of Dietary Lipids

Dietary lipids can produce rapid energy and short-chain fatty acids like cholesteryl esters can be utilized for ROS scavenging; for example, deficiency of dietary choline or dietary lipotropes (vitamin B_{12}, folate, pyridoxal, glycine, $PO_4{}^{3-}$, etc.) has been reported to result in ROS production, lipid peroxidation, tissue injury, malignancy, and death.[23-25]

Modulation of Arachidonic Acid Metabolism

Two aspects of arachidonic acid metabolism are strongly associated with carcinogenesis, and both are inhibited by antioxidants. The first is the prostaglandin synthetic pathway, and second is the prostaglandin H synthase and lipooxygenase activity. Retinoids are reported to suppress t phorbol ester mediated cyclooxygenase 2(COX2) induction in human oral epithelial cells.[26]

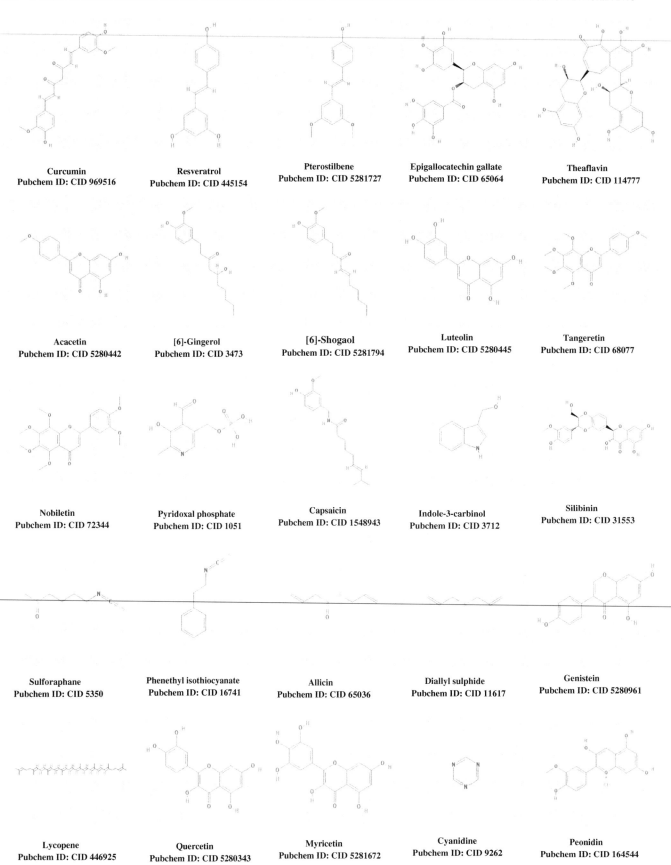

Curcumin
Pubchem ID: CID 969516

Resveratrol
Pubchem ID: CID 445154

Pterostilbene
Pubchem ID: CID 5281727

Epigallocatechin gallate
Pubchem ID: CID 65064

Theaflavin
Pubchem ID: CID 114777

Acacetin
Pubchem ID: CID 5280442

[6]-Gingerol
Pubchem ID: CID 3473

[6]-Shogaol
Pubchem ID: CID 5281794

Luteolin
Pubchem ID: CID 5280445

Tangeretin
Pubchem ID: CID 68077

Nobiletin
Pubchem ID: CID 72344

Pyridoxal phosphate
Pubchem ID: CID 1051

Capsaicin
Pubchem ID: CID 1548943

Indole-3-carbinol
Pubchem ID: CID 3712

Silibinin
Pubchem ID: CID 31553

Sulforaphane
Pubchem ID: CID 5350

Phenethyl isothiocyanate
Pubchem ID: CID 16741

Allicin
Pubchem ID: CID 65036

Diallyl sulphide
Pubchem ID: CID 11617

Genistein
Pubchem ID: CID 5280961

Lycopene
Pubchem ID: CID 446925

Quercetin
Pubchem ID: CID 5280343

Myricetin
Pubchem ID: CID 5281672

Cyanidine
Pubchem ID: CID 9262

Peonidin
Pubchem ID: CID 164544

Modulation of Hormonal and Growth Factor Activity

Chemicals may inhibit the growth associated with carcinogenesis by directly regulating the induction and activity of specific hormones and growth factors. The regulation can occur at the membrane receptors for growth factors or via cytoplasmic and nuclear receptors. Studies in MCF-7 cells derived from human breast cancer indicate the antiesterogenic effect of genistein, which may result from slowed translocation of genistein-bound receptors from the cytoplasm to the nucleus compared with that of oestradiol-bound receptors.[27]

Inhibition of Oncogene Activity

During the course of cell proliferation in carcinogenesis, numerous oncogenes are expressed abnormally, possibly functioning as intermediates in signal transduction pathway; for example, protein kinase. The citrus fruit constituent D-limonese and its cogener perillyl alcohol inhibit the progression of mammary tumors induced in rats.[28]

Alternative Biological Pathways

Alternative biological pathways have also been suggested for some forms of cancer, as for example, stomach cancer. Gastric cancer has generally been attributed to *Helicobacter pylori* infections, but this is also known to result from stress and trauma mediated by the cytokines. A sequence of pathological events has been elucidated: due to poor diet (low in proteins, fresh fruits, and vegetables) the inflammation leads normal gastric mucosa to sparse gastric mucus, which further progresses due to *H. pylori* infection, leading to free radical formation and cytokine activation. Depletion of antioxidants results in nitrosamine formation and mucus barrier destruction, which ultimately leads to tissue injury, mutations, and malignancy.[29] Although the *H. pylori* infection appears to be the most critical factor, this microorganism is highly resistant to antibiotic treatment. A more successful approach appears to be to administer ascorbic acid (may be from plant sources) to inhibit nitrosation, and subsequently, with the aid of antibiotics, to eliminate the microbial overgrowth and restore the normal gastric equilibrium.

Modulation of Signal Transduction

Cells respond to signals from extracellular stimuli via a complicated network of regulated events, collectively referred to as "signal transduction pathways." Stimulation of these pathways results in changes in transcriptional activity of genes. A recent idea about various cell-signaling pathways in ROS homeostasis is briefly discussed later.

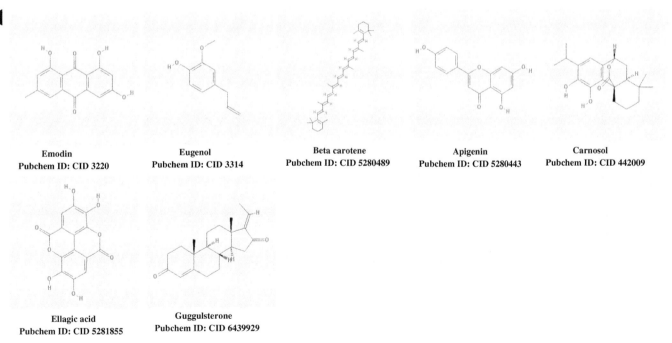

Emodin Pubchem ID: CID 3220 | **Eugenol** Pubchem ID: CID 3314 | **Beta carotene** Pubchem ID: CID 5280489 | **Apigenin** Pubchem ID: CID 5280443 | **Carnosol** Pubchem ID: CID 442009

Ellagic acid Pubchem ID: CID 5281855 | **Guggulsterone** Pubchem ID: CID 6439929

FIGURE 8.2 Chemical structures of some important bioactive compounds used as antioxidants for chemoprevention: curcumin, resveratrol, pterostilbene, epigallocatechin gallate, theaflavin, acacetin, [6]-gingerol, [6]-shogaol, luteolin, tangeretin, nobiletin, pyridoxal phosphate, capsaicin, indole-3-carbinol, silibinin, sulforaphane, phenethyl isothiocyanate, allicin, diallyl sulphide, genistein, lycopene, quercetin, myricetin, cyanidine, peonidin, emodin, eugenol, beta carotene, apigenin, carnosol, ellagic acid, guggulsterone. (See the color plate.)

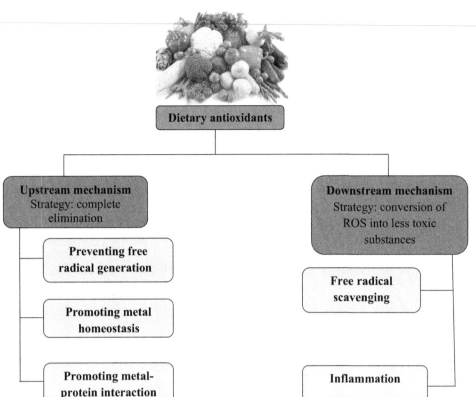

With regard to ROS homeostasis and redox regulation in cellular signaling, Ray et al. (2012)[30] proposed an in-depth mechanism of signal transduction pathways that could be targeted for cancer therapeutics. According to Ray et al. (2012),[30] "oxidative interphase" is the boundary between ROS and the signaling molecules they activate. The mechanism by which ROS generation alters protein function in cellular events is explained here.

Oxidation of the redox reactive cysteine (Cys) residue forms reactive sulfenic acid (-SOH) that can form disulfide bonds with nearby cysteines (-S-S-) or undergo further oxidation to sulfinic (-SO$_2$H) or sulfonic acid (-SO$_3$H). After some time with nearby nitrogen it forms sulfenamide. These oxidative modifications result in changes in the structure and functions of protein.[31] ROS regulate several signaling pathways, affecting a variety of cellular processes such as proliferation, differentiation, survival, metabolism, antioxidant and anti-inflammatory response, iron homeostasis, and DNA damage response. We will touch upon a few pathways associated with ROS homeostasis in this chapter, which follows several independent pathways as mentioned next, and which can serve the purpose of targeting cancer therapeutics (Figure 8.4).

ROS-RELATED SIGNALING PATHWAYS FOR TARGETED CANCER THERAPY

Regulation of MAPK Signaling Pathways by ROS

The mitogen-activated protein kinase (MAPK) and its associated kinases are evolutionarily conserved in eukaryotes and play pivotal roles in cellular responses to a wide variety of signals elicited by growth factors, hormones, and cytokines, in addition to genotoxic and oxidative stress. ROS-mediated regulation of MAPK is achieved by protein kinases and phosphatases in the MAPK signal cascade. Among the members of the MAPK cascades, apoptosis signal-regulated kinase 1(ASK1) is an upstream regulator that functions through a number of protein cascades leading to apoptosis through phosphorylation of signal proteins. Among many ASK1 associated proteins, the redox protein thioredoxin has been shown to constitutively interact with ASK1 and directly inhibit its kinase activity. ASK1 plays a pivotal role in promoting cell death under oxidative stress in mouse embryonic fibroblasts.[32]

Regulation of PI3K Signaling Pathways by ROS

Another signaling pathway that plays a key role in cell proliferation and survival in response to

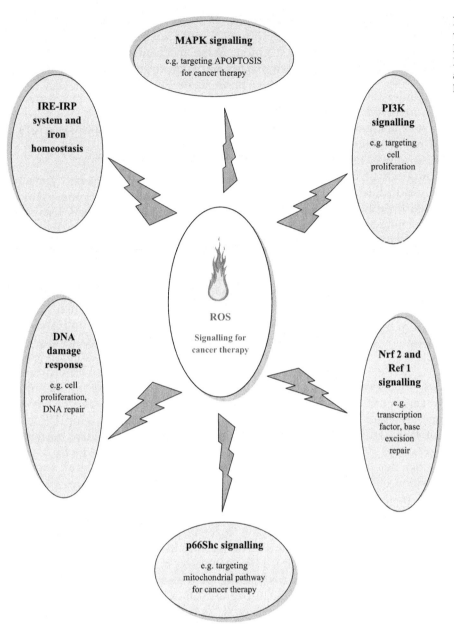

FIGURE 8.4 ROS-related signaling pathways for targeted cancer therapy include MAPK signaling, PI3K signaling, Nfr 2 and Ref 1 signaling, p66Shc signaling, DNA damage response, and IRE-IRP system and iron homeostasis. (See the color plate.)

growth factor, hormone, and cytokine stimulation is the phosphoinositide 3-kinase (PI3K) pathway. The PI3K is activated by various growth factors, such as Epidermal Growth Factor (EGF), Platelet-Derived Growth Factor (PDGF), Nerve Growth Factor (NGF), insulin, and Vascular Endothelial Growth Factor (VEGF). Through phosphatase and tensin homology (PTEN), the PI3K triggers the reversible reaction by the stimulation of the growth factor to stop ROS production. H_2O_2 was shown to oxidize and inactivate human PTEN through disulfide bond formation between the catalytic domain Cys-124 and Cys-71 residues.[33]

Nrf 2 and Ref 1 Mediated Redox Cellular Signaling

Redox regulation of transcription factors is significant in determining gene expression profile and cellular response to oxidative stress. Redox factor-1(Ref-1) is a multifunctional protein that not only regulates transcription factor activity, but also mediates base excision repair. Ref-1 was shown to be upregulated by genotoxic agents and oxidants, such as bleomycin and H_2O_2, and so protected cells from DNA and oxidative damage.[34] Ref-1 was shown to be involved in the transcriptional activation of PI3K-NFE2-like 2(Nrf2)-like antioxidant response element under oxidative stress.[35]

Regulation of p66Shc, Mitochondrial Oxidative Stress

The Shc adaptor protein family, encoded by the ShcA locus in mammalian cells, consisting of the p66Shc, is an important signaling molecule. The current understanding is that p66Shc is a proapoptotic protein involved in ROS production in mitochondria leading to mitochondrial damage and apoptosis under oxidative or genotoxic stress conditions, such as H_2O_2 or UV exposure. The molecular mechanism through which stress-activated p66Shc induces apoptosis has not been fully elucidated; however, p66Shc was shown to serve as a redox protein that produces H_2O_2 in mitochondria through interaction and electron transfer between p66Shc and cytochrome c, in which mutations in the redox centre of p66Shc (E132–E133 to Q132–Q133 in the CB domain) impaired opening of the mitochondrial permeability transition pore, thus negating the proapoptotic function of p66Shc.[36]

Regulation of IRE-IRP System and Iron Homeostasis by ROS

Iron is an essential element that plays a crucial role in cell proliferation and metabolism by serving as a functional constituent of various enzymes, including ribonucleotide reductase and cytochrome P450. However, when present in excess, free iron generates ROS via the Fenton reaction,[37] placing cells under deleterious oxidative stress. Therefore, tight regulation of iron homeostasis is crucial not only to maintain normal cellular function, but also to prevent iron-mediated oxidative stress. Iron-responsive element (IRE) and iron regulatory protein-1 and -2 (IRP1 and IRP2) are the examples of a cascade regulatory system that not only are regulated by cellular iron status but also regulated by ROS, in which cells elicit a defense mechanism against iron toxicity and iron-catalyzed oxidative stress.[37]

ROS and DNA Damage Response

Ataxia–telangiectasia mutated (ATM) and Ataxia–telangiectasia and Rad3-related (ATR) factors are the PI3K-like serine/threonine protein kinases activated under genotoxic stress conditions and phosphorylate various proteins involved in cell proliferation, cell death and survival, and DNA repair. These two signaling proteins were initially thought to be activated by a particular type of DNA damage, therefore serving in parallel signaling pathways; however, accumulating evidence suggests that the ATM- and ATR-pathways communicate and cooperate in response to DNA damage.[38]

On the basis of various signaling pathways discussed earlier, the National Cancer Institute of the United States has identified approximately 40 plant-based foods that exert cancer preventive effects including turmeric, green tea, red wine, ginger, soybean, cabbage, and cauliflower.[39] Among these, mechanisms of action of some prominent plants and plant products that show promise in cancer prevention are outlined next.

CAROTENOIDS (BETA CAROTENE AND LYCOPENE)

Carotenoids are lipophilic plant pigments typically containing a series of conjugated double bonds, which makes them susceptible to oxidative cleavage. This suggests that oxidation products have biological roles and therefore studying their biological effects is of great importance. Cleavage and oxidation of carotenoids can occur by (1) chemical cleavage and (2) enzymatic cleavage. A large number of carbonyl containing oxidation products are generated as a result of carotenoid oxidation and these can be further metabolized into the corresponding acids and alcohols. One such example is the formation of apocarotinoids. These oxidation products are suggested to carry biological activities through induction of apoptosis, or by modulating various transcription factors related to carcinogenesis prevention, for example, nuclear factor E2-related factor 2 (Nrf2), which activates the electrophile/antioxidant response element (EpRE/ARE); nuclear factor-κB (NF-κB); as well as nuclear receptors, such as the retinoic acid receptor (RAR), retinoid X receptor (RXR), peroxisome proliferator-activated receptor (PPAR), and estrogen receptor.[40]

Lycopene, another carotenoid that gives tomatoes and other fruits and vegetables their red color, has been of particular interest recently in regard to its role in cancer, especially in prostate cancer. Lycopene may react with oxygen free radicals by transfer of the unpaired electron leaving the lycopene in an excited triplet state. The excess energy is dissipated as heat, or by "bleaching" the lycopene, the free radical scavenging process starts. Lycopene has been demonstrated to scavenge reactive oxygen ($\bullet O_2$) and other free radicals like peroxyl radicals, and also is reported to interact with hydrogen peroxide and nitrogen dioxide to protect the cells from oxidative damage.[41] Lycopene has also been shown to have other possible anticancer activities, particularly relating to modulation of intercellular communication and alterations in intracellular signaling pathways. These include an upregulation in intercellular gap junctions, an increase in cellular differentiation, and alterations in phosphorylation of some regulatory proteins.[42] Lycopene has been demonstrated to be significantly more efficient than any carotene in inhibiting insulin-like growth factor type 1 (IGF1) induced proliferation of a number of tumor cell lines and in decreasing the occurrence of both spontaneous and chemically induced mammary tumors in animal models.[43]

GRAPES

The beneficial effects of grapes as an antioxidant for cancer therapeutics are believed to be due to the presence of major bioactive compounds such as resveratrol, catechins, anthrocyanins, phenolic acids, and procyanidins.[44] *In vitro* and *in vivo* studies have demonstrated that grape skin, seed, and pomace extracts possess potent free radical scavenging activities and transition metal ions chelating activities, which result in the reduction of physiological ROS.[45] The antioxidant properties of grapes have been attributed to their ability to induce cell cycle arrest and apoptosis in cancer cells.[46] The use of grape antioxidants is a promising strategy to inhibit growth of a broad range of cancer cells by targeting EGFR and its downstream pathways, inhibiting overexpression of COX and prostraglandin E2 receptors, or modifying estrogen receptor pathways.[47] Current research deals with resveratrol, one of the major bioactive compounds of grapes, to introduce the phytochemical into clinical trials. Cellular uptake, molecular targets, and *in vivo* biological networks demonstrated resveratrol as a Ca^{2+} mediated apoptosis inducing agent in cancer cells.[48]

GINGER

Ginger is one of the richest sources of antioxidants, which paved the way for its utilization to scavenge free radicals and allied health discrepancies. The bioactive molecules of ginger-like gingerols have shown antioxidant activity in various modules.[49] Ginger, owing to its functional ingredients like [6]-gingerol, [6]-shogaol, [6] paradol, and zerumbone, exhibits anti-inflammatory and antitumorigenic activities.[50,51] Ginger and its bioactive molecules are effective in controlling the extent of colorectal, gastric, ovarian, liver, and skin cancers.[52,53,54,55,50] Mechanisms of action of the major bioactive component [6]-gingerol for targeted cancer therapy include apoptotic potential,[56] inhibition of NF-κB activation,[57] inhibition of cell adhesion invasion motility and activities of MMP-2 and MMP-9,[58] TRAIL-induced NF-κB activation,[52] and inhibition of LTA (4) H activity.[50]

Mechanisms of action by [6]-shogaol include inhibition of AKT[51] and damaging microtubules[52] for cancer therapy.

SPINACH

Spinach is considered a beneficial source for various carotenoids and lipophilic active compounds (i.e., neoxanthin, lutein, zeaxanthin, and chlorophylls). Dietary intake of spinach extract has been reported to have beneficial effects on various types of cancer, such as ovarian,

prostate, lung, breast, and colon cancer.[59] A powerful water soluble natural antioxidant mixture (NAO) was obtained from spinach leaves and a number of antioxidative properties were tested both *in vitro* (DU145 and PC3) and *in vivo* (transgenic adenocarcinoma mouse prostate) models for its potent anticancer activity.[59]

FUTURE PERSPECTIVES

The disease states in which ROS signaling and toxicity have been implicated are areas of intensive research in regard to prevention and therapy. ROS regulates proliferative and apoptotic pathways, and aberrant regulation of proliferation and apoptosis is essential in tumorigenesis; therapeutic strategies exploiting the role of ROS in these pathways are being developed.[60] For the prevention of human cancer, agents that act through more than one mechanism of action are of particular importance. The method of combinational therapy is also the choice of cancer therapeutics. Unveiling the molecular mechanisms of disease pathogenesis and progression is therefore essential in providing relevant targets in order to develop innovative therapeutic strategies. Targeting ROS and relevant antioxidant mechanisms for targeted cancer therapy draws great attention in this regard by using different dietary antioxidants.

Even though the literature on cancer chemopreventive agents and diet constituents is rapidly expanding, the development of such compounds for clinical use is still in its infancy. There are a number of issues that should be addressed for developing any dietary antioxidants against cancer therapy. The concerns are dose and concentration selection, experimental verification and validation, side effects/adverse effects, feasibility of short-term acute and long-term chronic uses, and pharmacodynamic/pharmacokinetic properties. Therefore, these concerns related to either the bioactive part or the whole crude extract should be carefully tested and properly addressed before being considered for therapeutic use. In addition, with the advanced knowledge of nanobiotechnology, special methods can also be formulated for better adsorption, specificity, dose limitation of the dietary antioxidants, and drug bioavailability for targeted cancer therapy.

SUMMARY POINTS

- Food plant products, dietary antioxidants, and bioactive phytochemicals have effective roles in chemoprevention/anticancer drug formulation.
- Combination of environment, food, and lifestyle stress create imbalance of antioxidant defense and oxidants (ROS), resulting in breakdown of homeostasis and leading toward carcinogenesis.

- Oxidative damage to biomolecules such as lipids, proteins, and DNA results in an increased risk of diseases like cancer.
- Many dietary antioxidants target molecular pathways for inhibiting cancer development and growth.
- Molecular strategy targets upstream and downstream events toward achieving either complete elimination or sequestration of ROS.
- Understanding molecular mechanisms of targeted cancer therapy with ROS signaling pathways is important for anticancer drug formulation.

References

1. Howlader N, Noone AM, Krapcho M, Neyman N, Aminou R, Altekruse SF, Kosary CL, Ruhl J, Tatalovich Z, Cho H, Mariotto A, Eisner MP, Lewis DR, Chen HS, Feuer EJ, Cronin KA, editors. *SEER Cancer Statistics Review, 1975-2009 (Vintage 2009 Populations)*. Bethesda, MD: National Cancer Institute; 2012.
2. Greenwald P, McDonald SS. *Antioxidants and the prevention of cancer. Antioxidants in human health and disease.* Wallingford, UK: CABI Publishing; 1999.
3. Diplock AT. Antioxidants and disease prevention. *Food Chem Toxicol* 1996;**34**:1013–23.
4. Cozzi R, Ricordy R, Aglitti T, Gatta V, Perticone P, De Salvia R. Ascorbic acid and β-carotene as modulators of oxidative damage. *Carcinogenesis* 1997;**18**:223–8.
5. Krebs-Smith SM, Kantor LS. Choose a Variety of Fruits and Vegetables Daily: Understanding the Complexities. *J Nutr* 2001;**131**:487S–501S.
6. Colacino JA, Arthur AE, Dolinoy DC, Sartor MA, Duffy SA, Chepeha DB, et al. Pre treatment dietary intake is associated with tumor suppressor DNA methylation in head and neck squamous cell carcinomas. *Epigenetics* 2012;**7**:883–91.
7. Pan MH, Ho CT. Chemopreventive effects of natural dietary compounds on cancer development. *Chem Soc Rev* 2008;**37**:2558–74.
8. Basu TK, Temple N, Garg M. *Antioxidants in human health and disease.* Wallingford, UK: CABI Publishing; 1999.
9. Beliveau R, Gingras D. Role of nutrition in preventing cancer. *Can Fam Physician* 2007;**53**:1905–11.
10. Bush AI. Metals and neuroscience. *Curr Opin Chem Biol* 2000;**4**:184–91.
11. Demuro A, Mina E, Kayed R, Milton SC, Parker I, Glabe CG. Calcium dysregulation and membrane disruption as a ubiquitous neurotoxic mechanism of soluble amyloid Oligomers. *J Biol Chem* 2005;**280**:172–94.
12. Hanahan D, Weinberg RA. The hallmarks of cancer. *Cell* 2000;**100**:57–70.
13. Apel K, Hirt H. Reactive oxygen species: Metabolism, Oxidative Stress, and Signal Transduction. *Annu Rev Plant Biol* 2004;**55**:373–99.
14. Bergamini CM, Gambetti S, Dondi A, Cervellati C. Oxygen, reactive oxygen species and tissue damage. *Curr Pharm Des* 2004;**10**:1611–26.
15. Trush MA, Kensler TW. An overview of the relationship between oxidative stress and chemical carcinogenesis. *Free Radic Biol Med* 1991;**10**:201–9.
16. Chaudhary AK, Nokubo M, Marnett LJ, Blair IA. Analysis of the malondialdehyde-2′-deoxyguanosine adduct in rat liver DNA by gas chromatography/electron capture negative chemical ionization mass spectrometry. *Biol Mass Spectrom* 1994;**23**:457–64.
17. U.S. Department of Agriculture and U.S. Department of Health and Human Services. *Dietary Guidelines for Americans.* 7th ed. Washington, DC: U.S. Government Printing Office; 2010.
18. Perera FP. Molecular epidemiology: Insights into cancer susceptibility, risk assessment, and prevention. *J Natl Cancer Inst* 1996;**88**:496–509.
19. Liu RH. Potential Synergy of Phytochemicals in Cancer Prevention: Mechanism of Action. *J Nutr* 2004;**134**:3479S–85S.
20. Parke AL, Parke DV, Avery Jones F. Diet and nutrition in rheumatoid arthritis and other chronic inflammatory diseases. *J Clin Biochem Nutr* 1996;**20**:1–26.
21. Kensler TW, Trush MA, Guyton KZ. Free radicals as targets for cancer chemoprevention: Prospects and problems. In: Steele VE, Stoner GD, Kelloff GJ, Boone CW. editors. *Cellular and Molecular Targets for Chemoprevention.* Boca Raton, FL: CRC Press; 1992. p. 173–191.
22. Birchall JD. Silicon and the bioavailability of aluminium – Nutritional aspects. In: Food Nutrition, Toxicity Chemical, editors. London: Smith-Gordon; 1993. p. 215–26.
23. Vance DE. Phosphatidylcholine metabolism: masochistic enzymology, metabolic regulation and lipoprotein assembly. *Biochem Cell Biol* 1990;**68**:1151–65.
24. Lombardi B, Chandard N, Locker J. Nutritional model of hepatocarcinogenesis. Rat fed choline-devoid diet. *Dig Dis Sci* 1991;**36**:979–84.
25. Schrager TF, Newberne PM. Lipids, lipotropes and malignancy: A review and introduction of new data. In: Food Nutrition, Toxicity Chemical, editors. London: Smith-Gordon; 1993. p. 227–47.
26. Mestre JR, Subbaramaiah K, Sacks PG, Schantz SP, Tanabe T, Inoue H, et al. Retinoids suppress phorbol ester-mediated induction of cyclooxygenase-2. *Cancer Res* 1997;**57**:1081–5.
27. Martin PM, Horwitz KB, Ryan DS, McGuire WL. Phytoestrogen interaction with estrogen receptors in human breast cancer cells. *Endocrinology* 1978;**103**:1860–7.
28. Haag JD, Lindstrom MJ, Gould MN. Limonene-induced regression of mammary carcinomas. *Cancer Res* 1992;**52**:4021–6.
29. Parke DV. Sixty years of research into gastric cancer. *Toxicol Ecotoxicol News* 1997;**4**:132–7.
30. Ray PD, Huang BW, Tsuji Y. Reactive oxygen species (ROS) homeostasis and redox regulation in cellular signalling. *Cell Signal* 2012;**24**:981–90.
31. Roos G, Messens J. Protein sulfenic acid formation: from cellular damage to redox regulation. *Free Radic Biol Med* 2011;**51**:314–26.
32. Tobiume K, Matsuzawa A, Takahashi T, Nishitoh H, Morita K, Takeda K, et al. ASK1 is required for sustained activations of JNK/p38 MAP kinases and apoptosis. *EMBO Rep* 2001;**2**:222–8.
33. Kwon J, Lee SR, Yang KS, Ahn Y, Kim YJ, Stadtman ER, et al. Reversible oxidation and inactivation of the tumor suppressor PTEN in cells stimulated with peptide growth factors. *Proc Natl Acad Sci USA* 2004;**101**:16419–24.
34. Ramana CV, Boldogh I, Izumi T, Mitra S. Activation of apurinic/apyrimidinic endonuclease in human cells by reactive oxygen species and its correlation with their adaptive response to genotoxicity of free radicals. *Proc Natl Acad Sci USA* 1998;**95**:5061–6.
35. Iwasaki K, Mackenzie EL, Hailemariam K, Sakamoto K, Tsuji Y. Hemin-mediated regulation of an antioxidant-responsive element of the human ferritin H gene and role of Ref-1 during erythroid differentiation of K562 cells. *Mol Cell Biol* 2006;**26**:2845–56.
36. Pelicci G, Lanfrancone L, Grignani F, McGlade J, Cavallo F, Forni G, et al. A novel transforming protein (SHC) with an SH2 domain is implicated in mitogenic signal transduction. *Cell* 1992;**70**:93–104.
37. Ghio AJ. Disruption of iron homeostasis and lung disease. *Biochim Biophys Acta* 2009;**1790**:731–9.
38. Hurley PJ, Bunz F. ATM and ATR: Components of an integrated circuit. *Cell Cycle* 2007;**6**:414–7.
39. Surh YJ. Cancer chemoprevention with dietary phytochemicals. *Nat Rev Cancer* 2003;**3**:768–80.
40. Sharoni Y, Linnewiel-Hermoni K, Khanin M, Salman H, Veprik A, Danilenko M, et al. Carotenoids and apocarotenoids in cellular signaling related to cancer: A review. *Mol Nutr Food Res* 2012;**56**:259–69.

41. Woodall AA, Britton G, Jackson MJ. Carotenoids and protection of phospholipids in solution or in liposomes against oxidation by peroxyl radicals: relationship between carotenoid structure and protective ability. *Biochim Biophys Acta* 1997;**1336**:575–86.

42. Matsushima-Nishiwaki R, Shidoji Y, Nishiwaki S, Yamada T, Moriwaki H, Muto Y. Suppression by carotenoids of microcystin-induced morphological changes in mouse hepatocytes. *Lipids* 1995;**30**:1029–34.

43. Sharoni Y, Giron E, Rise M, Levy J. Effects of lycopene-enriched tomato oleoresin on 7, 12-dimethyl-benz[a]anthracene-induced rat mammary tumors. *Cancer Detect Prev* 1997;**21**:118–23.

44. Frayne RF. Direct analysis of the major organic components in grape must and wine using high performance liquid chromatography. *Am J Enol Viticult* 1986;**37**:281–7.

45. Sun Q, Prasad R, Rosenthal E, Katiyar SK. Grape seed proanthocyanidins inhibit the invasive potential of head and neck cutaneous squamous cell carcinoma cells by targeting EGFR expression and epithelial-to-mesenchymal transition. *BMC Complement Altern Med* 2011;**11**:134.

46. Aggarwal BB, Bhardwaj A, Aggarwal RS, Seeram NP, Shishodia S, Takada Y. Role of resveratrol in prevention and therapy of cancer: Preclinical and clinical studies. *Anticancer Res* 2004;**24**:2783–840.

47. Zhou HB, Chen JJ, Wang WX, Cai JT, Du Q. Anticancer activity of resveratrol on implanted human primary gastric carcinoma cells in nude mice. *World J Gastroenterol* 2005;**11**:280–4.

48. Subramanian L, Youssef S, Bhattacharya S, Kenealey J, Polans AS, Van Ginkel PR. Resveratrol: Challenges in Translation to the Clinic — A Critical Discussion. *Clin Cancer Res* 2010;**16**:5942–8.

49. Dugasani S, Pichika MR, Nadarajah VD, Balijepalli MK, Tandra S, Korlakunta JN. Comparative antioxidant and anti-inflammatory effects of [6]-gingerol, [8]-gingerol, [10]-gingerol and [6]-shogaol. *J Ethnopharmacol* 2010;**127**:515–20.

50. Jeong CH, Bode AM, Pugliese A, Cho YY, Kim HG, Shim JH, et al. [6]-gingerol suppresses colon cancer growth by targeting leukotriene a4 hydrolase. *Cancer Res* 2009;**69**:5584–91.

51. Hung JY, Hsu YL, Li CT, Ko YC, Ni WC, Huang MS, et al. [6] Shogaol, an active constituent of dietary ginger, induces autophagy by inhibiting the AKT/mTOR pathway in human non-small cell lung cancer A549 cells. *J Agric Food Chem* 2009;**57**:9809–16.

52. Ishiguro K, Ando T, Maeda O, Ohmiya N, Niwa Y, Kadomatsu K, et al. Ginger ingredients reduce viability of gastric cancer cells via distinct mechanisms. *Biochem Biophys Res Commun* 2007;**362**:218–23.

53. Sung B, Jhurani S, Ahn KS, Mastuo Y, Yi T, Guha S, et al. Zerumbone down-regulates chemokine receptor CXCR4 expression leading to inhibition of CXCL12-induced invasion of breast and pancreatic tumor cells. *Cancer Res* 2008;**68**:8938–44.

54. Brown AC, Shah C, Liu J, Pham JT, Zhang JG, Jadus MR. Ginger's (*Zingiber officinale Roscoe*) inhibition of rat colonic adenocarcinoma cells proliferation and angiogenesis in vitro. *Phytother Res* 2009;**23**:640–5.

55. Kim M, Miyamoto S, Yasui Y, Oyama T, Murakami A, Tanaka T. Zerumbone, a tropical ginger sesquiterpene, inhibits colon and lung carcinogenesis in mice. *Int J Cancer* 2009;**124**:264–71.

56. Nigam N, Bhui K, Prasad S, George J, Shukla Y. [6]-Gingerol induces reactive oxygen species regulated mitochondrial cell death pathway in human epidermoid carcinoma A431 cells. *Chem Biol Interact* 2009;**181**:77–84.

57. Rhode J, Fogoros S, Zick S, Wahl H, Griffith KA, Huang J, et al. Ginger inhibits cell growth and modulates angiogenic factors in ovarian cancer cells. *BMC Complement Altern Med* 2007;**7**:44.

58. Lee HS, Seo EY, Kang NE, Kim WK. [6]-Gingerol inhibits metastasis of MDA-MB-231 human breast cancer cells. *J Nutr Biochem* 2008;**19**:313–9.

59. Lomnitski L, Bergman M, Nyska A, Ben-Shaul V, Grossman S. Composition, efficacy, and safety of spinach extracts. *Nutr Cancer* 2003;**46**:222–31.

60. Antosiewicz J, Ziolkowski W, Kar S, Powolny AA, Singh SV. Role of reactive oxygen intermediates in cellular responses to dietary cancer chemopreventive agents. *Planta Med* 2008;**74**:1570–9.

CHAPTER

9

Oxidative Stress and Antioxidant Herbs and Spices in Cancer Prevention

R.I. Shobha, C.U. Rajeshwari, B. Andallu

Food Science and Nutrition Division, Sri Sathya Sai Institute of Higher Learning, Anantapur Campus, Anantapur, Andhra Pradesh, India

INTRODUCTION

Oxidative stress is defined as a discrepancy between production of free radicals and reactive metabolites, so-called oxidants or reactive oxygen species (ROS), and their eradication by defending mechanisms, referred to as antioxidants. This disparity leads to damage of important biomolecules and cells, with potential impact on the whole organism. ROS are products of a normal cellular metabolism and play vital roles in the stimulation of signaling pathways in plant and animal cells in response to changes in intra- and extracellular environmental conditions.[1] Most ROS are generated in cells by the mitochondrial respiratory chain. During endogenous metabolic reactions, aerobic cells produce ROS such as superoxide anion (O^{2-}), hydrogen peroxide (H_2O_2), hydroxyl radical ($OH^{\bullet}$), and organic peroxides as normal products of the biological reduction of molecular oxygen. Under hypoxic conditions, the mitochondrial respiratory chain also produces nitric oxide (NO), which can generate reactive nitrogen species (RNS), which can further generate other reactive species, for example, reactive aldehydes—malondialdehyde and 4-hydroxynonenal—by inducing excessive lipid peroxidation. Proteins and lipids are also significant targets for oxidative attack, and modification of these molecules can increase the risk of mutagenesis.[2]

Under a sustained environmental stress, ROS are produced over a long time, and thus significant damage may occur to cell structure and functions and may induce somatic mutations and neoplastic transformations. Indeed, cancer initiation and progression have been linked to oxidative stress by increasing DNA mutations or inducing DNA damage, genome instability, and cell proliferation.[3] Acting to protect the organism against these harmful pro-oxidants is a complex system of enzymatic antioxidants [superoxide dismutase (SOD), glutathione peroxidase (GPx), glutathione reductase, catalase, etc.] and nonenzymatic antioxidants (glutathione (GSH), vitamins C, D, E, etc.). However, the amount of these defensive antioxidant principles present under the normal physiological situation is sufficient only to cope with the physiological rate of free-radical generation. It is evident, therefore, that any further load of free radicals can tip the free-radical (pro-oxidant) and anti-free-radical (antioxidant) balance leading to oxidative stress, which may result in tissue injury and subsequent diseases.[4]

ANTIOXIDANTS IN DIET AND HUMAN HEALTH

The potential advantageous effects of antioxidants in protecting against disease have been used as an argument for recommending increasing intake of several nutrients above those derived by conventional methods. A dietary antioxidant can sacrificially scavenge ROS/RNS to stop radical chain reactions, considered as primary chain-breaking antioxidant or free radical scavenger (FRS), or it can inhibit the reactive oxidants from being formed in the first place, considered as secondary or preventive antioxidant. Primary antioxidants, when present in trace amounts, may either delay or inhibit the initiation step by inactivating or scavenging free radicals, thus inhibiting initiation and propagation reactions by reacting with peroxyl or alkoxyl radicals.[5]

Antioxidant efficiency is dependent on the ability of the FRS to donate hydrogen to the free radical. As the hydrogen-bond energy of the FRS decreases, the transfer

Cancer
http://dx.doi.org/10.1016/B978-0-12-405205-5.00009-X

© 2014 Elsevier Inc. All rights reserved.

of the hydrogen to the free radical is more energetically promising and rapid. The ability of FRS to donate hydrogen free radicals can be predicted from standard one-electron reduction potentials. Efficient FRS also produces radicals that do not react rapidly with oxygen to form peroxides. In foods, the efficiency of phenolic FRS also depends on additional factors such as volatility, pH sensitivity, and polarity.[6]

Carotenoids are one of the most widespread phytonutrients found mostly in the flowers, fruits, algae, and photosynthetic bacteria. Carotenoids have extensive applications as antioxidants and are, therefore, important for human health. The essential role of beta-carotene and others as the main dietary source of vitamin A includes its protective effects against serious disorders such as cancer, heart disease, and other disorders. Carotenoids, including xanthophylls (oxygen-containing carotenoids), are naturally occurring colored compounds that are abundant as pigments in plants. To date, about 500 to 600 specific carotenoids have been identified mostly from plants and algae. Carotenoids have the capacity to trap not only lipid peroxyl radicals, but also singlet oxygen species. The antioxidant capacity of carotenoids may also be related to the structure. A larger conjugated system such as astaxanthin is known to have higher antioxidant activity.[7]

Antioxidants such as vitamins C and E are also essential for protection against ROS. However, the majority of the antioxidant activity of herbs may be from compounds such as phenolic acids and flavonoids rather than from vitamins C, E, and β-carotene. Intake of controlled diets rich in herbs, fruits, and vegetables increased significantly the antioxidant capacity of plasma.[8]

DIETARY PHYTOCHEMICALS AS ANTIOXIDANTS

The "phyto" of the word phytochemicals is derived from the Greek word *phyto*, which means plant. Therefore, phytochemicals are plant chemicals and are bioactive nonnutrient plant compounds in fruits, vegetables, grains, and other plant foods. In recent years, many studies have shown that diets containing high content of phytochemicals can provide protection against various diseases. Approximately 90% of all cancer cases correlate with environmental factors, including one's dietary habits, and one-third of all cancer deaths are preventable by changing dietary habits.[9] These discoveries have rapidly augmented the consumer awareness of the probable benefits of naturally occurring compounds from plants in health promotion and maintenance, and researches in nutraceuticals, functional foods, and natural health products have been given top priority in recent years.

The phytochemicals found in herbs and other plant foods prevent and reverse many of the processes that underlie chronic diseases. These phytochemicals can be broadly classified as carotenoids, phenolics, alkaloids, nitrogen-containing compounds, and organosulfur compounds (Figure 9.1).[10]

The most important property of phytochemicals includes their role as antioxidants, where the compounds in plant foods provide protection against the often highly damaging oxidative processes in our bodies that are caused and perpetuated by free radicals. The unpaired electron common to all free radicals makes these molecules highly reactive and, in order to stabilize themselves, they steal electrons from other compounds and oxidize the targeted substances (e.g., proteins, fats, and DNA) that are all vulnerable to oxidation.[11]

The antioxidant phytochemicals include both enzymatic and nonenzymatic components that prevent radical formation, remove radical before damage can occur, repair oxidative damage, eliminate damaged molecules, and prevent mutations. A variety of sulfur-containing compounds and precursors in garlic also have antioxidant activity. Thus, due to aforementioned reasons, the protective effects of these phytochemicals found in fruits, vegetables, spices, and herbs were found not only for diabetes and cardiovascular diseases (CVDs), but also for other inflammatory diseases and cancer.[12]

ANTIOXIDANT PHYTOCHEMICALS AND CANCER PREVENTION

Cancer results from a multistage carcinogenesis process that involves three distinguishable but closely connected stages: initiation (normal cell→transformed or initiated cell), promotion (initiated cell→preneoplastic cell), and progression (preneoplastic cell→neoplastic cell; Figure 9.2). Initiation is an outcome of rather rapid and irreversible assault to the cell. The attack may be due to the initial uptake of a carcinogen and the subsequent stable genotoxic damage caused by its metabolic activation. Other causes of cancer initiation include oxidative stress, chronic inflammation, and hormonal imbalance.[13]

Generally, the leaf of a plant used in cooking may be referred to as a culinary herb, and any other part of the plant, often dried, as a spice. Spices can be the buds (cloves), bark (cinnamon), roots (ginger), berries (peppercorns), aromatic seeds (cumin), and even the stigma of a flower (saffron). Herbs and spices have a long history of both culinary use and of providing health benefits, as well as acting as preservatives. Herbs may act through several mechanisms to provide protection against cancer. Phytochemicals present in various herbs have also been proved to have great potential in combating cancer

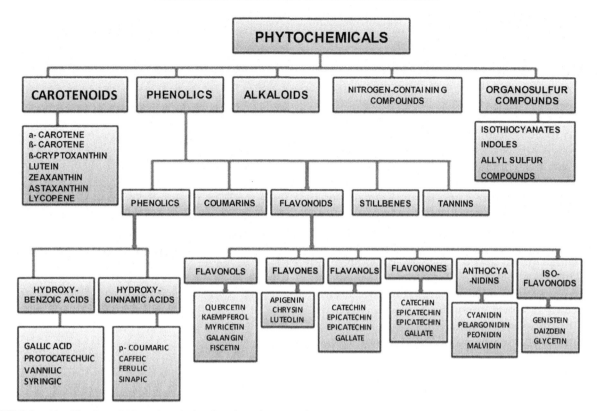

FIGURE 9.1 **Classification of Phytochemicals.** The phytochemicals found in herbs and other plant foods prevent and reverse many of the processes that underlie chronic diseases. These phytochemicals can be broadly classified as carotenoids, phenolics, alkaloids, nitrogen-containing compounds, and organosulfur compounds. *Source: Liu, 2004.*

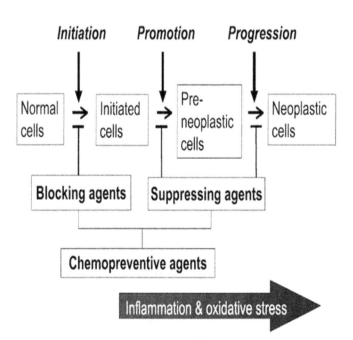

FIGURE 9.2 **Cancer—A Multistep Process.** Cancer results from a multistage carcinogenesis process that involves three distinguishable but closely connected stages: initiation (normal cell→transformed or initiated cell), promotion (initiated cell→preneoplastic cell), and progression (preneoplastic cell→neoplastic cell). *Source: Anand et al. (2008).*

processes (i.e., initiation, promotion, growth, and metastasis) and other chronic diseases that result from oxidative stress induced by free radicals.[11] Researchers have identified a host of cancer chemoprotective phytochemicals in many herbs. In addition, many herbs contain a variety of phytosterols, triterpenes, flavonoids, saponins, and carotenoids, which have been shown to be cancer chemopreventive.[11] These beneficial substances act as antioxidants and electrophile scavengers, stimulate the immune system, inhibit nitrosation and the formation of DNA adducts with carcinogens, inhibit hormonal actions and metabolic pathways associated with the development of cancer, and induce phase I or phase II detoxification enzymes.[14]

Chemoprevention is the use of chemical substances of either natural or synthetic origin to prevent, hamper, arrest, or reverse a disease. Several commonly used herbs such as coriander, cumin, garlic, ginger, mint, oregano, and turmeric have been identified by the National Cancer Institute as possessing cancer-preventive properties. Garlic is known to have antitumor properties, owing to its content of a wide variety of organic sulfides and polysulfides. Garlic is reported to enhance immune function by stimulating lymphocytes and macrophages to destroy cancer cells and is also reported to disrupt the metabolism of tumor cells. The inhibition of tumors by garlic seems to be the most effective when the tumor is small. Garlic can also inhibit the formation of nitrosamines, which are potent carcinogens, and also inhibits the formation of DNA adducts.[15]

Turmeric contains active phenolic components that inhibit cancer and also have antimutagenic activity and has been shown to suppress the development of stomach, breast, lung, and skin tumors. Its activity is largely due to curcumin, which has been shown to be an effective anti-inflammatory agent in humans. Curcumin has been found to increase expression of conjugation enzymes and has been shown to be one of the most potent inhibitors of Nuclear Factor-Kappa B (NF-κB), thereby exerting anti-inflammatory effects. Curcumin has been shown to decrease the proliferation of various cancer cells of the colon, the blood, the submandibular gland, and the liver by downregulating COX-2.[16] One of the novel molecular targets of the chemopreventive action of curcumin is β-catenin. The antitumor effect of curcumin was evidenced by its ability to decrease intestinal tumors in an animal model of familial adenomatous polyposis by reducing the expression of the oncoprotein β-catenin. Some human β-catenin /TCF target genes—including cyclin D, Matrix metalloproteinases (MMP 7), Osteopontin (OPN), Inter Leukin-8 (IL-8), and matrilysin—play a role in tumor promotion and progression. NF-kappa B repression and decreased β-catenin signaling are some of the mechanisms by which curcumin suppresses the promotion and progression of cancer[17,18] (Figure 9.3).[18]

Korean studies suggest that ginseng may lower the risk of cancer in humans. Ginseng seemed to be the most protective against cancer of the ovaries, larynx, pancreas, esophagus, breast, cervix, and thyroid, which may be attributed to the active ingredients present in ginseng such as ginsenosides.[19] Phytochemicals such as chlorogenic acid have also been found to protect against environmental carcinogen-induced carcinogenesis by their upregulation of phase II conjugating enzymes and suppression of ROS- mediated NF-κB, AP-1, and MAPK activation.[20]

Polyphenols such as quercetin, catechins, isoflavones, lignans, flavanones, ellagic acid, red wine polyphenols, and resveratrol induce a reduction of the number of tumors or of their growth. These polyphenols act as blocking agents at the initiation stage, influence the metabolism of procarcinogens by modulating the expression of cytochrome P_{450} enzymes involved in their activation to carcinogens, and also facilitate their excretion by increasing the expression of phase II conjugating enzymes. Polyphenols can form potentially toxic quinones in the body that are, themselves, substrates of these enzymes, which then activate these enzymes for their own detoxification and, thus, induce a general boosting of our defenses against toxic xenobiotics. Polyphenols may also limit the formation of initiated cells by stimulating DNA repair.[21]

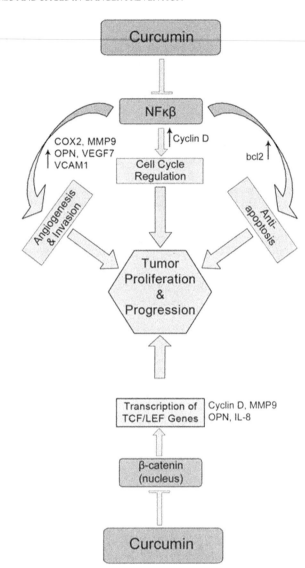

FIGURE 9.3   Schematic Representation of Chemopreventive Targets of Curcumin in Curtailing Tumor Proliferation and Progression. The antitumor effect of curcumin was evidenced by its ability to decrease intestinal tumors in an animal model of familial adenomatous polyposis by reducing the expression of the oncoprotein β-catenin. Some human β-catenin /TCF target genes—including cyclin D, MMP 7, OPN, IL-8, and matrilysin—play a role in tumor promotion and progression. NF-kappa B repression and decreased β-catenin signaling are some of the mechanisms by which curcumin suppresses the promotion and progression of cancer. *Source: Thangapazham et al. (2006).*

Polyphenols and flavonoids inhibit pro-inflammatory gene expression via inhibition of IκB, thus inhibiting NF-κB transactivation, as well as restoring transrepressive pathways through the activation of histone deacetylases (Figure 9.4).[22] In addition, expression of antioxidant genes such as GCL, Mn SOD, HO-1 via modulation of MAPK-ARE-Nrf 2 pathway are upregulated.[22]

In vitro anticancer studies have demonstrated that natural products of flavonoids like luteolin and quercetin have the power to inhibit the proliferation of cells in human carcinoma of larynx and sarcoma-180 cell lines.[23]

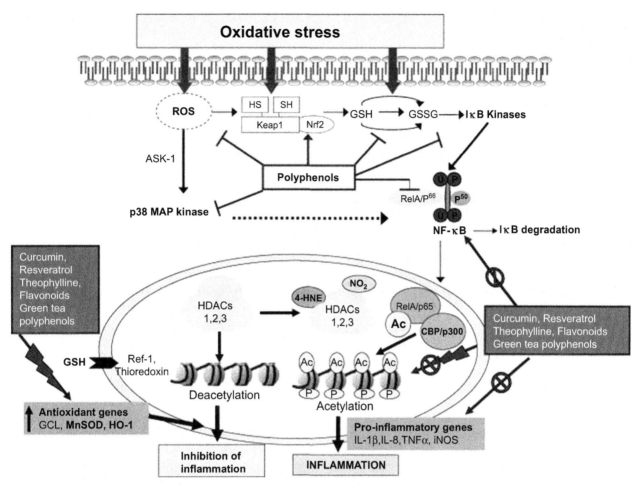

FIGURE 9.4 Dietary Polyphenols and Inhibition of Pro-Inflammatory Genes (IL-1β, TNFα, iNOS, NF-KB, etc.) Expression. Polyphenols and flavonoids inhibit pro-inflammatory gene expression via inhibition of IκB, thus inhibiting NF-κB transactivation, as well as restoring transrepressive pathways through the activation of histone deacetylases. Source: Rahman et al. (2006).

The nitrogenous phytochemicals, phenanthroindolizidine alkaloids [i.e., pergularinine and tylophorinidine isolated from *Pergularia pallida* (Roxb.) Wight and Arn (Asclepiadaceae)] inhibited the growth of *Lactobacillus leichmanni* cells by binding to thymidylate synthetase.[24]

Anthraquinone natural products like rubidianin, isolated from the alcoholic extract of *Rubia cordifolia*, demonstrated significant antioxidant activity in a dose-dependent manner. It prevented lipid peroxidation induced by ferrous sulphate and t-butylhydroperoxide. The antioxidant activity of rubidianin was found to be better than mannitol, vitamin E, and p-benzoquinone standards.[25]

Many phytochemicals from plants also affect lipoxygenase activity, which is also a cause for cancer. Lipoxygenase enzymes are found in a wide variety of plant and animal tissues. These enzymes have a nonheme iron serving as a catalytic center for the stereo and regiospecific dioxygenation of select carbon atoms in polyunsaturated fatty acids for their metabolism. Eighteen carbon chain fatty acids (e.g., linoleate) are the primary substrates of the plant lipoxygenases,

while the mammalian isozymes mainly catalyze the metabolism of fatty acids of 20 chain carbon length (e.g., arachidonate). The mammalian fatty acid, arachidonic acid, is metabolized by one of two enzyme pathways, cyclooxygenase (COX) or lipoxygenase (LOX) generating biologically active metabolites that are involved in carcinogenesis. It has been shown that the LOXs in particular are key regulators of cell survival and apoptosis in cells. It has also been shown that LOX is a regulator of human cancer development and it is overexpressed in a variety of tumors including breast, colorectal, and prostate cancer, and cancer cell lines. It has been reported that inhibition of oxidative enzymes such as 5-lipoxygenase and 12-lipoxygenase triggers tumor cell apoptosis, reduces tumor cell motility and invasiveness, or decreases tumor angiogenesis and growth.[26]

Onions and garlic are often used as food flavors and are used commonly in folk medicines in Asia and Africa. Belman et al.[27] studied the inhibitory effect of soybean lipoxygenase of onion and garlic constituents. The di-(1-propenyl) sulfide was the only irreversible inhibitor; diallyl trisulfide, allyl methyl trisulfide, and diallyl disulfide were competitive inhibitors; and 1-propenylpropyl sulfide and (E, Z)- 4,5,9 trithiadodeca-1,6,11-triene 9-oxide (ajoene) were mixed inhibitors. Sendl et al.[28] also studied lipoxygenase inhibitory activity of garlic.

Curry leaves, one of the spices used in Indian dishes, contain certain phytochemicals like carotenoids and fiber, which may contribute to the antioxidative and anticarcinogenic effects. A study conducted by Khanum et al.[29] on the antioxidant potential of curry leaves in rats treated with a known chemical carcinogen, dimethylhydrazine hydrochloride (DMH), revealed a significant increase in vitamin A content in the liver and no alteration in glutathione (GSH) content, a 50% decline in the micronuclei induced by DMH, and a 30% decrease in the activity of Γ-glutamyl transpeptidase. These results signify that curry leaves have high potential as a reducer of the toxicity of DMH.[29]

Ginger (Z. officinalis Roscoe, Zingiberaceae) is not only widely used as a dietary condiment but it has also been extensively utilized as a traditional oriental medicine. A study conducted by Abdullah et al.[30] provided evidence that ginger acts as a potent growth inhibitory compound in human colon adenocarcinoma cells and the study supports the possibility of chemopreventive potential of ginger in colon cancer cells. The cytotoxic effect could be as a result of the active component, gingerol, which has been reported to inhibit the growth of HCT 116 human colon colorectal and liver HepG2 cancer cells.[30] Azoxymethane-induced intestinal carcinogenesis in rats was significantly suppressed by dietary administration of gingerol. It was also reported that the inhibition mechanism of growth of colon cancer cells by ginger involved obstruction in both cell cycle progression in G0/G1 phase and apoptosis. Thus, it is clear that the antitumor effects on colon cancer cells were exerted by ginger by suppressing their growth, arresting the G0/G1-phase, reducing DNA synthesis, and inducing apoptosis.[31]

Saffron, used as a spice for flavoring and coloring food preparations, the main constituent of which is picrocrocin, has been reported to be useful in the treatment of numerous human diseases. Extracts of saffron have been shown to inhibit the formation of tumors and/or to retard tumor progression in a variety of experimental animal systems. The topical application of a saffron extract has been shown to inhibit both the initiation and the promotion of cancer by a common carcinogen, 7,12-dimethylbenz(a)anthracene (DMBA), which is used to induce skin cancer for experimental purposes. Oral administration of saffron extract inhibited the growth of mouse tumors that were derived from three different kinds of cancer cells

(S180, DLA, and EAC) and significantly increased (by two- to three-fold) the life span of treated tumor-bearing mice. It has also been found that naturally occurring saffron extract, in combination with two synthetic compounds, sodium selenite or sodium arsenite, may have a synergistic effect with saffron and may, therefore, have an important role in cancer chemoprevention.[32]

Cumin (Cuminum cyminum) has been a part of the diet, regularly used as a flavoring agent due to its pleasant flavor contributed by a major constituent, cuminaldehyde, in a number of ethnic cuisines. In a study conducted by Gagandeep et al.,[33] cancer chemopreventive potentials of different doses of a cumin seed-mixed diet were evaluated against benzo(a) pyrene [B(a)P]-induced forestomach tumorigenesis and 3-methylcholanthrene (MCA)-induced uterine cervix tumorigenesis. Results showed a significant inhibition of stomach tumor burden (tumors per mouse) by cumin. Cervical carcinoma incidence, compared with the MCA-treated control group (66.67%), reduced to 27.27% by a diet of 5% cumin seeds and to 12.50% by a diet of 7.5% cumin seeds. Levels of cytochrome P-450 (cyt P-450) and cytochrome b5 (cyt b(5) were significantly augmented by 2.5% dose of cumin seed diet. Lipid peroxidation measured as formation of malondialdehyde (MDA), showed significant inhibition by both doses of cumin. The results strongly suggest the cancer chemopreventive potentials of cumin seed and could be attributed to its ability to modulate carcinogen metabolism.[33]

The effect of red chili (Capsicum annum L.), cumin (Cuminum cyminum L.), and black pepper (Piper nigrum L.) on colon cancer induced in rats by a colon-specific carcinogen, 1,2-dimethylhydrazine (DMH), was investigated by Nalini et al.[34] The incidence and number of tumors in the colon were observed to be significantly higher in rats administered DMH and /or red chilies when compared to cumin+DMH and black pepper+DMH administered rats after the experimental period of 32 weeks. The levels of fecal bile acids and neutral sterols decreased in DMH and chili-supplemented rats whereas these levels increased significantly in rats supplemented with cumin and black pepper. Also, the levels of cholesterol, cholesterol/phospholipid ratio, 3-hydroxy-3-methylglutaryl-CoA reductase activity were decreased in cumin+DMH and black pepper+DMH-treated rats. Thus, these results confirm that red chili supplementation promotes colon carcinogenesis, whereas cumin and black pepper suppress colon carcinogenesis in the presence of the procarcinogen DMH.[34]

The chemical constituents of ginger, saffron, cumin, and black pepper are given in Figure 9.5.[11]

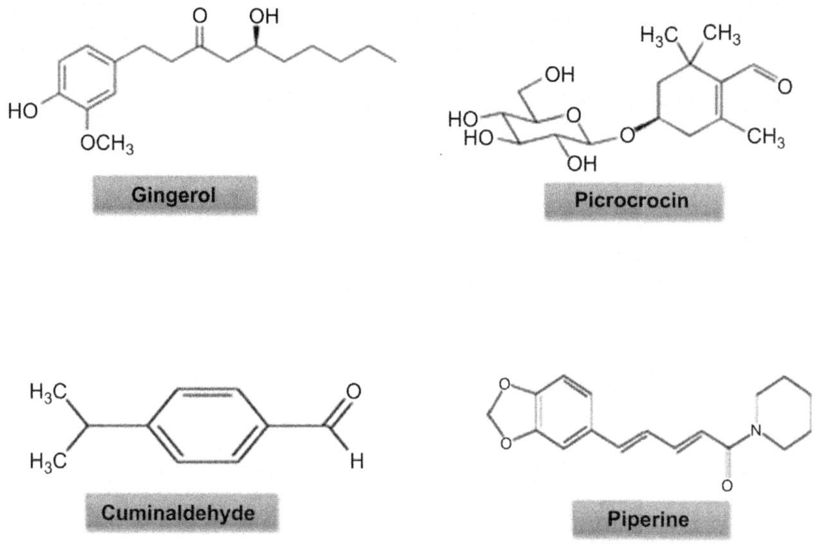

FIGURE 9.5 Bioactive Constituents of Ginger, Saffron, Cumin, and Black Pepper. The chemical constituents of ginger, saffron, cumin, and black pepper are given in the figure. *Source: Scott, 2006.*

TERPENOIDS AND CANCER PREVENTION

Terpenoids, the largest group of phytochemicals, traditionally used for medicinal purposes in India and China, are currently being explored as anticancer agents in clinical trials. Research has shown that terpenoids in plants increase tumor latency and decrease tumor multiplicity. Terpenoids in various herbs possess strong antioxidant activities. The isoprenoids are useful cancer chemopreventive agents as they suppress tumor growth by inhibiting HMG-CoA reductase.[35]

A large number of terpenoids exhibited cytotoxicity against a variety of tumor cells and also showed cancer preventive as well as anticancer efficacy in preclinical animal models. Epidemiological and experimental studies propose that monoterpenes may be helpful in the prevention and therapy of several cancers, including mammary, skin, lung, forestomach, colon, pancreatic, and prostate carcinomas. A large number of tri- terpenoids have been shown to curb the growth of a variety of cancer cells without exerting any toxicity in normal cells. Numerous preclinical efficacy studies have provided widespread indication that both naturally occurring and synthetic derivatives of triterpenoids possess chemopreventive and therapeutic effects against colon, breast, prostate, and skin cancers.[36] These triterpenoids and their derivatives act at various stages of tumor development; inhibit initiation and promotion of carcinogenesis; induce tumor cell differentiation and apoptosis; and suppress tumor angiogenesis, invasion, and metastasis through regulation of various transcription and growth factors as well as intracellular signaling mechanisms.

Currently, several phase clinical trials have been initiated to evaluate the chemopreventive as well as the anticancer efficacy of a number of triterpenoids.[37] 25-methoxyhispidol A, a novel triterpenoid isolated from the fruit of *Poncirus trifolata*, displayed antiproliferative effects against SK-Hep-1 cells. Apoptosis and G0/G1 phase arrest were suggested as the mechanisms of action. This correlated well with the down-regulation of cyclin D1, CDK4, c-myc, and retinoblastoma protein expressions, along with the upregulation of p21.[38] Zerumbone (ZER), a cytotoxic component isolated from the wild ginger, *Zingiber zerumbet* Smith, induced significant antiproliferative activity against HepG2 cells through mechanisms involving an elevation of the apoptotic process with increase in the level of pro-apoptotic protein Bax and decrease of anti-apoptotic protein Bcl-2 without involving p53.[39] Dietary isoprenic derivatives, including β-ionone, a cyclic isoprenoid present in grapes and wine, represent a promising class of chemopreventive agents (Figure 9.6).[36] β-ionone was found to inhibit hepatic preneoplastic lesions with a decrease in cell proliferation, inhibition of plasma cholesterol, and amelioration of DNA damage during the initial phases of hepatocarcinogenesis initiated with diethylnitrosamine (DENA) and promoted by 2-acetylaminofluorene (2-AAF) in rats.[40]

CHINESE HERBS AND CANCER PREVENTION

Many active compounds have been isolated from Chinese medicinal herbs and have been tested for anticancer effects (Figure 9.7).[41] Artesunate is a semisynthetic derivative of artemisinin, a natural product from the Chinese herb *Artemisia annua* L. exerted antimalarial activity, and, additionally, artemisinin and its derivatives are active against cancer cells. Artesunate induced apoptosis and necrosis and DNA breakage in a dose-dependent manner as shown by single-cell gel electrophoresis. This genotoxic effect was confirmed by

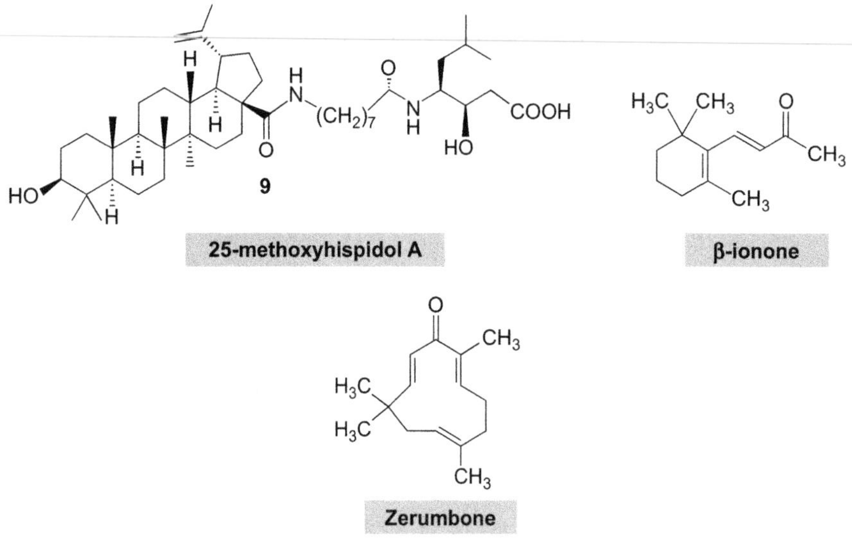

FIGURE 9.6   Structure of Terpenoids—25-Methoxyhispidol A, β-Ionone, and Zerumbone. Numerous preclinical efficacy studies have provided widespread indication that both naturally occurring and synthetic derivatives of triterpenoids possess chemopreventive and therapeutic effects against colon, breast, prostate, and skin cancer. *Source: Liby et al. (2007).*

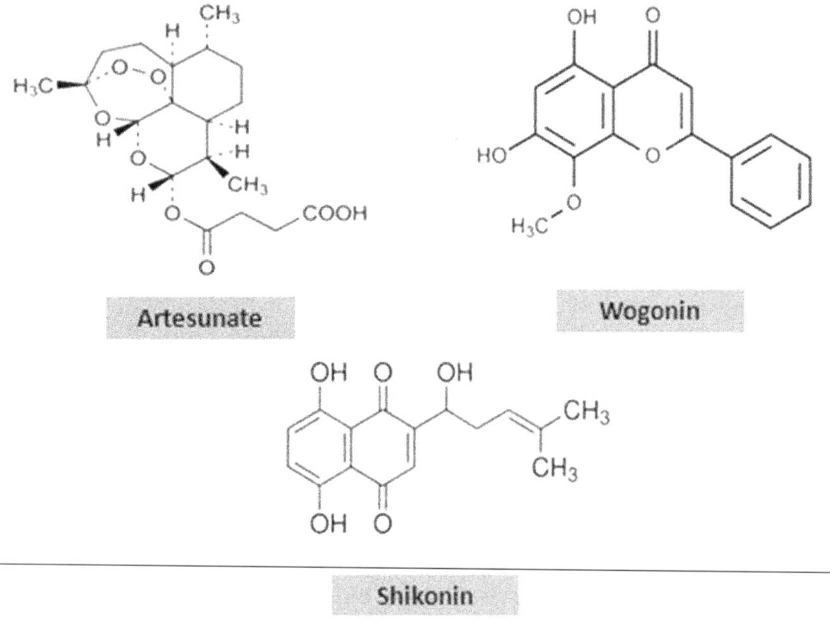

25-methoxyhispidol A

β-ionone

Zerumbone

FIGURE 9.7   Active Compounds in Chinese Medicinal Herbs. Many active compounds have been isolated from Chinese medicinal herbs and have been tested for anticancer effects. *Source: Tan et al. (2011).*

Artesunate

Wogonin

Shikonin

measuring the level of gamma-H2AX, which is considered to be an indication of DNA double-strand breaks (DSB). Polymerase beta-deficient cells were more sensitive than the wild-type to artesunate, indicating that the drug induces DNA damage that is repaired by base excision repair; irs1 and VC8 cells defective in homologous recombination (HR) due to inactivation of XRCC2 and BRCA2, respectively, were more sensitive to artesunate than the corresponding wild-type. This was also true for XR-V15B cells defective in nonhomologous end-joining (NHEJ) due to inactivation of Ku80. The data indicate that DSBs induced by artesunate are repaired by the HR and NHEJ pathways. They suggest that DNA damage induced by artesunate contributes to its therapeutic effect against cancer cells.[42]

Plants of the genus *Taraxacum*, commonly known as dandelions, have a history of use in Chinese, Arabian, and Native American traditional medicine to treat a variety of diseases including cancer. Three aqueous extracts prepared from the mature leaves, flowers, and roots investigated on tumor progression-related processes such as proliferation and invasion showed that the crude extract of dandelion leaf (DLE) decreased the growth of MCF-7/AZ breast cancer cells whereas the aqueous extracts of dandelion flower (DFE) and root (DRE) had no effect on the growth of either cell line. Furthermore, DRE was found to block invasion of MCF-7/AZ breast cancer cells while DLE blocked the invasion of LNCaP prostate cancer cells into collagen type I. Inhibition of invasion was further evidenced by decreased phosphorylation levels of Focal Adhesion Kinase (FAK) and src as well as reduced activities of matrix metalloproteinases, MMP-2 and MMP-9.[43]

Wogonin, one of the flavonoids isolated from *Scutellaria baicalensis* Georgi (Huangqin), a Chinese medicinal herb, with its dry herb weight consisting of up to 0.39 mg/100 mg of wogonin[44] has been extensively used in the management of various inflammatory diseases owing to its inhibition of nitric oxide (NO), prostaglandin E2, and pro-inflammatory cytokines production, as well as reduction of cyclooxygenase-2 (COX-2). Wogonin induces apoptosis through the mediation of Ca^{2+} and/or inhibition of NF-κB, shifting O^{2-} to H_2O_2 to some extent; H_2O_2, in turn, serves as a signaling molecule that activates phospholipase Cg. Wogonin may also directly activate the mitochondrial Ca^{2+} channel uniporter and enhance Ca^{2+} uptake, resulting in Ca^{2+} overload and mitochondrial damage. Wogonin is a good anticancer candidate due to its broad toxicities to various types of tumor cell lines and the low toxicities to normal tissues, as well as the synergistic effects.[45] Shikonin, a natural anthraquinone derivative isolated from the roots of *Lithospermum erythrorhizon* (Zicao), another Chinese herb, exerts antitumor effects mainly by inhibiting cell growth and inducing apoptosis. The fundamental molecular mechanisms differ with cell types and treatment methods. Shikonin modulates an estrogen enzyme by down-regulating the expression of steroid sulfatase, essential for estrogen biosynthesis, and also inhibits tumor invasion via the NF-κB signaling pathway in human high-metastatic adenoid cystic carcinoma cells. Consequently, shikonin may directly or indirectly inhibit or alter disease-related cellular targets in cancer.[46]

SUMMARY POINTS

- Oxidative stress is caused by an imbalance between reactive species and antioxidants and is implicated in a wide range of metabolic and degenerative disorders such as diabetes mellitus, arthritis, Parkinson's disease, Alzheimer's disease, and cancer.
- Cancer is a multistage process that involves three distinguishable interconnected stages: initiation, promotion and progression.
- Naturally occurring antioxidants from plants, phytochemicals, have been proven to have great potential in combating cancer and other chronic diseases that result from oxidative stress induced by free radicals.
- The phytochemicals from various herbs and spices, including polyphenols, flavonoids, flavonols, tannins, and terpenoids, act at various stages of tumor development, inhibit initiation and promotion of carcinogenesis, induce tumor cell differentiation and apoptosis, and suppress tumor angiogenesis.

- Thus, many of the phytochemicals act in different ways and thereby inhibit cancer initiation, promotion, and progression.

References

1. Durackova Z. Some current insights into oxidative stress. *Physiol Res* 2010;**59**:459–69.
2. Hussain SP, Hofseth LJ, Harris CC. Radical causes of cancer. *Nat Rev Cancer* 2003;**3**:276–85.
3. Visconti R, Grieco D. New insights on oxidative stress in cancer. *Curr Opin Drug Disco Dev* 2009;**12**:240–5.
4. Minelli A, Bellezza I, Conte C, Culig Z. Oxidative stress-related aging: A role for prostate cancer? *Biochim Biophys Acta* 2009;**1795**:83–91.
5. Roginsky V, Lissi EA. Review of methods to determine chain breaking antioxidant activity in food. *Food Chem* 2005;**92**:235–8.
6. Lee JH, Ozcelik B, Min DB. Electron donation mechanisms of β-carotene as a free radical scavenger. *J Food Sci* 2003;**68**(3):861–5.
7. Miki W. Biological function and activity of animal carotenoids. *Pure Appl Chem* 1991;**63**:141–6.
8. Cao G, Booth SL, Sadowski JA, Prior RL. Increases in human plasma antioxidant capacity after consumption of controlled diets high in fruits and vegetables. *Am J Clin Nutr* 1998;**68**:1081–7.
9. Willett WC. Diet, nutrition and avoidable cancer. *Environ Health Perspect* 1995;**103**(suppl. 8):165–70.
10. Liu RH. Potential synergy of phytochemicals in cancer prevention: Mechanism of action. *J Nutr* 2004;**134**:3479S–85S.
11. Scott K. How spices inhibit disease processes. *Medicinal seasonings*. South Africa: Medspice Press; 2006. p. 38–52.
12. Trichopoulos D, Willett WC. Nutrition and cancer. *Cancer Causes Control* 1996;**7**:1–180.
13. Anand P, Kunnumakara AB, Sundaram C, Harikumar BK, Tharakan ST, Lai OS, et al. Cancer is a preventable disease that requires major lifestyle changes. *Pharm Res* 2008;**25**:2097–116.
14. Kikuzaki H, Nakatani N. Antioxidant effects of some ginger constituents. *J Food Sci* 1993;**58**:1407–10.
15. Milner JA. Garlic: Its anticarcinogenic and antitumorigenic properties. *Nutr Rev* 1996;**54**:S82–6.
16. Du B, Jiang L, Xia Q, Zhong L. Synergistic inhibitory effects of curcumin and 5-fluorouracil on the growth of the human colon cancer cell line HT-29. *Chemotherapy* 2006;**52**:23–8.
17. Giles RH, van Es JH, Clevers H. Caught up in a Wnt storm: Wnt signaling in cancer. *Biochim Biophys Acta* 2003;**1653**:1–24.
18. Thangapazham RL, Sharma A, Maheshwari RK. Multiple molecular targets in cancer chemoprevention by curcumin. *AAPS J* 2006;**8**(3):E443–9.
19. Bruneton J. Pharmacognosy, phytochemistry, medicinal plants. In: Hatton CK, editor. *Translation of Pharmacognosie*. Paris: Lavoisier Publishers; 1995. p. 607–8.
20. Feng R, Lu Y, Bowman LL, Qian Y, Castranova V, Ding M. Inhibition of activator protein-1, NF-kappaB, and MAPKs and induction of phase 2 detoxifying enzyme activity by chlorogenic acid. *J Biol Chem* 2005;**280**:27888–95.
21. Yang CS, Landau JM, Huang MT, Newmark HL. Inhibition of carcinogenesis by dietary polyphenolic compounds. *Annu Rev Nutr* 2001;**21**:381–406.
22. Rahman I, Biswas SK, Kirkham PA. Regulation of inflammation and redox signaling by dietary polyphenols. *Biochem Pharmacol* 2006;**72**:1439–52.
23. Elangovan V, Ramamoorthy N, Balasubramanian S. Studies on the antiproliferative effect of some naturally occurring bioflavonoidal compounds against human carcinoma of larynx and sarcoma-180 cell lines. *Indian J Pharmacol* 1994;**26**:266–9.

24. Rao KN, Bhattacharya RK, Venkatachalam SR. Inhibition of thymidine synthase and cell growth by the phenanthroindolizidine alkaloids pergularinine and tylophorinidine. *Chem Biol Interact* 1997;**106**:201–12.

25. Tripathi YB, Sharma M, Manickam M. Rubidianin, a new antioxidant from *Rubia cordifolia*. *Indian J Biochem Biophys* 1997;**34**:523–6.

26. Pidgeon GP, Kandouz M, Meram A, Honn KV. Mechanisms controlling cell cycle arrest and induction of apoptosis after 12-lipoxygenase inhibition in prostate cancer cells. *Cancer Res* 2002;**62**:2721–7.

27. Belman S, Solomon J, Segal A, Block E, Barany G. Inhibition of soybean lipoxygenase and mouse skin tumor promotion by onion and garlic components. *J Biochem Toxicol* 1989;**4**:151–60.

28. Sendl A, Elbl G, Steinke B, Redl K, Breu W, Wagner H. Comparative pharmacological investigations of *Allium ursinum* and *Allium sativum*. *Planta Med* 1992;**58**:1–7.

29. Khanum F, Anilakumar KR, Sudarshana Krishna KR, Viswanathan KR, Santhanam K. Anticarcinogenic effects of curry leaves in dimethylhydrazine-treated rats. *Plant Foods Hum Nutr* 2000;**55**:347–55.

30. Abdullah S, Abidin SAZ, Murad NA, Makpol S, Ngah WZW, Yusof YAM. Ginger extract (*Zingiber officinale*) triggers apoptosis and G0/G1 cells arrest in HCT 116 and HT 29 colon cancer cell lines. *Afr J Biochem Res* 2010;**4**:134–42.

31. Yoshimi N, Wang A, Morishita Y, Tanaka T, Sugie S, Kawai K, et al. Modifying effects of fungal and herb metabolites on azoxymethane-induced intestinal carcinogenesis in rats. *J Cancer Res* 1992;**83**:1273–8.

32. Nair SC, Pannikar B, Pannikar KR. Antitumour activity of saffron (*Crocus sativus*). *Cancer Lett* 1991;**57**(2):109–14.

33. Gagandeep S, Dhanalakshmi S, Mendiz E, Rao AR, Kale RK. Chemopreventive effects of *Cuminum cyminum* in chemically induced forestomach and uterine cervix tumors in murine model systems. *Nutr Cancer* 2003;**47**(2):171–80.

34. Nalini N, Manju V, Menon VP. Effect of spices on lipid metabolism in 1,2-dimethylhydrazine-induced rat colon carcinogenesis. *J Med Food* 2006;**9**(2):237–45.

35. Yu SG, Abuirmeilah NM, Quershi AA, Elson CE. Dietary betaionone suppresses hepatic 3-hydroxy-3-methylglutaryl coenzyme A reductase activity. *J Agric Food Chem* 1994;**42**:1493–6.

36. Liby KT, Yore MM, Sporn MB. Triterpenoids and rexinoids as multifunctional agents for the prevention and treatment of cancer. *Nat Rev Cancer* 2007;**7**:357–69.

37. Rabi T, Gupta S. Dietary terpenoids and prostate cancer chemoprevention. *Front Biosci* 2008;**13**:3457–69.

38. Hong J, Min HY, Xu GH, Lee JG, Lee SH, Kim YS, et al. Growth inhibition and G1 cell cycle arrest mediated by 25-methoxyhispidol A, a novel triterpenoid, isolated from the fruit of *Poncirus trifoliata* in human hepatocellular carcinoma cells. *Planta Med* 2008;**74**:151–5.

39. Sakinah SA, Handayani ST, Hawariah LP. Zerumbone induced apoptosis in liver cancer cells via modulation of Bax/Bcl-2 ratio. *Cancer Cell Int* 2007;**7**:4–6.

40. de Moura Espíndola R, Mazzantini RP, Ong TP, de Conti A, Heidor R, Moreno FS. Geranylgeraniol and beta-ionone inhibit hepatic preneoplastic lesions, cell proliferation, total plasma cholesterol and DNA damage during the initial phases of hepatocarcinogenesis, but only the former inhibits NF-kappa B activation. *Carcinogenesis* 2005;**26**:1091–9.

41. Tan W, Lu J, Huang M, Li Y, Chen M, Wu G, et al. Anti-cancer natural products isolated from Chinese medicinal herbs. *Chinese Med* 2011;**6**:1–15.

42. Li PC, Lam E, Roos WP, Zdzienicka MZ, Kaina B, Efferth T. Artesunate derived from traditional Chinese medicine induces DNA damage and repair. *Cancer Res* 2008;**68**:4347–51.

43. Sigstedt SC, Hooten CJ, Callewaert MC, Jenkins AR, Romero AE, Pullin MJ, et al. Evaluation of aqueous extracts of *Taraxacum officinale* on growth and invasion of breast and prostate cancer cells. *Int J Oncol* 2008;**32**:1085–90.

44. Li C, Zhou L, Lin G, Zuo Z. Contents of major bioactive flavones in proprietary traditional Chinese medicine products and reference herb of *radix scutellariae*. *J Pharm Biomed Anal* 2009;**50**:298–306.

45. Baumann S, Fas SC, Giaisi M, Müller WW, Merling A, Gülow K, et al. Wogonin preferentially kills malignant lymphocytes and suppresses T-cell tumor growth by inducing PLCγ1- and Ca^{2+}-dependent apoptosis. *Blood* 2008;**111**:2354–63.

46. Min R, Zun Z, Min Y, Wenhu D, Wenjun Y, Chenping Z. Shikonin inhibits tumor invasion via down-regulation of NF-kappaB-mediated MMP-9 expression in human ACC-M cells. *Oral Dis* 2011;**17**:362–9.

The Indian Blackberry (Jamun), Antioxidant Capacity, and Cancer Protection

Farrukh Aqil, Radha Munagala, Jeyaprakash Jeyabalan, Thwisha Joshi, Ramesh C. Gupta

James Graham Brown Cancer Center and Department of Medicine, University of Louisville, Louisville, KY, USA

Inder P. Singh

National Institute of Pharmaceutical Education and Research, Punjab, India

INTRODUCTION

The adverse effects of oxidative stress on human health have become a serious issue. It has been known for some time that most degenerative diseases are associated with reactive oxygen species (ROS) such as superoxide anion radicals, hydroxyl radicals, and hydrogen peroxide.[1] Under stress conditions, our bodies produce more ROS than enzymatic and nonenzymatic antioxidants. The imbalance can ultimately lead to the development of degenerative diseases including cardiovascular diseases, cancers, neurodegenerative, and other inflammatory diseases.[2] There are endogenous as well as exogenous systems to protect cells from damage in animals. However, in many conditions the endogenous antioxidants are depleted and therefore exogenous supply of antioxidants becomes essential. Antioxidants play an important role in biological systems by suppressing the formation of ROS by reducing hydroperoxides ($ROO\bullet$) and H_2O_2, and scavenging free radicals among others.

Epidemiological data suggest that intake of antioxidants via increased consumption of fruits and vegetables contributes toward a reduced risk of certain types of cancers.[3] Plants have been the basis of traditional medicines throughout the world for thousands of years and continue to provide new remedies. Medicinal plants have been the subject of intense research due to their potential as sources of viable drugs or as lead compounds in drug development. A great deal of effort therefore has focused on using available experimental techniques to identify natural antioxidants from plants.

The upcoming antioxidants are generally comprised of different phenolics, flavonoids, isoflavones, flavones, and anthocyanins rather than the traditional vitamins C, E, and β-carotene.[4]

Dark-colored fruits, particularly berries, are highly protective because of their antioxidant, antiproliferative, and anti-inflammatory activities.[5,6] Because of their much higher antioxidant activity compared with other common fruits and vegetables, berries are beginning to gain much attention lately. The chemopreventive potential of berry bioactives could be mediated, at least in part, through their abilities to combat oxidative stress and inflammation.[6]

Syzygium cumini L. (family, Myrtaceae), commonly known as jamun or Indian blackberry, is native to India and is cultivated throughout the Asian subcontinent, Eastern Africa, South America, and Madagascar, and has also naturalized to Florida and Hawaii in the United States.[5,7,8] The fruits of this plant are oblong berries, deep purple or bluish in color with pinkish pulp (Figure 10.1A), and are widely consumed as a fruit particularly in India. In association to its dietary use, all parts of the tree (specifically the seeds) are used for the treatment of various diseases. Jamun seed is well known for its antidiabetic, antiscorbutic, diuretic, and antidiarrheal properties.[8-10] Studies in the past decade have shown jamun to possess antioxidant, antineoplastic,[5,7] radioprotective,[11] and chemopreventive effects. These protective properties have been attributed to the presence of diverse phytochemicals like anthocyanins, flavonoids, and terpenes of the jamun fruit.

Anthocyanins are brightly colored compounds responsible for much of the red, blue, and purple colors in

Cancer
http://dx.doi.org/10.1016/B978-0-12-405205-5.00010-6

© 2014 Elsevier Inc. All rights reserved.

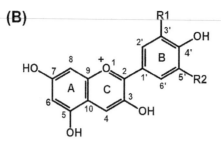

FIGURE 10.1 The photographs of Indian blackberry jamun fruits and seeds (A). Structures of various anthocyanidins in jamun fruit (B). *Photographs of jamun fruits were kindly provided by Dr. Manjeshwar S. Baliga.* (See the color plate.)

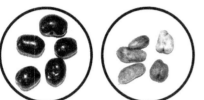

	R$_1$	R$_2$
Delphinidin	-OH	-OH
Cyanidin	-OH	-H
Peonidin	-OCH$_3$	-H
Petunidin	-OH	-OCH$_3$
Malvidin	-OCH$_3$	-OCH$_3$

fruits, vegetables, and ornamental crops. The anthocyanins found in foods are glycosides of mainly six aglycone anthocyanidins: delphinidin (Dp), cyanidin (Cy), petunidin (Pt), peonidin (Pe), malvidin (Mv), and pelargonidin (Pg). Jamun pulp contains all these anthocyanins, except Pg (Figure 10.1B). These compounds have been demonstrated to possess antioxidant, anti-inflammatory, anticancer, antidiabetic, cardioprotective, and neuroprotective effects.[5,12] In addition, anthocyanidins have been shown to affect xenobiotic-metabolizing enzymes, various transcription and growth factors, inflammatory cytokines, and subcellular signaling pathways of cancer cell proliferation, apoptosis, and tumor angiogenesis.[13,14] Recent reports indicate an inverse relationship between the dietary intake of antioxidant-rich foods and the incidence of human disease.[15]

In this chapter we discuss the method of extraction, phytochemical profile, antioxidant and anticancer properties of jamun as well as its major bioactive constituents, anthocyanins and ellagic acid.

EXTRACTION AND CHEMICAL COMPOSITION OF JAMUN PULP

Evidence continues to emerge suggesting the importance of anthocyanins for human health. Jamun is one of the richest sources of anthocyanins and ellagic acid/ellagitannins. The fruit pulp of jamun has been reported to contain glycones of five major anthocyanidins (Dp, Cy, Pt, Pe, and Mv) whereas its seed contains high amounts of ellagitannins and ellagic acid and virtually no anthocyanins.

Extraction methods for preparation of enriched extracts or isolation of pure compounds from berries can influence phytochemical profile. Various methods have been reported but most methods are laborintensive and lack reproducibility. We reported an efficient method for preparation of enriched extracts of the jamun pulp and seed using two stationary phases combined (XAD-761 and HP-20), which provided enhanced purity with greater yield of the anthocyanins and other phenolics.[5] A detailed schematic diagram for the extraction procedure is shown in Figure 10.2. Briefly, the method involved repeated extractions of jamun pulp and seed powder separately with acidified ethanol, followed by concentration using rotavapor, removal of free sugars and enrichment of phenolics by XAD-761/HP-20 column chromatography. The conversion of anthocyanins to anthocyanidins and ellagitannins to ellagic acid was achieved by hydrolysis with five volumes of 2N HCl, concentration by rotavapor, and finally purification by C18 column chromatography, which resulted in enriched anthocyanidins and ellagic acid, as confirmed by HPLC.[5] This approach of using a mixture of XAD-761/diaion HP-20 served as an effective means to enrich both anthocyanins and ellagitannins along with other polyphenolics. The pulp powder was found to contain 0.54% anthocyanins, 0.17% ellagic acid/ellagitannins, and 1.15% total phenolics. In contrast, jamun seed powder contained no detectable anthocyanins, but had higher amounts of ellagic acid/ellagitannins (0.5%) and total polyphenolics (2.7%). Anthocyanin content in the jamun pulp powder was higher (0.54%) than the content (0.23%) reported

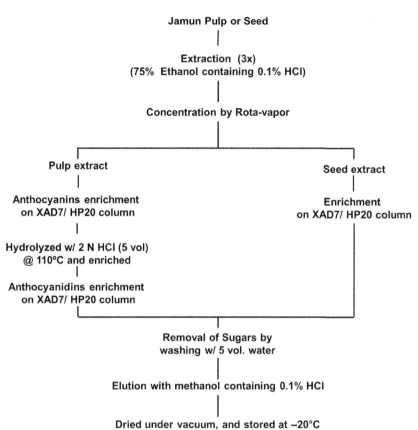

FIGURE 10.2 Scheme for the enrichment of antho-cyanins/anthocyanidins and other phenolics from pulp and seeds of jamun fruit. *Partially adopted from [5]*

TABLE 10.1 Phytochemicals of Enriched Jamun Seeds and Pulp Extracts

Dry Extract	Total Anthocyanins (mg/gm)	Total Phenolics (mg GAE/ gm)	Free EA
Pulp nonhydrolyzed	47.1 (4.7%)	142 (14.2%)	1.2%
Pulp hydrolyzed	45.7 (4.6%)	267 (26.7%)	3.8 %
Seed nonhydrolyzed	ND*	471 (47.1%)	13.8%
Seed hydrolyzed	ND*	483 (48.3%)	30.4%

by other researchers.[8] However, after hydrolysis, the enriched pulp extract yielded 0.23% of anthocyanidins. Further, HPLC profiles of the hydrolyzed samples suggest that while seed contains an abundance of ellagic acid/ellagitannins, pulp of the fruit contains ellagic acid/ellagitannins in addition to the anthocyanidins: Dp (20.3%), Cy (6.6%), Pt (24.2%), Pe (2.2%), and Mv (44.4%) (Figures 10.3 and 10.4).

Veigas et al.[8] described similar methodology with the use of Amberlite XAD-7 for the isolation of anthocyanins pigments from jamun pulp; however, these authors

reported only three anthocyanins as di-glucosides of Dp, Pt, and Mv by HPLC-ESI-MS.

Seeram and coworkers have shown the enrichment of jamun fruit pulp (Source: Gujarat, India) by successive extraction in cold hexane, followed by ethyl acetate and then methanol containing 0.1% hydrochloric acid. The methanol [~40% (w/w)] extract was enriched in anthocyanin content by XAD-16 Amberlite column chromatography. The final elution with acidic methanol (0.1% HCl) yielded enriched extract that contained 3.6% anthocyanins as cyanidin-3-glucoside equivalent, and as di-glucosides of five major anthocyanidins.[7]

Unlike jamun pulp, jamun seed is rich in ellagic acid/ellagitannins in addition to several unidentified polyphenolics (Table 10.1; Figure 10.3).[5] In addition, jamun seed has been reported to contain jambosine, gallic acid, ellagic acid, corilagin, 3,6-hexahydroxy diphenoylglucose, 1-galloylglucose, 3-galloylglucose, quercetin, β-sitoterol and 4,6-hexahydroxydiphenoylglucose.[10,16] De Brito et al.[17] showed the presence of di-glucosides of Dp, Cy, Pt, Pe, and Mv with the cumulative recovery of 0.77% in jamun pulp (Source: Ceará State, Brazil). In another study, Li et al. analyzed the chemical composition of freeze-dried jamun pulp and demonstrated that it contained di-glucosides of Dp, Cy, Pt, Pe, and Mv consistent with other reports.[7]

Vijayanand et al.[18] studied the volatile compounds of jamun by GC–MS analysis, which showed the presence of

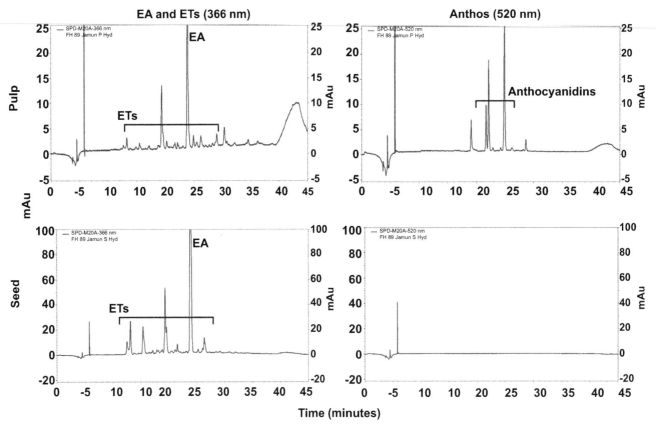

FIGURE 10.3   HPLC profiles of jamun pulp and seeds enriched in anthocyanidins and other phenolics. Hydrolyzed and enriched extracts (20 μl; 1 mg/ml) were injected on a reverse phase C18 (250 × 4.6 mm ID × 5 μm) column in a gradient of acetonitrile and 3.5% aqueous phosphoric acid and analyzed using photo diodarray detector in the range of 210–550 nm. Chromatograms in the first column represent ellagitannins (ETs) and free ellagic acid (EA) monitored at 366 nm. Chromatograms in the second column show anthocyanidins. mAu, milliabsorbance units.

almost 30 compounds. Essentially all these compounds were identified as a volatile mixture of *trans*-ocimene, *cis*-ocimene, β-myrcene, and α-terpineol as major entities. The mixture also included three esters, namely, dihydrocarvyl acetate, geranyl butyrate, and terpinyl valerate suggested to be responsible for the characteristic flavor of the jamun fruit. The complete phytochemical profile of jamun is shown in Table 10.2 and Figures 10.3 and 10.4.

Comparison of phytochemical profiles of various berries suggest that blueberry and bilberry also contain the same spectra of five anthocyanidins (Dp, Cy, Pt, Pe, and Mv) as jamun pulp; however, unlike jamun pulp, these berries contain only small quantities (<100 ppm) of ellagic acid/ellagitannins. On the other hand, blackberry and black raspberry contain Cy as essentially the only anthocyanidin along with high content of ellagic acid/ellagitannins (80–700 and 850 ppm, respectively), and strawberry contains Pg along with Cy and significant amounts of ellagic acid/ellagitannins (200 ppm).[19, 20] Thus, jamun appears to be the only berry that contains the diversity of five anthocyanidins along with an abundance of ellagic acid/ellagitannins.[5] In view of the known

anticarcinogenicity potential of these anthocyanidins in blueberry, and Cy together with ellagic acid/ellagitannins in black raspberry against estrogen-mediated breast cancer[21,22], this composition of anthocyanidins and ellagic acid/ellagitannins in jamun could potentially provide additive or synergistic effects.

ANTIOXIDANT POTENTIAL OF JAMUN AND ITS COMPONENTS

The array of antioxidants present in whole fruits and vegetables is considered to elicit higher activity than any individual constituents [23]. Antioxidant properties such as radical scavenging activity are important to counter the deleterious effects of free radicals in foods and in biological systems. Excessive formation of free radicals causes oxidation of lipids in foods, thereby decreasing quality of foods. On the other hand, free radicals generated in biological systems lead to stress and diseases. Hence, exploration for sources of natural plant antioxidants is crucial. In this section, the antioxidant potential of jamun extracts and its active constituents is discussed with special attention to anthocyanidins and EA.

TABLE 10.2 Phytochemical Profile of *Syzygium Cumini*

Plant Part	Phytochemical
Fruit pulp	Anthocyanins, diglucosides of delphinidin, petunidin, malvidin, peonidin, and cyanidin. Volatile oils (α-pinene, ß-pinene, ß-myrcene, cis-ocimene, trans-ocimene, terpinolene, linalool, 4-terpineol, α-terpineol, cis-dihydrocarvone, caryophyllene, α-humelene, cis-ß-farnesene, cis-α-farnesene, trans-α-farnesene, cis-nerolidol, geranyl butyrate, globulol, widdrol, torreyol, neocedranol, ß-bisabolol).[5,7,8,18]
Seeds	Ellagitannins, Jambosine, gallic acid, ellagic acid, corilagin, 3, 6-hexahydroxy diphenoylglucose, 1-galloylglucose, 3-galloylglucose, quercetin, β-sitoterol, and 4,6-hexahydroxydiphenoylglucose.[5,10,16]
Stem bark	Friedelin, friedelan-3-α-ol, betulinic acid, β-sitosterol, kaempferol, β-sitosterol-D-glucoside, gallic acid, ellagic acid, gallotannin, ellagitannin, and myricetine.[10,16]
Leaves	β-sitosterol, betulinic acid, mycaminose, crategolic (maslinic) acid, n-hepatcosane, n-nonacosane, n-hentriacontane, noctacosanol, n-triacontanol, n-dotricontanol, quercetin, myricetin, myricitrin and the flavonol glycosides myricetin 3-O-(4"-acetyl)-α- Lrhamnopyranosides. Essential oils (α-terpeneol, myrtenol, eucarvone, muurolol,α-myrtenal, 1, 8-cineole, geranyl acetone, α-cadinol, pinocarvone).[10,16,60]
Flowers	Oleanolic acid, ellagic acids, isoquercetin, quercetin, kampferol, and myricetin.[10]

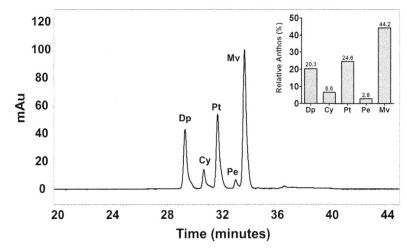

FIGURE 10.4 HPLC chromatogram of jamun anthocyanidins at 520 nm extracted from fruit pulp. Twenty μl of hydrolyzed pulp extract (1 mg/ml) was injected and analyzed using photo diodarray detector on a C18 reverse phase column in a gradient of 3.5% aqueous phosphoric acid and acetonitrile for 61 min. Bar diagram in inset represents relative (%) anthocyanidin content. Please note that solvent gradient used here was different than the gradient used in Figure 10.3 (see text). *Adapted with permission from [5]. Reprinted by permission of the publisher (Routledge, Taylor & Francis Ltd,* http://www.tandf.co.uk/journals).

There are several assays to measure the antioxidant potential of compounds, the 1,1-diphenyl-2-picrylhydrazine (DPPH) radical scavenging assay being the most extensively used antioxidant assay for plant samples. DPPH is a stable free radical that reacts with compounds able to donate a hydrogen atom. Numerous studies in the recent past have shown that the jamun fruit, seed, leaves, and bark possess antioxidant and free radical scavenging effects. Jamun pulp has high levels of Cy- and Dp- pigments that have a catechol (*ortho*-dihydroxyl) group and a pyrogallol (vicinal trihydroxyl) group, respectively. These structural features presumably contribute significantly to radical scavenging capacity of these compounds [23].

In search of the antioxidant compounds, Gracia-Alonso[24] analyzed 28 fruits and demonstrated that the fruits with greater antioxidant activity were all rich in anthocyanins, suggesting that these pigments could be contributing to the antioxidant activity. Anthocyanins

not only have free radical-scavenging properties, but also are observed to have high nitric oxide scavenging, ferric reducing power, and hydroxyl as well as superoxide radical scavenging activities as reviewed.[9,25]

The extracts of seeds, leaves, and stem bark of jamun have also been observed to be free radical scavengers in the DPPH and other scavenging assays.[26] Veigas et al. tested the anthocyanins-enriched extracts of jamun and demonstrated the antioxidant capacity of the extract using DPPH scavenging, ferric reducing power, and lipid peroxidation in rat brain.[8] The extract showed 78.2% DPPH scavenging at 2.5 ppm, while butylated hydroxylanisole (BHA) exhibited only 41.6% activity at the same concentration, thus proving anthocyanins to be more efficient scavengers of free radicals than well-known synthetic antioxidant BHA. The authors also demonstrated that 1 ppm of the extract was equivalent to 3.5 μM ascorbic acid, as estimated by reducing power assay. Nearly 94% inhibition of rat brain lipid peroxidation was achieved at 5 ppm. In another study Bajpai et al. observed that the hydromethanolic (50:50 v/v) extract of jamun seed was effective in scavenging (90.6%) free radicals as evaluated in the auto-oxidation of β-carotene and linoleic acid assays.[27]

The anthocyanins are glycosides of the anthocyanidins. The sugar that is usually bound to these anthocyanidins is glucose, arabinose, or galactose. We have recently demonstrated that anthocyanidins have comparatively higher antioxidant activity than anthocyanins (Figure 10.5). When analyzed in DPPH assay, the antioxidant activity was lower in jamun extract at anthocyanin level (25% reduction) compared with jamun extract at anthocyanidins level (46% reduction). However, ellagic acid/ellagitannins-enriched extract of jamun seeds showed higher activity compared to pulp extract.[5] Similarly, the concentrations of extracts that scavenged 50% of ABTS radicals during the 30-min incubation were in the following ascending order: seed extract (4.6 μg/ml), anthocyanidin-enriched pulp extract (11.8 μg/ml), and anthocyanin-enriched pulp extract (17.2 μg/ml). Similar trends were observed in ABTS radical scavenging assay and oxygen radical absorbance (ORAC) assay.[5]

Banerjee and colleagues demonstrated that the skin of jamun possesses antioxidant effects as demonstrated by the hydroxyl radical, superoxide radical, DPPH radical scavenging assays, and lipid peroxidation.[28] The presence of gallic acid, a strong antioxidant, has also been shown in the jamun extract. Antioxidant activity of gallic acid has been reported to be higher than the vitamin C and other phenolic constituents such as quercetin, epicatechin, catechin, rutin, and chlorogenic acid.[29] We have found high levels of ellagic acid in

jamun pulp as well as seed extracts. Ellagic acid has been studied extensively and has been shown to have strong antioxidant potential *in vitro* and *in vivo*.[30-32]

Jamun, because of its high levels of anthocyanins and ellagic acid/ellagitannins, is an ideal antioxidant fruit for human consumption and could help greatly to lower the risks of cancer and other degenerative diseases.

ANTIOXIDANT ACTIVITY OF INDIVIDUAL ANTHOCYANIDINS

As noted earlier, berry extracts and their individual constituents show great antioxidant activity in different systems including DPPH, ABTS, and H_2O_2 free radicals scavenging, ferric reducing power, and trolox equivalent antioxidant activity (ORAC) as reviewed.[25] In our laboratory, we have analyzed the five anthocyanidins (Dp, Cy, Pt, Pe, and Mv) present in jamun pulp for their antioxidant potential; the findings are discussed in the following.

Figure 10.6A illustrates a dose-dependent decrease in the concentration of DPPH radical due to the scavenging activity of the five major anthocyanidins. The scavenging activity of these anthocyanidins was in the following descending order: Dp > Pt > Cy > Pe > Mv, with an EC_{50} of 15.5, 21, 45, 46, 61, and 65 μg/ml. The trend was similar with respect to H_2O_2 scavenging effects (Figure 10.6B). Anthocyanidins also showed a concentration-dependent scavenging of ABTS radical (Figure 10.6C) with Dp and Cy (EC_{50} of 5.1 and 5.2 μg/ml) being most effective, followed by Pt, Pe, and Mv (EC_{50} 6–10 μg/ml). The higher antioxidant activity was dependent on the number of hydroxyl groups in the flavanoid structure of the anthocyanidins (Figure 10.1B). Previous studies have observed the effect of hydroxylation and methoxylation in anthocyanidins on radical scavenging ability in aqueous phase. Anthocyanidins that lack in *O*-diphenyl structure such as Mv, Pe, and Pt have lower DPPH scavenging activity compared to Dp or Cy. Cy and Dp pigments have an *ortho*-dihydroxyl group (catechol) and a vicinal trihydroxyl group (pyrogallol), respectively. These structural features have been shown associated with significant radical scavenging capacity of these compounds.[23]

However, Wang et al. did not observe any significant antioxidant activity of anthocyanidins related to the presence of hydroxyl groups.[33] Seyoum et al. suggested that the activity of flavonoids was dependent mainly on the position and/or pattern of hydroxylation rather than the number of hydroxyl groups, which could explain the activity behind another

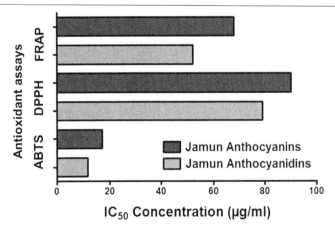

FIGURE 10.5 Antioxidant potential of anthocyanin- and anthocyanidin-enriched jamun pulp extracts. Horizontal bars represent the concentrations of extracts that demonstrate 50% of activity.

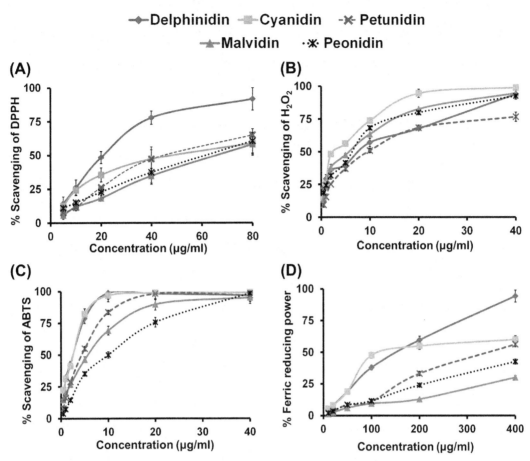

FIGURE 10.6 Antioxidant potential of individual anthocyanidins from *Syzygium cumini* in different assays: 2,2-diphenyl-1-picrylhydrazyl free radical scavenging activity (A); hydrogen peroxide scavenging activity (B); 2,2'-azino-bis(3-ethylbenzthiazoline-6-sulphonic acid) (ABTS) (C); scavenging activity and ferric ion reducing power (D). Each data point represents the mean of 3 experiments with standard deviation.

anthocyanidin, pelargonidin, with two hydroxyl groups.[34]

The reducing capacity of anthocyanidins was assayed using the FRAP method,[35] with modifications.[5] The FRAP method is a simple, accurate, and direct method for assessing total antioxidant activity of a given system, as no activity-changing interactions between antioxidants occur in the system.[36] Once again, the reducing capacity for anthocyanidins increased with the presence of hydroxyl groups in ring B (Cy ≈ Dp > Pt) whereas compounds lacking hydroxyl groups such as Pe and Mv showed no activity at the highest dose of 400 µg/ml (Figure 10.6D).

The ORAC assay measures specifically the ability of compounds to scavenge oxygen free radicals and as such is considered to be the closest to human physiology. As shown in Figure 10.7, the average antioxidant activities of the five anthocyanidins in µmol TE/g were Cy (6632) > Pt (4338) > Dp (3277) > Pe (2105) > Mv (1773). These findings clearly suggest that the individual anthocyanidins, which are bioactives of jamun and other berries, are potent antioxidants and could provide much required protection against insults such as oxidative stress and cellular damage.

JAMUN EXTRACTS INHIBIT OXIDATIVE DNA DAMAGE

Jamun polyphenolics have been shown to work as potential antioxidants as discussed earlier via different mechanisms, including direct scavenging of free radicals, reactive nitrogen species, and ROS that include hydroxyl and peroxyl radicals. We previously demonstrated by co-chromatography that the profile of polar oxidative adducts generated *in vitro* using either 4-hydroxy-estradiol (4-OHE$_2$)/CuCl$_2$ or H$_2$O$_2$/CuCl$_2$ was similar,[37] indicating that adducts in 4-OHE$_2$ and CuCl$_2$ reaction originated from oxidative mechanisms.

The potentials of jamun (pulp) aqueous and organic extracts were evaluated for their efficacy to inhibit 4-OHE$_2$/Cu^{2+}-induced oxidative DNA adducts in this

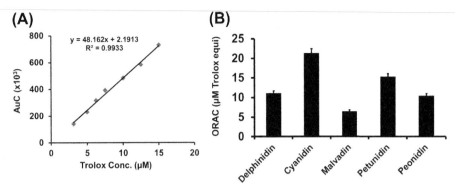

FIGURE 10.7 Oxygen radical absorbance capacity (ORAC) of individual anthocyanidins from *Syzygium cumini*. A: Line graph represents the linear plot of area under curve (AUC) versus different concentrations of trolox. B: ORAC values are equivalent to trolox (μM/g of extracts).

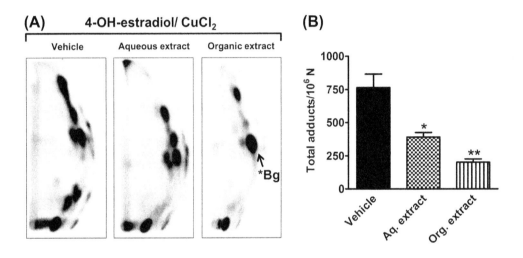

*Bg = Background spot

FIGURE 10.8 Representative autoradiographs of ^{32}P-postlabeled oxidative DNA adducts generated from Cu^{2+} (30 μM) catalyzed by redox cycling of 4-hydroxy estradiol (30 μM) in the presence of vehicle (1% DMSO), aqueous and organic extract of jamun pulp. Labeled adducts were separated by 2-directional PEI-cellulose thin-layer chromatography. (B) Graphical representation of the adduct levels at the indicated concentrations. Aqueous and organic extracts showed significant reduction in adduct levels (p = 0.016 and 0.0019, respectively). *Bg, background spot.

system. Although the data for the interaction of anthocyanins/anthocyanidins is scanty in regard to direct interaction with DNA, one of the jamun's active constituents, EA, has been shown to protect DNA from oxidative damage by covalent binding.[38] The other protective mechanisms include inhibition of ROS production and chelation of metal ions such as copper and iron.

Analysis of DNA damaged by the Cu^{2+}-catalyzed redox cycling of 4-OH-E_2 revealed several oxidative DNA adducts (Figure 10.8A). These adducts have been assigned as unidentified dinucleotides formed by an attack of hydroxyl radicals.[37] The formation of these adducts was significantly diminished in the presence of both organic (p = 0.0019) and aqueous (p = 0.016) extracts of jamun pulp, respectively (765 ± 226 versus 391 ± 69 and 203 ± 46 per 10^6 normal nucleotides; Figure 10.8B). Results from this study correlate well with previous studies from this laboratory, in which aqueous extracts of 0.5% each of mixed berries (strawberry, blueberry, blackberry, and red and black raspberry) or EA significantly reduced unidentified polar adduct burden induced by 4-OHE$_2$/CuCl$_2$.[30,39]

ANTICANCER EFFECTS OF JAMUN BIOACTIVES

Natural products have historically been a rich source of cancer drugs. Studies suggest that of all the antineoplastic drugs used, nearly one half of them originated from natural sources.[25] Chemoprevention refers to the use of agents to prevent, inhibit, delay, reverse, or retard tumorigenesis. Numerous phytochemicals derived from edible and medicinal plants have been reported to interfere with a specific stage of carcinogenic progressions. Commonly used drugs like vincristine, vinblastine, taxol, docetaxel, teniposide, etoposide, and campatothecin are all derived from plants.[40] Unfortunately, several of these chemotherapeutic compounds possess severe side effects by affecting the normal cells, thereby necessitating search for novel effective but nontoxic agents.

Berries are an excellent source of various bioactive polyphenolics, including anthocyanins, proanthocyanidins, ellagitannins, flavonols, and phenolic acids. The effects of individual berry bioactives on growth and apoptosis of human cancer cell lines have also been investigated. The bioactives from berries are reported to have many roles in cancer prevention including inhibition of the formation of carcinogen-induced DNA damage, protection against oxidative DNA damage, inhibition of carcinogen-induced tumorigenesis, and modulation of signaling pathways involved with cellular proliferation, inflammation, and cell cycle arrest.[14] In addition, berry bioactives also regulate xenobiotic-metabolizing enzymes, various transcription and growth factors, inflammatory cytokines, subcellular signaling pathways of apoptosis, and tumor angiogenesis. In the following we have summarized studies that have determined the mechanisms by which jamun and its constituents exert anticancer effects.

Studies indicate the potential of berries in cancer prevention and treatment. A series of papers from the Stoner group have shown that black raspberry and blackberry are highly effective against esophageal and colon cancer as reviewed.[41,42] Our own studies with black raspberry and blueberry have shown significant protection against estrogen-mammary carcinogenicity[21,22] as well as cigarette smoke-mediated lung cancer in mouse models.[43] Berry phytochemicals were suggested to sensitize tumor cells to chemotherapeutic agents by inhibiting pathways that lead to drug resistance and provide protection from therapy-associated toxicities.[44,45]

In a study that has wide clinical implications, Li et al.[7] demonstrated that the standardized jamun fruit extract possesses antiproliferative and proapoptotic effects in estrogen-dependent/aromatase positive (MCF-7aro), and estrogen-independent (MDA-MB-231) breast cancer cells. The effect of test extract was pronounced against MCF-7aro and MDA-MB-231 with IC_{50} of 27 µg/ml and 40 µg/ml, respectively. Most importantly, the authors showed that at equivalent concentrations the extract was relatively nontoxic as it did not induce cell death and apoptosis in the normal/nontumorigenic (MCF-10A) breast cell line ($IC_{50} > 100$ µg/ml).[7] In another study jamun extract has been shown to possess cytotoxic effects on cultured human cervical cancer cells, the HeLa (HPV-18 positive) and SiHa (HPV-16 positive). The extract caused a concentration-dependent cell death with the effect being more pronounced in the HeLa than SiHa cells.[46]

Parmar and colleagues (2010)[47] have shown the cancer chemopreventive properties of jamun extract in 7, 12-dimethylbenzanthracene-induced, croton oil-promoted two-stage skin carcinogenesis in Swiss albino mice. Daily dosing of 125 mg/kg b. wt. of the extract during pre- or postinitiation phases reduced the cumulative numbers of papillomas, tumor incidence, and increased latency period when compared to the control group. Similarly, Goyal et al. (2010)[48] have also observed that the administration of jamun extract (25 mg/kg b. wt., daily) was effective in preventing benzo[a]pyrene-induced forestomach carcinogenesis. It also reduced tumor incidence, tumor burden, and cumulative number of gastric carcinomas.

Recently, we demonstrated antiproliferative potential of anthocyanidin- and anthocyanin-enriched jamun extracts against human non-small-lung cancer cells (A549). A concentration-dependent response was observed with the jamun pulp at the concentrations tested (12.5–200 µg/ml; Figure 10.9). Interestingly, jamun pulp extract enriched at the anthocyanidin level showed strong antiproliferative activity with an IC_{50} value of 59 ± 4 µg/ml. However, anthocyanin-enriched pulp extract had a comparatively weaker effect. This suggests that anthocyanidins are a strong bioactive compared to their natural counterparts (anthocyanins).

Ding et al. showed that cyanidin-3-O-glucoside pretreatment dose-dependently decreased the expression of COX-2 and activities of AP-1, NF-κB, and TNF-α in cells treated with 12-O-tetradecanolyphorbol-13-acetate or ultraviolet light.[49] In another study, Baghchi and colleagues demonstrated that Opti-berry (an anthocyanin-rich preparation from wild blueberry, bilberry, cranberry, elderberry, raspberry, and strawberry) significantly inhibited both H_2O_2- and TNF-α-induced VEGF expression by human keratinocytes. VEGF is a key regulator of tumor angiogenesis.[12]

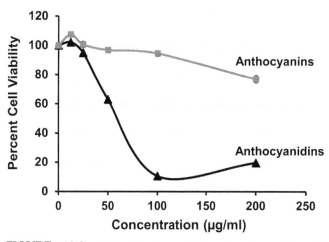

FIGURE 10.9 Antiproliferative activity of anthocyanin- and anthocyanidin-enriched jamun pulp extracts by MTT assay. Cells were treated with jamun extracts at 12.5–200 µg/ml concentrations for 72 h. Data are expressed as percentage of untreated cells, mean ± SD ($n = 3$). Each data point represents the mean of three experiments with standard deviation. *Adapted with permission from [5]. Reprinted by permission of the publisher (Routledge, Taylor & Francis Ltd, http://www.tandf.co.uk/journals).*

Although berry anthocyanidins (Dp, Cy, Pe, Pt, and Mv) have increasingly been explored for their anticancer effects, their combinatorial effects as a mixture, as present in blueberry, bilberry, and jamun, remained untested until recently. We showed that these anthocyanidins in combination synergistically suppress growth and invasive potential of human non-small-cell lung cancer cells.[14] The synergistic effect presumably resulted from the attack of the individual entities at distinct and overlapping targets associated with cell proliferation, apoptosis, and invasion.[14]

A series of papers from the Mukhtar group have shown Dp to induce apoptosis of human prostate cancer PC3 cells *in vitro* and *in vivo*.[50,51] They have also demonstrated that Dp treatment of HCT116 cells suppressed the NF-kB pathway, resulting in G2/M phase arrest and apoptosis. In cells *in vitro*, both glycosides and aglycones engage an array of antioncogenic mechanisms including antiproliferation, induction of apoptosis, and inhibition of activities of oncogenic transcription factors and protein tyrosine kinases as reviewed.[52]

EFFICACY OF JAMUN IN PREVENTING ESTROGEN-MEDIATED MAMMARY TUMORIGENESIS

Breast cancer is the most common female cancer diagnosed in the United States, and the second most common cause of cancer deaths in women.[53] This alone accounts for about 30% of new cancer cases among women with an estimated 230,480 cases of invasive breast cancer, of which 57,650 *in situ* cases were diagnosed in 2011.[54] Of the risk factors associated with breast cancer, cumulative estrogen exposure has the highest positive correlation to incidence. We have previously shown that ellagic acid (10 μM) and aqueous extracts of various berries (2%) were significantly effective in reducing 4-E_2-induced oxidative DNA damage.[30,39] In another study, a mixed berry diet (containing 0.5% each of blueberry, blackberry, red-raspberry, strawberry, and black raspberry) had a higher level of protection against E_2-induced oxidative DNA damage (8-oxodG) compared to control,[30] suggesting reduction of E_2-induced DNA damage as one possible mechanism for berries' chemopreventive action.[47] Likewise, as discussed earlier, jamun aqueous and organic extracts inhibit oxidative DNA damage. The higher activity of jamun's organic extract is presumably related to the presence of ellagic acid/ellagitannins in the extract. In another study, we showed diet supplemented with black raspberry, blueberry, and ellagic acid to significantly reverse the E_2-induced changes in the mRNA expression of several enzymes involved in E_2-metabolism.[21]

The use of chemical carcinogen-induced mammary tumors in rats as a preclinical model has been popular for the past few decades. We have previously refined the ACI rat model initially described by Shull and colleagues[55] by reducing the dose of E_2 from 27 to 9 mg and the surface area of the implants from 3 to 1.2 cm.[56] This model has been successfully used in several chemoprevention studies in our laboratory, including those involving berry intervention.[21,22,57] The mechanisms by which dietary berries prevent mammary tumors include modulation of E_2-metabolizing enzymes, reduction in DNA damage, decreased circulating prolactin levels, antiestrogenic, and antiproliferative actions.

We tested jamun pulp powder for modulation potential of cytochrome P450s (CYP 1A1 & CYP1B1), cell proliferation, and other biological markers using the ACI rat mammary tumorigenesis model. Diet supplemented with jamun powder (5%, w/w) did not affect diet intake or body weight gain, suggesting the jamun dose was well tolerated. E_2 treatment significantly enhanced the mammary tissue weights and pituitary prolactinomas. The latter effect was partly offset by dietary jamun treatment although it did not achieve significance. Jamun intervention significantly reduced the elevated level of plasma prolactin, a key factor in E_2-mediated mammary tumorigenesis and significantly reduced the high circulating level of estrogen. It also significantly offset E_2-associated increases in (1) mammary cell proliferation; (2) estrogen receptor-α; and (3) cyclin D1 (our unpublished data). Thus, these data from an animal model of breast cancer that is highly relevant to the human scenario suggests that jamun at 5% (w/w) dose can elicit significant reductions in E_2-mediated molecular markers responsible for mammary carcinogenesis.

BIOAVAILABILITY OF BERRY BIOACTIVES

Due to the potentially beneficial impact on human health, berries and their bioactives, anthocyanins and ellagic acid, have come into the focus of medicinal interest. However, data on the bioavailability of anthocyanins after oral intake are scarce and/or inconsistent. The bioavailability and metabolism of the anthocyanins and anthocyanins-containing fruits has been reviewed in various papers.[44, 58] Therefore, we have discussed this subject matter only in brief. Most of the studies published so far suggest low bioavailability of the anthocyanins and it has also been reported that these compounds are extensively metabolized in the tissues and by the colonic microflora. The gut microflora reduces the absorption and also degrades anthocyanins to release simple phenolics that conjugate in intestine and later in liver. However, a few reports contradict this understanding and suggest that anthocyanin glycosides remain intact when passing from the digestive tract into the blood circulation.[59]

Myrczylo et al. demonstrated the systemic levels and tissue distribution of cyanidin-3-glucoside (C3G) in C57BL6J mice receiving by gavage (500 mg/kg) or tail vein injection (1 mg/kg). The systemic bioavailability of parent C3G and total anthocyanins were 1.7 and 3.3%, respectively. In the same study, the authors also showed tissue bioaccumulation in liver, lung, heart, prostate, kidney, and bile.[58] Although anthocyanin bioavailability appears low, it could have been underestimated as some metabolites might have been ignored. Furthermore, the extraction efficiency from various organs varies widely, which may lead to underestimation of tissue levels based on the methodology used.

We have developed methods to detect berry anthocyanidins in the lung tissue of mice and rats administered anthocyanidins or dietary blueberry by modifying the published procedure.[58] Briefly, (1) lung tissue was homogenized, (2) anthocyanins were extracted in acidified acetonitrile, (3) the supernatant was evaporated and reconstituted in 50% methanol containing 2N HCL, (4) the extract was hydrolyzed to convert anthocyanins to anthocyanidins, (5) the latter were extracted in isoamyl alcohol, and (6) samples were analyzed by UPLC-PDA by reconstituting residue in acidified methanol immediately prior to analysis. In addition other precautions like use of silanizing glassware, reconstituting residue in acidified solvent immediately prior to use, and so on were exercised to minimize losses of anthocyanidins. The limits of detection for the various anthocyanidins (0.3–0.75 ng) and PCA (0.2 ng) were established. Data in Figure 10.10 show the presence of Dp (Peak 2′), Cy (peak 3′), PCA (peak 1′), and an unknown peak in

the lung tissues of rats given dietary blueberry (5%, w/w) for 5 weeks. When analyzed by LC-MS/MS, the lung tissue extract clearly showed the presence of Pe, Pt, and Mv due to significantly lower detection limits of these anthocyanidins by LC-MS (data not shown). Likewise, anthocyanidins were also detected in the lung tissue of nude mice given dietary highbush blueberry (5%, w/w) for 10 days.

AVAILABILITY OF JAMUN AS FUNCTIONAL FOODS

Jamun is an evergreen tropical tree in the flowering plant family Myrtaceae. The fruit, which is loaded with potentially bioactive compounds, is seasonal and available during monsoon season in India in the months of June and July. Thus, availability of this fruit throughout the year is a problem. Efforts have been made and jamun juice is available commercially. However, instability of jamun anthocyanins in an aqueous nonacidic environment remains a concern. Several studies have suggested that anthocyanins are highly unstable if not protected. Our studies suggest that anthocyanins and anthocyanidins can be stabilized for several months if stored at −20 °C under argon in amber bottles.

Isolation of anthocyanins and anthocyanidins in their purified form has been a challenging task. Although pure anthocyanidins are available from several commercial sources, these materials are prohibitively expensive, which may explain why only a handful of preclinical studies have been performed with pure compounds. Efforts are underway regarding availability of highly enriched anthocyanidins/anthocyanins as well as ellagitannins in the nutraceutical market. Procedures have been developed by us and others to prepare highly enriched jamun extracts and isolate pure anthocyanidins.

CONCLUSION

Studies have provided convincing evidence of the antioxidant potential and cancer preventive properties of jamun. The ability of jamun fruit to modulate several pathways in the carcinogenesis process is undoubtedly related to the fact that they contain an array of chemopreventive agents and the effect is due, at least in part, to the synergism among the anthocyanidins and with ellagic acid/ellagitannins. The method described in this chapter for the extraction provides an effective method for removing sugars as well as enriching a variety of polyphenolics, including anthocyanins and ellagic acid/ellagitannins. Jamun is the only berry that contains aglycones of five anthocyanidins and appreciable amounts of ellagic acid/ellagitannins. On the other hand, jamun seeds are rich in ellagic acid/ellagitannins in addition to other unidentified polyphenolics.

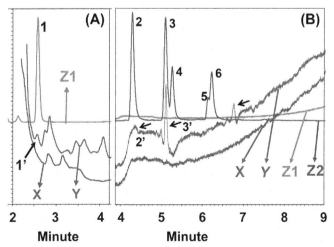

FIGURE 10.10 Detection of anthocyanidins at 520 nm (B) and 260 nm for protocatechuic acid (PCA) (A), a bioactive metabolite of cyanin in lung tissue of rats given dietary blueberry (5% w/w) for 5 weeks by HPLC-PDA. Chromatograms shown are composite: X = lung tissue from rats on control diet; Y = lung tissue from rats on blueberry diet. Reference compounds: Z1, peak 1 (PCA); Z2, reference anthocyanidins, peaks: 2 (Dp), 3 (Cy), 4 (Pt), 5 (Pe), and 6 (Mv). Solvent, gradient of 3.5% phosphoric acid in acetonitrile. Maps are cropped for presentation purposes. Reproduced from Aqil, F., Vadhanam, M.V., Jeyabalan, J., Cai, J. and Grupta RC. Detection of anthocyanins/anthocyanidins in animal tissues. J Agric Food Chem Under consideration, 2013. (See the color plate.)

Based on accumulating evidence, pure anthocyanidins as well as berry extracts enriched at anthocyanidin levels showed higher chemoprotective activities than berry extracts at anthocyanin level. Phytochemicals in both jamun pulp and seed extracts were found to be effective antioxidants in different *in vitro* assays. In addition, although there are several studies *in vitro* as well as *in vivo* systems to demonstrate anthocyanidins' robust protective activities, further studies should be designed to investigate the cancer-preventive potential of jamun in animal models and human subjects.

In conclusion, it is recommended that this area of research for berry fruits continue to be explored, and efforts should be made to make the berry bioactives available in large quantities to explore their protective effects in human studies.

SUMMARY POINTS

- Colored fruits, particularly berries, are highly chemoprotective because of their antioxidant, antiproliferative, and anti-inflammatory activities.
- The anthocyanidin-enriched jamun pulp extract showed significantly higher antiproliferative activity compared to anthocyanin-enriched extract.
- Jamun pulp contains five major anthocyanidins: delphinidin (20.3%), cyanidin (6.6%), petunidin (24.2%), peonidin (2.2%), and malvidin (44.4%).
- Jamun seems to represent the only berry that contains five anthocyanidins and appreciable amounts of ellagic acid/ellagitannins.
- Jamun seeds contain no detectable anthocyanins but are enriched in ellagic acid/ellagitannins.
- Jamun contains phytochemicals enriched in anthocyanins capable of significant antioxidant and antiproliferative potential.

Acknowledgements

Our work summarized in this article was supported by the US Public Health Sciences grant CA-118114, CA-152125, Kentucky Lung Cancer Research Program, Agnes Brown Duggan Endowment, and Hemsley Foundation.

References

1. Gilgun-Sherki Y, Melamed E, Offen D. Oxidative stress induced-neurodegenerative diseases: The need for antioxidants that penetrate the blood brain barrier. *Neuropharmacology* 2001;**40**(8):959–75.
2. Geronikaki AA, Gavalas AM. Antioxidants and inflammatory disease: Synthetic and natural antioxidants with anti-inflammatory activity. *Comb Chem High Throughput Screen* 2006;**9**(6):425–42.
3. Meyskens FL, Szabo E. Diet and cancer: The disconnect between epidemiology and randomized clinical trials. *Cancer Epidemiol Biomarkers Prev* 2005;**14**(6):1366–9.
4. Savikin K, Zdunic G, Jankovic T, Tasic S, Menkovic N, Stevic T, et al. Phenolic Content and Radical Scavenging Capacity of Berries and Related Jams from Certificated Area in Serbia. *Plant Foods Hum Nutr* 2009;**64**(3):212–7.
5. Aqil F, Gupta A, Munagala R, Jeyabalan J, Kausar H, Sharma RJ, et al. Antioxidant and Antiproliferative Activities of Anthocyanin/Ellagitannin-Enriched Extracts From *Syzygium cumini* L. (Jamun, the Indian Blackberry). *Nutr Cancer: An Int J* 2012;**64**(3):428–38.
6. Stoner GD, Wang LS, Casto BC. Laboratory and clinical studies of cancer chemoprevention by antioxidants in berries. *Carcinogenesis* 2008;**29**(9):1665–74.
7. Li L, Adams LS, Chen S, Killian C, Ahmed A, Seeram NP. *Eugenia jambolana* Lam. berry extract inhibits growth and induces apoptosis of human breast cancer but not non-tumorigenic breast cells. *J Agric Food Chem* 2009;**57**(3):826–31.
8. Veigas JM, Narayan MS, Laxman PM, Neelwarne B. Chemical nature, stability and bioefficacies of anthocyanins from fruit peel of *Syzygium cumini* Skeels. *Food Chem* 2007;**105**(2):619–27.
9. Baliga MS, Fernandes S, Thilakchand KR, D'Souza P, Rao S. Scientific Validation of the Antidiabetic Effects of *Syzygium jambolanum* DC (Black Plum), a Traditional Medicinal Plant of India. *J Altern Complement Med* 2012.
10. Sagrawat H, Mann AS, Kharya MD. Pharmacological potential of *Eugenia jambolana*: A review. *Pharmacognosy Mag* 2006;**2**:96–105.
11. Jagetia GC, Baliga MS. *Syzygium cumini* (Jamun) reduces the radiation-induced DNA damage in the cultured human peripheral blood lymphocytes: A preliminary study. *Toxicol Lett* 2002;**132**(1):19–25.
12. Bagchi D, Sen CK, Bagchi M, Atalay M. Anti-angiogenic, antioxidant, and anti-carcinogenic properties of a novel anthocyanin-rich berry extract formula. *Biochemistry (Mosc)* 2004;**69**(1):75–80 1 p preceding 75.
13. Hou DX, Fujii M, Terahara N, Yoshimoto M. Molecular Mechanisms Behind the Chemopreventive Effects of Anthocyanidins. *J Biomed Biotechnol* 2004;**2004**(5):321–5.
14. Kausar H, Jeyabalan J, Aqil F, Chabba D, Sidana J, Singh IP, et al. Berry anthocyanidins synergistically suppress growth and invasive potential of human non-small-cell lung cancer cells. *Cancer Lett* 2012;**325**(1):54–62.
15. Lobo V, Patil A, Phatak A, Chandra N. Free radicals, antioxidants and functional foods: Impact on human health. *Pharmacogn Rev* 2010;**4**(8):118–26.
16. Rastogi RM, Mehrotra BN. Compendium of Indian medicinal plants. Lucknow, India. *Cent Drug Res Inst* 1990:388–9.
17. de Brito ES, de Araujo MC, Alves RE, Carkeet C, Clevidence BA, Novotny JA. Anthocyanins present in selected tropical fruits: Acerola, jambolao, jussara, and guajiru. *J Agric Food Chem* 2007;**55**(23):9389–94.
18. Vijayanand P, Rao LJM, Narasimham P. Volatile flavour components of jamun fruit (*Syzygium cumini* L). *Flavour Fragrance J* 2001;**16**(1):47–9.
19. Stoner GD, Wang LS, Seguin C, Rocha C, Stoner K, Chiu S, et al. Multiple berry types prevent N-nitrosomethylbenzylamine-induced esophageal cancer in rats. *Pharm Res* 2010;**27**(6):1138–45.
20. Hager TJ, Howard LR, Prior RL. Processing and storage effects on the ellagitannin composition of processed blackberry products. *J Agric Food Chem* 2010;**58**(22):11749–54.
21. Aiyer HS, Gupta RC. Berries and ellagic acid prevent estrogen-induced mammary tumorigenesis by modulating enzymes of estrogen metabolism. *Cancer Prev Res (Phila)* 2010;**3**(6):727–37.

22. Aiyer HS, Srinivasan C, Gupta RC. Dietary berries and ellagic acid diminish estrogen-mediated mammary tumorigenesis in ACI rats. *Nutr Cancer* 2008;**60**(2):227–34.

23. Rice-Evans CA, Miller NJ. Antioxidant activities of flavonoids as bioactive components of food. *Biochem Soc Trans* 1996;**24**(3):790–5.

24. Garcia-Alonso M, de Pascual-Teresa S, Santos-Buelga C, Rivas-Gonzalo JC. Evaluation of the antioxidant properties of fruits. *Food Chem* 2004;**84**(1):13–8.

25. Baliga MS. Anticancer, Chemopreventive and Radioprotective Potential of Black Plum (*Eugenia Jambolana* Lam. *Asian Pac J Cancer Prev* 2011;**12**(1):3–15.

26. Jagetia GC, Baliga MS. The evaluation of nitric oxide scavenging activity of certain Indian medicinal plants in vitro: A preliminary study. *J Med Food* 2004;**7**(3):343–8.

27. Bajpai M, Pande A, Tewari SK, Prakash D. Phenolic contents and antioxidant activity of some food and medicinal plants. *Int J Food Sci Nutr* 2005;**56**(4):287–91.

28. Banerjee A, Dasgupta N, De B. In vitro study of antioxidant activity of *Syzygium cumini* fruit. *Food Chemistry* 2005;**90**(4):727–33.

29. Kim DO, Lee KW, Lee HJ, Lee CY. Vitamin C equivalent antioxidant capacity (VCEAC) of phenolic phytochemicals. *J Agric Food Chem* 2002;**50**(13):3713–7.

30. Aiyer HS, Vadhanam MV, Stoyanova R, Caprio GD, Clapper ML, Gupta RC. Dietary berries and ellagic acid prevent oxidative DNA damage and modulate expression of DNA repair genes. *Int J Mol Sci* 2008;**9**(3):327–41.

31. Moktar A, Ravoori S, Vadhanam MV, Gairola CG, Gupta RC. Cigarette smoke-induced DNA damage and repair detected by the comet assay in HPV-transformed cervical cells. *Int J Oncol* 2009;**35**(6):1297–304.

32. Seeram NP, Adams LS, Henning SM, Niu YT, Zhang YJ, Nair MG, et al. In vitro antiproliferative, apoptotic and antioxidant activities of punicalagin, ellagic acid and a total pomegranate tannin extract are enhanced in combination with other polyphenols as found in pomegranate juice. *J Nutr Biochem* 2005;**16**(6):360–7.

33. Wang CY, Wang SY, Yin JJ, Parry J, Yu LL. Enhancing antioxidant, antiproliferation, and free radical scavenging activities in strawberries with essential oils. *J Agric Food Chem* 2007;**55**(16):6527–32.

34. Seyoum A, Asres K, El-Fiky FK. Structure-radical scavenging activity relationships of flavonoids. *Phytochemistry* 2006;**67**(18):2058–70.

35. Benzie IFF, Strain JJ. The ferric reducing ability of plasma (FRAP) as a measure of antioxidant power: The FRAP assay. *Anal Biochem* 1996;**239**(1):70–6.

36. Song R, Kelman D, Johns KL, Wright AD. Correlation between leaf age, shade levels, and characteristic beneficial natural constituents of tea (*Camellia sinensis*) grown in Hawaii. *Food Chem* 2012;**133**(3):707–14.

37. Spencer WA, Vadhanam MV, Jeyabalan J, Gupta RC. Oxidative DNA Damage Following Microsome/Cu(II)-Mediated Activation of the Estrogens, 17 beta-Estradiol, Equilenin, and Equilin: Role of Reactive Oxygen Species. *Chem Res Toxicol* 2012;**25**(2):305–14.

38. Teel RW. Ellagic Acid Binding to DNA as a Possible Mechanism for Its Antimutagenic and Anticarcinogenic Action. *Cancer Lett* 1986;**30**(3):329–36.

39. Aiyer HS, Kichambare S, Gupta RC. Prevention of Oxidative DNA Damage by Bioactive Berry Components. *Nutr Cancer—An Int J* 2008;**60**:36–42.

40. DeVita VT, Lawrence TS, Hellman, Rosenberg's SA. Cancer: Principles & Practice of Oncology. *USA* 2004.

41. Mallery SR, Zwick JC, Pei P, Tong M, Larsen PE, Shumway BS, et al. Topical application of a bioadhesive black raspberry gel modulates gene expression and reduces cyclooxygenase 2 protein in human premalignant oral lesions. *Cancer Res* 2008;**68**(12):4945–57.

42. Wang LS, Stoner GD. Anthocyanins and their role in cancer prevention. *Cancer Lett* 2008;**269**(2):281–90.

43. Gupta RC, Stoner GD, Gairola CG. Inhibition of cigarette smoke-mediated lung tumorigenesis in A/J mice by dietary berries. *Proc Natl Acad Sci U S A* 2005;**46** #3895.

44. Seeram NP. Berry fruits for cancer prevention: Current status and future prospects. *J Agric Food Chem* 2008;**56**(3):630–5.

45. Kausar H, Jeyabalan J, Aqil F, Chabba D, Sidana J, Singh IP, et al. Berry Anthocyanidins Synergistically Suppresses Non-Small-Cell Lung Cancer Cell Growth and Metastasis and Enhances Sensitivity to the Chemotherapeutic Drug Paclitaxel. *J Thorac Oncol* 2011;**6**(6) S930–S931.

46. Barh D, Viswanathan G. *Syzygium cumini* inhibits growth and induces apoptosis in cervical cancer cell lines: A primary study. *Ecancermedicalscience* 2008;**2**:83.

47. Parmar J, Sharma P, Verma P, Goyal PK. Chemopreventive action of *Syzygium cumini* on DMBA-induced skin papillomagenesis in mice. *Asian Pac J Cancer Prev* 2010;**11**(1):261–5.

48. Goyal PK, Verma P, Sharma P, Parmar J, Agarwal A. Evaluation of Anti-Cancer and Anti-Oxidative Potential of *Syzygium Cumini* Against Benzo[a]pyrene (BaP) Induced Gastric Carcinogenesis in Mice. *Asian Pac J Cancer Prev* 2010;**11**(3):753–8.

49. Ding M, Feng R, Wang SY, Bowman L, Lu Y, Qian Y, et al. Cyanidin-3-glucoside, a natural product derived from blackberry, exhibits chemopreventive and chemotherapeutic activity. *J Biol Chem* 2006;**281**(25):17359–68.

50. Hafeez BB, Siddiqui IA, Asim M, Malik A, Afaq F, Adhami VM, et al. A dietary anthocyanidin delphinidin induces apoptosis of human prostate cancer PC3 cells in vitro and in vivo: involvement of nuclear factor-kappaB signaling. *Cancer Res* 2008;**68**(20):8564–72.

51. Syed DN, Suh Y, Afaq F, Mukhtar H. Dietary agents for chemoprevention of prostate cancer. *Cancer Lett* 2008;**265**(2):167–76.

52. Thomasset S, Teller N, Cai H, Marko D, Berry DP, Steward WP, et al. Do anthocyanins and anthocyanidins, cancer chemopreventive pigments in the diet, merit development as potential drugs? *Cancer Chemother Pharmacol* 2009;**64**(1):201–11.

53. Siegel R, Ward E, Brawley O, Jemal A. Cancer statistics, 2011: The impact of eliminating socioeconomic and racial disparities on premature cancer deaths. *CA Cancer J Clin* 2011;**61**(4):212–36.

54. American_Cancer_Society. Breast Cancer Facts & Figures 2011-2012. American Cancer Society, Inc. Atlanta: American Cancer Society.

55. Shull JD, Spady TJ, Snyder MC, Johansson SL, Pennington KL. Ovary-intact, but not ovariectomized female ACI rats treated with 17 beta estradiol rapidly develop mammary carcinoma. *Carcinogenesis* 1997;**18**(8):1595–601.

56. Ravoori S, Vadhanam MV, Sahoo S, Srinivasan C, Gupta RC. Mammary tumor induction in ACI rats exposed to low levels of 17beta-estradiol. *Int J Oncol* 2007;**31**(1):113–20.

57. Ravoori S, Vadhanam MV, Aqil F, Gupta RC. Inhibition of estrogen-mediated mammary tumorigenesis by blueberry and black raspberry. *J Agric Food Chem* 2012;**60**(22):5547–55.

58. Marczylo TH, Cooke D, Brown K, Steward WP, Gescher AJ. Pharmacokinetics and metabolism of the putative cancer chemopreventive agent cyanidin-3-glucoside in mice. *Cancer Chemother Pharmacol* 2009;**64**(6):1261–8.

59. Miyazawa T, Nakagawa K, Kudo M, Muraishi K, Someya K. Direct intestinal absorption of red fruit anthocyanins, cyanidin-3-glucoside and cyanidin-3,5-diglucoside, into rats and humans. *J Agric Food Chem* 1999;**47**(3):1083–91.

60. Shafi PM, Rosamma MK, Jamil K, Reddy PS. Antibacterial activity of *Syzygium cumini* and *Syzygium travancoricum* leaf essential oils. *Fitoterapia* 2002;**73**(5):414–6.

Preventive Effects of Broccoli Bioactives: Role on Oxidative Stress and Cancer Risk

Patrizia Riso, Cristian Del Bo'

Università degli Studi di Milano, DeFENS, Department of Food, Environmental and Nutritional Sciences, Division of Human Nutrition, Milan, Italy

Stefano Vendrame

Università degli Studi di Milano, DeFENS, Department of Food, Environmental and Nutritional Sciences, Division of Human Nutrition, Milan, Italy, Department of Food Science and Human Nutrition, University of Maine, Orono, ME, USA

List of Abbreviations

8-oxodG 8-oxo-7,8-dihydro-2'-deoxyguanosine
AhR aryl hydrocarbon receptor
ARE antioxidant response element
Bcl-2 B-cell lymphoma 2
CYP cytochrome
ER estrogen receptors
FPG formamidopyrimidine DNA glycosylase
GLs glucosinolates
GSH glutathione
GST glutathione-S-transferase
HDAC histone deacetylases
HO-1 heme oxygenase (decycling) 1
I3C indole-3-carbinol
IkB inhibitor kappa B
ITCs isothiocyanates
KEAP1 Kelch-like ECH-associated protein 1
MAPK mitogen-activated protein kinase
NF-kB Nuclear Factor kappa B
NQO NADPH quinone reductase
Nrf2 nuclear factor-erythroid-2-related factor 2
NUDT1 nucleoside diphosphate linked moiety X-type motif 1
OGG1 8-oxoguanine DNA glycosylase
PBMCs peripheral blood mononuclear cells
ROS reactive oxygen species
SERMs selective estrogen receptor modulators
SFN sulforaphane
UGTs UDP-glucuronosyl transferases
XRE xenobiotic response element

INTRODUCTION

Cruciferous vegetables belong to the botanical family known as Brassicaceae, a family of plants including the Brassica genus with a large number of accessions. They are characterized by a peculiar phytochemical composition; the most commonly consumed products within this class are broccoli, cauliflower, cabbage, and Brussels sprouts. Other brassica vegetables such as kale, watercress, turnip, mustard, and radish are consumed primarily in specific regions.[1]

A large number of studies, including cell culture, animal models, and epidemiological studies, have documented the importance of this class of vegetables for cancer prevention.[1] However, data from human intervention studies are relatively limited. The present chapter will discuss the available evidence on the role of cruciferous vegetables, with focus on broccoli, modulation of oxidative stress, and markers of cancer risk.[1]

CRUCIFEROUS VEGETABLES INTAKE AND CANCER RISK

Epidemiological evidence has linked brassica vegetable consumption to a reduced risk for a variety of cancers.[1-3] Most of the evidence comes from case-control studies, and appears to be most conclusive for an inverse association with lung and gastric cancers.[2]

Although it is extremely difficult to discriminate the contribution of a single food without incurring methodological biases and confounding factors when obtaining and evaluating epidemiological data, a few studies have attempted to focus specifically on the association between broccoli consumption and cancer risk.

Inverse associations between cruciferous vegetable intake and lung cancer have been repeatedly reported in case-control and cohort studies,[1] as recently reviewed

Cancer
http://dx.doi.org/10.1016/B978-0-12-405205-5.00011-8

© 2014 Elsevier Inc. All rights reserved.

by Lam and coworkers.[4] Concerning broccoli, a consumption more than once per week compared to no consumption has been inversely associated with lung cancer risk (OR 0.3, 95%CI 0.2,0.6).[5]

Inverse associations between cruciferous vegetable intake and gastric cancer risk have also been reported in different case-control studies,[1,6] and a study on a Japanese population reported a specific association with broccoli consumption (OR 0.60, 95%CI 0.34,1.08).[6]

Cruciferous vegetable intake has been repeatedly associated with reduced prostate cancer risk in case-control studies.[1,7] Broccoli intake specifically was associated with reduced prostate cancer risk in a population-based case-control study (OR 0.72, 95%CI 0.49,1.06).[7] In a prospective study, broccoli consumption more than once per week was associated with reduced metastatic prostate cancer risk versus consumption less than once per month (RR 0.55, 95%CI 0.34, 0.89).[8]

An inverse association between cruciferous vegetable intake and breast cancer has been found in few case-control studies, as reviewed by the International Agency on Cancer Research in 2004.[1] Regarding broccoli intake, one case-control study reported an inverse association with breast cancer risk with consumption of more than 625 g per month of broccoli versus a consumption of less than 305 g per month (OR 0.6, 95%CI 0.40,1.01) in premenopausal women; no association was found in postmenopausal women.[9]

Inverse associations between cruciferous vegetable intake and colorectal cancer have been reported in at least five case-control studies [1], but a pooled analysis of 14 prospective studies did not find significant differences between highest and lowest cruciferous vegetable intake.[10] A multicenter Japanese case-control study reported a strong inverse association between high broccoli consumption and colorectal cancer risk (OR 0.18, 95%CI 0.06,0.58).[6] An inverse association with the highest quartile of broccoli intake (OR 0.47, 95%CI 0.30,0.73) was also reported in a Singaporean case-control study.[11]

Very few studies are available for bladder cancer risk. Two prospective studies have found inverse associations between cruciferous vegetable intake and bladder cancer risk,[1,12] and one reported a specific inverse association with broccoli consumption (RR 0.61, 95%CI 0.42,0.87).[12]

Although several epidemiological studies have reported inverse associations between cruciferous vegetable intake and cancer risk, other studies have found no significant associations.[1-2] Of note, however, no study to date has ever found a positive association between cruciferous vegetable intake in general, or broccoli in particular, and cancer risk.

One possible explanation for the discrepancies in the epidemiological evidence on the protective effect of cruciferous vegetable consumption may be attributed to individual genetic makeup. In particular, as discussed later, polymorphisms in different isoforms of the glutathione-S-transferase (GST) family of phase II detoxifying enzymes have been extensively investigated in relation to the metabolism and biological activity of cruciferous vegetable bioactives. The null genotypes of the GSTM1 and GSTT1 do not produce a functional gene and may result in a lower enzyme activity and a higher risk of disease.

With regard to broccoli, case-control studies support this hypothesis in relation to different cancer sites. GSTM1 null individuals in the highest quartile of broccoli intake have been found to have a greater inverse association with colorectal cancer (OR = 0.36, 95%CI 0.19,0.68) compared with GSTM1 positive ones (OR = 0.76, 95% CI 0.40,0.99).[11] Another case-control study found the greatest inverse association with colon cancer in GSTM1 null individuals, with an OR of 0.23 for consumption of four or more servings/week of broccoli versus no intake (95%CI 0.10,0.54).[13] A large case-control study on lung cancer found the strongest inverse association with high broccoli and Brussels sprouts consumption versus low consumption (OR = 0.35, 95%CI 0.12,1.01) in individuals with GSTM1 and GSTT1 null genotype.[14] Conversely, GSTM1 positive individuals have been found to have the greatest reduction in prostate cancer risk associated with high broccoli intake (OR = 0.49; 95%CI 0.27,0.89).[7]

In view of the inconclusive results of very recent meta-analyses, more epidemiological studies are necessary to better elucidate the role of brassica vegetables. Moreover, a deeper understanding of the interactions between dietary intake, individual genotype, and metabolism may help clarify the potential contribution of brassica vegetable consumption to cancer prevention.

ANTIOXIDANTS AND OTHER BIOACTIVE COMPOUNDS IN BROCCOLI

Vegetables belonging to the brassica family are extremely heterogeneous in terms of composition as well as consumption patterns in different geographic areas, which could at least partly account for the apparent inconsistencies in findings from different studies investigating their protective effects.

The distinctive phytochemicals of brassica vegetables are glucosinolates (GLs, β-thioglucoside N-hydroxysulfates), a family of about 120 identified sulfur-containing compounds with a common glucone moiety and a variable aglucone side chain derived from one of eight amino acids giving aliphatic, aromatic, and indolil GLs.[1]

GLs undergo enzymatic conversion to biologically active isothiocyanates (ITCs), which are also responsible for the smell and taste of these vegetables. The conversion from GLs to their corresponding aglycones is catalyzed by the β-thioglucosidase myrosinase. The unstable aglycones spontaneously rearrange into ITCs, but in the

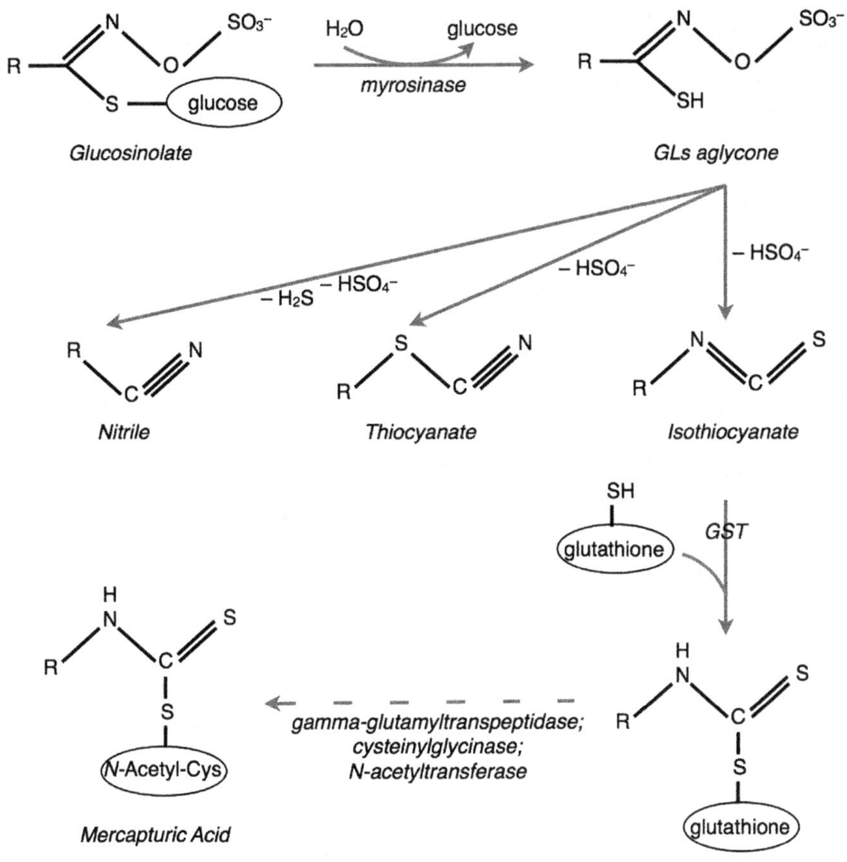

FIGURE 11.1 **Main Metabolic Pathway of Glucosinolates.** GL, glucosinolate; GST, glutathione-S-transferase. Glucosinolates are converted to their corresponding aglycones by the enzyme myrosinase, and subsequently catabolized to different products, including isothiocyanates. Isothiocyanates are conjugated *in vivo* to glutathione by GST and converted, through three enzymatic steps, to mercapturic acid for excretion.

presence of protein factors, they can also form nitriles or thiocyanates (Figure 11.1).

Myrosinase is compartmentalized in the vacuoles of vegetable cells to prevent it from coming in contact with its substrate under normal conditions. Following tissue injury, chewing, or cutting during food preparation, GLs are exposed to myrosinase and the conversion to ITCs can occur. Cooking can inactivate myrosinase, but GLs can still be converted into ITCs by the enzymatic activity of the intestinal microbiota, at least to some extent depending on interindividual variability.

The main GLs and their metabolic breakdown products are listed in Table 11.1. The GL content of broccoli has been reviewed by Latté et al.,[15] and is reported to be in the range of 12.8 to 20.9 μmol/g dry weight. Sixteen different GLs have been detected in broccoli, with glucoraphanin, glucoiberin, progoitrin, and the indole GLs glucobrassicin and neoglucobrassicin being the most abundant, but with extreme variability in the GLs profile between different cultivars[15] and different crops based on soil composition, weather conditions, and pathogens or wound-related stresses.[16] The enzymatic conversion of glucoraphanin by mirosinase generates sulforaphane (SFN, 1-isothiocyanato-4R-methylsulfinylbutane), the most abundant and widely investigated ITC in broccoli. Breakdown of glucobrassicin generates indole-3-carbinol (I3C), also extensively studied for its biological activity. Absorption of ITCs is influenced

TABLE 11.1 Main Glucosinolates in the Diet and Their Corresponding Isothiocyanates/Indoles

Food source	Glucosinolate (precursor)	Isothiocyanate/Indole
Broccoli	Glucoraphanin	Sulforaphane (SFN)
	Glucoiberin	Iberin
	Glucobrassicin	Indole-3-carbinol (I-3C)
Cabbage	Glucoiberin	Iberin
	Sinigrin	Allyl-Isothiocyanate (A-ITC)
	Glucobrassicin	Indole-3-carbinol (I-3C)
Brussels sprouts	Glucobrassicin	Indole-3-carbinol (I-3C)
	Sinigrin	Allyl-Isothiocyanate (A-ITC)
	Progoitrin	Goitrin
Cauliflower	Glucoiberin	Iberin
	Glucoraphanin	Sulforaphane (SFN)
	Glucobrassicin	Indole-3-carbinol (I-3C)
	Sinigrin	Allyl-Isothiocyanate (A-ITC)
Watercress	Gluconasturtiin	Phenethyl-Isothiocyanate (PEITC)
Horseradish	Sinigrin	Allyl-Isothiocyanate (A-ITC)

by different factors such as age, gender, dietary status, and eating behaviors, in particular the amount of time used for chewing. Studies on absorption of ITCs from different food sources in humans are increasing and demonstrate that they are rapidly metabolized and excreted and low levels are found in plasma.[17] For example the peak concentration of sulforaphane and its thiol conjugates after broccoli intake was shown to be less than 2 micromol/L, falling to levels of nanomol/L within a few hours.[18]

Most ITCs are metabolized *in vivo* through the mercapturic acid pathway. The first step of this pathway is the conjugation of ITCs with glutathione (GSH) by GSTs, to produce ditiocarbammate. These conjugates undergo further enzymatic modifications in the GSH portion to form nitro-acetylcisteine conjugates that are excreted in the urine (Figure 11.1).

Conversely, indole compounds can react with ascorbic acid in the acidic environment of the stomach, producing ascorbigen and a series of condensed products that may have biological activity.[1]

Broccoli is unique among brassica vegetables in that it is also a significant source of vitamins (e.g., C, K, folate), minerals (e.g., potassium, phosphorous, calcium, magnesium), and other bioactive compounds, including phenolics (flavonoids and hydroxycinnamic acids, and in particular quercetin derivatives, kaempferol derivatives and synapic acid derivatives) and carotenoids (mainly β-carotene and lutein), all of which potentially contribute together with ITCs to explain the health benefits associated with broccoli consumption.[1,19-20]

CANCER PREVENTIVE MECHANISMS OF ISOTHYOCIANATES AND INDOLES

GLs, and in particular their secondary breakdown products ITCs, indoles, and nitriles, are likely the major broccoli bioactive compounds involved in cancer protection, as they are involved in different but interconnected signaling pathways that play an important role in delaying cancer initiation and progression.[21] Many of these pathways are related to oxidative stress. This impairment between antioxidant and pro-oxidant species generates a stress response in cells that initiates a series of biological events that promote pathogens counteraction, adaptation, and survival. The consequences of oxidative stress can be both harmful and beneficial depending on its intensity, duration, and frequency. This phenomenon of biphasic dose response is known as hormesis.[22] The key conceptual features of hormesis are the disruption of homeodynamics, a modest overcompensation, and the reestablishment of homeodynamics.[22] ITCs, together with other bioactives including polyphenols, may represent an example of compounds able to cause stress-induced hormesis. In fact, they are perceived by the organism as xenobiotic components to be promptly metabolized and excreted, thus eliciting an oxidative stress response with the activation of several pathways. The main mechanisms of action of ITCs and indoles evaluated both *in vitro* and in animal models are reported in Figure 11.2.

MODULATION OF PHASE I AND PHASE II BIOTRANSFORMATION ENZYMES

The detoxification of xenobiotics, including carcinogens, requires their biotransformation by phase I and phase II enzymes. Phase I enzymes, such as those belonging to the cytochrome (CYP) P450 family, increase the reactivity of hydrophobic compounds, preparing them for reactions catalyzed by phase II enzymes. The reactions catalyzed by phase II enzymes, including NADPH quinone reductase (NQO), UDP-glucuronosyl transferases (UGTs), glutamate cysteine ligase, and GSTs increase the solubility in water and promote the elimination of the compounds.[23]

Some GLs breakdown products are involved in the modulation of phase I and phase II detoxification enzymes. Nitrile (i.e., crambene) and ITCs (i.e., SFN,

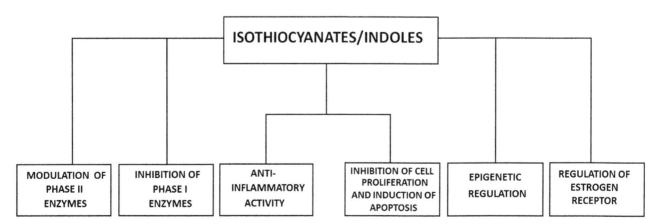

FIGURE 11.2 Mechanisms of Action of Isothiocyanates and Indoles in the Modulation of Signaling Pathways Involved in Cancer Prevention.

phenethyl isothiocyanate, and benzyl isothiocyanate) have been shown to induce phase II enzymes and activate antioxidant response element (ARE) through a modulation of the nuclear factor-erythroid-2-related factor 2 (Nrf2) pathway.[24] Under normal conditions Nrf2 is held in the cytoplasm by a specific inhibitory protein called Kelch-like ECH-associated protein 1 (KEAP1). Modification of KEAP1 cysteine residues induces a conformational change and a loss of Nrf2 binding, allowing Nrf2 to translocate to the nucleus where it heterodimerizes with specific cofactors and leads to the transcription of various genes through the regulatory regions of ARE.[21] The mechanism of upregulation of phase II enzyme is clearly dependent on intracellular uptake of ITCs and by the conjugation with GSH.[21] Studies documented that cells treated with a GSH-depleting agent could not accumulate ITCs, and the subsequent induction of phase II enzymes is blocked.[25]

Modulation of phase II enzymes through SFN has been documented in several *in vitro* and *ex vivo* studies.[26-28]

However, ITCs can play an important role in the inhibition of phase I enzymes, such as CYP450. This inhibition is an important mechanism to prevent the formation of carcinogenic metabolites. SFN demonstrated a competitive inhibitory effect of the CYP by preventing the formation of carcinogen-induced DNA-adducts in *in vitro* and *ex vivo* studies. For example, in rat hepatocytes SFN decreased the activity of CYP enzymes in a dose-dependent manner.[29]

On the contrary, indole compounds such as I3C and its metabolic products were shown to induce the expression of both phase I and II enzymes.[30] The mechanism of induction is driven by the binding of the aryl hydrocarbon receptor (AhR): this link induces the activation of the xenobiotic response element (XRE) and upregulates the expression of the enzymes involved in the process of biotransformation. These mechanisms are clearly documented in several *in vitro* studies.[31]

ANTI-INFLAMMATORY ACTIVITY

The mechanisms of chemoprevention of ITCs and indoles can be exerted through the modulation of several signaling pathways involved in the inflammatory response, including the direct and indirect inhibition of Nuclear Factor kappa B (NF-κB). NF-κB is a transcription factor and key mediator in the expression and activation of several pro-inflammatory mediators including cytokines, chemokines, and adhesion molecules. Generally, NFκB is retained in an inactive form in the cytoplasm by an inhibitor molecule, inhibitor kappa B (IκB).[32] When a pro-inflammatory stimulus or oxidative event occurs, the inhibitor molecule becomes phosphorylated and induces the degradation and liberation of the NF-κB complex in its active form. ITCs were shown to directly and indirectly inhibit NF-kB in *in vitro* and *in vivo* studies.[33-35] The role of ITCs in direct inhibition of NF-κB has been attributed to their capability to counteract phosphorylation, thus preventing IκB degradation and inhibiting transcriptional induction by NF-κB.[34] This inhibition decreases the expression of numerous NF-κB target genes, such as cyclooxygenase-2, interleukin 6, and tumor necrosis factor alpha, resulting in lower levels of inflammation.[34,36-37] Moreover, some evidence suggests that ITC, in particular SFN, can directly bind to a functional active subunit of NF-κB and inhibit its binding to DNA.[38]

Together with a direct inhibition, it has been hypothesized that ITCs can act indirectly in the inhibition of NF-κB activity through a mechanism that involves Nrf2. Activation of NF-κB is highly complex and mediated by many kinases, such as the mitogen-activated protein kinase (MAPK) cascade. Some studies have proposed a regulatory network for a coordinated modulation of Nrf2 and NF-κB through the MAPK cascade, based on common regulatory sequences in the transactivation domains of Nrf2 and NF-κB. ITCs seem to induce the activation of Nrf2 by causing a cross-talk between Nrf2 and NF-κB.[39] Although this potential inhibitory effect needs further investigation, it could at least partly explain the anti-inflammatory and anticancer activity of ITCs.

STIMULATION OF APOPTOSIS

Apoptosis is the process of programmed cell death; it plays a crucial role as a protective mechanism against cancer development by eliminating genetically damaged or excess cells that have been induced to divide. In cancer cells, the apoptosis process is impaired and cell proliferation is uncontrolled.[40] Several ITC and indole compounds have been shown to induce apoptosis through a number of targets, at different points in several pathways. Proteolytic activation of caspase-3, decreases in mitochondrial potential with subsequent release of apoptotic generating proteins cytochrome c and inhibitor of apoptosis family member Smac/DIABLO, activation of parallel MAPK cascades (such as extracellular signal-regulated kinase, c-Jun N-terminal kinase, and p38), down-regulation of antiapoptotic B-cell lymphoma 2 (Bcl-2), and upregulation of proapoptotic Bcl-2-associated X protein expression are only some of the pathways involved and regulated by ITCs and I3C in the apoptosis process.[39-40]

The mechanisms by which ITCs and indoles initiate intracellular signaling resulting in apoptosis is not completely understood. One of the plausible mechanisms is that exposure of cells to ITCs and indoles in high concentrations induces generation of reactive oxygen species (ROS) and consequently a depletion of intracellular GSH that may contribute to a series of signaling events leading to apoptosis.[31] Another plausible mechanism is that ITCs and indoles can interact with mitochondrial complex III and prevent the generation of ROS.[41] Alternatively, it has been recently reported that these compounds may induce apoptosis through inhibition of proteasome activity with marked accumulation of the tumor suppressor p53 in multiple myeloma cells, independent of ROS generation.[42]

REGULATION OF ESTROGEN RECEPTOR

Estrogens are hormones that function as signaling molecules interacting with cells in a variety of target tissues including brain, bone, liver, and heart. However, breast and uterus are two of the main estrogen targets. Estrogens act on target tissues by binding specific estrogen receptors (ER).[39] There are two forms of the estrogen receptor: α and β. In the absence of estrogen molecules, these estrogen receptors are inactive and have no influence on DNA.[39] Paradoxically, estrogens can be both beneficial and harmful molecules. The main beneficial effects of estrogens include their roles in programming breast and uterus for sexual reproduction, controlling cholesterol production, and preserving bone strength by helping to maintain the proper balance between bone mineralization and breakdown. Selective estrogen receptor modulators (SERMs) are a class of drugs that act by inhibiting the estrogen receptors of different target tissues, thus blocking cancer cell proliferation.

Several studies documented that many plant-derived micronutrients possess a SERM-like activity. ITCs and indoles have demonstrated to suppress ERα expression in several breast, prostate, and uterus cancer cells.[29] These compounds may also serve as modulators for the ratio of ER-α- to -ß-subtype concentrations. ER subtypes have been shown to have opposing actions on transcription factor activator protein-1-dependent genes, including those involved in mammary proliferation and cell growth.[43] ER-α generally enhances proliferation whereas ER-ß has been shown to reduce proliferation by stimulating cell cycle arrest.[39,44]

EPIGENETIC REGULATION

Epigenetic regulation of gene expression represents an important molecular mechanism connecting environmental factors with the genome; it consists of a modification of DNA that results in a change in gene expression or phenotype without altering the sequence of the genetic code itself. These modifications may have consequences on the health status throughout the life course. One of the most important pathways of epigenetic regulation is the modification of histone proteins, the major structural proteins of chromosomes. The acetylation of histones is a key factor for the process of gene expression. When acetyl groups are removed from histones through the action of the enzyme histone deacetylases (HDAC), access to DNA by transcription factors is restricted, and the process of transcription is suppressed. For these reasons the balance between acetylation and deacetylation is important in cancer prevention and it is often altered during the cancer process.[44]

Recent studies evidenced the potential epigenetic effect of ITCs, which inhibit the action of HDAC and make DNA accessible to transcription factors. Inhibition of HDAC has been shown for ITCs both in *in vitro* and *in vivo* studies.[44] However, this is not the only epigenetic mechanism in which ITCs are involved. ITCs also seem to exert an epigenetic effect via modulation of DNA methylation. The methylation of histones is a reaction mediated by the class of enzymes histone methyl-transferases, which provides a methyl group to be transferred to a lysine or arginine present at the N-terminus of histones. Some authors showed that SFN and other ITCs inhibited HDAC and down-regulated DNA methyltransferases, which led to site-specific CpG island demethylation in the telomerase reverse transcriptase gene in several cancer cell lines.[45-46]

OTHER MECHANISMS

Several studies indicate that ITCs can regulate angiogenesis and metastasis through different pathways. Recently, they have also been shown to modulate micro RNA expression with regulatory effects on cell proliferation, differentiation, apoptosis, Ras activation, p53 signaling, and NF-κB inhibition.[37]

Moreover, ITCs seem to play a crucial role as antibacterial compounds (e.g., the inhibitory effect on *Helicobacter pylori* growth), providing protection against proven risk factors for gastric cancer. Finally, ITCs may also modulate the human gut microbiota.[37]

ROLE OF BROCCOLI INTAKE ON MARKERS OF OXIDATIVE STRESS AND CANCER RISK IN HUMAN INTERVENTION STUDIES: THE CASE OF SMOKERS

As previously described, several studies have demonstrated an inverse association between brassica vegetable intake and cancer risk at several sites (e.g., lung cancer). Brassica vegetables, as a result of the presence of numerous bioactive compounds, could contribute to the reduction of free radical-related molecular damage and this could be particularly important in subjects with high exposure to continuous oxidative stress such as smokers.

Indeed, cigarette smoking has been clearly and unambiguously identified as a direct cause of cancers of the lung but also of the oral cavity, esophagus, stomach, pancreas, larynx, bladder, kidney, and leukemia.[47-48] A complex mixture of at least 4000 constituents and a variety of chemical products of tobacco combustion are inhaled both from mainstream and side stream smoke and a large amount of ROS as well as other substances that produce ROS are generated. ROS can cause oxidative damage to numerous type of cells as demonstrated *in vitro* and *in vivo*, and DNA represents a main target of damage.[49] In normal cells, it has been estimated that ROS can cause about one base modification per 130,000 bases in nuclear DNA. Such modifications can be of a different nature and their chemistry is highly complex. In conditions of oxidative stress, DNA damage can significantly increase due to an imbalance between oxidative product generation and the whole antioxidant defense system.

It is widely recognized that DNA integrity is fundamental for health maintenance, reducing the risk of cancer development. Genome instability caused by the numerous DNA damaging agents could have a catastrophic effect without an efficient DNA repair enzyme system and/or the activation of signals involved in the promotion of apoptosis in irretrievably damaged cells.

In general oxidative DNA damage involves the production of oxidized bases and strand breaks. These DNA lesions, including the promutagenic 8-oxo-7,8-dihydro-2'-deoxyguanosine (8-oxodG), could be implicated in early steps of cancer development.[50]

The analysis of intermediate endpoints to assess the chemopreventive ability of dietary regimens is particularly important and the use of appropriate methods (e.g., comet assay) to evaluate biomarkers of DNA damage and repair (Figure 11.3) represents an essential tool for the study of protective effects of food bioactives in human intervention studies.

As far as brassica vegetables are concerned, only a few studies investigated their potential effect on DNA damage and repair. In a recent human intervention study we found that the intake of a portion of broccoli for only 10 days could increase resistance to *ex vivo* H_2O_2-induced DNA strand breaks (evaluated through the comet assay) in peripheral blood mononuclear cells (PBMCs) of young smokers and nonsmokers. Conversely, oxidated purines (formamidopyrimidine DNA glycosylase-FPG-sensitive sites) were reduced significantly only in smokers.[51] At the same time a significant increase in plasma ITCs, but also in lutein, β-carotene, and folate concentrations was demonstrated, supporting a possible role of these molecules in the observed protective effect but also

underlying the importance of broccoli as a bioavailable source of numerous bioactives.[17]

As initially discussed, smokers have higher levels of oxidative stress, and thus we hypothesized that in these subjects the intake of brassica vegetables could be particularly advantageous determining a reduction of ROS deleterious effects as well as causing an induction of the endogenous defense system, for example GST activity, as suggested by other authors.[52] As mentioned previously GST represents a major group of detoxification enzymes and plays a central role in the elimination of major classes of tobacco carcinogens and ROS. All eukaryotic species possess multiple cytosolic and membrane-bound GST isoenzymes. The cytosolic enzymes are encoded by at least five gene families (designated class alpha, mu, pi, sigma, and theta GST). The level of expression of GST is a crucial factor in determining the sensitivity of cells to a broad spectrum of toxic chemicals. GST induction represents part of an adaptive response mechanism to chemical stress caused by electrophiles and many of the compounds that induce GST are themselves substrates for these enzymes; this is also the case for ITCs. Individual isoenzymes contribute to resistance to carcinogens, antitumor drugs, environmental pollutants, and products of oxidative stress.[53] They have overlapping activities but with different efficiency depending on the substrate. For example, GSTM1 seems particularly active in catalyzing conjugation reactions of ITCs with GSH; thus GSTM1 genotype is particularly interesting to investigate both ITCs absorption and cancer risk.

Since GST activity can depend on GST genotype, smokers enrolled in a subsequent intervention study with broccoli were characterized for their *GSTM1* genotype.[54] In fact it was demonstrated as a difference in ITCs metabolism depending on genotype: compared to GSTM1 positive subjects, GSTM1 null subjects excreted SFN metabolites at a faster rate during the first 6 h after consumption of broccoli, and at higher percentages after 24 h.[18] Moreover it was also hypothesized that GSTM1 null subjects could be more susceptible to DNA damage with respect to those with the positive genotype. Consequently, brassica vegetables intake in GSTM1 null smokers could be particularly advantageous to increase protection against DNA damage and improve the endogenous defense system.

To test this hypothesis a randomized crossover intervention study (Figure 11.4) was performed in which 27 male smokers were asked to consume 250 g of steamed broccoli or a standard diet for 10 days separated by a wash out period of 20 days.[54] The effect of such intervention on endogenous DNA damage (FPG sensitive sites) and on *ex vivo* induced DNA damage was evaluated through the comet assay. The repair activity of oxidized DNA lesions was also investigated through the repair assay and through the evaluation of the expression levels

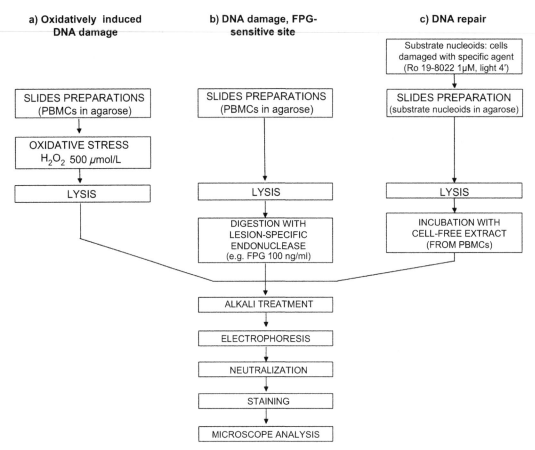

FIGURE 11.3 Evaluation of Endogenous, Oxidatively Induced DNA Damage (cell resistance against H_2O_2 insult) and Repair Activity in PBMCs Evaluated by Comet Assay. In addition, the comet assay can be modified to measure DNA repair activity toward a substrate DNA containing 8-oxodG. A treatment with Ro19-8022/white light, is used to generate 8-oxo-7,8-dihydroguanine. Cell extracts derived from PBMCs are incubated with substrate cells to measure the incision activity, as indicator of repair capacity. Through the assay, damaged DNA is recognized as a fluorescent core followed by a tail (i.e., comet), which is due to the electrophoretic migration of fragments. The quantification of damage is performed through image analysis systems or visual scoring and generally expressed as percentage DNA in tail ((DNA in tail/DNA head + tail)*100). H_2O_2, hydrogen peroxide; SSBs, single-strand breaks; PBMCs, peripheral blood mononuclear cells. The comet assay is a sensitive and rapid technique for quantifying and analyzing DNA damage in any eukaryotic cell population that can be obtained as a single-cell suspension. The damage detected is commonly referred to as strand breaks and alkali-labile sites. Cells, embedded in a thin agarose gel on microscope slides, can be also treated with H_2O_2 to induce DNA damage (i.e., to evaluate cell ability to counteract the oxidative insult). A common modification of the assay allows the measurement specifically of the level of 8-oxo-7,8-dihydro-20-deoxyguanosine (8-oxodG) as well as other altered purines by incorporating a digestion with the bacterial DNA repair enzyme formamidopyrimidine DNA glycosylase (FPG).

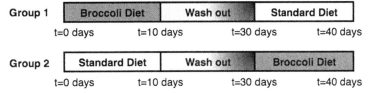

FIGURE 11.4 Effect of Broccoli Intake on Markers of Oxidative Stress and Cancer Risk in Human Volunteers: Experimental Design. Subjects were randomly divided into two groups of subjects and consumed a portion of broccoli or a standard diet, deprived of brassica vegetables, in the following order: Group 1 was assigned to the sequence broccoli diet/wash-out/standard diet, whereas group 2 followed the sequence standard diet/wash-out/broccoli diet. The two experimental treatments were separated by a 20 day wash-out period.

of repair and defense enzymes (i.e., 8-oxoguanine DNA glycosylase (OGG1), nucleoside diphosphate linked moiety X-type motif 1(NUDT1), and heme oxygenase (decycling) 1 (HO-1)).

OGG1 is an important glycosylase involved in the removal of 8-oxodG, while NUDTP1 removes 8-oxo-dGTP from the nucleotide pool to avoid its incorporation during replication or repair. Heme oxygenase 1 was also evaluated as a marker of an increased cell antioxidant protection in PBMCs. In fact the transcription factors for expression of HO-1 include the redox-sensitive Nrf2, which is also involved in the induction of phase II enzymes through the antioxidant responsive element by ITCs.[54]

Following broccoli intake a 41% reduction of FPG sensitive sites and 23% reduction of *ex vivo* induced DNA damage was found, and cell protection against hydrogen peroxide damage was higher in subjects with GSTM1 null genotype with respect to the positive one ($-27.6 \pm 13.6\%$ vs $-13.1 \pm 29.0\%$, respectively).

The effect of broccoli intake on DNA damage registered in our two intervention studies on smokers is reported in Figure 11.5.[51,54]

Other studies have reported protective effects of cruciferous vegetables intake, including Brussels sprouts, broccoli, and watercress, on biomarkers of oxidative damage to DNA in human intervention studies.[51,55–58] For example it was found that the intake of 300 g of steamed Brussels sprouts for 6 days was able to decrease endogenous and H_2O_2-induced DNA damage.[58] The same effect was found after 85 g of raw watercress for 8 weeks, particularly in smokers, and the effect was also associated with increased plasma carotenoid and vitamin C concentrations.[56]

With regard to the effect of brassica vegetables on the expression level and activity of DNA repair enzymes we could not find any effect of broccoli intake, probably due to a very high interindividual variability. Thus, more studies on a large sample of subjects are needed to investigate the potential role of broccoli on DNA repair.

ITCs may also act through the induction and activities of other antioxidant enzymes apart from GST; for example while no significant effect was demonstrated following a watercress intervention on *in vivo* PBMC gene expression of antioxidant enzymes such as glutathione peroxidase and superoxide dismutase (due to high individual variations), an increase in their activities in red blood cells was found particularly in GSTM1 null subjects.[59]

Taken together all the cited results suggest that the intake of brassica vegetables can improve protection against DNA damage through the effect of compounds with direct antioxidant action (e.g., carotenoids, polyphenols, vitamin C, etc.) but also by providing ITCs able to improve the endogenous defense system probably through the modulation of the expression and

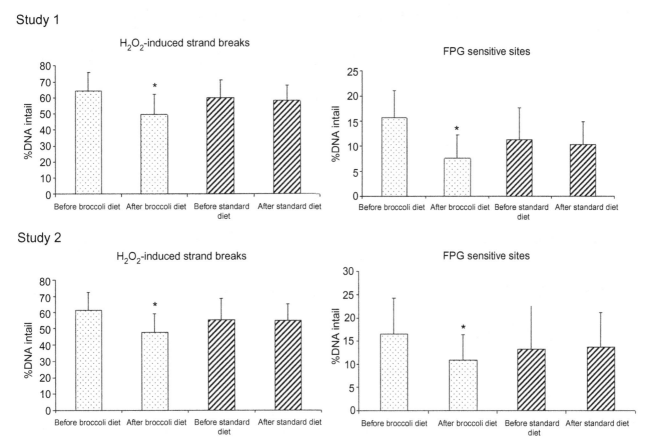

FIGURE 11.5 Effect of Broccoli Intake on Markers of DNA Damage in Smokers. Level of H_2O_2-induced DNA damage and FPG sensitive sites evaluated by comet assay: study 1 (Riso et al.[54]); study 2 (Riso et al.[51]).

activity of detoxification and antioxidant enzymes. Although some results are already available,[60] such effects needs to be further investigated in well-controlled human intervention studies with appropriate sample size.

CLOSING REMARKS AND CONCLUSIONS

To this day, available data suggest that brassica vegetable intake may represent a significant dietary tool for cancer prevention. However, research is still necessary in order to better understand, possibly through human intervention studies, the effect of the bioactives introduced with this class of vegetables on markers related to cancer development (including DNA damage and repair, changes in gene expression, and epigenetic mechanisms).

The concentration of ITCs may be a critical factor in determining their activity. It is increasingly clear that low physiological levels of ITCs such as SFN can produce only subtle effects on cell signaling; however, this is not purposeless as a long-term exposure can result in protective changes. In particular at low concentration, ITCs by inducing mild cellular stress activate the cytoprotective systems, including Keap1–Nrf2-mediated antioxidant response.[61] In contrast it is not excluded, and it is demonstrated *in vitro*, that strong and even toxic effects can be induced by high amounts of ITCs. For this reason, the use of supplements for preventive strategies is still not advisable.

Moreover results from *in vitro* studies indicate the possibility of a separate ITC chemoprevention pathway distinct from the classic Nrf2 pathway and with tumor cell-specific effects.[44] For example, exploration on HDAC activity modulatory effect of SFN rich broccoli represents an interesting issue for future research. Despite the fact that only a few human studies have been conducted, the obtained results are encouraging. In a pilot study a significant decrease in HDAC activity in PBMCs of healthy volunteers (with a concomitant increase in acetylated histones H3 and H4) was found 3 to 6 h after the ingestion of 68 g of broccoli sprouts, while HDAC activity was restored by 24 h, suggesting that this is a transient effect and that dietary HDAC inhibitors are weak ligands.[62] Thus, it is not surprising that in our study the regular intake of broccoli (providing lower amounts of ITCs than the cited study) for 10 days did not produce any change in HDAC activity in PBMCs of our healthy smoker and nonsmoker volunteers.[51] Interestingly, Clarke et al.[63] found that lower HDAC activity in PBMCs was present only in subjects consuming broccoli sprouts versus broccoli supplements for 7 days, and this was related to differences in the metabolism and excretion of glucoraphanin and glucoerucin metabolites in subjects consuming the two different sources.

Despite the increased understanding of absorption and metabolism of different ITCs, conclusions are not always univocal and this is particularly evident when GSTs involvement and polymorphism are also considered; thus future research is warranted.

Additionally a recent review well describes how ITCs, being electrophiles, are capable of covalently modifying proteins, and these modifications may cause functional changes in proteins resulting in downstream effects at the cellular level.[64] Very interesting results come from the study by Traka et al.,[65] who found that broccoli intervention up to 12 months in men with a previous diagnosis of high-grade prostatic intraepithelial neoplasia was associated with perturbation of transforming growth factor-β1, epidermal growth factor, and insulin signaling.[65] This was obtained by comparison of gene expression profiles of GSTM1 positive and null individuals who had consumed a broccoli-rich diet for six months, and by the paired analyses of gene expression profiles from biopsies obtained at 0 and 12 months from GSTM1 positive individuals who had consumed the broccoli-rich diet. The authors demonstrated that ITCs can readily form thioureas with signaling proteins in the plasma through covalently binding with the N-terminal residue, thus affecting transcription.

Finally, apart from remarking the importance of consuming dietary achievable amounts of bioactives for prevention purposes, it should also be emphasized that when discussing the potential anticancer properties of brassica vegetables we refer to whole foods that are sources of numerous different compounds, not just single bioactives. All these compounds can have additive or synergistic effects able to improve the endogenous defense system, and therefore the disease risk.

SUMMARY POINTS

- The intake of brassica vegetables has been inversely associated with cancer risk at several sites with cohort studies supporting inverse associations in particular for cancers of the digestive and respiratory tract. Results from recent meta-analyses suggest that further targeted epidemiological research is necessary to provide conclusive demonstration of the specific protective effect of brassica vegetables.
- The distinctive phytochemicals present in brassica vegetables are glucosinolates, a family of sulfur-containing compounds that are converted to biologically active compounds (i.e., isothiocyanates, and indoles) after ingestion. Broccoli is also a rich source of vitamins, minerals, and other bioactive compounds with antioxidant activity, including phenolics and carotenoids that could help explain the health benefits associated with broccoli consumption.

- Mechanistic studies document that isothyocianates, as other phytochemical compounds, seem to elicit an oxidative stress response crucial for the activation of multiple pathways involved in antioxidant and anticancer effects.
- Available dietary intervention studies with brassica vegetables seem to support their role in the reduction of ROS deleterious effects and in the modulation of cancer risk markers in human volunteers.
- In our studies, regular intake of broccoli for a short period was able to improve cell protection against DNA damage. We hypothesized a direct antioxidant action (e.g., exerted by carotenoids, polyphenols, vitamin C, etc.) and the modulation of endogenous defense systems mediated by isothyocianates (e.g., increased expression and activity of detoxification and antioxidant enzymes).
- Evidence exists that both isothiocyanate absorption and cancer risk could be dependent on GSTM1 genotype as reported by epidemiological and intervention studies. We demonstrated that protection against DNA damage following broccoli intake was higher in subjects with GSTM1 null compared to GSTM1 positive genotype, suggesting an important diet x genotype interaction.

References

1. International Agency for Research on Cancer. *Cruciferous vegetables, isothiocyanates and Indoles*, vol. 9. Lyon, France: IARC Handbooks of Cancer Prevention; 2004.
2. Kim MK, Park JH. Conference on Multidisciplinary approaches to nutritional problems. Symposium on Nutrition and health. Cruciferous vegetable intake and the risk of human cancer: epidemiological evidence. *Proc Nutr Soc* 2009;**68**:103–10.
3. Bosetti C, Filomeno M, Riso P, Polesel J, Levi F, Talamini R, et al. Cruciferous vegetables and cancer risk in a network of case-control studies. *Ann Oncol* 2012;**8**:2198–203.
4. Lam TK, Gallicchio L, Lindsley K, Shiels M, Hammond E, Tao XG, et al. Cruciferous vegetable consumption and lung cancer risk: A systematic review. *Cancer Epidemiol Biomark Prev* 2009;**18**:184–95.
5. Mettlin C. Milk drinking, other beverage habits, and lung cancer risk. *Int J Cancer* 1989;**43**:608–12.
6. Hara M, Hanaoka T, Kobayashi M, Otani T, Adachi HY, Montani A, et al. Cruciferous vegetables, mushrooms, and gastrointestinal cancer risks in a multicenter, hospital-based case-control study in Japan. *Nutr Cancer* 2003;**46**:138–47.
7. Joseph MA, Moysich KB, Freudenheim JL, Shields PG, Bowman ED, Zhang Y, et al. Cruciferous vegetables, genetic polymorphisms in glutathione S-transferases M1 and T1, and prostate cancer risk. *Nutr Cancer* 2004;**50**:206–13.
8. Kirsh VA, Peters U, Mayne ST, Subar AF, Chatterjee N, Johnson CC, et al. Prospective study of fruit and vegetable intake and risk of prostate cancer. *J Natl Cancer Inst* 2007;**99**:1200–9.
9. Ambrosone CB, McCann SE, Freudenheim JL, Marshall JR, Zhang Y, Shields PG. Breast cancer risk in premenopausal women is inversely associated with consumption of broccoli, a source of isothiocyanates, but is not modified by GST genotype. *J Nutr* 2004;**134**:1134–8.
10. Koushik A, Hunter DJ, Spiegelman Jr D, Kato I, Krogh V, Larsson SC, et al. Fruits, vegetables, and colon cancer risk in a pooled analysis of 14 cohort studies. *J Natl Cancer Inst* 2007;**99**:1471–83.
11. Lin HJ, Probst-Hensch NM, Louie AD, Kau IH, Witte JS, Ingles SA, et al. Glutathione transferase null genotype, broccoli and lower prevalence of colorectal adenomas. *Cancer Epidemiol Biomark Prev* 1998;**7**:647–52.
12. Michaud DS, Spiegelman D, Clinton SK, Rimm EB, Willett WC, Giovannucci EL. Fruit and vegetable intake and incidence of bladder cancer in a male prospective cohort. *J Natl Cancer Inst* 1999;**91**:605–13.
13. Slattery ML, Kampman E, Samowitz W, Caan BJ, Potter JD. Interplay between dietary inducers of GST and the GSTM-1 genotype in colon cancer. *Int J Cancer* 2000;**87**:728–33.
14. Brennan P, Hsu CC, Moullan N, Szeszenia-Dabrowska N, Lissowska J, Zaridze D, et al. Effect of cruciferous vegetables on lung cancer in patients stratified by genetic status: A mendelian randomisation approach. *Lancet* 2005;**366**:1558–60.
15. Appel KE, Lampen A. Health benefits and possible risks of broccoli—An overview. *Food Chem Toxicol* 2011;**49**:3287–309.
16. Fahey JW, Zalcmann AT, Talalay P. The chemical diversity and distribution of glucosinolates and isothiocyanates among plants. *Phytochemistry* 2001;**56**:5–51.
17. Riso P, Brusamolino A, Moro M, Porrini M. Absorption of bioactive compounds from steamed broccoli and their effect on plasma glutathione S-transferase activity. *Int J Food Sci Nutr* 2009 60S1: 56–71.
18. Gasper AV, Al-janobi A, Smith JA, Bacon JR, Fortun P, Atherton C, et al. Glutathione S transferase M1 polymorphism and metabolism of sulforaphane from standard and high-glucosinolate broccoli. *Am J Clin Nutr* 2005;**82**:1283–91.
19. Podsedek A. Natural antioxidants and antioxidant capacity of Brassica vegetables: A review. *LWT* 2007;**40**:1–11.
20. Singh J, Upadhyay AK, Prasad K, Bahadur A, Rai M. Variability of carotenes, vitamin C, E and phenolics in Brassica vegetables. *J Food Compos Anal* 2007;**20**:106–12.
21. Ferguson LR, Schlothauer RC. The potential role of nutritional genomic tools in validating high health foods for cancer control: Broccoli as example. *Mol Nutr Food Res* 2012;**56**:126–46.
22. Demirovic D, Rattan SI. Establishing cellular stress response profiles as biomarkers of homeodynamics, health and hormesis. *Exp Gerontol* 2013;**48**:94–8.
23. Iyanagi T. Molecular mechanism of phase I and phase II drug-metabolizing enzymes: Implications for detoxification. *Int Rev Cytol* 2007;**260**:35–112.
24. James D, Devaraj S, Bellur P, Lakkanna S, Vicini J, Boddupalli S. Novel concepts of broccoli sulforaphanes and disease: Induction of phase II antioxidant and detoxification enzymes by enhanced-glucoraphanin broccoli. *Nutr Rev* 2012;**70**:654–65.
25. Ye L, Zhang Y. Total intracellular accumulation levels of dietary isothiocyanates determine their activity in elevation of cellular glutathione and induction of Phase 2 detoxification enzymes. *Carcinogenesis* 2001;**22**:1987–92.
26. Abdull Razis AF, Bagatta M, De Nicola GR, Iori R, Ioannides C. Up-regulation of cytochrome P450 and phase II enzyme systems in rat precision-cut rat lung slices by the intact glucosinolates, glucoraphanin and glucoerucin. *Lung Cancer* 2011;**71**:298–305.
27. Konsue N, Ioannides C. Differential response of four human livers to modulation of phase II enzyme systems by the chemopreventive phytochemical phenethyl isothiocyanate. *Mol Nutr Food Res* 2010;**54**:1477–85.
28. Jana S, Mandlekar S. Role of phase II drug metabolizing enzymes in cancer chemoprevention. *Curr Drug Metab* 2009;**10**:595–616.
29. Barcelo S, Gardiner JM, Gescher A, Chipman JK. CYP2E1-mediated mechanism of anti-genotoxicity of the broccoli constituent sulforaphane. *Carcinogenesis* 1996;**17**:277–82.
30. Weng JR, Tsai CH, Kulp SK, Chen CS. Indole-3-carbinol as a chemoprevention and anti-cancer agent. *Cancer Lett* 2008;**262**:153–63.

31. Hayes JD, Kelleher MO, Eggleston IM. The cancer chemopreventive actions of phytochemicals derived from glucosinolates. Eur J Nutr 2008;47:73–88.

32. Gilmore TD. Introduction to NF-kappaB: players, pathways, perspectives. Oncogene 2006;25:6680–4.

33. Jeong WS, Kim IW, Hu R, Kong AN. Modulatory properties of various natural chemopreventive agents on the activation of NK-kappaB signalling pathway. Pharm Res 2004;21:661–70.

34. Youn HS, Kim YS, Park ZY, Kim SY, Choi NY, Joung SM, et al. Sulforaphane suppresses oligomerization of TLR4 in thiol-dependent manner. J Immunol 2010;184:411–9.

35. Wu L, Noyan Ashraf MH, Facci M, Wang R, Paterson PG, Ferrie A, et al. Dietary approach to attenuate oxidative stress, hypertension, and inflammation in the cardiovascular system. Proc Natl Acad Sci USA 2004;101:7094–9.

36. Clarke JD, Dashwood RH, Ho E. Multi-targeted prevention of cancer by sulforaphane. Cancer Lett 2008;269:291–304.

37. Karin M. Nuclear factor-kappaB in cancer development and progression. Nature 2006;441:431–6.

38. Heiss E, Herhaus C, Klimo K, Bartsch H, Gerhäuser C. Nuclear factor kappa B is a molecular target for sulforaphane-mediated anti-inflammatory mechanisms. J Biol Chem 2001;276:32008–15.

39. Navarro SL, Li F, Lampe JW. Mechanisms of action of isothiocyanates in cancer chemoprevention: An update. Food Funct 2011;2:579–87.

40. Keum YS, Jeong WS, Kong AN. Chemoprevention by isothiocyanates and their underlying molecular signalling mechanisms. Mutat Res 2004;555:191–202.

41. Xiao D, Powolny AA, Moura MB, Kelley EE, Bommareddy A, Kim SH, et al. Phenethyl isothiocyanate inhibits oxidative phosphorylation to trigger reactive oxygen species-mediated death of human prostate cancer cells. J Biol Chem 2010;285:26558–69.

42. Mi L, Gan N, Chung FL. Isothiocyanates inhibit proteosome activity and proliferation of multiple myeloma cells. Carcinogenesis 2010;32:216–23.

43. Paech K, Webb P, Kuiper GG, Nilsson S, Gustafsson J, Kushner PJ, et al. Differential ligand activation of estrogen receptors ERalpha and ERbeta at AP1 sites. Science 1997;277:1508–10.

44. Dashwood RH, Ho E. Dietary histone deacetylase inhibitors: from cells to mice to man. Semin Cancer Biol 2007;17:363–9.

45. Wang LG, Beklemisheva A, Liu XM, Ferrari AC, Feng J, Chiao JW. Dual action on promoter demethylation and chromatin by an isothiocyanate restored GSTP1 silenced in prostate cancer. Mol Carcinog 2007;46:24–31.

46. Meeran SM, Patel SN, Tollefsbol TO. Sulforaphane causes epigenetic repression of hTERT expression in human breast cancer cell lines. Plos One 2010;5:e11457.

47. Stämpfli MR, Anderson GP. How cigarette smoke skews immune responses to promote infection, lung disease and cancer. Nat Rev Immunol 2009;9:377–84.

48. Valavanidis A, Vlachogianni T, Fiotakis K. Tobacco smoke: Involvement of reactive oxygen species and stable free radicals in mechanisms of oxidative damage, carcinogenesis and synergistic effects with other respirable particles. Int J Environ Res Public Health 2009;6:445–62.

49. Storr S, Woolston C, Zhang Y, Martin S. Redox environment, free radical and oxidative DNA damage. Antioxid Redox Signal 2012;18(18):2399–408. http://dx.doi.org/10.1089/ars.2012.4920.

50. Evans MD, Dizdaroglu M, Cooke MS. Oxidative DNA damage and disease: Induction, repair and significance. Mutat Res 2004;567:1–61.

51. Riso P, Martini D, Møller P, Loft S, Bonacina G, Moro M, et al. DNA damage and repair activity after broccoli intake in young healthy smokers. Mutagenesis 2010;25:595–602.

52. Navarro SL, Chang JL, Peterson S, Chen C, King IB, Schwarz Y, et al. Modulation of human serum glutathione S-transferase A1/2 concentration by cruciferous vegetables in a controlled feeding study is influenced by GSTM1 and GSTT1 genotypes. Cancer Epidemiol Biomark Prev 2009;18:2974–8.

53. Hayes JD, Pulford DJ. The glutathione S-transferase supergene family: Regulation of GST and the contribution of the isoenzymes to cancer chemoprotection and drug resistance. Crit Rev Biochem Mol Biol 1995;30:445–600.

54. Riso P, Martini D, Visioli F, Martinetti A, Porrini M. Effect of broccoli intake on markers related to oxidative stress in healthy smokers and nonsmokers. Nutr Cancer 2009;61:232–7.

55. Murashima M, Watanabe S, Zhuo XG, Uehara M, Kurashige A. Phase 1 study of multiple biomarkers for metabolism and oxidative stress after one-week intake of broccoli sprouts. Biofactors 2004;22:271–5.

56. Gill CI, Haldar S, Boyd LA, Bennett R, Whiteford J, Butler M, et al. Watercress supplementation in diet reduces lymphocytes DNA damage and alters blood antioxidant status in healthy adults. Am J Clin Nutr 2007;85:504–10.

57. Verhagen H, Poulsen HE, Loft S, van Poppel G, Willems MI, van Bladeren PJ. Reduction of oxidative DNA damage in humans by Brussels sprouts. Carcinogenesis 1995;16:969–70.

58. Hoelzl C, Glatt H, Meinl W, et al. Consumption of Brussels sprouts protects peripheral human lymphocytes against 2-amino-1-methyl-6-phenylimidazol[4,5-b]pyridine (PhIP) and oxidative DNA-damage: Results of a controlled human intervention trial. Mol Nutr Food Res 2008;52:330–41.

59. Hofmann T, Kuhnert A, Schubert A, Gill C, Roland IR, Pool-Zobel BL, et al. Modulation of detoxification enzymes by watercress: In vitro and in vivo investigations in human peripheral blood cells. Eur J Nutr 2009;48:483–91.

60. Boddupalli S, Mein JR, Lakkanna S, James DR. Induction of phase 2 antioxidant enzymes by broccoli sulforaphane: Perspectives in maintaining the antioxidant activity of vitamins A, C, and E. Front Genet 2012;3:7.

61. Dinkova-Kostova AT, Holtzclaw WD, Cole RN, Itoh K, Wakabayashi N, Katoh Y, et al. Direct evidence that sulfhydryl groups of Keap1 are the sensors regulating induction of phase 2 enzymes that protect against carcinogens and oxidants. Proc Natl Acad Sci 2002;99:11908–13.

62. Myzak MC, Tong P, Dashwood WM, Dashwood RH, Ho E. Sulforaphane retards the growth of human PC-3 xenografts and inhibits HDAC activity in human subjects. Exp Biol Med 2007;232:27–34.

63. Clarke JD, Hsu A, Riedl K, Bella D, Schwartz SJ, Stevens JF, et al. Bioavailability and inter-conversion of sulforaphane and erucin in human subjects consuming broccoli sprouts or broccoli supplement in a cross-over study design. Pharmacol Res 2011;64:456–63.

64. Mi L, Gan N, Chung FL. Isothiocyanates inhibit proteasome activity and proliferation of multiple myeloma cells. Carcinogenesis 2011;32:216–23.

65. Traka M, Gasper AV, Melchini A, Bacon JR, Needs PW, Frost V, et al. Broccoli consumption interacts with GSTM1 to perturb oncogenic signalling pathways in the prostate. PLoS One 2008; 3:e2568.

Resveratrol and Lycopene in the Diet and Cancer Prevention

Vipin Arora, Anand Kamal Sachdeva, Prathistha Singh, Ankita Baveja, Kanwaljit Chopra, Anurag Kuhad

University Institute of Pharmaceutical Sciences, UGC-CAS, Panjab University, Chandigarh, India

List of Abbreviations

APAF-I Apoptotic protease activating factor 1
AKT Protein Kinase B (PKB)
ARE/ EpRE Antioxidant/electrophile response element
BAX Bcl-2–associated X protein
Bcl-2 B-cell lymphoma 2
Bcl-XL B-cell lymphoma-XL
bZIP Basic leucine zipper
Cdc Cell division cycle
CDK Cyclin-dependent kinases
COX-2 Cyclooxygenase-2
CYP Cytochrome p450
Egr-1 Early Growth Response-1
ERK Extracellular signal-regulated kinases
EGFR Epidermal Growth Factor Receptor
IFN Interferon
IGFR-1 Insulin-like growth factor receptor-1
IL Interleukin
Keap1 Kelch-like ECH-associated protein 1
MAPK Mitogen activated protein kinase
MMP Matrix metalloproteinase
mTOR mammalian target of rapamycin
NF-κB Nuclear factor kappa-light-chain-enhancer of activated B cells
Nrf2 Nuclear factor erythroid 2p45 related factor 2
iNOS Inducible nitric oxide synthase
PCNA Proliferating cell nuclear antigen
PDGF-BB Platelet derived growth factor-BB
pRB retinoblastoma protein
PI3K Phosphoinositol 3-kinase
PMA Phorbol 12-myristate13-acetate
PKCd Protein kinase C delta
ROS Reactive oxygen species
TEAC Trolox-equivalent antioxidant capacity
TGFβ Transforming growth factor
TNF-α Tissue necrosis factor-α
TRIAL TNF-related apoptosis-inducing ligand
UVB Ultraviolet B
VEGFR Vascular endothelial growth factor receptor

INTRODUCTION

Cancer is a leading cause of death worldwide, accounting for 7.6 million deaths in 2008. By 2030, global projection suggests that 22.3 million new cases of cancer will emerge and mortality will increase to 13.5 million deaths.[1] In the recent past, to arrest this global pandemic natural dietary agents such as fruits, vegetables, and spices have drawn a great deal of attention from both the scientific community and the general public, owing to their putative ability to suppress cancers[2] Dietary agents consist of a wide variety of biologically active compounds that are ubiquitous in plants, and the relationship between diet and cancer has been recognized throughout the recorded history. In 168 BCE, Galen, a Roman physician, wrote that unhealthy diet and bad climate were directly connected to cancer. In the Middle Ages (1676), Wiseman suggested that cancer might arise from "an error in diet," and recommended avoiding "salt, sharp and gross meats." The scientific studies of diet in relation to cancer risk started from the observations of Peyton Rous, when in 1914 he observed that restriction of food consumption delayed the development of cancer metastases in mice[3] and when Dr. Michael Sporn first coined the term "chemoprevention," referring to the possibility that natural forms of vitamin A could prevent the development and progression of epithelial cancer.[4] He probably did not imagine the impact that this new approach of chemoprevention would have had in cancer research. However, the study of diet and cancer risk reduction is complicated not only by the multistage, multifactorial nature of the disease, but also because of the inherent complexities of any diet. A diet is composed of a multitude of nutrients as well as nonnutritive components. In addition to the food itself, factors such as

Cancer
http://dx.doi.org/10.1016/B978-0-12-405205-5.00012-X

© 2014 Elsevier Inc. All rights reserved.

food preparation methods, portion sizes, nutrient synergy, the dietary pattern, and the role of physical activity on calorie balance are all a part of the nutritional picture that may affect cancer risk.[5]

In recent years, many studies have been carried out to investigate the potential cancer chemopreventive activities of natural antioxidants, in particular, dietary polyphenol. This chapter particularly deals with the chemopreventive properties of resveratrol and lycopene.

RESVERATROL

Resveratrol (3,5,4'-trihydroxystilbene) was first isolated from the roots of white hellebore (*Veratrum grandiflorum O. Loes*) in 1940,[6] and later, in 1963, from the roots of *Polygonum cuspidatum*, a plant used in traditional Chinese and Japanese medicine.[7] It is a naturally occurring phylloalexin produced by wide variety of plants such as grapes (Vitis *vinifera*), peanuts (Arachis *hypogaea*), and mulberries in response to stress, injury, ultraviolet (UV) radiation, and fungal infection (Botrytis *cinerea*). The root of the word resveratrol is a combination of the Latin prefix *Res*, meaning "which comes from," *veratr*, from the plant "Veratrum," and the suffix *ol*, indicating that it contains "alcohol" chemical groups. It exists in two isoforms; *trans*-resveratrol and *cis*-resveratrol where the *trans*-isomer is the more stable form (Figure 12.1). Resveratrol-glucuronide is the major form absorbed as compared to the very minute amounts of unconjugated resveratrol and resveratrol sulfate.

Frankel and coworkers (1993) were the first to demonstrate the ability of *trans*-resveratrol to function as an antioxidant.[8] Subsequently, Soleas and coworkers (1997) reported that resveratrol contributed significantly to the total antioxidant activity of wine as evaluated using the redox *in-vitro* assay.[9] Resveratrol is able to prevent the increase in ROS following exposure to oxidative agents such as tobacco-smoke condensate (TAR), H_2O_2, phorbol esters, ultraviolet radiation, and to decrease and scavenge ROS. It appears that resveratrol is an effective scavenger of hydroxyl, superoxide, and metal-induced radicals. Resveratrol exhibits a protective effect against lipid peroxidation in cell membranes and DNA damage

caused by ROS.[10] It is believed to be responsible for the so-called French Paradox, in which consumption of red wine has been shown to reduce the mortality rates from cardiovascular diseases and certain cancers.[11] In recent years, it got attention not only for its usefulness in French Paradox as a phytoestrogen agent,[12] but also for its anticancer[13] and antioxidant properties.[14]

Pharmacokinetics of Resveratrol

Bioavailability and metabolism of resveratrol have been widely studied; in rats and in humans its efficacy depends on its absorption and metabolism. In rats, resveratrol has been detected in the feces, urine, bile, and plasma as well as in kidneys, stomach, intestine, and liver, following oral administration. Resveratrol is efficiently absorbed in rats and in humans after oral administration. Resveratrol is metabolized by liver phase II enzymes leading to the production of mostly its glucoronide and sulphate metabolites. These metabolites include *trans*-resveratrol-3-*O*-glucuronide, *trans*-resveratrol-3-sulfate, *trans*-resveratrol-4'-sulphate, *trans*-resveratrol-3,5-disulfate, *trans*-resveratrol-3,4'-disulfate, and *trans*-resveratrol-3,4',5-trisulphate. As the parent structure has been discovered to remain intact in some target tissues, it could be that the phase I enzymes do not play any significant role in resveratrol metabolism. Despite its low bioavailability, resveratrol has been reported to exhibit its anticancer activity in rats even though there was no observed accumulation of this phytochemical in tumor tissues, where activity is needed. After biotransformation by detoxification enzymes, resveratrol and its conjugates are excreted in the urine and feces.[15,16]

Antioxidant Potential of Resveratrol

Excessive ROS accumulation may induce the oxidative modification of cellular macromolecules (lipid, proteins, and nucleic acids) with deleterious potential. In fact, DNA damage by ROS has been implicated in mutagenesis, oncogenesis, and aging. Oxidative lesions in DNA include base modifications, sugar damage, strand breaks, and basic sites. Since gene transcription can be regulated by oxidants, antioxidants, and other determinants of the intracellular redox state, ROS can also produce protein

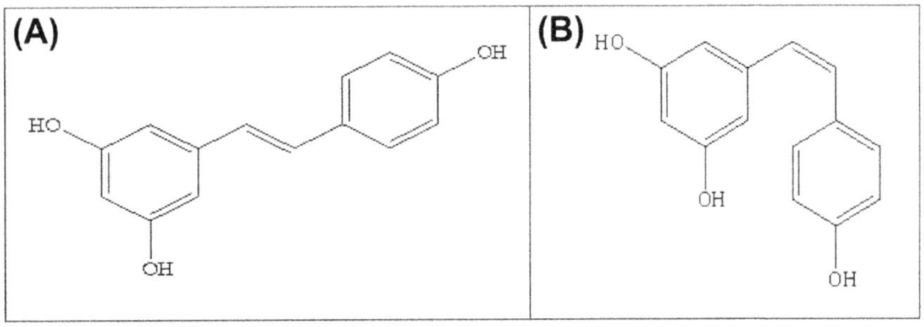

FIGURE 12.1 **Chemical Structure of Resveratrol.** 5-[2-(4-Hydroxy-phenyl)-vinyl]-benzene-1,3-diol / trans-Resveratrol 5-[2-(4-Hydroxy-phenyl)-vinyl]-benzene-1,3-diol / cis-Resveratrol

damage, inducing other types of mutations. Resveratrol is both a free radical scavenger and a potent antioxidant because of its ability to promote the activities of a variety of antioxidant enzymes (Figure 12.2). The ability of the polyphenolic compounds to act as antioxidants depends on the redox properties of their phenolic hydroxy groups and the potential for electron delocalization across the chemical structure. The common recognition of resveratrol as a natural antioxidant was clarified by Zini et al., (1999) who suggested three different antioxidant mechanisms: (1) competition with coenzyme Q and, to decrease the oxidative chain complex, the site of ROS generation; (2) scavenging $O_2\cdot^-$ radicals formed in the mitochondria; and (3) inhibition of lipid peroxidation induced by Fenton reaction products.[17]

To protect tissues against the deleterious effects of ROS, all cells possess numerous defense mechanisms that include enzymes such as SOD (superoxide dismutase), catalase, glutathione reductase, and glutathione peroxidase. Resveratrol can maintain the concentration of intracellular antioxidants found in biological systems. For instance, in a study by Losa (2003), stilbene appeared to maintain the glutathione content in peripheral blood mononuclear cells isolated *ex vivo* from a healthy human from oxidative damage caused by 2-deoxy-D-ribose.[18] In a previous study, in human blood platelets, resveratrol markedly decreased oxidation of thiol groups of proteins

in these cells. Similarly, resveratrol induced an increase in glutathione levels in a concentration-dependent manner in human lymphocytes activated with H_2O_2.

Resveratrol has been shown to increase glutathione peroxidase, glutathione S-transferase, and glutathione reductase. Scientific literature has reported the role of resveratrol in the regulation of NO production from vascular endothelium in the ischemic heart, brain, or kidney. Abnormally high concentrations of NO and its derivatives RNS have been associated with tumor growth and vascular invasion. Phytoalexin considerably diminished NO production upon the inducible isoform of NOS (iNOS expression), and it also induced an inhibitory effect on the iNOS enzyme activity.

Anticancer Potential of Resveratrol

Resveratrol prevents the initial DNA damage by two different pathways: (1) acting as an antimutagen through the induction of Phase II enzymes, such as quinine reductase, capable of metabolically detoxifying carcinogens by inhibiting COX and cytochrome P450, and (2) acting as an antioxidant through inhibition of DNA damage by ROS. It has been proposed that ROS derived from lipid peroxidation may function as tumor initiators. Leonard et al. (2003) have shown that resveratrol exhibits a protective effect against lipid peroxidation

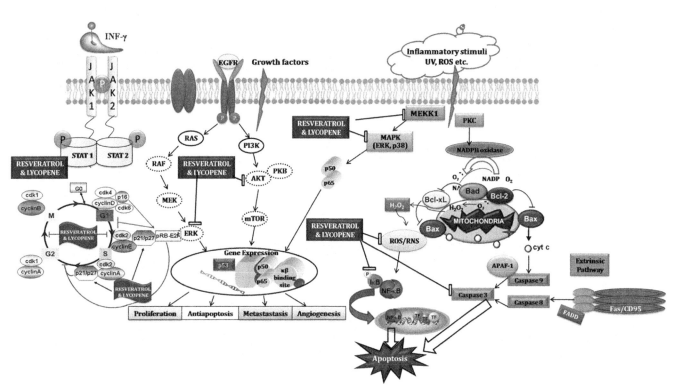

FIGURE 12.2 Antioxidant & Anticancer Properties of Resveratrol. Bcl-2: B-cell lymphoma 2; CDK: Cyclin-dependent kinases; ERK: Extracellular signal-regulated kinases; NF-κB: Nuclear factor kappa-light-chain-enhancer of activated B cells; PMA: Phorbol 12-myristate13-acetate; PKCd: Protein kinase C delta

in cell membranes and DNA damage caused by ROS.[10] The antipromotional properties of resveratrol can be partly attributed to its ability to enhance gap-junctional intercellular communications in cells exposed to tumor promoters such as PMA. The tumor-promoting activity mediated by PMA has also been associated with oxidative stress by increased production of O_2^- and H_2O_2, reduction of SOD activity, and interference with glutathione metabolism. In a model of PMA application to mouse skin, resveratrol induced the restoration of H_2O_2 and glutathione levels, and also myeloperoxidase, glutathione reductase, and SOD activities. The development of skin cancer is related to accumulative exposure to solar UVB as well as the nuclear transcription factor NF-κB, which plays a critical role in skin biology. NF-κB is involved in the inflammatory and carcinogenic signaling cascades, and resveratrol was able to block the damage caused by UVB exposure via its antioxidant properties blocking UVB-mediated NF-κB activation. Finally, resveratrol could inhibit tumor progression, partly by an inhibition of DNA polymerase and deoxyribonucleotide synthesis through its ability to scavenge the essential tyrosine radical of the ribonucleotide reductase and partly by inducing cell cycle arrest.

Chemoprevention and chemotherapy can be obtained by acting at multiple levels and by impairing combinatorial effects responsible for mutagenesis and mitogenesis. Among food-derived molecules that have been screened for the ability to inhibit or reverse such cellular processes, resveratrol is particularly interesting because it affects a broad range of intracellular mediators involved in the initiation, promotion, and progression of cancer (Figure 12.3). Many studies revealed a variety of resveratrol intracellular targets whose modulation gives rise to overlapping responses that lead to growth arrest and death.[15,16,19]

Alterations in Hormonal Activity

An alteration in estrogen and androgen levels leads to deregulated functioning of tissues and organs. Imbalanced hormonal stimulation may favor cell proliferation over differentiation and senescence, and may increase the risk of developing cancer. Consequently, hormone-dependent tumors (breast and prostate, but also others such as colon and lung) may be prevented by the daily intake of appropriate amounts of selective estrogen receptor modulators (SERMs). The polyphenol resveratrol can be considered as a dietary phytoestrogen with powerful beneficial effects on both ER (estrogen receptor)-expressing and nonexpressing human tumors. The chemical structure of resveratrol is similar to that of the synthetic estrogen diethylstilbestrol (4,4-dihydroxy-trans- diethylstilbene). Resveratrol belongs to the type I class of estrogens. It binds ERs in the low micromolar range with an affinity lower than that of estradiol; therefore, it behaves as a weak competitor. Despite the lower binding affinity, resveratrol may act as a super agonist in activating hormone receptor-mediated gene transcription. The super agonistic induction of gene expression is related to the promoter context and varies according to cell type.

Aside from super agonism, resveratrol also exerts an antiestrogen action by triggering parallel pathways that inhibit estrogen-induced cellular outcomes, such as proliferation, tumoral transformation, and progression. In the presence of estrogen, resveratrol exerted a mixed agonistic-antagonistic action in ER-positive breast cancer cells. Resveratrol binding to ERα suppressed the expression of a form of ER and regulated androgen receptor

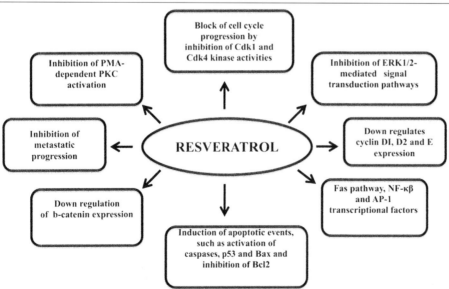

FIGURE 12.3 **Molecular Targets of Anti-cancer Properties of Resveratrol.** APAF-I: Apoptotic protease activating factor 1; AKT: Protein Kinase B (PKB); BAX: Bcl-2–associated X protein; Bcl-2: B-cell lymphoma 2; CDK: Cyclin-dependent kinases; ERK: Extracellular signal-regulated kinases; EGFR: Epidermal Growth Factor Receptor; IGFR-1: Insulin-like growth factor receptor-1; IL: Interleukin; MAPK: Mitogen activated protein kinase; MEK: Mitogen-activated protein kinase; mTOR: mammalian target of rapamycin; NF-κB: Nuclear factor kappa-light-chain-enhancer of activated B cells; pRB: retinoblastoma protein; PI3K: Phosphoinositol 3-kinase; PKC: Protein kinase C; ROS: Reactive oxygen species

signaling by repressing receptor coactivators and downstream gene transcription. It has been proposed that resveratrol inhibition of focal adhesion kinase (FAK) and protein kinase B (PKB/Akt) is responsible for inducing apoptosis in ER-positive breast cancer cells. Steroids (both estrogens and androgens) and their receptor complexes modulate mitogen-activated protein kinases (MAPKs) (by activating ERK, down-regulating Jun N-terminal kinases (JNK)) and thus downstream transcriptional events. Mitogen-activated protein kinases have been shown to also mediate upstream events induced by resveratrol, leading to downstream regulation of transcriptional factors. In breast cancer cells, 17α-estradiol reversed resveratrol-induced apoptosis even if estrogens and resveratrol acted similarly in stimulating MAPKs (specifically ERKs), either when administered alone or in combination (additive effect). This contrasting evidence may be explained by considering resveratrol ability to alter nongenotropic activities of steroid receptor complexes. Activated ER may rapidly stimulate the activity of G proteins and protein kinases, independently of gene transcriptional events.

Resveratrol may be considered as a natural SERM, although the balance between prosurvival genotropic and opposing nongenotropic activities is not clearly predictable due to the role of a broad array of intervening factors. At the low doses provided by dietary intake, resveratrol may act as a weak estrogen competitor, according to the receptor expression and hormonal status of tissues; it counteracts the proliferative effects of hormones and provides a balancing antitumoral activity. Tissue-specific expression of α and β ERs, cofactors regulating DNA binding, and different gene promoters all modulate the cellular response to resveratrol. In the absence of endogenous hormones and according to cellular specificity, the superagonistic activity of resveratrol may act in an opposite manner and prevent tissue senescence and apoptosis. When stress signals overcome proliferative signals, or when the latter are missing (absence of hormones), the polyphenol induced pathway may switch to apoptosis. Aggressive breast and prostate tumors often lose ER expression and become estrogen independent. However, resveratrol is also able to affect ER negative cells. Besides binding hormone receptors, resveratrol is able to trigger intracellular events that affect different metabolic pathways. Indeed, resveratrol exhibits antitumoral properties in a variety of cells that do not express steroid hormone receptors, as well as in cells treated with ER blockers (e.g., tamoxifen).[15,16,19]

Alterations in Metabolic Enzymes

Chemopreventive strategies include the inhibition of phase I enzymes responsible for activating xenobiotics and the induction of phase II enzymes that conjugate these activated compounds to endogenous ligands (e.g., glutathione). Resveratrol, being an exogenous lipophilic compound, can cross plasma membrane, be subjected to cellular metabolism, and possibly interacts with phase I enzymes. Resveratrol inhibited human recombinant CYP450 in vitro. Jang et al. (1997) found that resveratrol reduced the insurgence of preneoplastic lesions in mouse mammary gland cultures and decreased the incidence of tumor formation in mice treated with 7,12-dimethylbenz[a]anthracene (DMBA) used as tumor initiator, in combination with phorbol esters used as tumor promoter.[20] Since DMBA requires bioactivation by phase I enzymes CYP1A1, CYP1A2, and CYP1B1, the antitumoral activity of resveratrol in vivo includes prevention of the initiation phase of carcinogenesis by inhibiting phase I enzymes. Resveratrol was further shown to induce phase II enzymes such as UDP-glucuronyltransferase and NAD(P)H:quinone oxidoreductase in mouse epidermis.[15,16,19]

Modulation of Nitrosative Stress

High levels of endogenously produced NO may either induce apoptosis (by generating toxic reactive intermediates) or inhibit it, depending on the intracellular redox state. The promoter region of iNOS is controlled by transcription factors such as NF-κB, AP1 (Jun/Fos), CREB, and STATs. The expression of iNOS is often associated with the induction of tumor markers such as cyclooxygenase 2 (COX-2), vascular endothelial adhesion molecule 1 (VCAM1), and intercellular adhesion molecule 1 (ICAM1). Finally, iNOS is upregulated in cancer cells in relation to tumor progression. Resveratrol may favor antitumoral activity by impairing the process of angiogenesis. The compound was shown to inhibit endothelial cell migration and tube formation and to block oxygen radical formation and the related SRC-mediated expression of vascular endothelial cadherins. Resveratrol induced the expression of the endothelial isoform of NOS (eNOS), in association with the activation of p53 and p21 and cell cycle arrest in S/G2 phase.[15,16,19]

Modulation of Cyclooxygenase

Resveratrol was first discovered in roots for its ability to inhibit the activity of COXs, enzymes that catalyze the first committed steps of prostaglandin (PG) biosynthesis. Prostaglandins are known stimulators of cell proliferation and angiogenesis, and suppressors of immune surveillance. Cyclooxygenases also possess a hydroperoxidase activity that converts PGG2 to PGH2; this activity, which generates tyrosyl radicals, may be involved in the bioactivation of promutagens, as are phase I enzymes. Resveratrol noncompetitively inhibited both the COX and hydroperoxidase activities of COX-1 in vitro. This dual effect of resveratrol is unique, since classic nonsteroidal anti-inflammatory drugs only affect the COX activity. Since overexpression of COX-2 inhibits apoptosis and promotes metastasis of tumor cells, targeting COX-2 expression could be a promising chemotherapy strategy. Resveratrol inhibited human recombinant COX-2 in *in vitro* assays. Resveratrol discriminated between the two COX isoforms, being a poor inhibitor of COX-2 hydroperoxidase activity. Resveratrol was proposed in breast cancer cells to act by inhibiting the activation of the transcriptional factor NF-κB and upstream of COX-2 expression by inhibiting the activation of a cyclic-AMP-responsive element. In addition, in a similar cell model, micromolar concentrations of resveratrol blocked phorbol-ester-mediated translocation of protein kinase C to the membrane and activation of the COX-2 promoter.[15,16,19]

Alterations in Cell Cycle-Regulating Proteins

Experimental studies found that the resveratrol-induced cell cycle block did not cause apoptosis, was reversible, and was associated with cell differentiation in a variety of cancer cells. Cell sensitivity to apoptotic agents may change throughout cell cycle phases; therefore, resveratrol's ability to arrest cancer cells in S phase, which is the most vulnerable, may strongly increase its chemotherapeutic potential. Moreover, many authors reported that the induction of cell cycle arrest by resveratrol was followed by apoptotic cell death.

Various protein targets of resveratrol that play pivotal roles in cell cycle progression are exerted by p53 and by the retinoblastoma gene product (Rb). The p53 oncosuppressor is a DNA-binding protein that activates transcription of genes that induce cell cycle arrest. p53 is present at low levels in normally proliferating cells, and it accumulates, as a result of increased stability due to acetylation and phosphorylation, in response to stress and in senescence. p53 arrests the cell cycle in G1 by activating the cyclin inhibitor p21. The outcome of p53-arrested cells is apoptosis mediated by mitochondrial Bcl family proteins, resulting from the increased expression of proapoptotic factors and the activation of caspases. In a variety of cellular models, resveratrol strongly upregulated p53 and p21, imposing a checkpoint at G1/S transition, although the compound was also reported to act independently from p53 cellular status. This altered equilibrium led to the modulation of CDKs and cyclins, both at transcriptional and posttranscriptional levels, resulting in cell cycle arrest in a specific phase.

Resveratrol activated transcription of a whole set of p53- responsive genes (e.g., p21, p300/CBP, Apaf1, and BAK) related to cell cycle arrest and apoptosis, while it down-regulated tumor-associated antigens (e.g., PSA), NF-κB/p65, and Bcl2. Resveratrol not only increased p53 cellular content but also induced its posttranslational modification (phosphorylation and acetylation) required

for regulating gene transcription (such as p21 induction). Interesting observations have been reported on endothelial cell proliferation and angiogenesis, a process that plays an important role in tumor growth. In vascular endothelial cells, resveratrol activated p53 via serine phosphorylation but without increasing total concentration of p53 or p21; thus, resveratrol induced a block in DNA synthesis that was not followed by apoptosis. This partial activation of p53 reversibly blocked proliferation because washout of resveratrol restored normal cell cycle progression. The regulation of p53 by resveratrol has been proposed to occur via activation of MAPKs (specifically ERKs and p38). Mitogen-activated protein kinases family members are implicated in cellular proliferation and also in apoptosis. Resveratrol activated p38 and inhibited JNK in stimulated human cervical cancer cells, and it activated ERK1/2 in human breast cancer cells, human erythroleukemic cells, and human melanoma. Moreover, p53 activity was shown to be regulated by resveratrol via modulation of p300 expression. p300 is a transcription coactivator and member of the CREB-binding proteins family; it has acetyl-transferase activity that structurally alters many transcription factors, thereby exerting an antitumor role. Targets of p300 include transcriptional factors known to be regulated by resveratrol, such as NF-κB, p53 and EGR1. It is possible that p300 and p53 belong to an apoptotic loop of gene regulation that mediates resveratrol-induced apoptosis.

Cell Survival-Related Proteins

Natural polyphenols (e.g., catechins, epigallocatechins, theaflavins) bind Bcl2 and BclXl; the binding impaired the ability of these proteins to balance proapoptotic family members and thus it induced apoptosis. In colon carcinoma cancer cells, resveratrol induced mitochondrial translocation of proapoptotic Bcl2 family members (e.g., Bax, Bak) and initiated an apoptotic cascade. Resveratrol treatment also down-regulated the expression of antiapoptotic and increased that of proapoptotic Bcl2 members. Survivin, a member of the IAP family, directly inhibits apoptosis, and its expression is frequently high in cancer cells and correlates with resistance to chemotherapy. The survivin gene contains a cyclin-dependent element, and depression of this element allows the expression of survivin at transition phase G1/S; surviving levels remain high during mitosis since it participates in chromosome assembly at the mitotic spindle. Survivin expression is regulated by the formation or dissociation of different Rb/E2F complexes throughout the cell cycle, and it is switched off by an increase in p53/p21 complexes.

Besides regulating mitosis progression, survivin has a role in preventing apoptosis, possibly by impairing caspases activation and mitochondrial dysfunction. Resveratrol decreased survivin levels by enhancing its degradation as well as reducing its transcription; this was associated with decreased proliferation and sensitization to chemotherapy.

Interaction with Nuclear Factor Kappa Beta (NF-κB)

Resveratrol was shown to inhibit NF-κB activation; with some exceptions, this activation was associated with an antiproliferative action and with the induction of cell death. NF-κB controls the transcription of a variety of genes, including tumor-promoting COX-2, iNOS, matrix metalloprotease (MMP9), and endothelial adhesion molecules. The expression of these genes was reported to be down-regulated by resveratrol in different cell lines. In addition, dietary administration of resveratrol in DMBA-induced tumor-bearing rats reduced tumor growth and decreased transcription of NF-κB and of its regulated genes COX-2 and MMP in tumor tissues. Resveratrol was shown to inhibit NF-κB, AP1 and their target genes regulating the activity of the upstream MAPKs. Resveratrol down-regulated TNFα induction of NF-κB by blocking JNK and MEK activation, and it down-regulated phorbol myristate acetate (PMA)-induced NF-κB by blocking JNK and PKCγ activation. Similarly, resveratrol down-regulated AP1 induction by PMA and UV irradiation, possibly by upstream inhibition of Src and MAPKs. Resveratrol inhibition of tyrosine kinases such as Raf and Src may be considered an upstream event that opens access to multiple cascades. Studies either confirmed or refuted resveratrol inhibition of Src activity *in vitro*, questioning the possibility of a direct action. The potential of resveratrol as an antiangiogenic molecule can be related to its inhibition of NF-κB activation.

Sirtuins are a nicotinamide adenosine dinucleotide (NAD)-dependent class of deacetylases responsible for regulating the response to DNA damage and gene silencing processes of aging and survival. Resveratrol activated human sirtuin 1 (SIRT1) and this mediated RelA/p65 deacetylation, thus inhibiting TNFa-induced NF-κB transcription and sensitizing cells to apoptosis. Resveratrol enhanced the expression of p300, an acetylase that activates NF-κB and p53. Yeung et al. showed that resveratrol failed to inhibit and actually increased TNFα induction of NF-κB when SIRT1 was pharmacologically inhibited, suggesting that its activity may mediate resveratrol down-regulation of NF-κB; conversely, the overexpression of SIRT1 reversed TNFα-induced and possibly p300-mediated activation of NF-κB (Figure 12.2). From the reported data, it is possible to speculate that resveratrol controls two key gene transcriptional regulators: p300 and SIRT1. Mitogen activated protein kinases may be upstream mediators of Egr1 transcription that activates genes related to cell growth and differentiation. Resveratrol was shown to activate Egr1. Egr1 can

bind p21 promoter *in vivo*, and antisense Egr1 mRNA impaired resveratrol-induced p21 upregulation.[15,16,19]

Interaction with Sphingolipid

Ceramide is a sphingolipid mediator of intracellular signals, normally present in membranes in a complexed form as sphingomyelin or gangliosides. Ceramide interacts, either directly or indirectly, with a variety of intracellular targets, leading to differentiation, cell cycle arrest, and apoptosis. Resveratrol was shown to promote intracellular accumulation of ceramide in breast and prostate cancer cells. Resveratrol enhanced the *de novo* synthesis of this sphingolipid by increasing the activity of the rate-limiting enzyme.[15,16,19]

Evidences from Preclinical and Human Studies

It has been shown that the topical application of resveratrol on skin resulted in inhibition of UVB-radiation-induced tumor incidence and delay in the onset of skin tumorigenesis.[2] Topical application of resveratrol revealed that the molecule was able to down-regulate UV-mediated increases in critical cell cycle regulatory proteins, such as proliferating cell nuclear antigen (PCNA); cyclin-dependent kinase (cdk)-2, -4, and -6; cyclin- D1; and cyclin-D2. Further, resveratrol was also found to cause significant decreases in UVB-mediated upregulation of MAPK.[21] Treatment of human breast cancer MCF-7 cells with resveratrol also suppressed NF-κB activation and proliferation.[11] It was also demonstrated that resveratrol caused down-regulation of survivin, phosphosurvivin protein, and upregulation of proapoptotic Smac/ DIABLO protein in skin tumors and enhancement of apoptosis.[22] Resveratrol has been shown to induce apoptosis in LNCaP and DU145 prostate cancer cell lines through different PKC-mediated and MAPK-dependent pathways.[23] Treatment of androgen-sensitive prostate cancer cells (LNCaP) with resveratrol caused down-regulation of prostate-specific antigen and p65; these effects were associated with activation of p53, WAF1, p300/CBP, and APAF1.[24]

Resveratrol affected the growth of human breast cancer cell lines MCF7, MDA-MB-231, SK-BR-3, and Bcap-37 in a dose-dependent manner and has been shown to exert its growth-inhibitory/apoptotic effect on the breast cancer cells via the Akt caspase-9 pathway.[25] Treatment of human breast cancer MCF-7 cells with resveratrol also suppressed NF-κB activation and proliferation.[26]

Resveratrol was found to inhibit the growth of A549, EBC-1, and Lu65 lung cancer cells.[27] Resveratrol treatment of A549 cells resulted in a concentration-dependent induction of S phase arrest in cell cycle progression. Resveratrol significantly reduced the tumor volume, tumor weight, and metastasis to the lung in mice bearing highly metastatic Lewis lung carcinoma (LLC) tumors. In addition, resveratrol inhibited DNA synthesis, induced apoptosis, and decreased the S phase population in LLC cells. Resveratrol inhibited tumor-induced neovascularization in an *in vivo* model.[28] Resveratrol inhibited CYP1A1 expression in human HepG2 hepatoma cells, by preventing the binding of the aryl hydrocarbon receptor (AHR) to promoter sequences that regulate CYP1A1 transcription.[29] Resveratrol treatment caused induction of apoptosis as well as an increase in nuclear size and granularity in a concentration-dependent manner in human leukemia cell line HL-60 and the human hepatoma-derived cell line HepG2. Resveratrol also inhibited cell proliferation in a concentration- and time-dependent manner by interfering with different stages of the cell cycle.[30] Furthermore, it also modulated the NO/NOS system, by increasing iNOS and eNOS expression, NOS activity, and NO production. Inhibition of NOS enzymes attenuates its antiproliferative effect[31] (Table 12.1).[32]

LYCOPENE

Lycopene (w,w-carotene), one of more than 600 carotenoids synthesized by plants and photosynthetic microorganisms, is a tetraterpene hydrocarbon containing 40 carbon atoms and 56 hydrogen atoms with a molecular mass of 536.[33] Tomatoes and tomato products are the major dietary sources of lycopene. Other sources include watermelon, pink grapefruit, apricots, pink guava, and papaya.[34] Lycopene is the most abundant carotenoid in tomatoes (Lycopersicon *esculentum L.*) with concentrations ranging from 0.9 to 4.2 mg/100 g depending upon the variety. Tomato sauce and ketchup are concentrated sources of lycopene (33–68 mg/100 g) compared to unprocessed tomatoes.[33] Lycopene is an acyclic carotene with 11 conjugated double bonds, normally in the all-trans configuration (Figure 12.4). The long chromophore in the polyene chain accounts for the red color of lycopene (λ_{max} 472 nm) and also for its powerful antioxidant activity. It is able to react with singlet oxygen and various radical cations and has the highest TEAC (Trolox-equivalent antioxidant capacity) value of all

TABLE 12.1 Suggested Molecular Targets of Resveratrol in Different Anticancer Studies

S.No.	Cancer Cell Types	Mechanism
1.	Leukemia	Increases caspase 3, decreases Bcl-2, iNOS
	• Inhibits proliferation of chronic B lymphocytic leukemia	Increases caspases 9
	• Induces apoptosis in promyelocytic leukemia (HL-60 cells)	Decreases survivin
	• Induces apoptosis in adult T-cell leukemia	Arrests S phase, increases PARP cleavage, caspases
	• Inhibits growth of acute myeloid leukemia (AML)	Decreases NF-κB, IFN-A, IL-2, TNF, and IL-12
	• Inhibits proliferation of mitogen, IL-2, or alloantigen-induced splenic lymphocytes	Increases Bax, cytochrome c, caspases
	• Induces apoptosis in HL-60 cells	Increases ROS, caspases
	• Induces apoptosis in T-lymphoblastic leukemia	Increases caspases, PARP cleavage
	• Induces apoptosis in monocytic leukemia (THP-1) cells	Increases cytochrome and caspases
	• Induces apoptosis in U-937 cells	
2.	Breast	Mechanism is independent of ER status
	• Inhibits proliferation of breast epithelial (MCF-7, MCF-10F and MDA-MB-231) cells	Decreases TGF-α, IGF-1R, increases TGF-β
	• Inhibits growth of MCF-7 cells	Decreases ROS
	• Inhibits growth of MCF-7, T47D	Increases nSMase, ceramide, serine
	• Induces apoptosis of MDA-MB-231	Decreases CYP1A1
	• Inhibits growth of MCF-7 and T47D cells	
3.	Colon	Increases p53-independent apoptosis
	• Induces apoptosis of HCT116 cells	Decreases cyclin D1/Cdk4 complex, increases cyclin E and A
	• Induces apoptosis of Caco-2 and HCT-116 cells	Increases redistribution of Fas receptor in membrane rafts
	• Induce apoptosis SW480	Arrests G2 phase, decreases Cdk 7, increases Cdc2, arrests sub G0 phase
	• Induces apoptosis in (col-2) cancer cells	Arrests G0/G1 phase
	• Induces apoptosis in colon cancer cells	
4.	Ovarian and Endometria	Increases cyclin A; increases cyclin E; decreases Cdk2
	• Inhibits proliferation of endometrial adenocarcinoma cells	Increases NQO-1
	• Inhibits cell growth and induces apoptosis in ovarian cancer (PA-1) cells	Increases cytochrome c; increases caspases; increases autophagocytosis
	• Inhibited growth and induced death of five human ovarian carcinoma cells	Increases prostaglandin biosynthesis; arrests S phase
	• Inhibits proliferation in cervical tumor (HeLa and SiHa) cells	

carotenoids.[35] This reactivity of lycopene is the basis for its antioxidant activity in biological systems that might contribute to its efficacy as a chemoprevention agent.[33]

Antioxidant Potential of Lycopene

There are many proposed and studied mechanisms of lycopene action, both oxidative and nonoxidative. The nonoxidative mechanisms involve noncovalent binding interactions between lycopene and proteins and cell membranes, including growth factor signaling, modulation of hormone and immune responses, and enhancing gap junctional intercellular communication. These mechanisms may affect mutagenesis, carcinogenesis, and/or cell differentiation and proliferation. Specific examples include (1) carcinogen metabolism, (2) inhibition of hydroxymethylglutaryl coenzyme A (HMG-CoA) reductase and upregulation of LDL receptor activity in macrophages to decrease cholesterol, (3) repression of insulin-like growth factors to inhibit cellular proliferation, (4) gene function regulation, (5) modulation of the liver metabolizing enzyme cytochrome P450 2E1, (6) modulation of expression of cell cycle regulatory proteins to stop cell division at G0-G1 phase, (7) suppression of carcinogen-induced phosphorylation of regulatory proteins (p53 and RB anti-carcinogens), (8) inhibition of IL-6, androgens, and 5-lipoxygenase, and

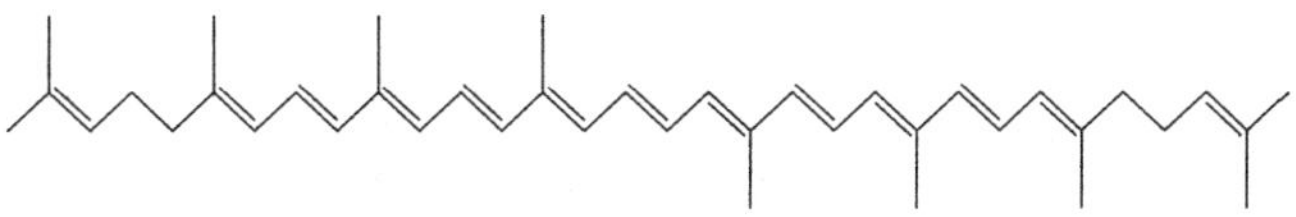

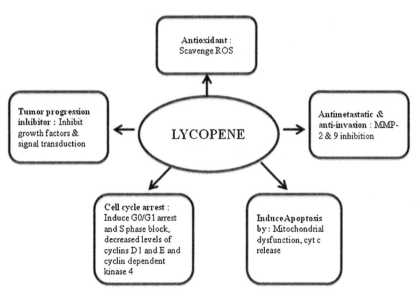

FIGURE 12.4 Chemical Structure of Lycopene. trans-Lycopene

(9) upregulation of connexion resulting in increased gap–junction communication. The antioxidative mechanisms of lycopene allow it to protect cellular biomolecules such as lipids, proteins, enzymes, lipoproteins, and DNA from oxidative modification, thus preventing development of chronic diseases resulting from oxidative stress.

Anticancer Potential of Lycopene

Cells must be self-sufficient and control systems to withstand endogenous and exogenous stresses. These controls sometimes place the cell at risk for corruption of pathways, which inevitably can lead to carcinogenesis. Mutations in DNA, genes, or cellular pathways can modify growth and differentiation, and migration or metastasis of cells. Lycopene has been reported to modulate many of these pathways resulting in reduced cellular carcinogenic properties (Figure 12.5). There are several mechanisms by which tomatoes, their products, and particularly lycopene exert inhibition of carcinogenesis; that is, scavenging of reactive oxygen species, enhancement of detoxification systems, suppression of cell cycle progression, as well as modulation of signal transduction pathways.[36]

Early studies suggested that the antioxidant effect of lycopene may reduce chronic diseases such as cancer. *In vitro*, lycopene is the most proficient carotenoid singlet oxygen quencher in nature.[37] At nonphysiological concentration, lycopene triggers apoptosis, activating the intrinsic pathway involving mitochondrial function, cytochrome C release, and exposure to Annexin V. The antiproliferative activity of lycopene is mostly identified with the ability of the molecule to block the cell cycle at G0/G1 phase.[38] Lycopene was shown to bind to PDGF-BB in human plasma. Lycopene can inhibit PDGF-BB-induced human Hs68 skin fibroblast

FIGURE 12.5 Antioxidant & Anticancer Properties of Lycopene. CDK: Cyclin-dependent kinases; MMP: Matrix metalloproteinase; ROS: Reactive oxygen species

migration, attenuate PDGF-BB-induced phosphorylation, and reduce PDGF-BB-induced signaling. PDGF-BB facilitates the growth, invasion, and metastasis of melanoma, and inhibition of these PDGF-BB effects can be a mechanism for arresting melanoma progression.[33] Lycopene also suppressed insulin-like growth factor-I-stimulated growth. Insulin-like growth factors are major autocrine/paracrine regulators of mammary and endometrial cancer cell growth.[34] Lycopene was shown to have antimetastatic and anti-invasion activity.

Evidences from Preclinical and Human Studies

Epidemiological and human clinical trials suggest that dietary intake of tomatoes or lycopene protects against cancers of the prostate, breast, cervix, ovary, endometrium, lung, bladder, oral cavity, esophagus, stomach, colon, and pancreas.[39] A case-control study in Italy involving 2706 cases of cancer of the oral cavity, pharynx, esophagus, stomach, colon, and rectum versus 2879 controls indicated that a high intake of tomatoes and tomato-based food was associated with reduced risk of digestive tract (especially stomach, colon, and rectal) cancers.[40] Serum lycopene is also associated with decreased risk of bladder cancer[41] and breast cancer.[42] Lycopene level showed inverse risk in cervical cancer.[43]

Lycopene's antiproliferative properties were investigated in a cell culture study in comparison with those of α- and β-carotene using endometrial, mammary, and lung human cancer cells. Their results revealed that lycopene inhibited the growth of the endometrial, breast, and lung cancer cells.[44] The study conducted by Salman and coworkers (2007) aimed to examine the *in vitro* effect of lycopene on the proliferation response and apoptosis on other malignant cell lines (i.e., HuCC, human colon carcinoma; EHEB, B chronic lymphocytic leukemia; K562, human erythroleukemia; and Raji, a Burkitt lymphoma cell line). Lycopene induced a dose-dependent inhibition of Raji and HuCC cell proliferation following incubation of K562 cells with increasing doses of lycopene; a dose-dependent reduction of ^{3}HTdR incorporation was observed. Incubation of EHEB cells with lycopene did not affect cell proliferation significantly [36] (Table 12.2).[45-48]

CONCLUSIONS

Carcinogenesis is an extremely complex process entailing numerous biochemical and molecular events. Consequently, screening dietary phytochemicals for potential chemopreventive activity requires a

TABLE 12.2 Suggested Molecular Targets of Lycopene in Different Anticancer Studies

Cancer Cell Types and Mechanism		Finding
Breast Inhibition of cell proliferation; cycle-arrested at G1/S phase; Antioxidant defense		Modulate cell cycle proteins such as beta tubulin, CK8/18, CK19, and heat shock proteins in MCF-7 breast cancer cells
Hepatic Induced G0/G1 arrest and S phase block; antimetastatic effect and inhibit cell growth.		Inhibition of toxic effects of aflatoxin B1; decreased the activities and protein expression of metalloproteinase-2 (MMP-2) and – 9 Increased the protein expression of nm23-H1 and the tissue inhibitor of MMP (TIMP)-1 and -2 Suppressed protein expression of Rho small GTPases Inhibited focal adhesion kinase-mediated signaling pathway, such as ERK/p38 and PI3K-Akt axis
Colon Suppress MMP-7 expression and leptin-mediated cell invasion Induces apoptosis in (col-2) cancer cells		Inhibit the phosphorylation of Akt; glycogen synthase kinase-3b (GSK-3b) inhibition; ERK 1/2 proteins and HT-29 cells inhibition Arrests G0/G1 phase
Gastric Enhances antioxidant enzyme activities and immunity function		Leukotriene inhibitor Increased blood IL-2, IL-4, IL-10, TNF-a levels and reduced the IL-6 level and enhanced blood IgA, IgG, and IgM levels; decreased MDA and increased blood and gastric antioxidant parameters (SOD, CAT, and GSH-Px)
Leukemia Trigger apoptosis		Apoptosis via a mitochondrial pathway. Redox state of cells was measured in terms of glutathione (GSH) content GSH depletion, suggesting that the loss of GSH may be a secondary consequence of ROS generation.
Lymphocytic leukemia **Erythroleukemia** **Burkitt lymphoma**	Antiproliferative and apoptotic effect	Exerted on the proliferation capacity of K562, Raji and HuCC lines; Increased apoptotic rate

systematic approach to identify promising foods for further investigation. From the studies reviewed in this chapter, it is clear that dietary antioxidants hold a great potential for prevention and treatment of cancer. With healthcare costs being a key issue today, the chemoprevention by dietary phytochemicals represents an inexpensive and a readily applicable approach to reduce cancer incidence.

SUMMARY POINTS

- Resveratrol and lycopene have shown encouraging anticancer activity.
- Antioxidant, anti-inflammatory inhibition of cell growth is attributed to anticancer activities of resveratrol.
- Lycopene exerts inhibition of carcinogenesis by scavenging of reactive oxygen species, suppression of cell cycle progression, as well as modulation of signal transduction pathways.

References

1. Bray F, Jemal A, Grey N, Ferlay J, Forman D. Global cancer transitions according to the Human Development Index (2008–2030): A population-based study. *Lancet Oncol* 2012;**13**(8):790–801.
2. Khan N, Afaq F, Mukhtar H. Cancer chemoprevention through dietary antioxidants: progress and promise. *Antioxid Redox Signal* 2008;**10**(3):475–510.
3. Glade MJ. Food, nutrition, and the prevention of cancer: A global perspective. American Institute for Cancer Research/World Cancer Research Fund, American Institute for Cancer Research, 1997. *Nutrition* 1999;**15**(6):523–6.
4. Sporn MB. Approaches to prevention of epithelial cancer during the preneoplastic period. *Cancer Res* 1976;**36**(7 PT 2):2699–702.
5. Guidelines on diet, nutrition, and cancer prevention: reducing the risk of cancer with healthy food choices and physical activity. The American Cancer Society 1996 Advisory Committee on Diet, Nutrition, and Cancer Prevention. *CA Cancer J Clin* 1996;**46**(6):325–41.
6. Takaoka M. Of the phenolic substances of white hellebore (Veratrum grandiflorum Loes. fil.). *J Faculty of Sci Hokkaido Imperial University* 1940;**3**:1–16.
7. Nonomura S, Kanagawa H, Makimoto A. Chemical Constituents of Polygonaceous Plants. I. Studies on the Components of KO-J O-KON. (POLYGONUM CUSPIDATUM SIEB. ET ZUCC. *Yakugaku zasshi* 1963;**83**:988.
8. Frankel EN, Waterhouse AL, Kinsella JE. Inhibition of human LDL oxidation by resveratrol. *Lancet* 1993;**341**(8852):1103–4.
9. Soleas GJ, Tomlinson G, Diamandis EP, Goldberg DM. Relative contributions of polyphenolic constituents to the antioxidant status of wines: development of a predictive model. *J Agric Food Chem* 1997;**45**(10):3995–4003.
10. Leonard SS, Xia C, Jiang BH, Stinefelt B, Klandorf H, Harris GK, et al. Resveratrol scavenges reactive oxygen species and effects radical-induced cellular responses. *Biochem Biophys Res Commun* 2003;**309**(4):1017–26.
11. Surh YJ. Cancer chemoprevention with dietary phytochemicals. *Nat Rev Cancer* 2003;**3**(10):768–80.
12. Kopp P. Resveratrol, a phytoestrogen found in red wine. A possible explanation for the conundrum of the 'French paradox'? *Eur J Endocrinol* 1998;**138**(6):619–20.
13. Atten MJ, Godoy-Romero E, Attar BM, Milson T, Zopel M, Holian O. Resveratrol regulates cellular PKC alpha and delta to inhibit growth and induce apoptosis in gastric cancer cells. *Invest New Drugs* 2005;**23**(2):111–9.
14. Kim YA, Lim SY, Rhee SH, Park KY, Kim CH, Choi BT, et al. Resveratrol inhibits inducible nitric oxide synthase and cyclooxygenase-2 expression in beta-amyloid-treated C6 glioma cells. *Int J Mol Med* 2006;**17**(6):1069–75.
15. Signorelli P, Ghidoni R. Resveratrol as an anticancer nutrient: molecular basis, open questions and promises. *J Nutr Biochem* 2005;**16**(8):449–66.
16. de la Lastra CA, Villegas I. Resveratrol as an antioxidant and pro-oxidant agent: mechanisms and clinical implications. *Biochem Soc Trans* 2007;**35**(Pt 5):1156–60.
17. Zini R, Morin C, Bertelli A, Bertelli AAE, Tillement JP. Effects of resveratrol on the rat brain respiratory chain. *Drugs under Exp Clin Res* 1999;**25**(2-3):87–97.
18. Losa GA. Resveratrol modulates apoptosis and oxidation in human blood mononuclear cells. *Eur J Clin Invest* 2003;**33**(9):818–23.
19. Udenigwe CC, Ramprasath VR, Aluko RE, Jones PJ. Potential of resveratrol in anticancer and anti-inflammatory therapy. *Nutr Rev.* 2008;**66**(8):445–54.
20. Jang M, Cai L, Udeani GO, Slowing KV, Thomas CF, Beecher CW, et al. Cancer chemopreventive activity of resveratrol, a natural product derived from grapes. *Science* 1997;**275**(5297):218–20.
21. Reagan-Shaw S, Afaq F, Aziz MH, Ahmad N. Modulations of critical cell cycle regulatory events during chemoprevention of ultraviolet B-mediated responses by resveratrol in SKH-1 hairless mouse skin. *Oncogene* 2004;**23**(30):5151–60.
22. Aziz MH, Reagan-Shaw S, Wu J, Longley BJ, Ahmad N. Chemoprevention of skin cancer by grape constituent resveratrol: relevance to human disease? *Faseb J* 2005;**19**(9):1193–5.
23. Shih A, Zhang S, Cao HJ, Boswell S, Wu YH, Tang HY, et al. Inhibitory effect of epidermal growth factor on resveratrol-induced apoptosis in prostate cancer cells is mediated by protein kinase C-alpha. *Mol Cancer Ther* 2004;**3**(11):1355–64.
24. Narayanan BA, Narayanan NK, Re GG, Nixon DW. Differential expression of genes induced by resveratrol in LNCaP cells: P53-mediated molecular targets. *Int J Cancer* 2003;**104**(2):204–12.
25. Li Y, Liu J, Liu X, Xing K, Wang Y, Li F, et al. Resveratrol-induced cell inhibition of growth and apoptosis in MCF7 human breast cancer cells are associated with modulation of phosphorylated Akt and caspase-9. *Appl Biochem Biotechnol* 2006;**135**(3):181–92.
26. Banerjee S, Bueso-Ramos C, Aggarwal BB. Suppression of 7,12-dimethylbenz(a)anthracene-induced mammary carcinogenesis in rats by resveratrol: role of nuclear factor-kappaB, cyclooxygenase 2, and matrix metalloprotease 9. *Cancer Res* 2002;**62**(17):4945–54.
27. Kubota T, Uemura Y, Kobayashi M, Taguchi H. Combined effects of resveratrol and paclitaxel on lung cancer cells. *Anticancer Res* 2003;**23**(5A):4039–46.
28. Kimura Y, Okuda H. Resveratrol isolated from *Polygonum cuspidatum* root prevents tumor growth and metastasis to lung and tumor-induced neovascularization in Lewis lung carcinoma-bearing mice. *J Nutr* 2001;**131**(6):1844–9.
29. Ciolino HP, Daschner PJ, Yeh GC. Resveratrol inhibits transcription of CYP1A1 in vitro by preventing activation of the aryl hydrocarbon receptor. *Cancer Res* 1998;**58**(24):5707–12.
30. Stervbo U, Vang O, Bonnesen C. Time- and concentration-dependent effects of resveratrol in HL-60 and HepG2 cells. *Cell Prolif* 2006;**39**(6):479–93.

31. Notas G, Nifli AP, Kampa M, Vercauteren J, Kouroumalis E, Castanas E. Resveratrol exerts its antiproliferative effect on HepG2 hepatocellular carcinoma cells, by inducing cell cycle arrest, and NOS activation. *Biochim Biophys Acta* 2006;**1760**(11):1657–66.

32. Aggarwal BB, Bhardwaj A, Aggarwal RS, Seeram NP, Shishodia S, Takada Y. Role of resveratrol in prevention and therapy of cancer: preclinical and clinical studies. *Anticancer Res.* 2004;**24**(5A):2783–840.

33. van Breemen RB, Pajkovic N. Multitargeted therapy of cancer by lycopene. *Cancer Lett* 2008;**269**(2):339–51.

34. Omoni AO, Aluko RE. The anti-carcinogenic and anti-atherogenic effects of lycopene: A review. *Trends Food Sci Technol* 2005;**16**(8):344–50.

35. Rice-Evans CA, Sampson J, Bramley PM, Holloway DE. Why do we expect carotenoids to be antioxidants in vivo? *Free Radic Res* 1997;**26**(4):381–98.

36. Salman H, Bergman M, Djaldetti M, Bessler H. Lycopene affects proliferation and apoptosis of four malignant cell lines. *Biomed Pharmacother* 2007;**61**(6):366–9.

37. Di Mascio P, Kaiser S, Sies H. Lycopene as the most efficient biological carotenoid singlet oxygen quencher. *Arch Biochem Biophys* 1989;**274**(2):532–8.

38. Russo M, Spagnuolo C, Tedesco I, Russo GL. Phytochemicals in cancer prevention and therapy: truth or dare? *Toxins* 2010;**2**(4):517–51.

39. Ford N, Erdman Jr JW. *Lycopene and Cancer*. Springer: Carotenoids and Human Health; 2013 193–214.

40. Franceschi S, Bidoli E, La Vecchia C, Talamini R, D'Avanzo B, Negri E. Tomatoes and risk of digestive-tract cancers. *Int J Cancer* 1994;**59**(2):181–4.

41. Helzlsouer KJ, Comstock GW, Morris JS. Selenium, lycopene, alpha-tocopherol, beta-carotene, retinol, and subsequent bladder cancer. *Cancer Res* 1989;**49**(21):6144–8.

42. Dorgan JF, Sowell A, Swanson CA, Potischman N, Miller R, Schussler N, et al. Relationships of serum carotenoids, retinol, alpha-tocopherol, and selenium with breast cancer risk: Results from a prospective study in Columbia, Missouri (United States). *Cancer Causes Control* 1998;**9**(1):89–97.

43. Sengupta A, Das S. The anti-carcinogenic role of lycopene, abundantly present in tomato. *Eur J Cancer Prev* 1999;**8**(4):325–30.

44. Levy J, Bosin E, Feldman B, Giat Y, Miinster A, Danilenko M, et al. Lycopene is a more potent inhibitor of human cancer cell proliferation than either α-carotene or β-carotene. *Nutr Cancer* 1995.

45. Karas M, Amir H, Fishman D, Danilenko M, Segal S, Nahum A, et al. Lycopene interferes with cell cycle progression and insulin-like growth factor I signaling in mammary cancer cells. *Nutr Cancer* 2000;**36**(1):101–11.

46. Ghosh J, Myers CE. *Molecular Mechanisms of Prostate Cancer Cell Death Triggered by Inhibition of Arachidonate 5-Lipdxygenase: Involvement of Fas Death Receptor-Mediated Signals*. Springer: Eicosanoids and Other Bioactive Lipids in Cancer, Inflammation, and Radiation Injury, 5; 2003 415–420.

47. Seren S, Lieberman R, Bayraktar UD, Heath E, Sahin K, Andic F, et al. Lycopene in cancer prevention and treatment. *Am J Ther* 2008;**15**(1):66–81.

48. Kazim Sahin D, Kucuk O. *Lycopene in Cancer Prevention*. Springer: Natural Products; 2013 3875–3922.

13

Iron, Oxidative Stress, and Cancer

Mi-Kyung Sung

Department of Food and Nutrition, Sookmyung Women's University, Seoul, Korea

Yun-Jung Bae

Department of Food and Nutritional Sciences, Hanbuk University, Gyeonggi, Korea

List of Abbreviations

4-OHE 4-hydroxyestradiol
BC breast cancer
CRC colorectal cancer
DMBA dimethylbenz[a]anthracene
EC endometrial cancer
EMT epithelial-mesenchymal transition
FHC ferritin heavy chain
FPN ferroportin
HH hereditary hemochromatosis
HIF-1 hypoxia inducible factor-1
HO-1 heme oxygenase-1
HRT hormone replacement therapy
MMP-1 matrix metalloproteinase-1
NASH nonalcoholic steatohepatitis
RCC renal cell carcinomas
ROS reactive oxygen species
RR ribonucleotide reductase
SOD superoxide dismutase
TIBC total iron-binding capacity
UIBC unsaturated iron binding capacity
VEGF-1 vascular endothelial growth factor 1

INTRODUCTION

Iron is an essential micronutrient playing a key role in oxygen supply. A number of cellular enzymes possess iron as a cofactor to maintain their functions. Cancer is a multistage disease involving both genetic and environmental risk factors. Excess iron has been suggested as a risk factor in cancer development mostly due to increased production of reactive oxygen species (ROS), which induces genetic damages initiating and accelerating abnormal cell growth. Recent studies have also suggested that iron depletion may be effective therapeutic tools in cancer. Excess body iron pool is induced either by heritable genetic diseases or by excess iron intake. In this review, we discussed epidemiological and experimental evidences on the relationship between iron overload and cancer. Molecular mechanisms involved in excess-iron-induced tumor initiation, promotion, and proliferation are suggested based on current evidences, and possible application of iron chelation in cancer therapy is briefly discussed.

EPIDEMIOLOGICAL EVIDENCES

Dietary Iron Intake and Cancer

Dietary iron is obtained from either heme iron or nonheme iron. Heme iron is present in animal foods as a component contained in the center of a large heterocyclic organic ring called porphyrin, which makes up 40% of the iron in meat, poultry, and fish. Nonheme iron is derived from animal tissue and plant foods such as fruits, vegetables, grains, and nuts. Heme iron is more bioavailable and readily absorbable compared to nonheme iron.

Epidemiological evidences have suggested that iron intake is positively related to the cancer risk. A most plausible mechanistic explanation is the increased pro-oxidant load, which may ultimately lead to more oxidative stress and DNA damage. Also, animal food products, major sources of dietary iron, are sources of carcinogenic compounds such as N-nitroso compounds and heterocyclic amines. However, the linkage between heme iron or red meat intake and cancer needs to be concluded attentively because heme iron content of meat varies by meat type and methods of preparation. A recent case-control study indicated that colorectal adenoma risk was significantly related to the intakes of pan fried meat and red meat, while no association was found with grilled meat intake.[1] Also, micronutrient contents

© 2014 Elsevier Inc. All rights reserved.

of heme iron containing foods possibly modify the association between heme iron intake and cancer risk. Zinc is a trace element involved in metallothionein synthesis, which is thought to inhibit free radical production and as a consequence may reduce the frequency of harmful gene mutations. A previous study reported that the mixed effects of prooxidant iron and antioxidant zinc may negate associations between cancer and consumption of iron- and zinc-rich foods, such as meats.[2] These evidences suggest that more research is needed to verify the association between iron intake and cancer.

Colorectal Cancer

Previous studies have shown that the population having Western-style dietary habits are at high risk for colorectal cancer (CRC).[3] However, study results on the relationship between dietary iron intake and CRC are inconsistent (Table 13.1). In a large population-based case-control study in Canada, higher risk of CRC was observed with higher iron intake (OR = 1.34; in the Q5 vs. Q1, 95% CI 1.01–1.78).[4] In contrast, other case-control studies found no association between dietary iron intake and CRC risk.[1,5]

Few cohort studies have investigated the association between dietary iron or red meat intake and CRC risk. Analyses of a large, US prospective cohort study with 2,719 CRC cases ascertained in 300,948 men and women followed for an average of 7 years showed that CRC risk was positively associated with intakes of total iron, heme iron, red meat, processed meat, and heterocyclic amines.[6] However, several prospective cohort studies observed no association between dietary iron intake and CRC risk.[7,8]

Meanwhile, studies have suggested that the effect of heme iron may depend on the location of tumors in the colon because there are embryologic, morphologic, physiologic, and biochemical differences between right- and left-sided colon.[9] In the Western Australian Bowel Health Study, dietary iron was associated with decreased risk of left-sided CRC (p for trend = 0.044; OR = 0.78 in the Q5 vs. Q1, 95% CI 0.56–1.09), whereas OR of right-sided CRC showed no association.[5]

As mentioned previously, the association between iron intake and CRC risk can be modified by the concurrent intake of minerals such as copper and zinc. Experimental evidences on the relevance of oxidative damage due to metal toxicity and carcinogenicity have been especially strong for both iron and copper.[10] A case-control study showed the significant positive associations between dietary iron (OR = 2.2; 95% CI 1.1–4.7) and copper (OR = 2.4; 95% CI 1.3–4.6) intake and CRC risk.[11] On the other hand, zinc exhibits antioxidant functions and main food sources of zinc include those with high heme iron content. It is reported that zinc plays a role in protecting

chemically induced colonic preneoplastic progression in an animal model.[12]

Alcohol consumption is known to disrupt iron homeostasis. A previous study reported that the positive association between heme iron intake and CRC risk was significantly stronger among alcohol drinkers than that of nondrinkers.[2] In a prospective study, the association of heme iron intake with CRC risk was not modified by alcohol consumption although a slightly elevated risk of CRC by the increased heme iron intake was observed among alcohol drinkers (≥ 10 g/day).[7]

Breast Cancer

Iron has been demonstrated as a potential risk factor in breast cancer (BC). Although excessive intake of iron may predispose to mammary carcinogenesis due to the fact that free iron works as a catalyst for ROS generation, epidemiological evidences are much less conclusive (Table 13.2). In a population-based case-control study in Shanghai, China, it was found that the consumption of animal-derived (largely heme) iron was positively associated with increased risk of BC (p for trend < 0.01; OR = 1.49 in the Q4 vs. Q1, 95% CI 1.25–1.78) after adjusting confounding factors.[13] The observed effect was similar in pre- and postmenopausal women. In contrast, a nested case-control study in Shanghai reported that the intake of dietary iron was not associated with risk of BC (p for trend = 0.81; OR = 0.96 in the Q4 vs. Q1, 95% CI 0.53–1.77).[14]

Few cohort studies have investigated the association between iron intake and BC risk. A large prospective cohort study with 2,545 BC cases ascertained in 49,654 women aged 40 to 59 at enrollment showed no association between intakes of total dietary iron, heme iron, meat iron, or red meat iron and BC risk after a 16.4 year follow-up.[15] In addition, there was no association between iron intake and BC risk within strata defined by alcohol consumption or hormone replacement therapy (HRT) use. In contrast, in an 8-year follow-up prospective cohort study with 1,205 BC cases ascertained in the Prostate, Lung, Colorectal, and Ovarian (PLCO) Cancer Screening Trial, dietary iron was associated with increased risk of BC (p for trend = 0.03; HR = 1.25 in the Q5 vs. Q1, 95% CI 1.02–1.52), whereas total iron and heme iron from meat showed no association.[16] Also, Lee et al.[17] reported that dietary heme intake was positively associated with risk of BC among drinkers who consumed 20g/d (p for interaction < 0.01).

An individual's genetic make-up involved in the metabolism of ROS may modify the role of iron in breast carcinogenesis. In a nested case-control study of postmenopausal women from the American Cancer Society Prevention II Nutrition Cohort, women with genotypes resulting in potentially higher levels of iron-generated

TABLE 13.1 Epidemiological Evidences on the Association of Iron and Meat Intake with Colon Cancer Risk

Case-Control Study

Study	Population		Cases/Controls	OR (95% CI)	P trend
Sun et al., 2012	Canada	CRC	1,760/2,481	Iron (OR =1.34, Q5 vs. Q1, 1.01–1.78)	0.168
Van Lee et al., 2011 (Western Australian Bowel Health Study)	Western Australia	CRC	854/958	Iron (CRC: OR = 0.86, Q5 vs. Q1, 0.64–1.15) (Right-sided CRC: OR = 1.03, Q5 vs. Q1, 0.68–1.52) (Left-sided CRC: OR = 0.78, Q5 vs. Q1, 0.56–1.09)	0.702 0.044
Cross et al., 2011 (Prostate, Lung, Colorectal and Ovarian Cancer Screening Trial)	USA	Colorectal adenoma	356/396	Total dietary iron (OR = 0.98, Q4 vs. Q1, 0.62–1.55) Heme iron (OR = 1.46, Q4 vs. Q1, 0.94–2.29) Red meat (OR = 1.59, Q4 vs. Q1, 1.02–2.49)	0.98 0.08 0.03
Ferrucci et al., 2009 (CONCeRN study)	USA	Colorectal adenoma	158/649 (Women)	Total iron (OR = 0.91, Q4 vs. Q1, 0.52–1.59) Dietary iron (OR = 1.11, Q4 vs. Q1, 0.57–2.16) Iron from meat (OR = 1.71, Q5 vs. Q1, 0.94–3.14) Heme iron from meat (OR = 1.50, Q4 vs. Q1, 0.83–2.73)	0.34 0.77 0.17 0.32
Cross et al., 2006 (a-tocopherol, b-carotene cancer prevention study)	USA	Colorectal adenoma	130/260	Dietary iron (OR = 0.4, Q4 vs. Q1, 0.1–1.1)	0.06
Senesse et al., 2004	France	CRC	171/309	Iron (OR = 2.2, Q4 vs. Q1, 1.1–4.7)	0.02

Cohort Study

Study	Population		Cases/Participants	RR/HR (95% CI)	P trend
Zhang et al., 2011 (Nurses' Health Study and Health Professionals Follow-up Study)	USA	CRC (22y)	2,114/173,229	Dietary iron (RR = 1.09, Q5 vs. Q1, 0.93–1.30) Heme iron (RR = 1.10, Q5 vs. Q1, 0.93–1.30)	0.37 0.51
Cross et al., 2010	USA	CRC (7y)	2,719/300,948	Dietary iron (HR = 0.75, Q5 vs. Q1, 0.65–0.87) Total iron (HR = 0.75, Q5 vs. Q1, 0.66–0.86) Heme iron (HR = 1.13, Q5 vs. Q1, 0.99–1.29) Red meat (HR = 1.24, Q5 vs. Q1, 1.09–1.42) Processed meat (HR = 1.16, Q5 vs. Q1, 1.01–1.32)	<0.001 <0.001 0.022 <0.001 0.017
Balder et al., 2006 (Netherlands Cohort Study)	Netherlands	CRC (9.3y)	1,535/120,852	Heme iron (Men: RR = 1.16, Q5 vs. Q1, 0.87–1.55) (Women: RR = 1.22, Q5 vs. Q1, 0.89–1.68)	0.27 0.22
Lee et al., 2004 (Iowa Women's Health Study)	USA	Colon cancer (15y)	438/34,708 (Postmenopausal women)	Heme iron (Proximal colon cancer: RR = 2.18, Q5 vs. Q1, 1.24–3.86) (Distal colon cancer: RR=0.90, Q5 vs. Q1, 0.45–1.81)	0.01 0.77

TABLE 13.2   Epidemiological Evidences on the Association of Iron and Meat Intake with Breast Cancer Risk

Case-Control Study				
Study	Population		Cases/Controls	OR (95% CI)	P trend
Kallianpur et al., 2008 (Shanghai Breast Cancer Study)	China	BC	3,452/3,474	Animal-derived iron (OR = 1.49, Q4 vs. Q1, 1.25–1.78)	< 0.01
Moore et al., 2009	China	BC	248/1,040	Dietary iron (OR = 0.96, Q4 vs. Q1, 0.53–1.77)	0.81

Cohort Study				
Study	Population		Case/Participant	RR/HR (95% CI)	P trend
Kabat et al., 2007	Canada	BC(16.4y)	2,545/49,654	Total iron (HR = 0.97, Q5 vs. Q1, 0.85–1.10)	0.63
				Heme iron (HR = 1.03, Q5 vs. Q1, 0.90–1.18)	0.25
				Meat iron (HR = 1.09, Q5 vs. Q1, 0.96–1.24)	0.26
				Red meat (HR = 0.98, Q5 vs. Q1, 0.86–1.12)	0.91
Ferrucci et al., 2009 (Prostate, Lung, Colorectal, and Ovarian Cancer Screening Trial)	USA	BC(5.5y)	1,205/52,158	Total iron (HR = 1.08, Q5 vs. Q1, 0.90–1.30)	0.58
				Dietary iron (HR = 1.25, Q5 vs. Q1, 1.02–1.52)	0.03
				Heme iron (HR = 1.16, Q5 vs. Q1, 0.95–1.42)	0.37
				Meat heme iron (HR = 1.12, Q5 vs. Q1, 0.92–1.38)	0.59

oxidative stress may be at increased risk of BC, and this association may be most relevant among women with high iron intake.[18] These results indicate that the association between dietary iron and BC risk varies depending on the exogenous estrogens, alcohol consumption, other dietary factors, and genetic variation.

Endometrial Cancer

Heme iron intake has been shown to be positively associated with the risk of diabetes and obesity, both of which are suspected or established risk factors for endometrial cancer (EC).[19] However, epidemiological studies have shown mixed results (Table 13.3). A recent population-based case-control study in Shanghai, China showed that the animal-derived iron intake was positively associated with EC risk (p for trend < 0.01; OR = 1.86 in the Q4 vs. Q1, 95% CI 1.22–2.85), predominantly after menopause (p for trend < 0.01; OR = 2.02 in the Q4 vs. Q1, 95% CI 1.15–3.55) and in postmenopausal women with BMI ≥ 25 kg/m^2 (p for trend = 0.002; OR = 3.25 in the Q4 vs. Q1, 95% CI 1.41–7.50).[20] However, the Canadian National Breast Screening Study, a prospective cohort study, reported no association of meat intake or any of the dietary iron-related variables with the risk of EC.[21] In contrast, a prospective cohort study with 720 EC cases ascertained in the Swedish Mammography Cohort of 21 years of follow-up, heme iron intake (p for trend = 0.02; RR = 1.24 in the Q4 vs. Q1, 95% CI 1.01–1.53), total iron intake (p for trend = 0.009; RR = 1.31 in the Q4 vs. Q1, 95% CI 1.07–1.61) and liver consumption (p for trend = 0.01; RR = 1.29 in ≥ 100 g/wk vs. < 100g/wk, 95% CI 1.06–1.56) were associated with increased risk of EC.[22] Also, a meta-analysis observed random-effects dose-response summary estimates of 1.26 (95% CI 1.03–1.54) per 100 g of total meat/d, 1.51

(95% CI 1.19–1.93) per 100 g of red meat/d, 1.03 (95% CI 0.32–3.28) per 100 g of poultry/d.[23] The authors suggested that meat consumption, particularly red meat, increases EC risk.

Biochemical Indices of Body Iron Status and Cancer

Iron absorption is tightly regulated to maintain body iron pool homeostasis (Table 13.4). The efficiency of iron absorption is influenced by various factors, especially the dietary source of iron and body iron storage. Among various iron indices available, serum/plasma ferritin has been acknowledged as the most accurate indicator of body iron stores, although it can be elevated with many clinical conditions, including liver disease, infection, and malignant neoplasms. While serum iron is considered as a less specific marker for body iron stores, transferrin saturation has been shown to reflect tissue-iron stores. Transferrin is the major iron binding protein in circulation and its level increases as serum iron availability decreases. Total iron-binding capacity (TIBC) is commonly assessed in clinical settings as an indicator of iron load and is an indicator of transferrin concentration. Unlike other serum iron indices, such as ferritin, TIBC is not affected by the inflammatory conditions, and does not exhibit day-to-day variation in measurements. Unsaturated iron binding capacity (UIBC) is also a measure of the availability of iron binding sites on transferrin, and also increases as iron availability decreases.

In a nested case-control study within the α-Tocopherol, ß-Carotene Cancer Prevention Study cohort, it was reported that serum ferritin, serum iron, and transferrin saturation were all inversely associated with colon cancer risk specifically (OR = 0.2, 95% CI 0.1–0.7; OR = 0.2,

TABLE 13.3 Epidemiological Evidences on the Association of Iron and Meat Intake with Endometrial Cancer (EC) Risk

Case-Control Study					
Study	**Population**		**Cases/Controls**	**OR (95% CI)**	**P trend**
Kallianpur et al., 2009	China Shanghai Breast Cancer Study	EC	1,165/1,147	Animal iron (OR = 1.86, Q4 vs. Q1, 1.22-2.85)	< 0.01
Cohort Study					
Study	**Population**		**Case/Participant**	**RR/HR (95% CI)**	**P trend**
Genkinger et al., 2012	Swedish mammography cohort	EC(21y)	720/60,825	Total iron (RR = 1.25, Q4 vs. Q1, 1.02–1.54) Heme iron (RR = 1.24, Q4 vs. Q1, 1.01–1.52) Liver iron (RR = 1.29, ≥100g/wk vs. '100g/wk, 1.06–1.56) Red meat iron (RR = 1.06, ≥600g/wk vs. '100g/wk, 0.68–1.66)	0.05 0.03 0.01 0.11
Kabat et al., 2008	Canadian National Breast Screening	EC(8y)	1,205/34,148	Red meat iron (HR = 0.86, Q5 vs. Q1, 0.61–1.22) Total iron (HR = 0.90, Q5 vs. Q1, 0.64–1.26) Meat iron (HR = 0.79, Q5 vs. Q1, 0.57–1.16) Heme iron (HR = 0.82, Q5 vs. Q1, 0.59–1.16)	0.75 0.22 0.37 0.22

TABLE 13.4 Normal Blood Levels of Iron Status Indices

Indices	Normal Ranges
Iron	Male: 80–180 μg/dl Female: 60–160 μg/dl Newborn: 10–250 μg/dl Child: 50–120 μg/dl
Total iron binding capacity	250–460 μg/dl
Transferrin	Adult male: 215–365 μg/dl Adult female: 250–380 μg/dl Newborn: 130–275 μg/dl Child: 203–360 μg/dl
Transferrin saturation	Male: 20–50% Female: 15–50%

Source: Modified from Pagana, K.D., and Pagana, T. J. (2003) Mosby's Diagnostic and Laboratory Test Reference, 6th ed. Mosby, Inc., St. Louis.

95% CI 0.1–0.9; OR = 0.1, 95% CI 0.02–0.5), whereas serum UIBC was positively associated with colon cancer risk (OR = 4.7, 95% CI 1.4-15.1).[24] Another case-control study reported that strong inverse associations were found between serum ferritin and proximal colon cancer risk.[25] On the other hand, the Prostate, Lung, Colorectal and Ovarian (PLCO) Cancer Screening Trial found that serum TIBC and UIBC were inversely associated with colorectal adenoma (OR = 0.57, in the Q4 vs. Q1, 95% CI 0.37–0.88; OR = 0.62, in the Q4 vs. Q1, 95% CI 0.40–0.95), while serum levels of ferritin, iron, and transferrin saturation were not associated with colorectal adenoma incidence.[26] Also, a nested case-control study in Shanghai showed no association of serum levels of ferritin with BC risk[27], whereas the Japanese atomic bomb survivor

study found that serum level of ferritin level was positively associated with increased risk of BC.[27]

IRON HOMEOSTASIS AND CANCER DEVELOPMENT: MECHANISTIC VIEWS

Iron as a Source of Reactive Oxygen Species and Its Role in Carcinogenesis

Reactive Oxygen Species (ROS) and Carcinogenesis

ROS in living organisms are formed during metabolic processes including mitochondrial respiratory chain where superoxide radical is frequently produced from the electron transport chain in mitochondria (Figure 13.1). Superoxide radical subsequently forms hydrogen peroxide, hydroxyl, and peroxyl radicals. The cytochrome P450 metabolic pathways, inflammatory processes, and xanthine oxidase pathway are other important endogenous sources of cellular ROS. Besides these sources, metal ions, radiations, and chemicals are exogenous ROS inducers. Due to their highly toxic nature, ROS produced in the cellular environment are quickly removed by endogenous and exogenous defense mechanisms. Under normal circumstances, cellular oxidation-reduction potential homeostasis is maintained by an antioxidant defense system. However, excess ROS formation, prolonged oxidative stress, or defects in the defense system disrupt the oxidation-reduction homeostasis.

Excess ROS have been shown to be closely related to carcinogenic processes. Carcinogenesis is a multistep process composed of tumor initiation, promotion, and

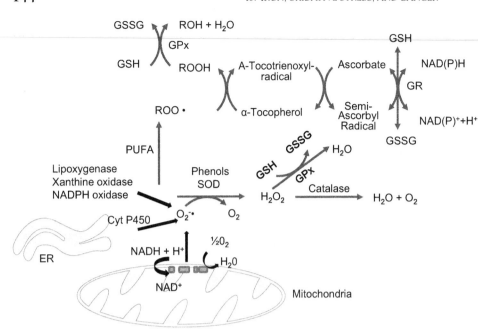

FIGURE 13.1 Generation of superoxide radical and antioxidant defense system. Superoxide radicals are generated during normal cellular metabolism. Endogenous antioxidant defense enzymes include superoxide dismutase (SOD), catalase, and glutathione peroxidase (GPx). Exogenous antioxidants are tocopherols, ascorbic acid, and phenols derived from dietary sources. *From Sung & Bae (2013) Dietary Antioxidants and Rheumatoid Arthritis. In: Watson RR and Preedy VR (eds.) Bioactive Food as Dietary Interventions for Arthritis and Related Inflammatory Diseases, pp. 515–527. San Diego: Academic Press. Reprinted with permission from Elsevier.* (See the color plate.)

progression. Permanent and heritable damages in cellular DNA are key mechanisms involved in tumor initiation. ROS cause oxidative DNA damage inducing gene mutation and subsequent aberrant gene expression possibly through base modifications, strand breaks, and DNA-protein cross linkages.[28] A most common DNA lesion, 8-oxo-7,8-dihydro-2′-deoxyguanosine, induces base alterations leading to mutations in *RAS* oncogenes and *p53* tumor suppressor genes. Under normal circumstances, mutated genes are repaired by several different protective mechanisms; however, sustained oxidative stress disables the repair process, causing mutations in genes involved in the repair process, which can cause the proliferation of mutated cells leading to carcinogenesis. Also ROS act as a secondary messenger in cellular signaling pathways that regulate cellular processes including cell proliferation, differentiation, and survival. An excellent review on the redox regulation in cellular signaling has been recently published elsewhere.[29] Oxidative stress also induces epigenetic alterations in human carcinogenesis.[28] Epigenetic changes due to alterations in DNA methylation pattern are key events involved in the development of human cancer. Especially the hydroxyl-radical-induced base modification inhibits necessary DNA methylation by DNA methyltransferase, causing global hypomethylation.[30] Oxidative stress also induces hypermethylation in the promoter region of tumor suppressor genes, resulting in gene silencing.[56]

Iron-Induced ROS Formation

Metals have shown to induce DNA damage in both a direct and indirect manner, and therefore they have been implicated as potential carcinogens.[31] Although precise mechanisms of action are not completely understood, metals promote the formation of ROS, and thereby cause damages in DNA and other cellular macromolecules. Iron in a biological system is mostly present either as free ions or as a component of heme. Depending on the oxidation status, iron is found in two forms as either ferrous ions (Fe^{2+}) or ferric ions (Fe^{3+}). Under normal physiological conditions, iron is present in stable ferric form while ferric iron is converted into ferrous ion when the reduced form of iron is required for biological processes such as the synthesis of heme and storage of iron in the form of ferritin.

In a biological system, the superoxide radical formed during cellular metabolism is rapidly converted into hydrogen peroxide by the action of superoxide dismutase (SOD). Either a superoxide or a hydrogen peroxide readily interacts with a ferrous iron to form a hydroxyl radical via the Fenton-type reaction.

$$Fe^{2+} + H_2O_2 \rightarrow Fe^{3+} + \cdot OH + OH^- \text{ (Fenton-type reaction)}$$

Not only free iron but heme iron is subjected to H_2O_2 induced oxidation. The oxidized heme is reverted to reduced form by the action of antioxidants. However, persistent oxidative stress leading the depletion of antioxidants induces further oxidation and releases iron from heme by protein denaturation followed by the Fenton-type reaction.

Iron Overload as a Risk Factor of Cancer: Evidences from Experimental Studies and Plausible Mechanisms of Action

Previous epidemiological studies have suggested that excess iron intake is positively related to the risk of different types of cancer (refer to Chapter 8). Hemochromatosis is a well-known genetic disease caused by iron overload resulting in liver cirrhosis or hepatocellular carcinoma due to tissue injury. A number of animal models showed that inhalational, intraperitoneal, or intramuscular administration caused iron-induced tumor formation at different sites.[32] The most plausible mechanisms of action include (1) oxidative damages to DNA, lipid and protein initiating carcinogenesis and (2) excess oxidative stress-induced impairments in antioxidant defense mechanisms and related signal transduction pathways promoting abnormal cell growth.

LIVER CANCER

Iron overload has been suggested to play a role in hepatic carcinogenesis. Liver is a major iron storage site. Under normal physiological conditions, iron-bound ferritin is stored in the liver, while excess iron entry decomposes ferritin protein releasing iron ion exerting toxic effects on hepatocytes.

A total of 38 patients were studied to elucidate the association between hepatic oxidative damage and the presence of nonalcoholic steatohepatitis (NASH).[33] Hepatic oxidative damage measured by 8-oxodG was significantly higher in patients with NASH compared to that of the simple steatosis. Oxidative damage was significantly correlated with iron overload and clinical severity of hepatic steatosis in NASH patients. The iron reduction therapy performed using biweekly treatment with phlebotomy and dietary iron restriction significantly reduced 8-oxodG levels from 20.7 to 13.8 positive cells/10^5 μm^2, with improvement in liver function markers.

More recently, iron overload has been suggested to be involved in liver carcinogenesis. Swiss albino mice were subjected to a liver procarcinogen p-dimethylaminoazobenzine (p-DAB) treatment for 7 weeks. Oxidative stress markers, mineral accumulation in the liver, and metallothine, a metal binding stress protein, were measured to determine the involvement of metal ion-induced oxidative stress in liver carcinogenesis. Results indicated that p-DAB induced iron accumulation in the liver, and the expression of metallothionein was upregulated in association with elevated oxidative stress. The authors suggested that p-DAB increased free radicals causing excess hemolysis, which possibly accelerates liver iron deposition again increasing the risk of oxidative damage in the liver followed by neoplastic responses.

Cellular iron homeostasis is shown to be an important regulator of ROS production and subsequent ROS level determines cellular potential to migrate. A proteomic analysis of transforming growth factor-β1-induced epithelial-mesenchymal transition (EMT) in murine hepatocytes indicated ferritin heavy chain (FHC) is one of the significantly altered proteins.[34] FHC possesses a ferroxidase function to a controllable iron pool, suggesting that available free iron may be an important factor to induce EMT and thereby cancer cell migration.

Due to its high toxic effects, iron homeostasis is tightly controlled by regulating iron absorption, transport, and storage at a cellular and systemic level. Liver cirrhosis and hepatocellular carcinomas are frequent consequences of hereditary hemochromatosis (HH), which occurs due to genetic defects in genes involved in iron absorption, transport, and storage.[35] There are four different types of genetic defects in HH. Type 1 HH is caused by mutations in gene *HFE* (a gene related to major histocompatibility complex class I proteins), which suppresses the expression of hepcidin. Hepcidin is a hepatic peptide hormone facilitating the degradation of iron transporter through the binding with ferroportin (FPN). Type 2 HH involves mutations in *HAMP* (hepcidin autimicrobial peptide) or *HFE2*, which is also known as *HJV* (hemojuvelin). HH type 2 develops cardiac and endocrine disorders. HAMP gene produces hepcidin autimicrobial peptide, a hormone regulated by the expression of SLC40A1. Chronic hepatitis virus infection is also known to suppress the production of HAMP thereby increasing the unregulated iron transportation and the risk of hepatic carcinogenesis. HJV is a membrane GPI-linked protein and BMP coreceptor through which hepcidin transcription signal is transmitted. Type 3 HH is caused by a mutation in transferrin R2 (*TFR2*), which is related to the expression of hepcidin as well as iron transportation. Type 4 HH is due to a mutation in ferroportin (*FPN*), which is also known as *SLC40A1* (the gene encoding duodenal iron-regulated transporter). The mutant ferroportin is known to be resistant to hepcidin-induced internalization causing iron overload in hepatocytes[36,37] due to the inability of hepcidin to induce ferroportin degradation.[36]

KIDNEY CANCER

Although random genomic damages are generally accepted as ROS-induced oxidative damages, a recent study suggested that ferric nitrilotriacetate (Fe-NTA) administration caused deletion and single nucleotide substitutions at G:C sites in the renal cell carcinomas (RCC) of experimental animals.[38] Genomic sites specifically fragile to iron-induced oxidative stress have been suggested as $p15^{INK4B}$ and $p16^{INK4A}$ tumor suppressor genes.[39] Another study suggested that Fe-NTA-induced RCC contains genomic amplification of *ptprz1*, a tyrosine phosphatase that is associated with β-catenin pathway activation.[40] Another *in vivo* study indicated that repeated administration of ferric nitrilotriacetate induced global genomic alterations, preferentially Cdkn2a/2b deletion and a Met amplification in renal tumor tissue samples.[41] The authors suggested that rapid regeneration of tubular cells under chronic oxidative stress by ferric nitrilotriacetate administration increased abnormal DNA replication and chromosomal missegregation leading the development of renal cell carcinomas.

BREAST CANCER

Iron concentration in breast cancer tissue was five times higher compared to the level in adjacent normal tissue,[42] suggesting a possible relationship between iron and mammary tumor growth. Previous animal studies showed that iron depletion suppresses chemical carcinogen-induced mammary tumor development.[43–45] Female Sprague-Dawley rats were fed a diet with a low (2ppm), adequate (120 ppm), or excess (1200 ppm) amount of iron ($FeSO_4 \cdot 7H_2O$) for 32 weeks.[43] At the age of 50 days, animals were intraperitoneally administered with 1-methyl-1 nitrosourea (MNU) at a concentration of 25 mg/kg body weight. Results indicated that excess iron supply promoted mammary tumor formation while iron depletion did not suppress the growth of tumors compared to adequate iron supplementation. To provide the evidence of direct tumor growth stimulatory effects of iron, female Sprague-Dawley rats received a single 5 mg/kg intraperitoneal injection of dimethylbenz[a]anthracene (DMBA) at the age of 55 days.[44] After 8 days of DMBA administration, animals were subjected to the subcutaneous injection of iron sulfate (50 μmol/kg, 2 time/week for 53 weeks). Study results indicated that iron supplementation showed higher tumor frequency and larger tumor size. Another study indicated that iron-deficient diet suppressed DMBA-induced mammary tumor development compared to the diet containing an appropriate amount of iron (50 mg/kg) while iron repletion attenuated the suppressive effects of iron depletion indicating that sufficient iron is required for tumor growth.[45]

A recent in vitro study suggested that iron-induced ROS generation was facilitated by the increased heme oxygenase-1 (HO-1) activity resulting in increased expression of matrix metalloproteinase-1 (MMP-1) in human breast cancer cells.[46] MMPs are a group of proteinases involving cancer cell migration and angiogenesis that indicate a possible involvement of iron overload with cancer metastasis.

Several studies suggested that the increased risk of breast cancer with iron overload is related to estrogen metabolism. Breast tumor tissues exhibit elevated DNA damage possibly due to the formation of redox active estrogen metabolites.[47] Endogenous estradiol generates 2-hydroxy estradiol and 4-hydroxy estradiol (4-OHE) by the action of cytochrome 1A1 and 1B1, and these metabolites are shown to form hydrogen peroxide in normal or malignant breast epithelial cells.[47] In our previous study, the expression of DNA repair protein was suppressed by 4-OHE in MCF-10A breast epithelial cells.[48] Excess iron may replace zinc in zinc finger of estrogen receptor producing more free radicals. Alcohol, a well-known risk factor of breast cancer triggers the release of iron ion from ferritin, which may explain a part of the mechanisms of action.

A more recent study, however, suggested that both iron depletion and iron overload contribute to breast cancer development.[49] Menopause-induced iron accumulation explains the increased risk of breast cancer in postmenopausal women possibly through oxidative stress-induced activation of mitogen-activated protein kinase.[49]

OTHER TYPES OF CANCER

The AICR report suggested that red meat consumption is significantly associated with the risk of colon cancer. Two studies provided evidence of elevated iron deposition in colon carcinoma tissues compared to that of the adenocarcinomas.[50] When a total of 54 colon cancer and 59 colon adenocarcinomas were subjected to histological examination and Perl' Prussian staining to reveal ferric iron deposition, frequent iron deposition was observed in the stroma of large adenomas, pedunculated adenomas, and adenocarcinomas suggesting iron deposition is an event secondary to intralesional hemorrhage in these tissues.[51] Few studies provided mechanistic explanations for the role of iron in colon carcinogenesis and more studies are needed to explain the association between colon cancer and iron overload.

Male non-small cell lung cancer patients (n = 125) were studied as to whether iron status parameters are associated with cancer survival.[52] Also the expression of transferrin receptor 1 and ferritin expression was analyzed and compared between tumor and normal tissues. Results indicated that 90% of the patients had lower serum ferritin iron content compared to the normal reference values. Serum iron content was positively correlated with survival while tumor tissues have elevated expression of ferritin and transferrin receptor 1.

In a prostate cancer cell line (DU145), TNF-alpha triggers JNK activation, which increases the degradation of ferritin liberation iron ion. UVB radiation is known as a cancer-initiating factor. In human dermal fibroblasts, UVB-induced ROS generation increases JNK2 activity and the expression of c-jun, MMP-1, and MMP-3 mRNA, stimulating the cell growth.[53]

Iron and Immune Dysregulation

Macrophage activation is a critical surveillance mechanism and plays a role in tumor surveillance. The lymphokine-dependent tumoricidal activity was significantly inhibited in the presence of ferritin. Ferritin pretreatment also inhibited phytohaemagglutinin or concanavalin A–stimulated lymphocyte function against K-562 cells.[54] Although no precise mechanisms are suggested, the inhibition of effector molecules such as ROS and cytolytic factor by iron may play a role.

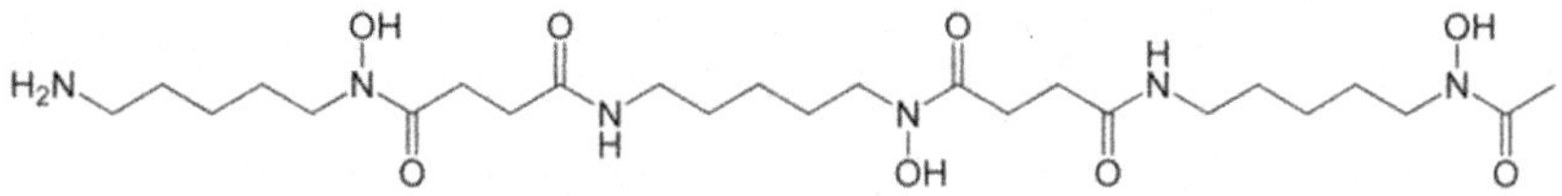

FIGURE 13.2   The structure of deferoxamine. *From* http://en.wikipedia.org/wiki/File:Deferoxamine-2D-skeletal.png

Iron Chelation in Cancer Treatment

Apart from iron overload as a risk factor in cancer development, iron is an essential nutrient required to maintain cell growth and function. Several iron chelators have been suggested as effective cancer therapeutic agents with high selectivity against neoplastic cells. Due to its essentiality in cell proliferation, iron depletion induces cell cycle arrest and apoptosis possibly through acting as a cofactor in enzymes involved in energy metabolism and DNA synthesis.[55] A previous study using lymphocytes indicated that deferoxamine (Figure 13.2), a hydroxylamine produced by *Streptomyces pilosus*, binds ferric iron resulting in cell cycle arrest at S phase by inhibiting ribonucleotide reductase (RR).[55] Iron is a cofactor of RR, which converts ribonucleotides into deoxyribonucleotides for DNA synthesis, and high selectivity of iron chelators against tumor cells may be due to greater activity of RR in tumor cells compared to the activity in normal cells.[56] Efficacy of iron chelators to suppress tumor cell growth was evidenced in a number of clinical trials.[57] Fe chelation not only inhibits DNA synthesis by suppressing the activity of RR, but regulates the expression of proteins involved in cell cycle progression. These proteins include cyclin D1, cyclin D2, cyclin D3, CDK2, p21, and p53.[55]

The involvement of iron depletion in cancer cell angiogenesis and metastasis has been discussed. Hypoxia inducible factor-1 (HIF-1) is a transcriptional factor regulating the expression of genes involved in angiogenesis including vascular endothelial growth factor 1 (VEGF-1), endocrine gland derived VEGF-1, leptin, and transforming growth factor β3. Under the condition of adequate oxygen supply, HIF-1 is decomposed by binding with active VHL protein. However, hypoxic conditions suppress the activation of VHL protein leading the overexpression of angiogenesis genes. It has been suggested that Fe depletion mimics the condition of hypoxia.[58] A recent study showed that the oral intake of iron chelator deferasirox inhibits DMS53 lung carcinoma growth in nude mice[59] possibly through the increased expression of the metastasis suppressor protein *N-myc* downstream-regulated gene 1 and cyclin-dependent kinase inhibitor p21.[CIP1/WAF1] Deferasirox also increased the expression of apoptotic molecules including cleaved caspase-3 and cleaved polu (ADP-ribose) polymerase 1. Another study showed that iron chelators desferrioxamine or di-2-pyridylketone-4,4-dimethyl-3-thiosemicarbozone regulated mitogen-activated protein kinase pathway thereby inducing cell cycle arrest and apoptosis in different cancer cell lines.[60] Antitumor iron chelators include desferrioxamine, Thiosemicarbazones, Tachpyridine, ICL670A, Di-2-pyridylketone thiosemicarbazone series, 2-Benzoylpyridine thiosemicarbazone series and Thiohydrazone series.

SUMMARY POINTS

- Iron increases the production of reactive oxygen species.
- ROS are involved in tumor initiation and promotion.
- Dietary iron intake is positively associated with colorectal cancer.
- Hepatic iron accumulation is a possible cause of liver cancer.
- Genetic defects in iron homeostasis are associated with cancer risk.
- Iron chelators are possible cancer therapeutics.

References

1. Ferrucci LM, Sinha R, Graubard BI, Mayne ST, Ma X, Schatzkin A, et al. Dietary meat intake in relation to colorectal adenoma in asymptomatic women. *Am J Gastroenterol* 2009;**104**:1231–40.
2. Lee DH, Anderson KE, Harnack LJ, Folsom AR, Jacobs Jr DR. Heme iron, zinc, alcohol consumption, and colon cancer: Iowa Women's Health Study. *J Natl Cancer Inst* 2004;**96**:403–7.
3. Giovannucci E, Rimm EB, Stampfer MJ, Colditz GA, Ascherio A, Willett WC. Intake of fat, meat, and fiber in relation to risk of colon cancer in men. *Cancer Res* 1994;**54**:2390–7.
4. Sun Z, Zhu Y, Wang PP, Roebothan B, Zhao J, Dicks E, et al. Reported intake of selected micronutrients and risk of colorectal cancer: results from a large population-based case-control study in Newfoundland, Labrador and Ontario, Canada. *Anticancer Res* 2012;**32**:687–96.
5. van Lee L, Heyworth J, McNaughton S, Iacopetta B, Clayforth C, Fritschi L. Selected dietary micronutrients and the risk of right- and left-sided colorectal cancers: a case-control study in Western Australia. *Ann Epidemiol* 2011;**21**:170–7.
6. Cross AJ, Ferrucci LM, Risch A, Graubard BI, Ward MH, Park Y, et al. A large prospective study of meat consumption and colorectal cancer risk: An investigation of potential mechanisms underlying this association. *Cancer Res* 2010;**70**:2406–14.
7. Zhang X, Giovannucci EL, Smith-Warner SA, Wu K, Fuchs CS, Pollak M, et al. A prospective study of intakes of zinc and heme iron and colorectal cancer risk in men and women. *Cancer Causes Control* 2011;**22**:1627–37.
8. Balder HF, Vogel J, Jansen MC, Weijenberg MP, van den Brandt PA, Westenbrink S, et al. Heme and chlorophyll intake and risk of colorectal cancer in the Netherlands cohort study. *Cancer Epidemiol Biomark Prev* 2006;**15**:717–25.
9. Iacopetta B. Are there two sides to colorectal cancer? *Int J Cancer* 2002;**101**:403–8.
10. Buzard GS, Kasprzak KS. Possible roles of nitric oxide and redox cell signaling in metal-induced toxicity and carcinogenesis: a review. *J Environ Pathol Toxicol Oncol* 2000;**19**:179–99.

11. Senesse P, Meance S, Cottet V, Faivre J, Boutron-Ruault MC. High dietary iron and copper and risk of colorectal cancer: A case-control study in Burgundy, France. *Nutr Cancer* 2004;**49**:66–71.

12. Chadha VD, Goel A, Dhawan D. Regulatory role of zinc on the biokinetics and biodistribution of (65)Zn during the initiation of experimentally induced colon cancer. *Nutr Cancer* 2011;**63**:212–7.

13. Kallianpur AR, Lee SA, Gao YT, Lu W, Zheng Y, Ruan ZX, et al. Dietary animal-derived iron and fat intake and breast cancer risk in the Shanghai Breast Cancer Study. *Breast Cancer Res Treat* 2008;**107**:123–32.

14. Moore AB, Shannon J, Chen C, Lampe JW, Ray RM, Lewis SK, et al. Dietary and stored iron as predictors of breast cancer risk: A nested case-control study in Shanghai. *Int J Cancer* 2009;**125**:1110–7.

15. Kabat GC, Miller AB, Jain M, Rohan TE. Dietary iron and heme iron intake and risk of breast cancer: a prospective cohort study. *Cancer Epidemiol Biomark Prev* 2007;**16**:1306–8.

16. Ferrucci LM, Cross AJ, Graubard BI, Brinton LA, McCarty CA, Ziegler RG, et al. Intake of meat, meat mutagens, and iron and the risk of breast cancer in the Prostate, Lung, Colorectal, and Ovarian Cancer Screening Trial. *Br J Cancer* 2009;**101**:178–84.

17. Lee DH, Anderson KE, Harnark LJ, Jacobs Jr DR. Dietary iron intake and breast cancer: The Iowa Women's Health Study. *Proc Amer Assoc Cancer Res* 2004;**45**. Available from: http://www.aacrmeetingabstracts.org/cgi/content/abstract/2004/1/535-c.

18. Hong CC, Ambrosone CB, Ahn J, Choi JY, McCullough ML, Stevens VL, et al. Genetic variability in iron-related oxidative stress pathways (Nrf2, NQ01, NOS3, and HO-1), iron intake, and risk of postmenopausal breast cancer. *Cancer Epidemiol Biomark Prev* 2007;**16**:1784–94.

19. Rajpathak S, Ma J, Manson J, Willett WC, Hu FB. Iron intake and the risk of type 2 diabetes in women: A prospective cohort study. *Diabetes Care* 2006;**29**:1370–6.

20. Kallianpur AR, Lee SA, Xu WH, Zheng W, Gao YT, Cai H, et al. Dietary iron intake and risk of endometrial cancer: a population-based case-control study in Shanghai, China. *Nutr Cancer* 2010;**62**:40–50.

21. Kabat GC, Miller AB, Jain M, Rohan TE. Dietary iron and haem iron intake and risk of endometrial cancer: a prospective cohort study. *Br J Cancer* 2008;**98**:194–8.

22. Genkinger JM, Friberg E, Goldbohm RA, Wolk A. Long-term dietary heme iron and red meat intake in relation to endometrial cancer risk. *Am J Clin Nutr* 2012;**96**:848–54.

23. Bandera EV, Kushi LH, Moore DF, Gifkins DM, McCullough ML. Consumption of animal foods and endometrial cancer risk: a systematic literature review and meta-analysis. *Cancer Causes Control* 2007;**18**:967–88.

24. Cross AJ, Gunter MJ, Wood RJ, Pietinen P, Taylor PR, Virtamo J, et al. Iron and colorectal cancer risk in the alpha-tocopherol, beta-carotene cancer prevention study. *Int J Cancer* 2006;**118**:3147–52.

25. Kato I, Dnistrian AM, Schwartz M, Toniolo P, Koenig K, Shore RE, et al. Iron intake, body iron stores and colorectal cancer risk in women: A nested case-control study. *Int J Cancer* 1999;**80**:693–8.

26. Cross AJ, Sinha R, Wood RJ, Xue X, Huang WY, Yeager M, et al. Iron homeostasis and distal colorectal adenoma risk in the prostate, lung, colorectal, and ovarian cancer screening trial. *Cancer Prev Res (Phila)* 2011;**4**:1465–75.

27. Stevens RG, Cologne JB, Nakachi K, Grant EJ, Neriishi K. Body iron stores and breast cancer risk in female atomic bomb survivors. *Cancer Sci* 2011;**102**:2236–40.

28. Ziech D, Franco R, Pappa A, Panayiotidis MI. Reactive oxygen species (ROS)-induced genetic and epigenetic alterations in human carcinogenesis. *Mutat Res* 2011;**711**:167–73.

29. Ray PD, Huang BW, Tsuji Y. Reactive oxygen species (ROS) homeostasis and redox regulation in cellular signaling. *Cell Signal* 2012;**24**:981–90.

30. Toyokuni S. Molecular mechanisms of oxidative stress-induced carcinogenesis: from epidemiology to oxygenomics. *IUBMB Life* 2008;**60**:441–7.

31. Kawanishi S, Hiraku Y, Murata M, Oikawa S. The role of metals in site-specific DNA damage with reference to carcinogenesis. *Free Radic Biol Med* 2002;**32**:822–32.

32. Toyokuni S. Role of iron in carcinogenesis: cancer as a ferrotoxic disease. *Cancer Sci* 2009;**100**:9–16.

33. Fujita N, Miyachi H, Tanaka H, Takeo M, Nakagawa N, Kobayashi Y, et al. Takei Y., Iron overload is associated with hepatic oxidative damage to DNA in nonalcoholic steatohepatitis. *Cancer Epidemiol Biomark Prev* 2009;**18**:424–32.

34. Zhang KH, Tian HY, Gao X, Lei WW, Hu Y, Wang DM, et al. Ferritin heavy chain-mediated iron homeostasis and subsequent increased reactive oxygen species production are essential for epithelial-mesenchymal transition. *Cancer Res* 2009;**69**:5340–8.

35. Chen J, Chloupkova M. Abnormal iron uptake and liver cancer. *Cancer Biol Ther* 2009;**8**:1699–708.

36. Schimanski LM, Drakesmith H, Merryweather-Clarke AT, Viprakasit V, Edwards JP, Sweetland E, et al. In vitro functional analysis of human ferroportin (FPN) and hemochromatosis-associated FPN mutations. *Blood* 2005;**105**:4096–102.

37. Liu XB, Yang F, Haile DJ. Functional consequences of ferroportin 1 mutations. *Blood Cells Mol Dis* 2005;**35**:33–46.

38. Jiang L, Zhong Y, Akatsuka S, Liu YT, Dutta KK, Lee WH, et al. Deletion and single nucleotide substitution at G: C in the kidney of gpt delta transgenic mice after ferric nitrilotriacetate treatment. *Cancer Sci* 2006;**97**:1159–67.

39. Tanaka T, Iwasa Y, Kondo S, Hiai H, Toyokuni S. High incidence of allelic loss on chromosome 5 and inactivation of p15INK4B and p16INK4A tumor suppressor genes in oxystress-induced renal cell carcinoma of rats. *Oncogene* 1999;**18**:3793–7.

40. Liu YT, Shang D, Akatsuka S, Ohara H, Dutta KK, Mizushima K, et al. Chronic oxidative stress causes amplification and overexpression of ptprz1 protein tyrosine phosphatase to activate beta-catenin pathway. *Am J Pathol* 2007;**171**:1978–88.

41. Akatsuka S, Yamashita Y, Ohara H, Liu YT, Izumiya M, Abe K, et al. Fenton reaction induced cancer in wild type rats recapitulates genomic alterations observed in human cancer. *PLoS One* 2012;**7**:e43403.

42. Ionescu JG, Novotny J, Stejskal V, Latsch A, Blaurock-Busch E, Eisenmann-Klein M. Increased levels of transition metals in breast cancer tissue. *Neuro Endocrinol Lett* 2006;**27**(Suppl. 1):36–9.

43. Thompson HJ, Kennedy K, Witt M, Juzefyk J. Effect of dietary iron deficiency or excess on the induction of mammary carcinogenesis by 1-methyl-1-nitrosourea. *Carcinogenesis* 1991;**12**:111–4.

44. Diwan BA, Kasprzak KS, Anderson LM. Promotion of dimethyl-benza.anthracene-initiated mammary carcinogenesis by iron in female Sprague-Dawley rats. *Carcinogenesis* 1997;**18**:1757–62.

45. Hrabinski D, Hertz JL, Tantillo C, Berger V, Sherman AR. Iron repletion attenuates the protective effects of iron deficiency in DMBA-induced mammary tumors in rats. *Nutr Cancer* 1995;**24**:133–42.

46. Kim DH, Kim JH, Kim EH, Na HK, Cha YN, Chung JH, et al. 15-Deoxy-Delta12,14-prostaglandin J2 upregulates the expression of heme oxygenase-1 and subsequently matrix metalloproteinase-1 in human breast cancer cells: Possible roles of iron and ROS. *Carcinogenesis* 2009;**30**:645–54.

47. Fussell KC, Udasin RG, Smith PJ, Gallo MA, Laskin JD. Catechol metabolites of endogenous estrogens induce redox cycling and generate reactive oxygen species in breast epithelial cells. *Carcinogenesis* 2011;**32**:1285–93.

48. Lee EJ, Oh SY, Kim MK, Ahn SH, Son BH, Sung MK. Modulatory effects of alpha- and gamma-tocopherols on 4-hydroxyestradiol induced oxidative stresses in MCF-10A breast epithelial cells. *Nutr Res Pract* 2009;**3**:185–91.

49. Jian J, Yang Q, Dai J, Eckard J, Axelrod D, Smith J, et al. Effects of iron deficiency and iron overload on angiogenesis and oxidative stress—a potential dual role for iron in breast cancer. *Free Radic Biol Med* 2011;**50**:841–7.

50. Labropoulou EP, Datsis A, Kekelos S, Rogdakis A, Katsiki E, Labropoulou CM, et al. The presence and significance of iron in neoplasia of the colorectum. *Tech Coloproctol* 2004;**8**(Suppl. 1):s211–3.

51. Han HS, Lee SY, Seong MK, Kim JH, Sung IK, Park HS, et al. Presence of iron in colorectal adenomas and adenocarcinomas. *Gut Liver* 2008;**2**:19–22.

52. Kukulj S, Jaganjac M, Boranic M, Krizanac S, Santic Z, Poljak-Blazi M. Altered iron metabolism, inflammation, transferrin receptors, and ferritin expression in non-small-cell lung cancer. *Med Oncol* 2010;**27**:268–77.

53. Antosiewicz J, Ziolkowski W, Kaczor JJ, Herman-Antosiewicz A. Tumor necrosis factor-α-induced reactive oxygen species formation is mediated by JNK 1-dependent ferritin degradation and elevation of labile iron pool. *Free Rad Biol Med* 2007;**43**:265–70.

54. Matzner Y, Hershko C, Polliack A, Konijn AM, Izak G. Suppressive effect of ferritin on in vitro lymphocyte function. *Br J Haematol* 1979;**42**:345–53.

55. Le NT, Richardson DR. The role of iron in cell cycle progression and the proliferation of neoplastic cells. *Biochim Biophys Acta* 2002;**1603**:31–46.

56. Takeda E, Weber G. Role of ribonucleotide reductase in expression in the neoplastic program. *Life Sci* 1981;**28**:1007–14.

57. Donfrancesco A, Deb G, De Sio L, Cozza R, Castellano A. Role of deferoxamine in tumor therapy. *Acta Haematol* 1996;**95**:66–9.

58. Callapina M, Zhou J, Schnitzer S, Metzen E, Lohr C, Deitmer JW, et al. Nitric oxide reverses desferrioxamine- and hypoxia-evoked HIF-1alpha accumulation-implications for prolyl hydroxylase activity and iron. *Exp Cell Res* 2005;**306**:274–84.

59. Lui GY, Obeidy P, Ford SJ, Tselepis C, Sharp DM, Jansson PJ, et al. The iron chelator, deferasirox, as a novel strategy for cancer treatment: oral activity against human lung tumor xenografts and molecular mechanism of action. *Mol Pharmacol* 2013;**83**:179–90.

60. Yu Y, Richardson DR. Cellular iron depletion stimulates the JNK and p38 MAPK signaling transduction pathways, dissociation of ASK1-thioredoxin, and activation of ASK1. *J Biol Chem* 2011;**286**: 15413–27.

Role of Black Chokeberries in Breast Cancer: A Focus on Antioxidant Activity

Beata Olas

Department of General Biochemistry, Faculty of Biology and Environmental Protection,
University of Lodz, Lodz, Poland

List of Abbreviations

8-isoPGF$_{2\alpha}$ 8-epi-prostaglandin F$_{2\alpha}$
O$_2$$^{-\bullet}$ superoxide anion radical
ORAC oxygen radical absorbing capacity
ROS reactive oxygen species
TAS total antioxidant level
TEAC trolox equivalent antioxidant capacity
WF fresh weight

OXIDATIVE STRESS IN BREAST CANCER

Oxidative stress causes damage in different molecules such as proteins, lipids, and DNA.[1–4] Oxidative stress is also found in cancer cells, but unfortunately, the mechanism(s) responsible for its induction is unknown. They may include oncogenic signals, mutations in mitochondrial DNA, intensive metabolism, inflammation, and cytokine action.[5] Overproduction of reactive oxygen species (ROS) has been demonstrated in different cancer cells: prostate, melanoma, or pancreatic cells; and also liver, bladder, colon, ovarian, and breast cancers.[6] Afanas'ev[6] suggests that the ROS overproduction in cancer cells is a major factor in their survival and resistance to chemotherapy, and the enhanced oxidative stress is responsible for aging development. It is interesting that an additional ROS generation induced by prooxidants may induce apoptosis in cancer cells.[6] Moreover, oncology therapy (radiotherapy or chemotherapy) induces the generation of reactive oxygen species not only in cancer cells, but also in normal cells.[7–13] ROS produced during cancer chemotherapy are often a source of serious side effects (cardiotoxicity, nephrotoxicity, ototoxicity, and hematological toxicity).[14,15]

The oxidative stress, detected by using various biomarkers (including the level of carbonyl groups and 3-nitrotyrosine in proteins, the total antioxidant level (TAS), the activity of antioxidative enzymes), had been observed in breast cancer patients before and during different phases of treatment.[7–13,16–23] Results of Kedzierska et al.[11] have shown that the total level of glutathione (an important physiological antioxidant) in plasma isolated from breast cancer patients (before and during treatment, during different phases of chemotherapy (doxorubicin and cyclophosphamide)) was lower than in healthy persons.

Experiments of other authors have demonstrated that treatment of breast cancer patients was also associated with a decrease of level of glutathione and an increase of lipid peroxidation in plasma.[24] Significantly lowered levels of antioxidative enzymes (superoxide dismutase, glutathione peroxidase, and glutathione reductase) were observed in breast cancer patients when compared to control subjects.[18–20] In Kasapovic et al.[18] and Rajneesh et al.[19] researchers showed that the level of plasma antioxidant activity was also significantly lower, whereas plasma level of malondialdehyde (MDA; a marker of lipid peroxidation) was significantly higher in patients than control. The clinical studies indicated that the patients with invasive breast cancer have extremely high levels of urinary 8-epi-prostaglandin F$_{2\alpha}$ (8-isoPGF$_{2\alpha}$), a marker of lipid peroxidation induced by free radicals[8,25] that indicates that the oxidative stress takes place. Rossner et al.[25] measured the urinary level of 15-F(2t)-isoprostane, and they observed a statistically significant trend in breast cancer risk with an increase of 15-F(2t)-isoprostane level. They suggest that the increased level may reflect with oxidative stress associated with recent treatment of patients (oncology therapy), but according to experiments by Kedzierska et al.,[8] can reflect with samples from patients before surgery, and breast cancer patients without radiotherapy or chemotherapy. Other experiments indicate that

© 2014 Elsevier Inc. All rights reserved.

serum levels of 8-hydroxydeoxyguanosine (8-OHdG) and superoxide dismutase activities were higher in patients with breast cancer, whereas glutathione peroxidase activity was lower in these patients. Results of Rossner et al.[25] also suggest that oxidative DNA damage may be associated with breast cancer risk.

It has been established that the oxidative/nitrative stress may stimulate changes of hemostasis in breast cancer patients (before and during the treatment).[13,26,27] The changes in protein tyrosine nitration in blood platelets isolated from breast cancer patients may modulate tyrosine phosphorylation (an important process for signal transduction) and platelet activation. According to the results obtained by Kedzierska et al.,[26] the oxidative stress (measured by the level of superoxide anion radicals ($O_2^{-\bullet}$)) in patients with breast cancer (before and during chemotherapy) may be an influence in increasing blood platelet aggregation (induced by strong physiological platelet agonist, thrombin) in patients with breast cancer before and after the surgery and also after various phases of the chemotherapy. Moreover, results of these authors showed a correlation between the $O_2^{-\bullet}$ generation and changes of platelet aggregation in breast cancer patients before and during various forms of treatment (the surgery and the chemotherapy (doxorubicin and cyclophosphamide)). Kedzierska et al.[26] suggest that the production of $O_2^{-\bullet}$ in blood platelets (activated by thrombin) obtained from breast cancer patients may induce the changes of platelet aggregation, which may contribute to thrombosis in these patients.

Several recent studies have also demonstrated platelet hyperactivity (measured by increased expression of P-selectin on the surface of blood platelets isolated from breast cancer patients).[14] Observations of Kim et al.[15] have found that doxorubicin can induce procoagulant activity; and what is more, ROS production is involved in this process. Moreover, results of Kim et al.[15] showed that doxorubicin can induce cytotoxicity in blood platelets through ROS generation and also decrease glutathione or protein thiol levels. Increased complication of thrombotic disorders is well documented in cancer patients treated with other ROS-generating chemotherapeutic drugs (bleomycin or cisplatin).[28] Results described by Trachootham et al.[29] and Panis et al.[30] showed that breast cancer patients subjected to chemotherapy (with doxorubicin) present immediate systemic oxidative stress and oxidative damages in red cells.

DIETARY ANTIOXIDANTS AND THEIR ROLE IN CANCER

The latest research confirmed low antioxidant status and increased oxidative stress in cancer patients before and during treatment.[1,2,13] Dietary antioxidants

(minerals, vitamins, and phenolic compounds) may protect against reactive oxygen species–induced damage of cells. An increased dietary intake of antioxidants is associated with a reduced risk of different diseases, including cancer.[2,31] The Mediterranean diet is also associated with a reduced risk of cancer. The biological mechanisms for cancer prevention associated with this diet have been related to the favorable action of balanced ration of omega 3 and 6 essential fatty acids and high amounts of antioxidants, including polyphenols present in vegetables and fruits. On the other hand, the consumption of low levels of antioxidants in the form of fruits and vegetables has been demonstrated to more than double the incidence of cancer.[2,31]

Diet and nutrition have been studied in relationship with breast cancer risk.[24,32–34] Experiments of Rockenbach et al. [24] showed that breast cancer diagnosis and treatments (surgery, chemotherapy, and radiation therapy) were associated with dietary intake changes and increased body weight, the body mass index, and oxidative stress. Some foods and nutrients have been suggested to increased the risk for breast cancer. On the other hand, dietary antioxidants (present in vegetables and fruits, including black chokeberries) may reduce the risk of breast cancer.[7–12,35,36] Moreover, antioxidant supplements (e.g., vitamin E and glutamine) may decrease side effects associated with treatment. According to Butalla et al.[37] regular carrot juice (8 oz of fresh BetaSweet or Balero orange carrot juice) intake induces an increase of plasma total carotenoid concentrations in breast cancer patients (increased by 1.65 and 1.38 µM for the BetaSweet and Balero carrot juice, respectively), and reduces oxidative stress. In this experiment, 69 breast cancer patients consumed fresh carrot juice for three weeks. The protective effect of other dietary supplements (resveratrol, vitamin C, lycopene, and anthocyanins (Ixor®)) was observed in reducing skin toxicity induced by radiotherapy in breast cancer patients. These compounds were used at a dose of two tablets a day, starting from 10 days before the radiotherapy until 10 days after the end of treatment.[38] The studies of Zhang et al.[39] demonstrated anticancer activities of tea epigallocatechin-3-gallate in breast cancer patients under radiotherapy.

ACTIVITY OF BLACK CHOKEBERRIES IN BREAST CANCER

Aronia melanocarpa (A. melanocarpa), native to eastern North America and East Canada, has become popular in Eastern Europe and Russia. The genus name *Aronia* (Rosaceae family, Maloideae subfamily) has been replacing the common name, chokeberry. Two species can be distinguished: *Aronia melanocarpa* [Michx.] Elliot (black

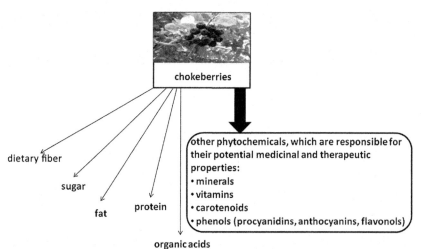

FIGURE 14.1 Chemical compounds of chokeberries, which are responsible for their potential medical and therapeutic properties [42; modified]. (See the color plate.)

chokeberry, Aronia noir) and *Aronia arbutifolia* [L.] Elliot (red chokeberry, Aronia rouge).[40–43] The plants have no problems with pests.[42,44] *A. melanocarpa* fruits have been traditionally used by Potawatomii Native Americans to cure colds.[43] At present, *A. melanocarpa* fruits are used for syrup, jam, juice, jellies, tea, wine, liqueur, and commercial extract (Aronox® by Agropharm (Poland)) production.[35,42,45] The components of chokeberries are dependent on various factors: fertilization, cultivar, maturation of fruits, harvest date, and location.[41,42,46,47]

Chokeberries contain very different chemical compounds,[42] including dietary fiber (to 5.62 g/100 fresh weight (FW)[48]), sugar 16 to 18%,[41] protein (to 0.7 g/100 g FW),[48] fat (to 0.14 g/100 gFW),[48] organic acids (1–1.5%),[48] minerals (to 440 mg/100g[48]), vitamins (B$_1$, B$_2$, B$_6$, C, niacin, and panthothenic acid), and other phytochemicals (carotenoids and phenols: procyanidins, anthocyanins, and flavonols; Table 14.1).[42] Figure 14.1 shows chemical compounds of chokeberries. Products from chokeberries are also a source of valuable phytochemicals, because the fruits of *A. melanocarpa* are one of the richest plant sources of phenolic substances.

The procyanidins, which are oligomeric and polymeric (epi)catechins formed from the association of several monomeric units, were identified as the major class of phenolic compounds in *A. melanocarpa* fruits.[42,49] In *A. melanocarpa* fruits, the anthocyanidins represent about 25% of total phenols.[42,49] The anthocyanins are water-soluble plant pigments (for the dark blue and even black color of the fruits). This class of phenolic compound has different biological activity: antioxidant, anti-inflammatory, antiplatelet, antimicrobial, hepatoprotective, gastroprotective, antiviral, and other activities.[42,45,50–52] Moreover, *A. melanocarpa* anthocyanins exhibit immunomodulatory activity in patients with breast cancer.[52]

TABLE 14.1 Chemical Compounds of Chokeberries [42; modified]

Chemical Compounds	
Glucose and fructose	66–176 g/kg FW
Dietary fiber	56 g/kg FW
Fat	0.14% FW
Protein	0.7% FW
Vitamin C	13–270 mg/kg FW
Folate	200 µg/kg FW
Vitamin B$_1$	180 µg/kg FW
Vitamin B$_2$	200 µg/kg FW
Vitamin B$_6$	280 µg/kg FW
Niacin	3000 µg/kg FW
Pantothenic acid	2790 µg/kg FW
Tocopherols	17.1 mg/kg FW
Vitamin K	242 µg/kg FW
Carotenoids	48.6 mg/kg FW
Na	26 mg/kg FW
K	2180 mg/kg FW
Ca	322 mg/kg
Mg	162 mg/kg FW
Fe	9.3 mg/kg FW
Zn	1.47 mg/kg FW
Citric acid	2.1 g/kg FW
Malic acid	13.1 g/kg FW
Phenolic constituents	2010–6902 mg/100 g FW

Research of other authors also indicates the immunomodulatory effect of *A. melanocarpa* in the combination with apple pectin in patients with breast cancer undergoing radiotherapy.[53] The experiment comprised 42 women (19–25 years of age) receiving 20 ml of chokeberry concentrate in combination with 15 g of apple pectin twice daily during radiation therapy. Experiments of Lala et al.[54] demonstrate the protective property of the extract from *A. melanocarpa* in colon carcinogenesis in model system *in vitro*, and indicate multiple mechanism of action. *A. melanocarpa* extract has also anticancer effects, mediated by the increase of tumor suppressor genes as well as by reduction of oxidative stress and resulting DNA damage important for the proliferation of cancer cells.[42,54] Another study demonstrated that *A. melanocarpa* extract stimulates apoptosis of human HT-29 colon cancer cells.[42,55,56] *A. melanocarpa* juice also inhibits Caco-2 cell proliferation by causing G2/M cell cycle arrest.[42] It was found that the extract from chokeberry

was active against the sensitive leukemic cell line HL60 and retained the activity against multidrug resistant sublines *in vitro*.[57]

Most of the effects of *A. melanocarpa* anthocyanins and procyanidins are due to their antioxidative activity.[50] *A. melanocarpa* juice exhibits the highest antioxidant capacity among the phenol-rich beverages (four times higher than blueberry juice, cranberry juice, or red wine; Tables 14.2, 14.3).[42,58]

Results of Gasiorowski et al.[59] showed the antimutagenic activity of anthocyanins isolated from *A. melanocarpa* fruits. These properties of anthocyanins are exerted by their ROS scavenging action and by the inhibition of enzymes activating promutagens and converting mutagens to DNA-reacting derivatives.[59] Gosiorowski et al.[59] found that *A. melanocarpa* anthocyanins inhibited mutagenic effects of 2-aminofluorene and benzo[a] pyrene. Olas et al.[35] and Kedzierska et al.[7,9,12] observed that *A. melanocarpa* commercial extract (Aronox®; Table 14.4) may have a protective action on oxidative/nitrative stress (measured by the level of various biomarkers of this process such as the amount of carbonyl groups and 3-nitrotyrosine in proteins, the generation of superoxide anion radicals, or the amount of glutathione in plasma and in blood platelets isolated from breast cancer patients before surgery, after surgery, and during different phases of chemotherapy (doxorubicin 60 mg/m^2 intravenous injection (iv) day 1 (d1) + cyclophosphamide 600 mg/m^2 (iv) (d1); the chemotherapy cycled every 21 days). Radical scavenging inhibitory effect of this extract was about 35% (for breast cancer patients before surgery), about 9% (for breast cancer patients after surgery), and about 29% (for breast cancer patients after I phase of chemotherapy).[12]

Other experiments indicate that in the presence of *A. melanocarpa* commercial extract (Aronox®), changes in amount of different low-molecular-weight thiols, including glutathione, cysteine, and cysteinylglycine, which are physiological free radical scavengers (determined by high-performance liquid chromatography) in plasma from breast cancer patients during different phases of treatment (before or after the surgery and patients after different phases of chemotherapy (doxorubicin and cyclophosphamide)) were significantly reduced.[11] Moreover,

TABLE 14.2 Antioxidant Capacity (measured as ORAC) of Chokeberries and Other Fruits [42; modified]

Fruit	ORAC (Oxygen Radical Absorbing Capacity) [μmol of Trolox equivalents/g]
Chokeberry	160.2
Blackberry	About 55.7
Strawberry	15.4–20.6
Jostaberry	About 28.1
Gooseberry	17.0–33.0
Cranberry	10.4–18.5
Orange	About 7.5
Apple	About 2.2
Red grape	About 7.4
White grape	About 4.5

TABLE 14.3 Antioxidant Capacity (measured as TEAC) of Chokeberry Juices and Other Fruit Juices [42; modified]

Fruit juice	TEAC (Trolox Equivalent Antioxidant Capacity) [μmol/ml]
Chokeberry	65–70
Blueberry	About 15.0
Cranberry	About 10.4
Orange	About 4.2
Apple	About 3.6

TABLE 14.4 Phenolic Compounds of Chokeberries Commercial Extract (Aronox®)[7,12,35]

Phenolic Compounds	
Phenolic acids	149.2 mg/g of extract
Anthocyanins	110.7 mg/g of extract
Flavonoids	49.7 mg/g of extract

TABLE 14.5 Biological Effects of Chokeberry Commercial Extract (Aronox®) in Breast Cancer Patients

Biological Effects of Chokeberries Commercial Extract (Aronox®) in Breast Cancer Patients	References
Antioxidant activity of Aronox® in blood platelets and plasma: • Protection effect of Aronox® against protein oxidation and nitration • Inhibition of superoxide anion radicals production in resting blood platelets and blood platelets activated by thrombin	[7,9,12,35]
Reduction of changes in amount of different low-molecular-weight thiols (glutathione, cysteine, and cysteinylglycine) in plasma	[11,36]
Decreases the level of homocysteine in plasma	[11,36]
Reduction of changes in antioxidative enzymes activity in plasma	[11]
Reduction of changes in hemostasis (the fibrin polymerization and lysis)	[27]

tested commercial extract reduced modifications of antioxidative enzymes activity (superoxide dismutase, glutathione peroxidase, and glutathione reductase) in plasma from patients during different phases of treatment, but this effect was not statistically significant.[11] Aronox® also significantly decreased the level of homocysteine in plasma from patients with invasive breast cancer (at all tested groups),[11] and reduced the changes of hemostasis (the fibrin polymerization and lysis) in breast cancer patients caused by the surgery and various phases of the chemotherapy in model system *in vitro* (Table 14.5).[27]

It is very important that there is no information about any unwanted and toxic properties of black chokeberries, juice, jam, wine, and other commercial products.[42] The daily dose of chokeberry product should correspond to an anthocyanin content of 1600 to 3000 mg contained in about 700 to 1400 ml juice per day.[60] It seems that aronia extract and other aronia products (which may be not only a food supplement in healthy persons) may also play an important function in the nonpharmacologic prevention of breast cancer patients before and during different phases of treatment.

SUMMARY POINTS

- Oxidative stress is observed in breast cancer patients.
- Dietary antioxidants may reduce the risk of breast cancer.
- *Aronia melanocarpa* (*A. melanocarpa*, black chokeberry) fruits are a source of valuable phytochemicals.
- Black chokeberries are one of the richest plant sources of phenolic compounds.
- Black chokeberry extract and other commercial products may reduce the risk of breast cancer and the oxidative stress in breast cancer patients.

Acknowledgements

Supported by grant 506/810 from University of Lodz, Poland.

References

1. Valko M, Rhodes CJ, Moncol J, Izakovic M, Mazur M. Free radicals, metals and antioxidants in oxidative stress-induced cancer. *Chem Biol Interact* 2006;**160**:1–40.
2. Fuchs-Tarlovsky V. Role of antioxidants in cancer therapy. *Nutrition* 2013;**29**(1):15–21.
3. Klaunig J, Kamendilus LM. The role of stress in carcinogenesis. *Annu Rev Pharmacol Toxicol* 2004;**44**:239–67.
4. Ziech D, Franco R, Georgkalikas AG, Georgakila S, Malamou-Mitsi V, Schoneveld O, et al. The role of reactive oxygen species and oxidative stress in environmental carcinogenesis and biomarker development. *Chem Biol Interact* 2010;**188**:334–9.
5. Karihtala P, Soini Y. Reactive oxygen species and antioxidant mechanisms in human tissues and their relation to malignancies. *APMIS* 2007;**115**:81–103.
6. Afanas'ev I. Reactive oxygen species signaling in cancer: Comparison with aging. *Aging Dis* 2011;**2**:219–30.
7. Kedzierska M, Olas B, Wachowicz B, Stochmal A, Oleszek W, Jeziorski A, et al. An extract from berries of *Aronia melanocarpa* modulates the generation of superoxide anion radicals in blood platelets from breast cancer patients. *Planta Med* 2009;**75**:1405–9.
8. Kedzierska M, Olas B, Wachowicz B, Jeziorski A, Piekarski J. The lipid peroxidation in breast cancer patients. *General Physiol Biophys* 2010;**29**:208–10.
9. Kedzierska M, Olas B, Wachowicz B, Stochmal A, Oleszek W, Jeziorski A, et al. The nitrative and oxidative stress in blood platelets isolated from breast cancer patients; the protectory action of *Aronia melanocarpa* extract. *Platelets* 2010;**21**:541–8.
10. Kedzierska M, Malinowska J, Glowacki R, Olas B, Bald E, Jeziorski A, et al. The elevated homocysteine stimulates changes of hemostatic function of plasma isolated from breast cancer patients. *Mol Cell Biochem* 2011;**355**:193–9.
11. Kedzierska M, Glowacki R, Czernek U, Szydłowska-Pazera K, Potemski P, Piekarski J, et al. Changes in plasma thiol levels induced by different phases of treatment in breast cancer; the role of commercial extract from black chokeberry. *Mol Cell Biochem* 2013;**372**(1-2):47–55.

12. Kedzierska M, Olas B, Wachowicz B, Glowacki R, Bald E, Czernek U, et al. Effects of the commercial extract of aronia on oxidative stress in blood platelets isolated from breast cancer patients after the surgery and various phases of the chemotherapy. *Fitoterapia* 2012;**83**:310–7.

13. Kedzierska M, Olas B, Wachowicz B, Jeziorski A, Piekarski J. Relationship between thiol, tyrosine nitration and carbonyl formation as biomarkers of oxidative stress and changes of hemostatic function of plasma isolated from breast cancer patients before surgery. *Clin Biochem* 2012;**45**:231–6.

14. Blann AD, Gurney D, Wadley M, Bareford D, Stonelake P, Lip GYH. Increased soluble P-selectin in patients with haematological and breast cancer: A comparison with fibrinogen, plasminogen activator inhibitor and von Willebrand factor. *Blood Coagul Fibrinol* 2001;**12**:43–50.

15. Kim EJ, Lim KM, Kim KY, Bae ON, Noh JY, Chung SM, et al. Doxorubicin-induced platelet cytotoxicity: A new contributory factor for doxorubicin-mediated thrombocytopenia. *J Thromb Haemost* 2009;**7**:1172–83.

16. Gonenc A, Erten D, Aslan S, Akinci M, Simsek B, Torun M. Lipid peroxidation and antioxidant status in blood and tissue of malignant breast tumor and benign breast disease. *Cell Biol Int* 2006;**30**:376–80.

17. Chandramathi S, Suresh K, Anita ZB, Kuppusamy UR. Comparative assessment of urinary oxidative indices in breast and colorectal cancer patients. *J Cancer Res Clin Oncol* 2009;**135**:319–23.

18. Kasapović J, Pejić S, Todorović A, Stojilijkovic V, Pajovic SB. Antioxidant status and lipid peroxidation on the blood of breast cancer patients in different ages. *Cell Biochem Funct* 2008;**26**:723–30.

19. Rajneesh CP, Manimaran A, Sasikala KR, Adaikappan P. Lipid peroxidation and antioxidant status in patients with breast cancer. *Singapore Med J* 2008;**49**:640–3.

20. Sharhar S, Normah H, Fatimah A, Fadilah RN, Rohi GA, Amin I, et al. Antioxidant intake and status, and oxidative stress in relation to breast cancer risk: A case-control study. *Asian Pac J Cancer Prev* 2008;**9**:343–50.

21. Hamo-Mahmood I, Abdullah KS, Abdullah MS. Total antioxidant status in women with breast cancer. *Pak J Med Sci* 2009;**4**:609–12.

22. Himmetoglu S, Dincer Y, Ersoy YE, Bayraktar B, Celik V, Akcay T. DNA oxidation and antioxidant status in breast cancer. *J Investig Med* 2009;**57**:720–3.

23. Mannello F, Tonti GA, Medda V. Protein oxidation in breast microenvironment: Nipple aspirate fluid collected from breast women contains increased protein carbonyl concentration. *Cell Oncol* 2009;**31**:383–92.

24. Rockenbach G, Di Pietro PF, Ambrosi C, Boaventura BC, Vieira FG, Crippa C, et al. Dietary intake and oxidative stress in breast cancer: Before and after treatments. *Nutr Hosp* 2011;**26**:737–44.

25. Rossner P, Gammon MD, Terry MB, Agrawal M, Zhang FF, Teitelbaum SL, et al. Relationship between urinary 15-F2t-isoprostane and 8-oxodeoxyguanosine levels and breast cancer risk. *Cancer Epidemiol Biomarkers Prev* 2006;**15**:639–44.

26. Kedzierska M, Czernek U, Szydłowska-Pazera K, Potemski P, Piekarski J, Jeziorski A, et al. The changes of blood platelet activation in breast cancer patients before the surgery, after the surgery and various phases of the chemotherapy. *Platelets* 2013;**24**(6):462–8.

27. Kedzierska M, Malinowska J, Kontek B, Kołodziejczyk-Czepas J, Czernek U, Potemski P, et al. Chemotherapy as the modulator of biological activity of plasma isolated from breast cancer patients; the protective properties extract from black chokeberry. *Food Chem Toxicol* (sub.) 2013;**53**:126–32.

28. Weijl NI, Rutten MF, Zwinderman AH, Keizer HJ, Nooy MA, Rosendaal FR, et al. Thromboembolic events during chemotherapy for germ cell cancer: A cohort study and review of the literature. *J Clin Oncol* 2000;**18**:2169–78.

29. Trachootham D, Alexandre J, Huang P. Targeting cancer cells by ROS-mediated mechanisms: A radical therapeutic approach? *Nat Rev Drug Discov* 2009;**8**:579–91.

30. Panis C, Herrera ACSA, Victorino VJ, Campos FC, Freitas LF, De Rossi T, et al. Oxidative stress and hematological profiles of advanced breast cancer patients subjected to paclitaxel or doxorubicin chemotherapy. *Breast Cancer Res Treat* 2012;**133**(1):89–97.

31. Romieu I. Diet and breast cancer. *Salud Publica Mex* 2011;**53**: 430–9.

32. Hakimuddin F, Paliyath G, Meckling K. Selective cytotoxicity of a red grape wine flavonoid fraction against MCF-7 cells. *Breast Cancer Res Treat* 2004;**85**:5–79.

33. Greenlee H, Hershman DL, Jacobson JS. Use of antioxidant supplements during breast cancer treatment: A comprehensive review. *Breast Cancer Res Treat* 2009;**115**:437–52.

34. Thomson CA, Thompson PA. Dietary patterns, risk and prognosis of breast cancer. *Future Oncol* 2009;**5**:1257–69.

35. Olas B, Wachowicz B, Nowak P, Kedzierska M, Tomczak A, Stochmal A, et al. Studies on antioxidant properties of polyphenol-rich extract from berries of *Aronia melanocarpa* on blood platelets. *J Physiol Pharmacol* 2008;**59**:823–35.

36. Olas B, Kedzierska M, Wachowicz B, Stochmal A, Oleszek W, Jeziorski A, et al. Effect of aronia on thiol levels in plasma of breast cancer patients. *Cent Eur J Biol* 2010;**5**:38–46.

37. Buttala AC, Crane TE, Patil B, Wertheim BC, Thompson P, Thomson CA. Effects of a carrot juice intervention on plasma carotenoids, oxidative stress, and inflammation in overweight breast cancer survivors. *Nutr Cancer* 2012;**64**:331–41.

38. Di Franco R, Calvanese M, Murino P, Manzo R, Guida C, Di Gennaro D, et al. Skin toxicity from external beam radiation therapy in breast cancer patients: Protective effects of resveratrol, lycopene, vitamin C and anthocianin (Ixor®). *Radiat Oncol* 2012;**30**:7–12.

39. Zhang G, Wang Y, Zhang Y, Wan X, Li J, Liu K, et al. Anticancer activities of tea epigallocatechin-3-gallate in breast cancer patients under radiotherapy. *Curr Mol Med* 2012;**12**:163–76.

40. Hardin JW. The enigmatic chokeberries (*Aronia*, Rosaceae). *Bull Torrey Bot Club* 1973;**100**:178–84.

41. Strigl AW, Leitner E, Pfannhauser W. Die schwarze Apfelbeere (*Aronia melanocarpa*) als naturliche Farbstoffquelle. *Dtsch Lebensmitt Rundsch* 1995;**91**:177–80.

42. Kulling SE, Rawel HM. Chokeberry (*Aronia melanocarpa*) – A review on the characteristic components and potential health effects. *Planta Med* 2008;**74**:1625–34.

43. Kokotkiewicz A, Jeremicz Z, Luczkiewicz M. Aronia plants: A review of traditional use, biological activities, and perspectives for modern medicine. *J Med Food* 2010;**13**:255–69.

44. Scott RW, Skirvin RM. Black chokeberry (*Aronia melanocarpa* Michx.): A semi-edible fruit with no pests. *J Am Pomol Soc* 2007;**61**:135–7.

45. Olas B, Wachowicz B, Tomczak A, Erler J, Stochmal A, Oleszek W. Comparative anti-platelet and antioxidant properties of polyphenol-rich extracts from berries of *Aronia melanocarpa*, seeds of grape and bark of *Yucca schidigera* in vitro. *Platelets* 2008;**19**:70–7.

46. Jeppson N, Johansson R. Changes in fruit quality in black chokeberry (*Aronia melanocarpa*) during maturation. *J Hortic Sci Biotechnol* 2000;**75**:340–5.

47. Skupien K, Oszmianski J. The effect of mineral fertilization on nutritive value and biological activity of chokeberry fruit. *Agric Food Sci* 2007;**16**:46–55.

48. Tanaka T, Tanaka A. Chemical components and characteristics of black chokeberry. *J Jpn Soc Food Sci Technol* 2001;**48**:606–10.

49. Oszmianski J, Wojdylo A. *Aronia melanocarpa* phenolics and their antioxidant activity. *Eur Food res Technol* 2005;**221**:809–13.

50. Kahkonen MP, Hopia AI, Vuorela HJ, Rauha JP, Pihlaja K, Kujala TS, et al. Antioxidant activity of plant extracts containing phenolics compounds. *J Agric Food Chem* 1999;**47**:3954–62.

51. Valcheva-Kuzmanova S, Marazova K, Krasnaliev I, Galunska B, Borisova P, Belcheva A. Effect of *Aronia melanocarpa* fruit juice on indamethacin-induced gastric mucosal damage and oxidative stress in rats. *Exp Toxicol Pathol* 2005;**56**:385–92.

52. Valcheva-Kuzmanova SV, Belcheva A. Current knowledge of *Aronia melanocarpa* as a medicinal plant. *Folia Med* 2006;**48**:11–7.

53. Yaneva MP, Botushanova AD, Grigorov LA, Kokov JL, Todorova EP, Krachanova MP. Evaluation of the immunomodulatory activity of Aronia in combination with apple pectin in patients with breast cancer undergoing postoperative radiation therapy. *Folia Med* 2002;**44**:22–5.

54. Lala G, Mailk M, Zhao C, He J, Kwon Y, Giusti MM, et al. Anthocyanin-rich extracts inhibit multiple biomarkers of colon cancer in rats. *Nutr Cancer* 2006;**54**:84–93.

55. Malik M, Zhao C, Schoene N, Guisti MM, Moyer MP, Magnuson BA. Anthocyanin-rich extract from *Aronia melanocarpa* E induces a cell cycle block in colon cancer but not normal colonic cells. *Nutr Cancer* 2003;**46**:186–96.

56. Zhao C, Giusti MM, Malik M, Moyer MP, Magnuson BA. Effects of commercial anthocyanin-rich extracts on colonic cancer and nontumorigenic colonic cell growth. *J Agric Food Chem* 2004;**52**:6122–228.

57. Skupien K, Kostrzewa-Nowak D, Oszmianski J, Tarasiuk J. In vitro antileukaemic activity of extracts from chokeberry (*Aronia melanocarpa* [Michx] Elliot) and mulberry (Morus alba L.) leaves against sensitive and multidrug resistant HL60 cells. *Phytother Res* 2008;**22**:689–94.

58. Seeram NP, Aviram M, Zhang Y, Henning SM, Feng L, Dreher M, et al. Comparison of antioxidant potency of commonly consumed polyphenol-rich beverages in The United States. *J Agric food Chem* 2008;**56**:1415–22.

59. Gasiorowski K, Szyba K, Brokos B, Kolaczynska B, Jankowiak – Wlodarzyk M, Oszmianski J. Antimutagenic activity of anthocyanins isolated from *Aronia melanocarpa* fruits. *Cancer Let* 1997; **119**:37–46.

60. Chrubasik C, Li G, Chrubasik S. The clinical effectiveness of chokeberry: A systematic review. *Phytother Res* 2010; **24**:1107–14.

Curcumin, Oxidative Stress, and Breast Cancer

Gloria M. Calaf

Instituto de Alta Investigación, Universidad de Tarapaca, Arica, Chile and Center for Radiological Research, Columbia
University Medical Center, New York, NY, USA

List of Abbreviations

βE2 17β-Estradiol
αE2 17α-Estradiol
αEE 17 α –Ethinylestradiol
8-iso-PGF$_{2α}$ 8-isoprostaglandin F2α
ROS Reactive Oxygen Species
RNS Reactive Nitrogen Species
(NAD(P)H Nicotinamide Adenine Dinucleotide Phosphate
8-OHdG 8-hydroxy-20-deoxyguanosine
EGFR Epidermal Growth Factor Receptor
LET Linear Energy Transfer
Estrogen 17β-estradiol
ER Estrogen receptor
NFκB Nuclear Factor κ B
PGE2 Prostaglandin E2
8-iso-PGF2α 8-isoprostaglandin F2α
PARP Poly (ADP-Ribose) Polymerases
HE Hematoxylin and Eosin
MnSOD or SOD2 Mn Superoxide Dismutase
GDP Guanidine Diphosphate
GTP Guanidine Triphosphate
GRF Growth-Regulating Factor
COX-2 Cyclooxygenase-2

INTRODUCTION

Breast cancer, the most frequent spontaneous malignancy diagnosed in women in the western world, is a classic model of hormone-dependent malignancy. There are substantial evidences that breast cancer risk is associated with prolonged exposure to female hormones, since onset of menarche, late menopause, and hormone replacement therapy are associated with greater cancer incidence.[1] Breast cancer progression follows a complex multistep process that depends on various exogenous (diet, breast irradiation) and endogenous (age, hormonal imbalances, proliferative lesions, and family history of breast cancer) factors.[2] Such cancer may have its genesis and cell growth influenced by hormonal factors since about one third of breast cancers is responsive to endocrine therapies. Estrogen has generally been considered beneficial, based on a variety of hormonal effects; however, in the past 15 to 20 years, epidemiological studies have increasingly pointed to an increased breast risk associated with them. The potential carcinogenic activity of estrogen-containing medications in humans has not been recognized for many years.

Authors[3–5] have shown that estrogen administration, a risk factor for humans, increases with continuous doses of estrogen and with the length of treatment. Slightly elevated levels of circulating estrogen are also a risk factor for breast cancer. Several studies have demonstrated strong relationships between endogenous estrogen levels and breast cancer risk.[6–9] This role of endogenous estrogen in human breast carcinogenesis is supported by risk factors of breast cancer such as high serum or urine estrogen levels.[6]

Estrogen is associated with carcinogenesis in humans and animals,[10–15] and the exact effect of estrogen in breast cancer remains unclear at this time. Estrogen has been implicated in a variety of cancers. Since that time, many more reports on tumor induction by estrogen have been published, and many rodent tumor models have been reviewed.[9] The evidence for the carcinogenic activity of estrogen in animals has been reported by those groups. This conclusion was based on numerous experiments related to the administration to rodents of oral or subcutaneous estrogen, which resulted in an increased incidence of mammary tumors.[16,17]

ESTROGENS (17β-ESTRADIOL) AND OXIDATIVE STRESS

Oxidative stress plays a crucial role in estrogen-induced carcinogenesis.[16,17] It has been suggested that oxidative stress resulting from metabolic activation of carcinogenic estrogen plays a critical role in estrogen-induced carcinogenesis. In hamsters, a high incidence of malignant kidney tumors occurred in intact and castrated

© 2014 Elsevier Inc. All rights reserved.

males and in ovariectomized females.[16] Hamsters were implanted with 17β-estradiol (βE2), 17α-estradiol (αE2), 17 α–ethinylestradiol (αEE), menadione, a combination of αE2 and αEE, or a combination of αEE and menadione for 7 months. The βE2-treated group developed kidney tumors and showed more than a two-fold increase in 8-iso-PGF$_{2α}$ levels compared with controls. Kidneys of hamsters treated with a combination of menadione and αEE showed increased 8-iso-PGF$_{2α}$ levels compared with control. It was concluded in this work that chemicals known to produce oxidative stress or a potent estrogen with poor ability to produce oxidative stress were nontumorigenic in hamsters when given as single agents; however, when given together they induced renal tumors.

OXIDATIVE STRESS

Free radical production is ubiquitous in all organisms and it is enhanced in many diseases due to carcinogen exposure, such as under conditions of stress, that seems to contribute widely to cancer development.[13,15,18] A free radical is any chemical species capable of independent existence, possessing one or more unpaired electrons. Chemical reactions seem to occur in every cell including oxidation and reduction of molecules. These reactions can lead to the production of free radicals that react with organic substrates such as lipids, proteins, and DNA. Through oxidation free radicals cause damage to these molecules, disturbing their normal function, and may therefore contribute to a variety of diseases.

Oxidative stress occurs under pathologic circumstances and leads to an overproduction of highly reactive oxygen species (ROS), which can induce chemical changes in DNA, proteins, and lipids. An antioxidation system, which consists of enzymatic antioxidants and nonenzymatic antioxidants, defends against oxidative stress. Biological free radicals are thus highly unstable molecules that have electrons available to react with various organic substrates. The presence of free radicals and nonradical reactive molecules derived from free radicals at high concentrations is dangerous to living organisms because of their ability to damage cell organelles. Nitrogen monoxide, superoxide anions, and related ROS and reactive nitrogen species (RNS) also play important modulating roles in certain signal transduction pathways.[18] Several ROS-mediated reactions protect the cell from oxidative stress and serve to stabilize redox homeostasis. Chemical compounds capable of generating potential toxic oxygen species can be referred to as pro-oxidants. In a normal cell, there is an appropriate pro-oxidant–antioxidant balance. However, this balance can be shifted toward the pro-oxidants when production of oxygen species is increased greatly such as following ingestion of certain chemicals or drugs, or with diminished levels of antioxidants.[13]

Authors[13] have introduced the concept of oxidative stress, that is, the dissolution of the pro-oxidant–antioxidant equilibrium. Oxidative stress is basically caused by two mechanisms: (1) reduced concentration of antioxidants (e.g., due to mutated antioxidant enzymes, toxins, or the reduced intake of natural antioxidants) and (2) increased number of oxygen/nitrogen/carbon-based reactive species. Following the oxidative stress, the activity of a given signal transduction molecule is either reduced or increased; additionally the function of the molecule may also change. Large amounts of ROS may be generated in one of two ways by a significant stimulation of NAD (P) H oxidases or from the mitochondrial respiratory chain. In the mitochondria ROS are the unwanted by-products of oxidative metabolism. Severity of the stress seems to affect the cell through the amount of ROS produced and biochemical status of the cell such as activity of antioxidative and other enzymes, antioxidant content, pH, integrity of membranes, redox characteristics, and others.

Clinical, epidemiological, and experimental findings have provided evidence of a role for free radicals in the etiology of cancer.[14] Furthermore, generation of hydrogen peroxide by oxidative metabolism of estrogens has been documented. Thus an increase in hydroxy free radical damage to DNA has been identified in human mammary tissue of breast cancer patients compared with controls,[15] which are induced by estrogen. It has been reported that the superoxide generated by this redox system could be converted to hydrogen peroxide and subsequently to hydroxyl radical, which can cause oxidative damage in DNA.[16] It has been generally accepted that active oxygen produced under stress is a detrimental factor, which causes lipid peroxidation and most interestingly oxidative damage to DNA. Both an imbalance of nutrients and ROS generation can alter antioxidant activity of cells and apoptosis.[17] Oxidative stress can be induced by decreasing the ability of a cell to scavenge reactive ROS or by a shortage of antioxidants. Cells utilize enzymatic and nonenzymatic compounds—the so-called antioxidants—to defend themselves against oxidative stress. The term antioxidant can serve as a label for any substance whose presence, even at low concentrations, delays or inhibits the oxidation of a substrate.[18] There are several molecules that play a role in antioxidant defense; these are either endogenous (internally synthesized) or exogenous (nutritional substances). Antioxidants can be divided into two groups depending on their mechanism of action. They can be either chain breaking antioxidants or preventive antioxidants.

Correlations between the extents of oxidative DNA damage in different tissues are important indicators of the individual oxidative stress levels in different physiological systems. Specific biomarkers show that oxidative damage can characterize cell damage *in vitro*. 8-hydroxy-20-deoxyguanosine (8-OHdG) is characteristic for DNA damage.[14] Multiple environmental mutagens and carcinogens are known to react with components of DNA with or without metabolic activation and form adducts with adenine, guanine, cytosine, and thymine. Among these targets, guanine is known to be one of the most important. Even though background levels of oxidative damage to DNA exist, oxidative stress can lead to an increase in such damage, which has been described in various pathological conditions, such as carcinogenesis. 8-OHdG is one of the most frequent lesions among over 20 known base modifications caused by oxygen radicals.[18] It is possible that formation of 8-OHdG is proportional to that of other DNA modifications, regardless of organisms, type of ROS, and conditions of exposure. Therefore, measurement of 8-OHdG should be very useful in estimating oxygen. Elevated levels of 8-OHdG, a marker of hydroxyl radical-induced damage in DNA, have been detected in DNA from kidneys of hamsters chronically administered with diethylstilbestrol.[11] It has been shown that there is concurrent damage not only to lipids but also to DNA during lipid oxidation.

CURCUMIN AS AN ANTIOXIDANT

Oxidative stress is one of the important pathogenic factors of cancer development. Among the antioxidants, curcumin (1, 7-bis (4-hydroxy-3-methoxyphenyl)-1,6-heptadiene-3,5-dione; diferuloylmethane) is a well-known major dietary natural yellow pigment derived from the rhizome of the herb *Curcuma longa* (Zingiberaceae). It is also named turmeric and is a perennial herb belonging to the ginger family, native to India and Southeast Asia. It measures up to 1 m high with a short stem and tufted leaves. Curcumin is present in extracts of the plant. Curcuminoids are responsible for the yellow color of turmeric and curry powder. It is a pigment of turmeric that is used for imparting color and flavor to foods. This nonnutritive phytochemical is pharmacologically safe, considering that it has been consumed as a dietary spice, at doses up to 100 mg/day, for centuries.[19] In the United States curcumin is used as a coloring agent in cheeses, spices, mustard, cereals, pickles, potato flakes, soups, ice creams, and yogurts. The most active component is curcumin, which makes up 2 to 5% of the total spice in turmeric. It has been shown to be a potent anti-inflammatory, antioxidant, anticarcinogenic, and chemopreventive agent.

This phytochemical has also been shown to suppress the proliferation of numerous types of tumor cells by down-regulating c-myc[20], cyclin D1[21], activator protein-1 (AP-1)[22], phosphatidylinositol-3-kinase /AKT signaling[23], and epidermal growth factor receptor (EGFR)

signaling.[24] An imbalance of nutrients and generation of ROS can alter antioxidant activity of cells and induce apoptosis.[17] Curcumin has been previously shown to prevent the formation of many chemically induced cancers including breast cancer in mice[25,26] and rats.[27,28] It has been shown[20-30] to prevent cancer in colon, skin, stomach, and duodenum following oral administration. Curcumin has been shown to have chemopreventive and chemotherapeutic effects by blocking tumor initiation induced by benzo[a]pyrene and 7,12 dimethylbenz[a] anthracene;[25,26,28] it suppressed phorphol ester-induced tumor promotion;[25,26] and suppressed carcinogenesis of the skin,[25,26,31,32] forestomach,[26] and colon[29,33] in mice. Curcumin and its analogues are known to protect against peroxidation damage. In the present study the effect of curcumin as an antioxidant was analyzed by measuring several parameters related to oxidative stress.

To gain insight into the effects of curcumin on oxidative stress an established *in vitro* experimental breast cancer model (Alpha model) was used.[34-37] Such a model was developed with the immortalized human breast epithelial cell line, MCF-10F[37] that was exposed to low doses of high LET (linear energy transfer) α particles (150 keV/μm) of radiation, values comparable to α particles emitted by radon progeny, and subsequently cultured in the presence or absence of 17β-estradiol (estrogen). This model consisted of human breast epithelial cells in different stages of transformation: (1) a control cell line, named MCF-10F, (2) an estrogen-treated cell line, named Estrogen, (3) a malignant cell line, named Alpha3, (4) a malignant and tumorigenic cell line, named Alpha5, and (5) Tumor2 derived from cells originated from a tumor after injection of Alpha5 cell line injected in the nude mice. The present study evaluated whether curcumin had any effect on oxidative stress in human breast epithelial cells transformed by the effect of radiation in the presence of estrogen. The Alpha model gives an opportunity to study phenotypic and molecular alterations induced by an increased amount of oxidative stress.[36,37]

EFFECT OF CURCUMIN ON A MULTIFUNCTIONAL NUCLEAR TRANSCRIPTION FACTOR AND MANGANESE SUPEROXIDE DISMUTASE (MNSOD) PROTEIN EXPRESSION

It is known that curcumin interferes with the transcription activation induced by transcription factors such as nuclear factor-κB (NFκB), resulting in the negative regulation of various cell cycle control genes and oncogenes.[38-40] The nuclear NFκB complex containing p65 (Rel A) and p50 consists of closely related proteins that act as a multifunctional nuclear transcription factor.[41] This complex acts as a transcription factor that regulates the expression of multiple genes that promote carcinogenesis. Upstream activators of nuclear NFκB include various cellular stressors such as ROS, carcinogens, tumor promoters, apoptosis inducers, and cytokines. Curcumin may inhibit NFκB in experimental conditions and may also regulate DNA binding in pancreatic cancer cells, inducing apoptosis, therefore decreasing cell survival.[42,43] Activation of NFκB has been implicated in resistance of cancer cells to radiotherapy and chemotherapy.[38] It has also been implicated in growth control and G0/G1 to S-phase transition. These studies showed that NFκB protein expression was altered by curcumin since this compound is directly related to specific oxidative stress pathways. Figure 15.1A shows the effect of curcumin (30 μM) on NFκB (50kDa subunit) protein expression in the Alpha model by protein analyses. Results indicated that curcumin decreased NFκB protein expression of MCF-10F, Alpha3, Alpha5, and Tumor2 cell lines in comparison to their counterparts without curcumin. The graph corresponds to the relative grade of luminescence of protein expression level.

Under oxidative stress conditions, superoxide anions are produced and converted to hydrogen peroxide through a specific antioxidant system, and then to water to complete the detoxification pathway.[44] Abnormal levels of MnSOD in cancer have been documented to play a critical role in the survival of cells. MnSOD is a nuclear encoded mitochondrial antioxidant enzyme that catalyzes the conversion of superoxide radicals into molecular oxygen and hydrogen peroxide, which are further reduced into water by peroxide metabolizing enzyme systems, mainly catalase, an endogenous antioxidant that neutralizes hydrogen peroxide by converting it into water and oxygen. Thus, it has been reported that MnSOD2 expression was upregulated in response to oxidative stress in various types of cells and tissues by toxic stimuli and treatments, such as ionizing radiation and ultraviolet light.[45] Figure 15.1B shows the effect of curcumin on MnSOD (SOD-2) protein expression in the five cell lines under study. The control MCF-10F cell line responded to the curcumin effect by decreasing its expression as indicated by Western blot analysis. Curcumin decreased SOD-2 protein expression in the control MCF-10F, Alpha3, and Tumor2 cell lines; however, MnSOD was slightly increased in the Alpha5 cell line. The MCF-10F and Alpha5 cell lines had a higher background of SOD-2 expression than the other two cell lines. On the other hand, Alpha3 and Tumor2 and curcumin-treated cell lines had very low SOD-2 protein expression. Representative images of Figures 15.2A and B show the effects of curcumin (15 μM) analyzed either by peroxidase (2A) or immunofluorescent (2B) techniques on SOD-2 protein expression of MCF-10F, Estrogen, Alpha3, Alpha5, and Tumor2 cell lines. SOD-2 protein expression

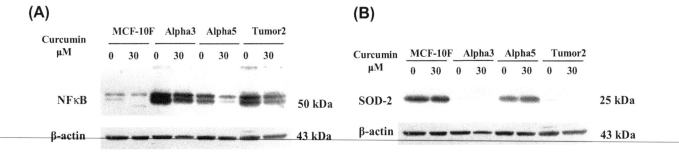

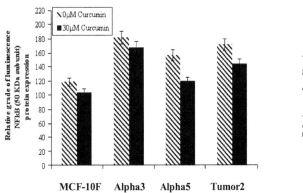

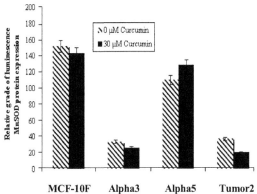

FIGURE 15.1   Effect of Curcumin on NFκB and MnSOD Protein Expression in a Breast Cancer Model. Western blot analysis of MCF-10F, Alpha3, Alpha5, and Tumor2 cell lines treated with 30 μM of curcumin on (A) NFκB and (B) SOD-2 protein expression. β-actin was used as a loading control. The graphs represent the relative grade of luminescence of protein expression level.

can be appreciated in mitochondria of those treated cell lines by peroxidase techniques. A decrease of such expression was observed by the effect of curcumin in all the cell lines.

Oxidative damage can be measured by an agent that recognizes cell damage. High levels of hydrogen peroxide favors ROS formation and an increase of these molecules may play an important role in carcinogenesis.[46] An observed excess of hydrogen peroxide can be an indication of the imbalance in redox control that causes oxidative cell damage *in vitro*. Thus, hydrogen peroxide can be useful to recognize oxidative stress in the cells. It has been reported that curcumin is a good scavenger of hydrogen peroxide at high concentrations (over 27 µM), but at low concentrations it activates the Fenton reaction to increase the production of hydrogen peroxide.[47] Hydrogen peroxide levels were measured by the Amplex™ Red Hydrogen Peroxide Assay kit in MCF-10F, Estrogen, Alpha3, Alpha5, and Tumor2 treated and nontreated cell lines with curcumin (15 µM). It was interesting to find that curcumin decreased the formation of hydrogen peroxide in all the cells of this model when compared with their nontreated counterparts (data not shown).[35,36]

Catalase is a peroxisome specific marker protein that belongs to the catalase family. It is an important regulator of oxidative stress and protector of the cells from hydrogen peroxide. Catalase and MnSOD are the principal defense against intracellular oxidative stress.[48] Figure 15.2C

corresponds to representative images of catalase protein expression by immunoperoxidase staining. It shows the effects of curcumin (15 µM) on MCF-10F, Estrogen, Alpha3, Alpha5, and Tumor2 cell lines. Catalase protein expression was decreased in Alpha3 and Alpha5 cell lines when treated with curcumin in comparison to their counterpart. It was interesting to see that the Tumor2 cell line did not show any expression, indicating the low defense against oxidative stress of cells present in tumors.

EFFECTS OF CURCUMIN ON LIPID PEROXIDATION

It has been generally accepted that active oxygen produced under stress is a detrimental factor that causes lipid peroxidation.[49] This is a natural metabolic process under normal conditions; however, it is an important factor in the patho-physiological functions of numerous diseases.[50] Specific biomarkers can recognize oxidative cell damage *in vitro*. The best biomarkers of lipid peroxidation are the isoprostanes such as the 8-isoprostaglandin F2α (8-iso-PGF2α), malondialdehyde that show lipid damage.[51–53] It is a good marker to determine the effects of an antioxidant, such as curcumin. As an end product it provides a useful tool to monitor oxidative stress in human organisms. Authors showed that lipid peroxidation is known to be a free radical-mediated reaction leading to cell membrane damage, and the inhibition of

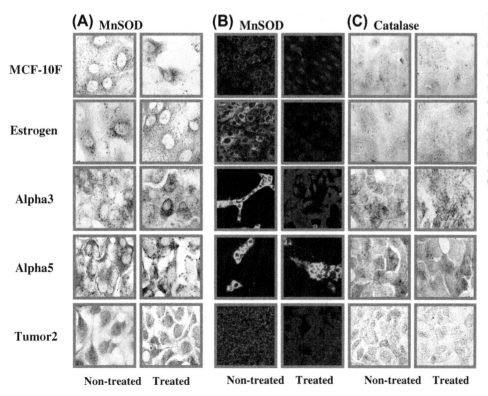

FIGURE 15.2 Effect of Curcumin on MnSOD and Catalase Protein Expression in a Breast Cancer Model. Representative images of the effect of 15 µM curcumin on MnSOD protein expression analyzed by either (A) peroxidase or (B) immunofluorescent techniques in the MCF-10F, Estrogen, Alpha3, Alpha5, and Tumor2 cell lines. (C) Catalase protein expression analyzed by peroxidase technique in the MCF-10F, Estrogen, Alpha3, Alpha5, and Tumor2 cell lines. (See the color plate.)

peroxidation by curcuminoids is mainly attributed to the scavenging of the reactive free radicals involved in such a process.[54] It is one of the major isoprostanes formed *in vivo* and is initially generated *in situ* from esterified arachidonic acid in phospholipids and released in the free form into the circulation.

One of the most useful aspects of measuring 8-iso-PGF$_{2\alpha}$ to assess oxidative stress is the fact that they can be accurately and specifically measured since they are present in all biological tissues or fluids studied. In animals, the levels of 8-iso-PGF$_{2\alpha}$ are increased in oxidative stress models and reduced by dietary anti-oxidant supplementation. DNA damage caused by per-oxidation of lipids may be implicated in tumorigenesis. Such studies demonstrated that curcumin has a diverse range of molecular targets, confirming the concept that it acts upon numerous biochemical and molecular cas-cades. Evidence is increasing that isoprostanes, a novel class of prostaglandin-like compounds produced upon peroxidation of lipoproteins, play a causative role in carcinogenesis. Measurement of isoprostane concentra-tions is likely to have an important diagnostic potential to assess oxidative stress in several disorders, such as carcinogenesis.

Studies of the modulation of isoprostanes by anti-oxidant nutrients are becoming available. For exam-ple, authors reported that Vitamin C administered to heavy smokers significantly reduced 8-iso-PGF$_{2\alpha}$ excretion in urine.[25,26] It has been generally accepted that active oxygen produced under stress is a detrimen-tal factor that causes oxidative damage to DNA.[49–57] There are specific biomarkers that can recognize oxidative cell damage *in vitro* such as isoprostanes. Levels of 8-iso-prostaglandin F$_{2\alpha}$ (8-iso-PGF2α) can indicate lipid damage and it is an end product that provides a useful tool to monitor oxidative stress in human organisms. Enzyme levels of 8-iso-PGF2α of benign and malignant breast lesions derived from biopsy specimens can be seen in Figure 15.3. Results indicated that malignant tissues had significantly greater 8-iso-PGF2α (pg/ml) than the normal coun-terpart tissue from the same patients. DNA damage caused by peroxidation of lipids may be implicated in tumorigenesis[58] and it can be suggested as an excellent marker for malignant progression in breast cancer.

Authors[57] have shown that lifetime administration of curcumin reduced COX-2 expression in murine intestinal adenomas by decreasing the oxidative DNA adducts. Aberrant COX-2 expression has been detected in many human malignancies including breast cancer. These authors measured the oxidative DNA adduct lev-els that reflected direct antioxidant effects of curcumin. Several studies have indicated this substance may exert its effect by specifically inhibiting COX-2 enzyme activity and preventing the conversion of arachidonic

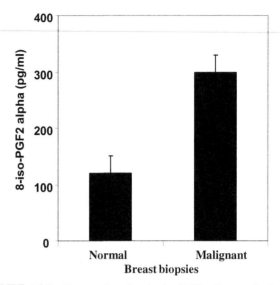

FIGURE 15.3 Enzyme Levels of 8-iso-PGF2α Present in Benign Lesions and Malignant Breast Carcinomas. Graph that represents 8-iso-PGF2α enzyme levels (pg/ml) present in benign and malignant lesions derived from breast specimens and measured by an immunoas-say kit. (*Assay Designs, Ann Arbor, MI*)

acid to prostaglandin during prostaglandin synthesis. Therefore, the suppression of prostaglandin synthesis through selective inhibition of COX-2 has been sug-gested as a strategy applicable to the development of chemopreventive substances. In normal cells, COX-2 gene is highly inducible by signals that activate the NFκB pathway. In contrast, many types of cancer cells possess high basal levels of COX-2, due to permanent activation of NFκB in these cells followed by expres-sion of the COX-2 gene.[57] The downstream product of COX-2 enzymatic activity is prostaglandin E2 (PGE2), which serves as an important stimulus for induction of several cell signaling pathways, including the NFκB pathway that subsequently regulates cell proliferation and motility. Indeed, inhibition of COX-2 enzymatic activity by specific pharmacological inhibitors is an effective tool for controlling cancer progression. It was interesting to find decreased COX-2 protein expression in the curcumin-treated MCF-10F and the pretumori-genic Alpha5 cell lines.[35,36] However, it did not have any effect in the curcumin-treated Alpha3 and Tumor2 cell lines in comparison to their counterparts.

Figure 15.4A corresponds to a cross-section of breast specimen with noninvasive ductal carcinoma (Case #1; HE; left side) and isolated cells positive for ER alpha (right side) that were originated from such tissue. Figures 15.4B and C correspond to cells from passage 3 that were cultured with 30 μM curcumin for 2 days. These studies indicated decreased NFκB (Figure 15.4B) and COX-2 (Figure 15.4C) protein expression by the effect of this substance.

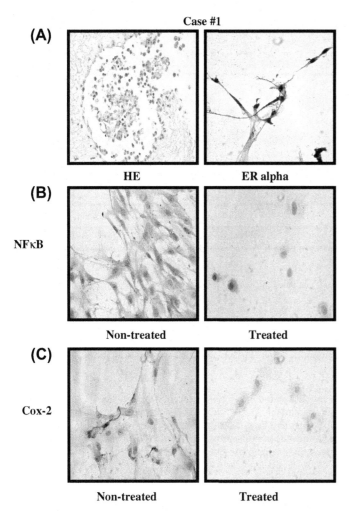

(A)

Case #1

HE   ER alpha

(B)

NFκB

Non-treated   Treated

(C)

Cox-2

Non-treated   Treated

FIGURE 15.4 **Effect of Curcumin on NFκB and COX-2 Protein Expression in Isolated Cells Derived from Specimens with Breast Carcinomas.** Cross-section of breast specimen with noninvasive ductal carcinoma (Case #1) (Hematoxylin–eosin) (A; left side). Protein expression of isolated cells from such tissue after 3 passages of Estrogen receptor alpha (Right side); NFκB, (B) and COX-2, (C) in nontreated and treated cells with 30 µM curcumin for 2 days in culture. (See the color plate.)

OXIDATIVE STRESS AND GENOMIC INSTABILITY

Ras proteins must first release bound GDP mediated by GRF to reach their active GTP bound state. Curcumin increased GRF1 protein expression of MCF-10F and E cell lines with or without curcumin preventing the binding. The cell lines Alpha3, Alpha5, and Tumor2 had increased protein expression.[35,36] However, curcumin decreased such expression increasing the binding. Results also showed that Rho-A protein expression decreased in the malignant cell lines Alpha3, Alpha5, and Tumor2 with 15 µM curcumin but not in MCF-10F and estrogen. Previously, our laboratory demonstrated[37] that the parental MCF-10F cell line exposed to double

doses of alpha-particle radiation and treated with estrogen showed a more complex pattern of allelic imbalance when compared to cell lines treated with a single dose of radiation without estrogen. Progressive changes were observed due to allelic alterations induced by irradiation with either a single or double dose of alpha particles in the presence or absence of estrogen that were expressed either in the form of loss of heterozygosity/microsatellite instability and by parallel phenotypic changes, such as anchorage independence or invasive capabilities.[34] Consequently, the doses of radiation of the cell lines after irradiation directly influenced these alterations, and this genetic effect was more deleterious when given in combination with estrogen. The *c-Ha-ras* oncogene, mapped to 11p15.5, acquired transforming capacity either by single point mutation (s) in codon 12 or 61, resulting in the expression of an aberrant gene product.

To determine the effect of curcumin on c-Ha-ras, Rho-A and SOD-2 protein expression in tissues of breast cancer specimens were analyzed for hormone receptor status and Her 2 neu (antibodies against HER-2) by immunocytochemistry. Cross-section of breast specimen with invasive ductal carcinoma, Grade II can be observed in Figure 15.5A (Case #2). Isolated cells from such tissue were considered a triple negative patient since cells were negative for ER alpha, ER beta, and ErbB-2 protein expression. Such cells were cultured for 10 passages and then cultured in the presence of 30 µM curcumin for 2 days. Cross-section of breast specimen with lobular carcinoma, Grade III can be seen in Figure 15.5B (Case #3). Isolated cells from such tissue were positive for ER alpha, ER beta, and ErbB-2 protein expression after 10 passages. Then cells were cultured in the presence of 30 µM curcumin for 2 days.

Isolated cells derived from breast specimen with ductal carcinoma indicated positive c-Ha-ras and Rho-A protein expression by the effect of 30 µM of curcumin after three passages and 2 days in culture as seen in Figure 15.6A (Case #1). Those isolated cells derived from invasive ductal carcinoma, Grade II as seen in Figure 15.6B (Case #2) showed decreased protein expression of SOD-2 and Rho-A after 30 µM curcumin after 10 passages and 2 days in culture. Those isolated cells derived from lobular carcinoma, Grade III as seen in Figure 15.6C (Case #3) show decreased protein expression of c-Ha-ras and Rho-A in isolated cells derived from such tissue after three passages and 2 days in culture. These results indicated that 30 µM curcumin induced apoptosis and decreased in cell number. These figures are representative images of studies performed in numerous breast specimens.

On the other hand, previous studies demonstrated that curcumin had a significant inhibitory effect not only on cell growth but in colony formation in breast

(A)   Case #2   **(B)**   Case #3

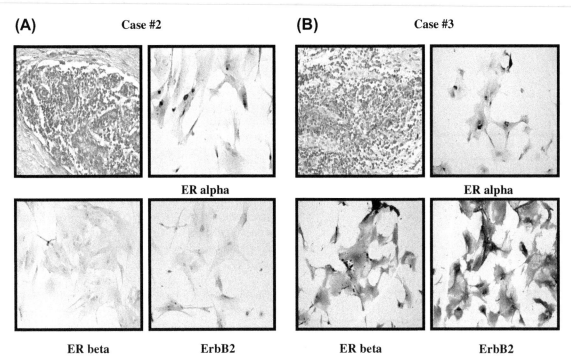

FIGURE 15.5   **Effect of Curcumin on ER Alpha, ER Beta, and ErbB-2 Protein Expression in Isolated Cells Derived from Specimens with Breast Carcinomas.** (A) Cross-section of breast specimen with invasive ductal carcinoma, grade II (Case # 2). Protein expression of positive ER alpha, ER beta, and ErbB-2 of isolated cells derived from such tissue after 10 passages and 2 days in culture in the presence of curcumin. (B) Cross-section of breast specimen with lobular carcinoma, grade III (Case #3). Protein expression of negative ER alpha, ER beta, and ErbB-2 of isolated cells derived from such tissue after 10 passages and 2 days in culture in the presence of curcumin. (See the color plate.)

carcinogenesis by activating DNA damage. Cell cycle analysis of these cell lines showed that curcumin altered S and G2/M phases by reducing the proportion of cells in G0/G1 phase and increasing the number of G2/M phase cells indicating that curcumin imposed a strong G2/M checkpoint as compared to untreated cells but less in Tumor2.[35] The effect was more pronounced in Alpha3 and Alpha5 than in Tumor2 cell lines. That curcumin seem to inhibit cell proliferation, induce apoptosis, and promote accumulation of cells in the G2/M phase of the cell cycle has been previously demonstrated.[35] Although cell cycle modulators are designed to target cancer cells, some of these can also be applied for a different purpose, as to protect normal cells against the lethality of chemotherapy. In some cell types like thymocytes, curcumin induced apoptosis-like changes whereas in many other normal and primary cells curcumin either stimulates or inhibits proliferation. Inhibition of both proliferation and apoptosis of T lymphocytes by curcumin led other authors[20] to conclude that the inhibition of cell proliferation by curcumin was not always associated with programmed cell death. Interestingly, curcumin has been found to inhibit proliferation of normal, nonselective, as well as malignant cells, although its apoptotic effect is more profound in malignant cells.

Previous results showed that PARP-1 was cleaved upon curcumin treatment in malignant and tumorigenic cells.[35] Cleaved PARP-1 protein was increased in the presence of curcumin in the control MCF-10F and Estrogen cell lines, as well as in the malignant Alpha3 and Tumor2 cell lines, inducing apoptosis. However, it was not reduced in Alpha5 cell lines, indicating that PARP-1 was not cleaved upon curcumin treatment in premalignant cell lines. The γ-H2AX protein expression was overexpressed in MCF-10F, Alpha5, and Tumor2 cell lines upon curcumin treatment but there was no expression in the Estrogen and Alpha3 cell lines. Since DNA double-strand breaks induce histone H2AX phosphorylation it can be concluded that PARP-1 and H2AX conferred cellular protection against radiation and estrogen-induced transformation when curcumin was present. It can be concluded that targeting either PARP-1 or histone H2AX may have an effective therapeutic value as antioxidants for cancer prevention. Moreover, it can be concluded that curcumin acted upon oxidative stress affecting genomic instability in human breast epithelial cells transformed by the effect of radiation in the presence of estrogen.

In conclusion, since curcumin had an inhibitory effect only on malignant cell lines it could be used as an important substance for prevention of breast

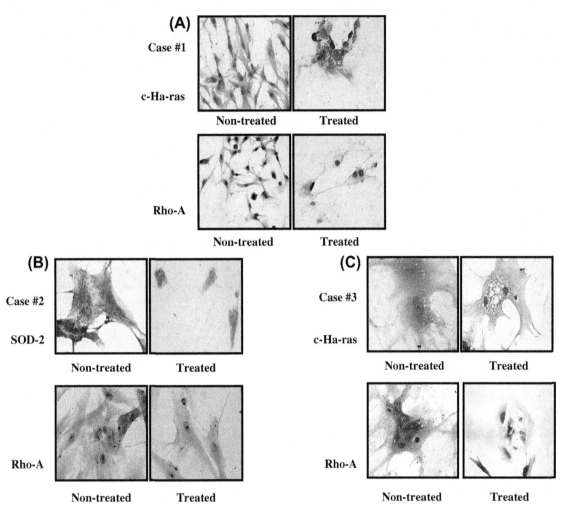

FIGURE 15.6 **Effect of Curcumin on c-Ha-ras and Rho-A and SOD-2 Protein Expression in Isolated Cells Derived from Specimens with Breast Carcinomas.** (A) Isolated cells derived from breast specimen with ductal carcinoma after 3 passages (Case #1). Protein expression of c-Ha-ras and Rho-A in nontreated and treated cells with 30 µM curcumin for 2 days in culture. (B) Isolated cells derived from breast specimen with invasive ductal carcinoma, grade II (Case #2) after 10 passages. Protein expression of SOD-2 and Rho-A in nontreated and treated with 30 µM curcumin after 10 passages and 2 days in culture. (C) Isolated cells derived from breast specimen with lobular carcinoma, grade III. Protein expression of c-Ha-ras and Rho-A in nontreated and treated (30 µM) curcumin in isolated cells derived from such tissue after 10 passages and 2 days in culture. (See the color plate.)

carcinogenesis. On the other hand, this breast cancer model can be considered as an important tool for monitoring the effects of natural dietary compounds on multiple protein expressions that could be key factors in signaling pathways in carcinogenesis.

SUMMARY POINTS

- Oxidant stress plays a crucial role in estrogen-induced carcinogenesis
- Role for free radicals in the etiology of cancer
- Specific biomarkers of oxidative damage to characterize cell damage *in vitro* as DNA damage
- Oxidative stress, an important pathogenic factor of cancer development
- Among antioxidants, curcumin (1, 7-bis (4-hydroxy-3-methoxyphenyl)-1,6-heptadiene-3,5-dione;

diferuloylmethane) is a major dietary natural yellow pigment derived from the rhizome of the herb *Curcuma longa* (Zingiberaceae)
- Curcumin and its analogues are known to protect against peroxidation damage
- Curcumin as an antioxidant measured by several parameters related to oxidative stress
- Curcumin on oxidative stress in an established *in vitro* experimental breast cancer model
- Specific biomarkers of lipid peroxidation are the isoprostanes such as the 8-isoprostaglandin F2α
- Curcumin as an important substance for prevention of breast carcinogenesis

Acknowledgements

The financial support received from grant FOND-ECYT #.1080482 and # 1120006 and MINEDUC, UTA

(GMC) is greatly appreciated. I am sincerely thankful for the technical assistance of Guiliana Rojas Ordóñez, Richard Ponce Cusi, and Georgina Vargas Marchant.

References

1. Henderson BE, Pike MC, Ross RK. Epidemiology and risk factors. In: Bonadonna GJ, editor. *Breast Cancer: Diagnosis and Management.* New York: John Wiley and Sons Ltd.; 1984.

2. Krieger N. Exposure, susceptibility and breast cancer risk: A hypothesis regarding exogenous carcinogens, breast tissue development, and social gradients, including black/white differences, in breast cancer incidence. *Breast Cancer Res Treat* 1989;**13**(3):205–23.

3. Bernstein L. The epidemiology of breast cancer. *LOWAC J* 1998;**1**:7–13.

4. Feigelson HS, Henderson BE. Estrogens and breast cancer. *Carcinogenesis* 1996;**17**:2279–84.

5. Bernstein L, Ross RK, Pike MC, Brown JB, Henderson BE. Hormone levels in older women: A study of post-menopausal breast cancer patients and healthy population controls. *Br J Cancer* 1990;**61**:298–302.

6. Toniolo PG, Levitz M, Zeleniuch-Jacquotte A, Banerjee S, Koenig KL, Shore RE, et al. A prospective study of endogenous estrogens and breast cancer in post-menopausal women. *J Natl Cancer Inst* 1995;**86**:1076–82.

7. Berrino F, Muti P, Micheli A, Bolelli G, Krogh V, Sciajno R, et al. Serum sex hormone levels after menopause and subsequent breast cancer. *J Natl Cancer Inst* 1996;**88**:291–6.

8. Li J, Li SA, Klicka JK, Parsons JA, Lam LKT. Relative carcinogenic activity of various synthetic and natural estrogens in the Syrian hamster kidney. *Cancer Res* 1983;**43**:5200–4.

9. International Agency for Research on Cancer. *Monographs on the Evolution of Carcinogenic Risks to Humans: Hormonal Contraception and Postmenopausal Hormone Therapy,* vol. 72. France: IARC, Lyon; 1999.

10. Cavalieri EL, Stack DE, Devanesan PD, Todorovic R, Dwivedy I, Higginbotham S, et al. Molecular origin of cancer: Catechol estrogen-3,4-quinones as endogenous tumor initiators. *Proc Natl Acad Sci USA* 1997;**99**:10937–42.

11. Highman B, Greenman DL, Norvell MJ, Farmer J, Shellenberger TE. Neoplastic and preneoplastic lesions induced in female C3H mice by diets containing diethylstilbestrol or 17β estradiol. *J Environ Pathol Toxicol* 1980;**4**:81–5.

12. Shull DJ, Spady TJ, Snyder MC, Johansson S, Pennington KL. Ovary intact, but not ovariectomized female ACI rats treated with 17β-estradiol rapidly develop mammary carcinoma. *Carcinogenesis* 1997;**18**:1595–601.

13. Sies H. Oxidative stress: Oxidants and antioxidants. *Exp Physiol* 1997;**82**:291–5.

14. Ames BN, Shigenaga MK, Hagen TM. Oxidants, antioxidants, and the degenerative diseases of aging. *Proc Natl Acad Sci USA* 1993;**90**(17):7915–22.

15. Malins DC, Holmes EH, Polissar NL, Gunselman SJ. The etiology of breast cancer. Characteristic alteration in hydroxyl radical-induced DNA base lesions during oncogenesis with potential for evaluating incidence risk. *Cancer* 1993;**71**(10):3036–43.

16. Li JJ, Li SA. Estrogen-induced tumorigenesis in hamsters: Roles for hormonal and carcinogenic activities. *Arch Toxicol* 1984;**55**:110–8.

17. Bhat HK, Calaf GM, Hei TK, Loya T, Vadgama JV. Critical role of oxidative stress in estrogen-induced carcinogenesis. *Proc Natl Acad Sci USA* 2003;**100**(7) 3913–3818.

18. Halliwell B. Oxygen and nitrogen are pro-carcinogens. Damage to DNA by reactive oxygen, chlorine and nitrogen species: Measurement, mechanism and the effects of nutrition. *Mutat Res* 1999;**443**:37–52.

19. Ammon HP, Wahl MA. Pharmacology of *Curcuma longa. Planta Med* 1991;**57**(1):1–7.

20. Aggarwal BB, Kumar A, Bharti AC. Anticancer potential of curcumin: Preclinical and clinical studies. *Anticancer Res* 2003;**23**:363–98.

21. Mukhopadhyay A, Banerjee S, Stafford LJ, Xia C, Liu M, Aggarwal BB. Curcumin-induced suppression of cell proliferation correlates with down-regulation of cyclin D1 expression and CDK4-mediated retinoblastoma protein phodphorilation. *Oncogene* 1994;**21**:8852–61.

22. Surh YJ, Han SS, Keum YS, Seo HJ, Lee SS. Inhibitory effects of curcumin and capsaicin on phorbol ester-induced activation of eukaryotic transcription factors, NF-kappaB and AP-1. *Biofactors* 2000;**12**:107–12.

23. Korutla L, Kumar R. Inhibitory effect of curcumin on epidermal growth factor receptor kinase activity in A431 cells. *Biochim Biophys Acta* 1994;**1224**:597–600.

24. Hussain AR, Al-Rasheed M, Manogaran PS, Al-Hussein KA, Platanias LCA, Kuraya K, et al. Curcumin induces apoptosis via inhibition of PI3′-kinase/AKT pathway in acute T cell leukemias. *Apoptosis* 2006;**11**:245–54.

25. Huang MT, Newmark HL, Frenkel K. Inhibitory effects of curcumin on tumorigenesis in mice. *J Cell Biochem* 1997;**27**:26–34.

26. Huang MT, Lou YR, Ma W, Newmark HL, Reuhl KR, Conney AH. Inhibitory effects of dietary curcumin on forestomach, duodenal, and colon carcinogenesis in mice. *Cancer Res* 1994;**54**(22):5841–7.

27. Pereira MA, Grubbs B, Barnes LH. Effects of phytochemicals, curcumin and quercetin upon azoxymethane-induced colon cancer and 7, 12-dimethylbenz[a]anthracene-induced mammary cancer in rats. *Carcinogenesis* 1996;**17**:1305–11.

28. Singletary K, MacDonald C, Wallig M, Fisher C. Inhibition of 7,12-dimethylbenz[a]anthracene (DMBA)-induced mammary tumorigenesis and DMBA-DNA adduct formation by curcumin. *Cancer Lett* 1996;**103**(2):137–41.

29. Rao CV, Rivenson A, Simi B, Reddy BS. Chemoprevention of colon carcinogenesis by dietary curcumin, a naturally occurring plant phenolic compound. *Cancer Res* 1995;**55**(2):259–66.

30. Kawamori T, Lubet R, Steele VE, Kelloff GJ, Kaskey RB, Rao CV, et al. Chemopreventive effect of curcumin, a naturally occurring anti-inflammatory agent, during the promotion/progression stages of colon cancer. *Cancer Res* 1999;**59**(3):597–601.

31. Limtrakul P, Lipigorngoson S, Namwong O, Apisariyakul A, Dunn FW. Inhibitory effect of dietary curcumin on tumorigenesis in mice. *J Biochem* 1997;(Suppl. 27):26–34.

32. Lu YP, Chang RL, Lou YR, Huang MT, Newmark HL, Reuhl KR, et al. Effect of curcumin on 12-O-tetradecanoylphorbol-13-acetate-and ultraviolet B light-induced expression of c-jun and c-fos in JB6 cells and in mouse epidermis. *Carcinogenesis* 1994;**15**(10):2363–70.

33. Kim JM, Araki DJ, Park CB, Takasuka N, Baba-Toriyama H, Ota T, et al. Chemo preventive effect of carotenoids and curcumins on mouse colon carcinogenesis after 1,2 dimethylhydrazine initiation. *Carcinogenesis* 1998;**19**(1):81–5.

34. Calaf GM, Hei TK. Establishment of a radiation- and estrogen-induced breast cancer model. *Carcinogenesis* 2000;**21**:769–76.

35. Calaf GM, Echiburú-Chau C, Wen G, Balajee AS, Roy D. Effect of curcumin on irradiated and estrogen-transformed human breast cell lines. *Int J Oncol* 2012;**40**:436–4236.

36. Calaf GM, Echiburú-Chau C, Roy D, Chai Y, Wen G, Balajee AS. Protective role of curcumin in oxidative stress of breast cells. *Oncol Reports* 2011;**26**:1029–35.

37. Roy D, Calaf GM, Hei TK. Allelic imbalance at 11p15.5-15.4 correlated with *c-Ha-Ras* mutation during radiation-induced neoplastic transformation of human breast epithelial cells. *Int J Cancer* 2003;**103**:730–7.

38. Aggarwal BB, Surh YH, Shishodia S. The molecular targets and therapeutics of curcumin in health and disease. *Adv Exp Biol* 2007; **995** (Springer Publ).

39. Aggarwal BB, Shishodia S, Takada Y, Banerjee S, Newman RA, Bueso-Ramos CE, et al. Curcumin suppresses the paclitaxel-induced nuclear factor-kappa B pathway in breast cancer cells and inhibits lung metastasis of human breast cancer in nude mice. *Clin Cancer Res* 2005;**11**:7490–8.

40. Baeuerle PA. The inducible transcription activator NF-kappa B: regulation by distinct protein subunits. *Biochim Biophys Acta* 1991;**1072**:63–80.

41. Brennan P, O'Neill LA. Inhibition of nuclear factor kappaB by direct modification in whole cells—Mechanism of action of nordi-hydroguaiaritic acid, curcumin and thiol modifiers. *Biochem Pharmacol* 1998;**55**:965–73.

42. Siebenlist U, Franzoso G, Brown K. Structure, regulation and function of NF-kappa B. *Annu Rev Cell Biol* 1994;**10**:405–55.

43. Anand P, Sundaram C, Jhurani S, Kunnumakkara AB, Aggarwal BB. Curcumin and cancer: An 'old-age' disease with an 'age-old' solution. *Cancer Lett* 2008;**267**:133–64.

44. Basile V, Ferrari E, Lazzari S, Belluti S, Pignedoli F, Imbriano C. Curcumin derivatives: Molecular basis of their anti-cancer activity. *Biochem Pharmacol* 2009;**78**:1305–15.

45. Wispe JR, Clark JC, Burhans MS, Kropp KE, Korfhagen TR, Whitsett JA. Synthesis and processing of the precursor for human mangano-superoxide dismutase. *Biochim Biophys Acta* 1989;**994**: 30–6.

46. Klaunig JE, Kamendulis LM. The role of oxidative stress in carcinogenesis. *Annu Rev Pharmacol Toxicol* 2004;**44**:239–67.

47. Kunchandy E, Rao MN. Effect of curcumin on hydroxyl radical generation through Fenton reaction. *Int J Pharmaceut* 1989;**57**: 173–6.

48. Ahn J, Gammon MD, Santella RM, Gaudet MM, Britton JA, Teitelbaum SL, et al. Associations between breast cancer risk and the catalase genotype, fruit and vegetable consumption, and supplement use. *Am J Epidemiol* 2005;**162**:943–52.

49. Urata Y, Yoshito I, Hiroaki M, Shinji G, Takehiko K, Junji Y, et al. 17β estradiol protects against oxidative stress-induced cell death through the glutathione/glutaredoxin-dependent redox regulation akt in myocardiac H9c2 cells. *J Biol Chem* 2006;**281**(19):13092–102.

50. Roberts II LJ, Morrow JD. Measurement of F2-isoprostanes as an index of oxidative stress *in vivo. Free Radic Biol Med* 2000;**28**:505–13.

51. Morrow JD, Minton TA, Badr KF, Roberts II LJ. Evidence that the F2-isoprostane, 8-epi-prostaglandin F2a, is formed in vivo. *Biochim Biophys Acta* 1994;**1210**:244–8.

52. Sodergren E, Cederberg J, Basu S, Vessby B. Vitamin E supplementation decreases basal levels of F(2)-isoprostanes and prostaglandin (2alpha) in rats. *J Nutr* 2000;**130**:10–4.

53. Wong YT, Ruan R, Tay FE. Relationship between levels of oxidative DNA damage, lipid peroxidation and mitochondrial membrane potential in young and old F344 rats. *Free Radic Res* 2006;**40**:393–402.

54. Priyadarsini I. Free radical reactions of curcumin in membrane models. *Free Radical Biol Med* 1997;**23**(6):838–43.

55. Valko M, Izakovic M, Mazur M, Rhodes CJ, Telser J. Role of oxygen radicals in DNA damage and cancer incidence. *Mol Cell Biochem* 2004;**266**:37–56.

56. Marnett LJ. Oxyradicals and DNA damage. *Carcinogenesis* 2000;**21**:361–70.

57. Williams CS, Mann M, Dubois RN. The role of cyclooxygenases in inflammation cancer and development. *Oncogene* 1999;**18**:7908–16.

Antioxidant Vitamins and Genetic Polymorphisms in Breast Cancer

Daehee Kang

Department of Preventive Medicine, Seoul National University College of Medicine, Seoul, Republic of Korea

Sang-Ah Lee

Department of Preventive Medicine, Kangwon National University School of Medicine, Gangwon-do, Republic of Korea

List of Abbreviations

ADP Adenosine diphosphate
CAT Catalase
CBS cystathionine-beta-synthase
CpG C-phosphate-G
CPS-II Cancer Prevention Study II Nutrition cohort
DCH Diet, Cancer and Health cohort
dTMP Deoxythymidylate
E3N Etude Epidémiologique auprès de femmes de la Mutuelle Générale de l'Education Nationale
eNOS Endothelial nitric oxide synthase
EPIC European Prospective Investigation into Cancer and Nutrition
ER/PR Estrogen receptor/Progesterone receptor
FTHFD formyltetrahydrofolate dehydrogenase
GPX Glutathione peroxidase
IWHS Iowa Women's Health Study
MDC Malmö Diet and Cancer
MnSOD Manganese superoxide dismutase
MPO Myeloperoxidase
MTHF Methylenetetrahydrofolate
MTHFD1 methylenetetrahydrofolate dehydrogenase1
MTR Methionine synthase
MTRR Methionine synthase reductase
NAD Nicotinamide adenine dinucleotide
NADPH Nicotinamide adenine dinucleotide phosphate
NHS Nurses' Health Study
PARPs Poly-ADP-ribose polymerase
PLCO Prostate, Lung, Colorectal, and Ovarian Cancer Screening Trial
ROS Reactive oxygen species
RR Risk ratio
SHMT1 Serine hydroxymethyltransferase1
SWHS Shanghai Women's Health Study
TYMS Thymidylate synthase
V Valine
WHS Women's Health Initiative Observational Study

INTRODUCTION

As current knowledge of diet-related carcinogenesis is still limited, so individual variability in the potential relationship between dietary constituents and breast cancer risk should continue to clarify the role of gene-nutrient interactions in the etiology of breast cancer. In particular, interaction studies have great potential for investigating relevant mechanisms, identifying susceptible populations/individuals, and making practical use of their results to develop preventive strategies beneficial to breast cancer incidence.

The role of diet for the risk of breast cancer is of great interest as a potentially modifiable risk factor. Dietary factors are of particular interest in the context of breast cancer including fat intake, fruits and vegetables, antioxidant vitamins, glycemic index/load, isoflavone, green tea, heterocylic amines, and so on. A role for diet in breast cancer etiology may be ascribed to the antioxidant properties of selected nutrients, influence on DNA repair, DNA mutations, DNA adducts, metabolic detoxification, stimulation of growth factors, and potential antiestrogenic influence of some nutrients.

Future studies on gene-diet interactions are also warranted and perhaps particularly important in the field of diet and breast cancer because most of the existing evidence has not revealed strong associations with risk. It is possible that the beneficial or harmful effects of dietary exposures are restricted to a subgroup of women defined by specific genetic characteristics with direct relevance to the biological pathways underling the associations

Cancer
http://dx.doi.org/10.1016/B978-0-12-405205-5.00016-7

© 2014 Elsevier Inc. All rights reserved.

under investigation. Studies of gene-diet interactions should be grounded solidly in biological plausibility (biological relationship of gene and dietary exposure of interest and demonstration of functional significance) and must have sufficient statistical power to detect suspected modifying effects of a candidate genetic polymorphism on the association between dietary exposure and breast cancer risk.

EFFECT OF ANTIOXIDANT VITAMINS ON BREAST CANCER INCIDENCE

Among the prospective epidemiologic studies conducted on diet and breast cancer to date, there is no association that is consistent, strong, and statistically significant, except for regular alcohol consumption, overweight, and weight gain. Most prospective studies do not provide evidence of an association between folate intake and breast cancer risk (Table 16.1) and the results from the Nurses' Health Study are only suggestive of an inverse association with circulating folate levels.[1] High intake of folate as well as circulating levels may be associated with a lower risk of breast cancer among moderate to high alcohol-drinkers (Table 16.2). There are several possible explanations for this apparent lack of association between diet and breast cancer incidence:

- No causal association. Diet during adulthood may not be an important predictor of breast cancer risk.
- Measurement error by self-administered diet. It is possible that a modest or moderate association such as a true RR of 1.5 or 0.7 cannot be detected with our current diet assessment methods.
- Variation in diet. Because the variation in diet studied needs to exceed the measurement error, diet studies may need to take advantage of the international variation in diet when exploring the association with cancer.
- Timing of dietary assessment. Most cohort studies have focused on adult intake assessed at baseline among adults aged 40 and older. The human breast may be most susceptible to dietary influences during early life, in particular before puberty.
- Follow-up. The average follow-up time in cohort studies is 8 to 9 years. This time period may be too short to capture the role of diet as an initiator or initial preventive factor of cancer.
- Subgroups of women characterized by ER/ PR status or genetic, epigenetic, or hormonal status. Stratification of breast cancer by specific characteristic should be considered.
- It is possible that food toxins might be ingested with regular diet and could counteract some beneficial effect of food such as fruits and vegetables on breast cancer risk.

- Food and nutrients have interactive effects that are difficult to capture in epidemiological studies. Having metabolic profiles of individuals through a metabolomics approach could provide a more integrated evaluation about the impact of diet on breast cancer.
- Gene-diet interaction is particularly important in the field of diet and breast cancer because most existing evidence has not revealed strong association with risk. Large genomewide association studies are on their way and will provide further insight on potential gene–diet interaction.
- Diet may also interact with genetic predisposition via epigenetic mechanisms. Exposure during adulthood can influence methylation and epigenetic malfunction appears to play an important role in cancer development.

ANTIOXIDANT VITAMINS AND GENOMIC INTEGRITY: DEVELOPMENTAL AND DEGENERATIVE CORRELATES

Vitamins A, B_{12}, B_6, B_3, C, and E and folic acid are critical in preventing oxidative lesions and both single- and double-stranded DNA breaks (Table 16.3). They are also factors that can influence embryogenesis as well as being etiologically important in chronic adulthood disease. Although the intricate relationship between folate and DNA damage has been shown earlier and is shown in Figure 16.1, other vitamins have similar relevance when considering the elaboration, maintenance, and expression of DNA (Table 16.3).

Folate participates in DNA metabolism in the synthesis of purines and thymidilate and is a methyl donor for DNA methylation reactions. Low levels of folate may result in a disruption of DNA repair and replication processes and in abnormal methylation and gene expression. Folate is a central player in one-carbon metabolism, including the methylation of DNA. It is converted to 5,10-methylenetetrahydrofolate (MTHF), the metabolites of which are important for the synthesis of DNA components, guanine and adenine. Importantly, the product of MTHF reductase is a methyl donor for the conversion of homocyseine to methionine, which ultimately serves in the form of S-adenosylmethionine for a variety of reactions, including methylation of DNA. Polymorphism in MTHF reductase has been shown to interact with B vitamins to moderate breast cancer risk.[2] Because folate deficiency may lead to global hypomethylation, there is a strong mechanistic basis for the role of alterations in dietary intake levels of folate in the initiation and

TABLE 16.1 Dietary Antioxidants Intake and Breast Cancer Risk in Prospective Studies

Antioxidant Vitamins	Cohort	Country	No. of Subject Cases	No. of Subject Total	Results
Folate	SWHS	Chinese	718	73237	No association, even stratified by hormone-receptor status or menopausal status
	DCH	Denmark*	1072	27296	No association
	CPS II	USA*	3898	70656	No association
	MDC	Sweden*	392	11699	Decreased risk (Q1–Q5, HR = 0.56, 95% CI = 0.35–0.95)
	E3N	France*	1812	62739	Decreased risk (Q1–Q5, RR = 0.78, 95% CI = 0.67–0.90)
	PLCO	USA*	691	15400	No association
	NHS	USA	3797	88744	No association, even stratified by hormone-receptor status
	IWHS	USA	1875	34393	No association, even stratified by hormone-receptor status
Beta-carotene	DCH	Denmark*	1072	27296	No association
	EPIC	EU**	1480	118437	No association
	WHS	USA*	2879	84805	Decreased risk among women with ER+/PR+ ($Q_{1vs.5}$, RR = 0.78, 95% CI = 0.66–0.94)
	Netherlands	Netherland	650	62573	No association
Vitamin B_2	SWHS	Chinese	718	73237	No association, even stratified by hormone-receptor status or menopausal status
	IWHS	USA*	1586	35973	No association
Vitamin B_3	SWHS	Chinese	718	73237	Increased risk among women with ER+/PR+ ($Q_{1vs.4}$, HR = 1.62, 95% CI = 1.07–2.46)
Vitamin B_6	SWHS	Chinese	718	73237	No association, even stratified by hormone-receptor status or menopausal status
	CPS II	USA*	3898	70656	No association
	IWHS	USA*	1586	35973	No association
Vitamin B_{12}	SWHS	Chinese	718	73237	No association, even stratified by hormone-receptor status or menopausal status
	CPS II	USA*	3898	70656	No association
	IWHS	USA*	1586	35973	No association
Vitamin C	DCH	Denmark*	1072	27296	No association
	EPIC	EU**	1480	118437	No association
	UK Dietary Cohort	UK	707	2851	No association
	WHS	USA*	879	35266	No association
	NHS	USA	1439	90933	No association
	Netherlands	Netherland	650	62573	No association
Vitamin E	DCH	Denmark*	1072	27296	No association
	EPIC	EU**	1480	118437	No association
	WHS	USA*	879	35266	No association
	NHS	USA	1439	90933	No association
	Netherlands	Netherland	650	62573	No association

SWHS, Shanghai Women's Health Study; DCH, Diet, Cancer and Health cohort; CPS-II, Cancer Prevention Study II Nutrition cohort; E3N, Etude Epidémiologique auprès de femmes de la Mutuelle Générale de l'Education Nationale; EPIC, European Prospective Inverstigation into Cancer and Nutrition; IWHS, Iowa Women's Health Study; MDC, Malmö Diet and Cancer; NHS, Nurses' Health Study; PLCO, Prostate, Lung, Colorectal, and Ovarian Cancer Screening Trial; WHS, Women's Health Initiative Observational Study

* *postmenopausal only*

** *premenopausal only*

TABLE 16.2   Blood Level of Antioxidants and Breast Cancer Risk in Prospective Studies

| Antioxidant Vitamins | Cohort | No. of Subject | | Association |
		Cases	Controls	
Folate	NHS	712	712	Decreased risk ($Q_{1vs.5}$, RR = 0.69, 95% CI = 0.49–0.98)
	WHS	848	848	No association
	Washington County serum bank	195	195	No association, even stratified by menopausal status
Beta-carotene	Prospective study in Guernsey	39	78	No association
Vitamin B_6	NHS	712	712	Decreased risk ($Q_{1vs.5}$, RR = 0.68, 95% CI = 0.49–0.96)
	WHS	848	848	No association
Vitamin B_{12}	NHS	712	712	No association
	WHS	848	848	No association
	Washington County serum bank	195	195	No association, even stratified by menopausal status
Vitamin E	Prospective study in Guernsey	39	78	No association

WHS, Women's Health Initiative Observational Study; NHS, Nurses' Health Study

TABLE 16.3   The Major Source and Function of Antioxidant Vitamins and Genomic Integrity

Vitamin (unit)	RDA[c]	Dietary Sources	Genomic Integrity
Folic acid (µg)	F:180 M:200	Green vegetables, orange juice, nuts, legumes, grain products	DNA metabolism in the synthesis of purines and thymidilate. Methyl donor for DNA methylation reactions. Player in one-carbon metabolism, including methylation of DNA. Synthesis of DNA components (guanine and adenine) throughout the production of 5, 10- methylenetetrahydrofolate (MTHF).
B_{12} (µg)	F&M:2	Animal products	Folate is trapped at the level of 5-methyltetrahydrofolate because B_{12}-dependent methionine synthase activity is compromised. Regulated DNA expression.
B_2 (Riboflavin) (mg)	F: 1.3 M: 1.7	Dairy products, meats, eggs, green leafy vegetables, spinach	Improve folate-dependent pathways such as de novo methionine biosynthesis needed to maintain genomic CpG methylation patterns, regeneration of tetrahydrofolate for nucleotide biosynthesis.
B_6 (mg)	F: 1.6 M: 2.0	Meat, chicken, banana, sunflower	Regulation of the methionine cycle and the balance between homocysteine remethylation with B_{12} and folate. Regulation of steroid hormones, because it helps to remove the hormone-receptor complex from DNA binding and hence terminates hormone action.
B_3 (Niacin) (NE)	F: 15 M: 19	Nuts, meats, chicken, liver, mushroom	The transfer of ADP-ribose moieties from niacin derived NAD to arginine, lysine, or asparagines residues of nucleoproteins involved in DNA repair and replication is catalyzed by poly-ADP-ribose polymerase (PARPs).
C (ascorbic acid) (mg)	F&M :60	Fruits and vegetables	Vitamin C can neutralize reactive oxygen species (ROS) and hence protect nuclear and membrane structures. Mega dose of vitamin C may actually potential radical formation and compromise the integrity of genomic mechanism, as with other antioxidants such as β-carotene.
E(TE[b])	F: 8 M: 10	Vegetable oils, nuts, seeds, asparagus, margarine	Protect PUFA within the cell membrane and plasma lipoproteins from lipid peroxidation. Modulates two important signal transduction pathways (phosphatidylinisitol 3-kinase and protein kinase C).
A (RE[a])	F : 800 M : 1000	Liver, carrot, spinach, tomato, pumpkin	β-carotene in particular has an important role as a free radical scavenger, and it protects DNA and membrane structures.

[a] RE: retinol equivalents
[b] TE: tocopherol equivalents
[c] Healthy Adults Ages 19–50, F: Female, M: Male

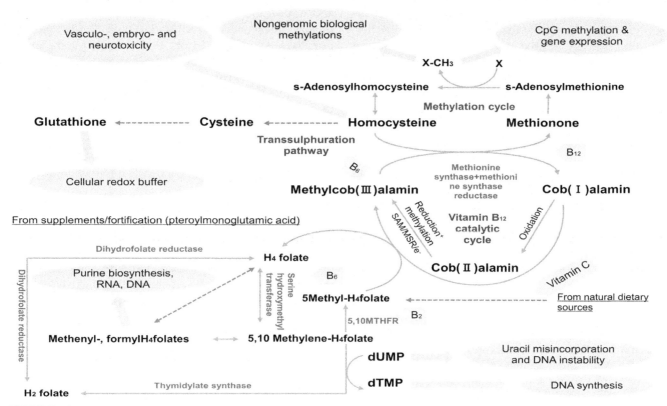

FIGURE 16.1 The schematic of the folate-dependent process and the cooperation with the other vitamins in the elaboration, maintenance, and expression of DNA. (See the color plate.)

progression of breast cancer by acting on DNA integrity, synthesis, repair, and methylations. Several reports have suggested that folate may confer protection in early carcinogenesis and also promote cancer growth later during the neoplastic process, including that of mammary tumors.[3,4] This is a plausible scenario because, as the disease progresses, the organization of the cell nucleus, including the genome, may be modified.[5] Therefore, one-carbon metabolism-mediated DNA methylation triggered by folate might affect genes that are protecting against breast cancer development as well as genes that are promoting tumor development and progression depending on whether or not nuclear organization renders these genes accessible for epigenetic modification.

Folate is trapped at the level of 5-methyltetrahydrofolate because B_{12}-dependent methionine synthase activity is compromised (Figure 16.1). This "functional folate deficiency" leads to a reduction in the folate vitamers required for dTMP synthesis and reduces the availability of *de novo* methyl groups required for CpG methylation and hence regulated DNA expression. B_{12} deficiency acts in synergy with folate deficiency to adversely affect DNA integrity, but studies have also identified the discrete influence of B_{12} in this process.[6] Supplementation with vitamin B_2 may improve the activity of the variant form of C677T-MTHFR and therefore improve folate-dependent pathways such as

de novo methionine biosynthesis needed to maintain genomic CpG methylation patterns, as well as regeneration of tetrahydrofolate for nucleotide biosynthesis. B_6 along with B_{12} and folate help to regulate the methionine cycle and the balance between homocysteine remethylation and its tranulphuration to cysteine. Cystathionine β-synthase and γ-cystathionase are both tranulphuration pathway enzymes and both B_6 dependent. Homocysteine is thought to be embryotoxic, and thus, dietary B_6, B_{12}, and folate may critically influence the level of homocysteine exposed to the early embryo, having an impact on reproductive success. B_6 is also important in the regulation of steroid hormones because it helps to remove the hormone-receptor complex from DNA binding and hence terminates hormone action. Thus, when B_6 is deficient, there is an increased sensitivity to circulating hormone.

The transfer of ADP-ribose moieties from B_3 (niacin) derived NAD to arginine, lysine, or asparagines residues of nucleoproteins involved in DNA repair and replication is catalyzed by poly-ADP-ribose polymerase (PARPs). Five or more different PARPs exist, and in the region of a double-strand DNA break, hundreds of ADP-ribose molecules are polymerized per minute. It is therefore not surprising that a deficiency of vitamin B_3 leads to DNA instability. Outside of its crucial role as a coenzyme for copper-dependent hydroxylases and α-ketoglutarete-linked iron-containing hydroxy-ases, vitamin C has generalized antioxidant properties, including salvage of the spent tocopheroxyl radical back to α-tocopherol (vitamin E). Vitamin C can neutralize reactive oxygen species (ROS) and, hence, protect nuclear and membrane structures: it is worth noting that at high levels, vitamin C, as with other antioxidants such as β-carotene, may actually potentiate radical formation and compromise the integrity of genomic mechanisms.

Vitamin E protects polyunsaturated fatty acids within the cell membrane and plasma lipoproteins from lipid peroxidation. It reduces lipid peroxide radicals to unrective fatty acids. As a consequence of peroxyl-radical scavenging, vitamin E is converted into the stable tocopheroxyl radical. Both vitamin C and selenium enzyme glutathione peroxidase can salvage this compound back into α-tocopherol. It is this role that classifies selenium as an antioxidant. α-Tocopherol also modulates two important signal transduction pathways that are centered on phosphatidylinisitol 3-kinase and protein kinase C. These pathways alter cell proliferation, platelet aggregation, and NADPH-oxidase activation. α-Tocopherol also regulates genes independent of these kinase pathways. γ-Tocopherol also has some gene regulatory properties.[7]

INTERACTION BETWEEN GENETIC POLYMORPHISM AND ANTIOXIDANT VITAMINS ON BREAST CANCER RISK

A growing body of literature focuses on the effect of genetic polymorphisms on the associations between various dietary exposures and breast cancer risk. Such investigations hold promise to detect associations between dietary exposures and breast cancer risk because they are focused on examining such links among women who are considered genetically susceptible or not susceptible to the beneficial or harmful effects of these exposures.

The recognition of genetic and biological variability in nutrient requirements contributed to the development of extensive studies on gene–nutrient interactions.[8] It seems very useful to analyze individual genotypes with a specific focus on common genetic polymorphisms modifying the bioavailability, metabolism, affinity, and activity of several dietary constituents.

Although most of the gene–diet interactions have focused on dietary fats, one of the most solidly established is the *MTHFR* gene, which has been comprehensively reviewed.[9,10]. Findings from observational studies are informative and are of great potential value, but their clinical validity must be confirmed by intervention studies. Ideally, studies designed to test gene–diet interactions should involve prior selection of participants based on the genotype of interest. Most of the currently reported intervention studies are still using retrospective

and opportunistic analyses of subjects participating in dietary intervention studies designed for nongenetic purposes. As expected from their cost and complexity, the number of participants in these studies is very small and subject to errors and spurious associations and interactions. Similar to observational studies, most of the interventional reports have focused on well-known candidate genes, and none of the novel loci identified from genomewide association studies have yet made their way to the literature, although it is anticipated that reports about new genes will trickle into the literature in the near future.

The modification of a phenotype related to a genotype, particularly by dietary habits, could support the notion that some of the inconsistencies in findings from molecular epidemiologic studies could be due to differences in the populations studied and do not account for the underlying characteristics mediating the relationship between genetic polymorphisms and the actual phenotypes. Given the evidence that diet can modify cancer risk, gene–diet interactions in cancer etiology would be anticipated. However, much of the evidence in this area comes from observational epidemiology, which limits the causal inference. Thus, the investigation of these interactions is essential to gain a full understanding of the impact of genetic variation on health outcomes.

This report reviews current approaches to gene–diet interactions in epidemiological studies. Characteristics of gene and dietary factors are divided into four categories:

1. One-carbon metabolism-related gene polymorphisms and dietary factors on breast cancer risk
   The gene–environment interaction studies have focused on "low-methyl" diet and genotype. Folate deficiency contributes to chromosomal instability and may increase susceptibility to radiation-induced DNA damage.[11] Thus, folate deficiency may contribute to carcinogenesis through several biological mechanisms and these mechanisms may be differentially important to cancer etiology. Two mechanisms have been proposed by which folate deficiency could affect malignancy: (1) it can cause DNA hypomethylation and proto-oncogene activation and (2) it can induce uracil misincorporation during DNA synthesis, leading to catastrophic DNA repair, DNA strand breakage, and chromosome damage, although human evidence in support of these mechanisms is limited.[12,13]
   The relationships between folate status and the *MTHFR* genotype have been examined in respect to breast cancer risk in Chinese women.[14] Although there was no difference in the distribution of the *MTHFR C677T* genotype among cases and controls, there was a significant inverse association of breast

cancer risk with dietary folate intake for each of the genotypes that appeared to be stronger for those carrying the TT version of the gene. Eight recent studies reported the interactive effect between folate intake and the one-carbon metabolism-related gene on breast cancer (Table 16.4). There are gene–diet interactions between folate, vitamin B_2, B_6, B_{12}, and methionine as dietary factors and *MTHFR, MTR,*

TABLE 16.4 Interactive Effect between One-Carbon Metabolism-Related Gene Polymorphisms and Antioxidant Vitamins on Breast Cancer Risk

Author (Population)	Study	SNPs	Diets/Nutrients	Results SNPs	Diet/Nutrients	Interaction
Ericson et al., (Sweden).[26]	Malmo Diet and Cancer Cohort	*MTHFR*	Folate intake	rs1801133 (OR, 1.34 for women above 55 yr, OR, 1.29 for below 55 yr)	N/A	Folate intake and MTHFR (rs1801131)CC Genotype (p = 0.03)
Suzuki et al., (Japan)[27]	HERPACC	*MTHFR; MTR; MTRR; TS(2R)*	Folate intake	*MTHFR* (p = 0.048) for postmenopausal *MTRR* (p = 0.041) for premenopausal	Inverse assoc (P = 0.010)	Lowest category of folate and MTRR GG genotype (p = 0.008)
Stevens et al., (USA)[28]	CPS-II Nutrition Cohort	*MTHFR; MTR; MTRR; SHMT1; MTHFD1; TYMS; DHFR; FTHFD; CBS*	Dietary folate, total folate, alcohol, and Methionine	*MTHFD1* (rs1950902, P = 0.048) *FTHFD*(rs2276731 and rs2002287, p = 0.022 and 0.034, respectively)	N/A	Total folate intake and *MTHFR* (p = 0.03) Methionine and *FRHFD*(rs2002287) genotype (p = 0.035)
Xu et al., (USA)[29]	Long Island Breast Cancer Study (LIBCSP)	*DHFR*	Multivitamin use	No association	N/A	19bp +/+vs. −/− OR = 1.52 for vitamin user only
Shrubsole et al., (China)[14]	Shanghai Breast Cancer Study	*MTRR* *MTR*	Folate Methionine Vitamin B_2,B_6	No association	Folate (p = 0.02) MET(p < 0.01) Vit. B_6 (p = 0.03)	No interactive effect
Chen et al., (USA)[2]	LIBCSP	*MTHFR*	Dietary folate Total folate Vitamins B_1, B_2, B_6 Niacin	*MTHFR* (rs1801133 and rs1801131) (p trend = 0.03 and 0.03, respectively)	Vitamin B_1 (p = 0.002), B_2 (p = 0.05) and B_6 (p = 0.03) for no supplement user	No interactive effect However, low-risk (CC genotype and high folate intake) vs. high risk group OR, 1.83. Similar associations were observed vitamins B_1, B_2, B_6, and niacin (OR, 2.06, 1.88, 2.05, and 2.36, respectively)
Marchand et al., (USA)[30]	Multiethnic Cohort Study	*MTHFR*	Folate Ethanol	No association	N/A	No interactive effect
Maruti et al., (Japan).[31]	Vitamins and Lifestyle cohort	*MTHFR*	Folate, Methionine, Vitamins B2, B6, B12	CC vs. TT (OR, 1.6; 95% CI = 1.1–2.5)(p trend = 0.04)	N/A	Folate intake and MTHFR (p = 0.02), Vitamin B_2 from diet plus supplement and MTHFR (p = 0.01)
Lissowska et al., (Poland).[32]	Population-based case control	*MTHFR; MTR; MTRR; CBS; SLCI9AI*	Folate Vitamins B2, B6 B12, Methionine Ethanol	MTR (rs1805087, p = 0.013)	No association for all selected nutrients	Lowest category of vitamin B_2 and MTR GG Genotype (p = 0.05)

MTRR, SHMT1, MTHFD1, TYMS, FTHFD, and *CBS* as the one-carbon metabolism-related genes. Among these gene-diet combinations, there was an interactive effect between *MTHFR* and folate, *MTRR* and folate, and *MTR* and vitamin B_2 on breast cancer risk.

2. Oxidative stress-related gene polymorphisms and dietary factors on breast cancer risk
   Reactive oxygen species (ROS) cause oxidative damage to biomolecules such as DNA, proteins, and lipids, and can cause cellular alteration that may lead to tumorigenesis.[15] Oxidative stress occurs but leads to imbalance between the antioxidant defense system and production of ROS. Fruits and vegetables are rich sources for a number of antioxidants, such as carotenoids, tocopherols, and ascorbic acid, in which these compounds can decrease the oxidative load.[16] ROS are also endogenously generated by enzymes including myeloperoxidase (MPO) and endothelial nitric oxide synthase (eNOS)[17,18] or neutralized by enzymes including manganese superoxide dismutase (MnSOD), catalase (CAT), and glutathione peroxidase (GPX).[19] Variability in these enzymes and environmental exposures may determine the level of oxidative stress in the organism and play a role in cancer risk.
   Ambrosone et al.[20] reported that premenopausal women who were homozygous for the *A* allele and with low intake of fruits and vegetables had increased risk of breast cancer. The MnSOD, key component of the mitochondrial antioxidant defenses, catalyzed the dismutation of the superoxide anione into oxygen. The *MnSOD* genotype that has the alanine (*A*) allele rather than the valine (*V*) allele appears to enhance transport of the protein into the mitochondrial matrix and may lead to greater MnSOD activity.[21] Since 2000, 14 studies have examined the interactive effect between oxidative stress related gene polymorphisms and diet/nutrients on breast cancer risk, and Table 16.5 summarizes those results. Only four studies observed an interactive effect; between *MPO* [22,23], *NOS3* [22], *CAT* [22], and *MGMT* [24] as the related genetic factor, and fruit and vegetable intake including β-carotene and vitamin C as dietary factors. Li et al.[22] observed borderline significant interactions between vegetable and fruit intake and the common *CAT* CC genotype. This genotype was inversely associated with breast cancer risk only among women with higher consumption of fruits and vegetables. Similarly, the low-risk *NOS* T allele and the *HO-1* S allele and MM genotype were also found to be protective among women with high fruit and/or vegetable intake, particularly when the total number of low-risk alleles was jointly considered.

3. Research priorities for gene–diet interaction approach
   Methodologies are currently lacking for specification of hypotheses, clarification of functional effects, and statistical analysis relating to such complex gene–environment pathways. This area of research must be a priority if advancements in the understanding of disease etiology are to be achieved. Finally, Sharp and Little[25] suggested research priorities for the gene–diet interaction approach. In the context of breast cancer prevention, research on gene–environment interactions seems promising. Future research on gene–environment interactions should be performed in large prospective design settings to exclude recall bias and gain enough statistical power. A significance level was either barely achieved or attenuated when other risk factors were taken into account in multivariate analysis. Assessment of gene–environment interactions only included small population of individuals and generated wide confidence intervals, and testing of multiple hypotheses increased the possibility that some results arose purely by chance.

TABLE 16.5 Interactive Effect between Oxidative Stress-Related Gene Polymorphisms and Antioxidant Vitamins on Breast Cancer Risk

Author (Population)	Study	SNPs	Diets/Nutrients	Results SNPs	Diet/Nutrients	Interaction
Li et al. (USA)[22]	CPS-II Nutrition Cohort	CAT; MPO; HO-1; NOS3	Vegetables and fruit	No association	N/A	Vegetables, fruit, and CAT (p = 0.04), NOS3 (p = 0.005), MPO (p = 0.02), and total no. of low-risk alleles (p = 0.006)
He et al. (USA)[33]	Nurses' Health Study Cohort	MPO COMT	Fruit and vegetables Carotenoids Vitamin C Vitamin E	No association	N/A	No interactive effect MPO CC genotype was associated with non significant 42% reduction in risk among women with high fruit intake
Lee et al. (China)[34]	Shanghai Breast Cancer Study	GSTP	Cruciferous vegetable (isothicyanate)	Ile/Ile vs. Val/Val (OR, 1.50; 95% CI = 1.12–1.99)	Isothicyanate (p < 0.001)	No interactive effect the association of GSTPI and cruciferous vegetable was predominantly seen among premenopausal women (OR, 2.08)
McCullough et al. (USA)[35]	CPS II Nutrition Cohort	VDR (FOK1, Taq1, Apa1, Bsm1, Poly (A) tail); Gc protein l CYP24A1	Total Ca Total Vitamin D	No association	N/A	Total Ca intake and polymorphisms near the 3' end of the VDR (p = 0.01)
Ahn et al. (USA)[36]	LIBCSP	GSTA1xB	Cruciferous Yellow, Leafy vegetables	No association	N/A	No interactive effect
Shen et al. (USA)[37]	LIBCSP	XRCC1	Fruit and vegetables β-Carotene Vitamin C Vitamin E α-Carotene	No association	N/A	Fruit, vegetables, and XRCC1 (Arg194Trp genotype (p = 0.04)
Shen et al. (USA)[24]	LIBCSP	MGMT	Fruit and vegetables β-Carotene Vitamins C, E α-Carotene	No association	N/A	The association between fruit and vegetable intake or α-Carotene and breast cancer risk was apparent among women with at least one variant G allele for codon 143 (OR, 0.6, in both)
Gaudet et al. (USA)[38]	LIBCSP	MnSOD	Fruit and vegetables	No association	N/A	Women carrying the Ala allele and high fruit and vegetables had a significant 37% reduction of BC risk (OR, 0.63)
Ahn et al. (USA)[39]	LIBCSP	CAT	Fruit and vegetables β-Carotene Vitamins C, E	TT/TC vs. CC (OR, 0.83; 95% CI = 0.69–1.00)	N/A	Interactive effect between fruit and vitamin C (p = 0.02 and 0.03, respectively) and CATCC genotype

(Continued)

TABLE 16.5   Interactive Effect between Oxidative Stress-Related Gene Polymorphisms and Antioxidant Vitamins on Breast Cancer Risk—cont'd

Author (Population)	Study	SNPs	Diets/Nutrients	Results SNPs	Diet/ Nutrients	Interaction
Ahn et al. (USA)[23]	LIBCSP	MPO	Fruit and vegetables Carotenoids Vitamins C, E	GG vs. GA/AA (OR, 0.82;95% CI = 0.73–1.04)	N/A	Fruit and vegetable intake and MPO genotype (p = 0.04)
Egan et al. (US)[40]	Collaborative Breast Cancer Study	MnSOD	Fruit and vegetables	Val/Val vs. Val/Ala (OR, 1.9; 95% CI = 1.0-3.4) among premenopausal	N/A	No interactive effect
M. Inoue et al. (China)[41]	Singapore Chinese Health Study	MTHFR; TYMS	Green tea	No association	No association	High activity MTHFR/ TYMS genotypes and weekly/daily green tea vs. less frequent intake OR, 0.66
Yuan et al. (Singapore)[42]	Singapore Chinese Health Study	ACE	Green tea	(TT and/or DD) or AT/AA and ID/II (OR, 0.5; 95% CI = 0.4-0.9)	No association	AT/AA and ID/II ACE and green tea intake (p = 0.01)
Ahn et al. (USA)[36]	Population-based case control study	CAT	Fruit and vegetables	% difference: = 29.9 (p = 0.0001)	No association	No interactive effect
Slanger et al. (Germany)[43]	Case-control study	MnSOD	Fruit and vegetables β-Carotene Vitamin C	No association	N/A	No interactive effect
Ambrosone et al. (USA)[20]	Case-control study	MnSOD	Fruit and vegetables Carotenoids Vitamins C, E	Val/Val+Val/Ala vs. Ala/Ala (OR, 3.5; 95% CI = 1.8-6.8) among premenopausal	N/A	The deleterious effect of the MnSOD poluymorphisms was most pronounced among premenopausal women with low intake of fruit and vegetables (OR, 6.0)
Ambrosone et al. (USA)[44]	Case-control study	GSTM1	Fruit and vegetables Carotenoids Vitamins C, E	No association	N/A	No interactive effect
Lee et al. (Korea)[45]	Case-control study	MTHFR	Green vegetables White vegetables, fruits	No association	Green vegetables (p < 0.001), White vegetables (p = 0.015), Fruits (p = 0.784)	Low green veg intake and CC+CT(OR, 3.8; 95% CI = 2.1-6.9), low green veg intake and TT(OR, 5.6; 95% CI = 1.2-26.3)
Wu et al. (Asian-American)[46]	Case-control study	COMT	Tea intake	No association	N/A	At least one low activity COMT allele and tea intake (OR, 0.5; 95% CI = 0.3-0.8)

References

1. Zhang S, Hunter DJ, Hankinson SE, Giovannucci EL, Rosner BA, Colditz GA, et al. A prospective study of folate intake and the risk of breast cancer. *JAMA* 1999;**281**:1632–7.

2. Chen J, Gammon MD, Chan W, Palomeque C, Wetmur JG, Kabat GC, et al. One-carbon metabolism, MTHFR polymorphisms, and risk of breast cancer. *Cancer Res* 2005;**65**(4):1606–14.

3. Stolzenberg-Solomon RZ, Chang SC, Leitzmann MF, Johnson KA, Johnson C, Buys SS, et al. Folate intake, alcohol use, and postmenopausal breast cancer risk in the Prostate, Lung, Colorectal, and Ovarian Cancer Screening Trial. *Am J Clin Nutr* 2006;**83**(4):895–904.

4. Lin J, Lee IM, Cook NR, Selhub J, Manson JE, Buring JE, et al. Plasma folate, vitamin B-6, vitamin B-12, and risk of breast cancer in women. *Am J Clin Nutr* 2008;**87**(3):734–43.

5. Lelievre SA. Tissue polarity-dependent control of mammary epithelial homeostasis and cancer development: An epigenetic perspective. *J Mammary Gland Biol Neopl* 2010;**15**(1):49–63.

6. Ingelman-Sundberg M. The human genome project and novel aspects of cytochrome P450 research. *Toxicol Appl Pharmacol* 2005;**207**(Suppl 2.):52–6.

7. Azzi A, Gysin R, Kempna P, Munteanu A, Negis Y, Villacorta L, et al. Vitamin E mediates cell signaling and regulation of gene expression. *Ann New York Acad Sci* 2004;**1031**:86–95.

8. Fairweather-Tait SJ. Human nutrition and food research: Opportunities and challenges in the post-genomic era. Philosophical transactions of the Royal Society of London Series B. *Biol Sci* 2003;**358**(1438):1709–27.

9. Cummings AM, Kavlock RJ. Gene-environment interactions: A review of effects on reproduction and development. *Crit Rev Toxicol* 2004;**34**(6):461–85.

10. Friso S, Choi SW. Gene-nutrient interactions in one-carbon metabolism. *Curr Drug Metabolism* 2005;**6**(1):37–46.

11. Beetstra S, Thomas P, Salisbury C, Turner J, Fenech M. Folic acid deficiency increases chromosomal instability, chromosome 21 aneuploidy and sensitivity to radiation-induced micronuclei. *Mutat Res* 2005;**578**(1-2):317–26.

12. Duthie SJ. Folic acid deficiency and cancer: Mechanisms of DNA instability. *Br Med Bull* 1999;**55**(3):578–92.

13. Ames BN. DNA damage from micronutrient deficiencies is likely to be a major cause of cancer. *Mutat Res* 2001;**475**(1-2):7–20.

14. Shrubsole MJ, Gao YT, Cai Q, Shu XO, Dai Q, Jin F, et al. MTR and MTRR polymorphisms, dietary intake, and breast cancer risk. *Cancer Epidemiol Biomark Prev* 2006;**15**(3):586–8.

15. Cooke MS, Evans MD, Dizdaroglu M, Lunec J. Oxidative DNA damage: Mechanisms, mutation, and disease. *FASEB J* 2003;**17**(10):1195–214.

16. de Zwart LL, Meerman JH, Commandeur JN, Vermeulen NP. Biomarkers of free radical damage applications in experimental animals and in humans. *Free Radic Biol Med* 1999;**26**(1-2):202–26.

17. Lancaster Jr JR, Xie K. Tumors face NO problems? *Cancer Res* 2006;**66**(13):6459–62.

18. Arnhold J. Properties, functions, and secretion of human myeloperoxidase. *Biochem Biokhim* 2004;**69**(1):4–9.

19. Gilbert DL. *Reactive Oxygen Species in Biological System: An Interdisplinary Approach* 1999.

20. Ambrosone CB, Freudenheim JL, Thompson PA, Bowman E, Vena JE, Marshall JR, et al. Manganese superoxide dismutase (MnSOD) genetic polymorphisms, dietary antioxidants, and risk of breast cancer. *Cancer Res* 1999;**59**(3):602–6.

21. Shimoda-Matsubayashi S, Matsumine H, Kobayashi T, Nakagawa-Hattori Y, Shimizu Y, Mizuno Y. Structural dimorphism in the mitochondrial targeting sequence in the human manganese superoxide dismutase gene. A predictive evidence for conformational change to influence mitochondrial transport and a study of allelic association in Parkinson's disease. *Biochem Biophys Res Commun* 1996;**226**(2):561–5.

22. Li Y, Ambrosone CB, McCullough MJ, Ahn J, Stevens VL, Thun MJ, et al. Oxidative stress-related genotypes, fruit and vegetable consumption and breast cancer risk. *Carcinogenesis* 2009;**30**(5):777–84.

23. Ahn J, Gammon MD, Santella RM, Gaudet MM, Britton JA, Teitelbaum SL, et al. Myeloperoxidase genotype, fruit and vegetable consumption, and breast cancer risk. *Cancer Res* 2004;**64**(20):7634–9.

24. Shen J, Terry MB, Gammon MD, Gaudet MM, Teitelbaum SL, Eng SM, et al. MGMT genotype modulates the associations between cigarette smoking, dietary antioxidants and breast cancer risk. *Carcinogenesis* 2005;**26**(12):2131–7.

25. Sharp L, Little J. Polymorphisms in genes involved in folate metabolism and colorectal neoplasia: A huge review. *Am J Epidemiol* 2004;**159**(5):423–43.

26. Ericson U, Sonestedt E, Ivarsson MI, Gullberg B, Carlson J, Olsson H, et al. Folate intake, methylenetetrahydrofolate reductase polymorphisms, and breast cancer risk in women from the Malmo Diet and Cancer cohort. *Cancer Epidemiol Biomark Prev* 2009;**18**(4):1101–10.

27. Suzuki T, Matsuo K, Hirose K, Hiraki A, Kawase T, Watanabe M, et al. One-carbon metabolism-related gene polymorphisms and risk of breast cancer. *Carcinogenesis* 2008;**29**(2):356–62.

28. Stevens VL, McCullough ML, Pavluck AL, Talbot JT, Feigelson HS, Thun MJ, et al. Association of polymorphisms in one-carbon metabolism genes and postmenopausal breast cancer incidence. *Cancer Epidemiol Biomark Prev* 2007;**16**(6):1140–7.

29. Xu X, Gammon MD, Wetmur JG, Rao M, Gaudet MM, Teitelbaum SL, et al. A functional 19-base pair deletion polymorphism of dihydrofolate reductase (DHFR) and risk of breast cancer in multivitamin users. *Am J Clin Nutr* 2007;**85**(4):1098–102.

30. Le Marchand L, Haiman CA, Wilkens LR, Kolonel LN, Henderson BE. MTHFR polymorphisms, diet, HRT, and breast cancer risk: The multiethnic cohort study. *Cancer Epidemiol Biomark Prev* 2004;**13**(12) 2071-7.

31. Maruti SS, Ulrich CM, Jupe ER, White E. MTHFR C677T and postmenopausal breast cancer risk by intakes of one-carbon metabolism nutrients: A nested case-control study. Breast cancer research:. *BCR* 2009;**11**(6) R91.

32. Lissowska J, Gaudet MM, Brinton LA, Chanock SJ, Peplonska B, Welch R, et al. Genetic polymorphisms in the one-carbon metabolism pathway and breast cancer risk: A population-based case-control study and meta-analyses. *Int J Cancer J Int du Cancer* 2007;**120**(12):2696–703.

33. He C, Tamimi RM, Hankinson SE, Hunter DJ, Han J. A prospective study of genetic polymorphism in MPO, antioxidant status, and breast cancer risk. *Breast Cancer Res Treat* 2009;**113**(3):585–94.

34. Lee SA, Fowke JH, Lu W, Ye C, Zheng Y, Cai Q, et al. Cruciferous vegetables, the GSTP1 Ile105Val genetic polymorphism, and breast cancer risk. *Am J Clin Nutr* 2008;**87**(3):753–60.

35. McCullough ML, Stevens VL, Diver WR, Feigelson HS, Rodriguez C, Bostick RM, et al. Vitamin D pathway gene polymorphisms, diet, and risk of postmenopausal breast cancer: A nested case-control study. Breast cancer research:. *BCR* 2007;**9**(1) R9.

36. Ahn J, Nowell S, McCann SE, Yu J, Carter L, Lang NP, et al. Associations between catalase phenotype and genotype: Modification by epidemiologic factors. *Cancer Epidemiol Biomark Prev* 2006;**15**(6):1217–22.

37. Shen J, Gammon MD, Terry MB, Wang L, Wang Q, Zhang F, et al. Polymorphisms in XRCC1 modify the association between polycyclic aromatic hydrocarbon-DNA adducts, cigarette smoking, dietary antioxidants, and breast cancer risk. *Cancer Epidemiol Biomark Prev* 2005;**14**(2):336–42.

38. Gaudet MM, Gammon MD, Santella RM, Britton JA, Teitelbaum SL, Eng SM, et al. MnSOD Val-9Ala genotype, pro- and anti-oxidant environmental modifiers, and breast cancer among women on Long Island, New York. *Cancer Causes Control: CCC* 2005;**16**(10): 1225–34.

39. Ahn J, Gammon MD, Santella RM, Gaudet MM, Britton JA, Teitelbaum SL, et al. Associations between breast cancer risk and the catalase genotype, fruit and vegetable consumption, and supplement use. *Am J Epidemiol* 2005;**162**(10):943–52.

40. Egan KM, Thompson PA, Titus-Ernstoff L, Moore JH, Ambrosone CB. MnSOD polymorphism and breast cancer in a population-based case-control study. *Cancer Letters* 2003;**199**(1):27–33.

41. Inoue M, Robien K, Wang R, Van Den Berg DJ, Koh WP, Yu MC. Green tea intake, MTHFR/TYMS genotype and breast cancer risk: The Singapore Chinese Health Study. *Carcinogenesis* 2008;**29**(10):1967–72.

42. Yuan JM, Koh WP, Sun CL, Lee HP, Yu MC. Green tea intake, ACE gene polymorphism and breast cancer risk among Chinese women in Singapore. *Carcinogenesis* 2005;**26**(8):1389–94.

43. Slanger TE, Chang-Claude J, Wang-Gohrke S. Manganese superoxide dismutase Ala-9Val polymorphism, environmental modifiers, and risk of breast cancer in a German population. *Cancer Causes Control: CCC* 2006;**17**(8):1025–31.

44. Ambrosone CB, Coles BF, Freudenheim JL, Shields PG. Glutathione-S-transferase (GSTM1) genetic polymorphisms do not affect human breast cancer risk, regardless of dietary antioxidants. *J Nutr* 1999;**129**(2S Suppl) 565S-8S.

45. Lee SA, Kang D, Nishio H, Lee MJ, Kim DH, Han W, et al. Methylenetetrahydrofolate reductase polymorphism, diet, and breast cancer in Korean women. *Exp Mol Med* 2004;**36**(2):116–21.

46. Wu AH, Tseng CC, Van Den Berg D, Yu MC. Tea intake, COMT genotype, and breast cancer in Asian-American women. *Cancer Res* 2003;**63**(21):7526–9.

CHAPTER

17

Dietary Antioxidants in Prostate Cancer

Chris Hamilton, Lucas Aidukaitis, Pingguo Liu, Richard Robison,
Kim O'Neill

Brigham Young University, Department of Microbiology and Molecular Biology, Provo, UT, USA

List of Abbreviations

ROS reactive oxygen species
RNS reactive nitrogen species
PSA prostate specific antigen
LNCaP Lymph Node Carcinoma of the Prostate
CARET Beta-Carotene and Retinol Efficacy Trial
ATBC Alpha-Tocopherol Beta-Carotene Study
SELECT Selenium and Vitamin E Cancer Prevention Trial
PLCO Prostate, Lung, Colorectal, and Ovarian Study
NK3.1 NK3 homeobox 1 gene
HIF-1 Hypoxia-Inducible Factor 1 gene

INTRODUCTION

In 2008, the department of health estimated that nearly 1 million men were diagnosed and over 250,000 men died from prostate cancer worldwide.[1] The prostate gland is a small gland located behind the bladder in males. It is primarily responsible for secreting an alkaline fluid that makes up a portion of semen. Prostate cancer is strongly correlated with age and is related to other risk factors such as race, smoking, and a high-fat diet.[2] It has been shown that one of the major causes of prostate cancer is accumulation of certain mutations possibly as a result of long-term exposure to oxidative stress on the tissue.[3] A major by-product of aerobic respiration is a group of compounds known as reactive oxygen species (ROS) and reactive nitrogen species (RNS). ROS/RNS are highly reactive compounds that damage cellular structures such as cell membranes, DNA, and other organelles. They can also react with molecules such as proteins, lipids, and carbohydrates circulating in blood to alter the normal function and reactivity of these molecules. Environmental conditions such as smoking, alcohol intake, and lack of exercise can increase oxidative stress and lead to DNA damage.

The human body has a group of antioxidant defense systems that act to neutralize and control the damaging effects of naturally and environmentally occurring oxidative metabolites (Figure 17.1). Such systems include the glutathione redox system and the antioxidant response elements.[2] These systems work to maintain homeostasis in the presence of oxidative metabolites. Overexposure to oxidative stress can lead to an imbalance between the defense system's ability to clear this stress, and the accumulation of free radicals. When oxidative stress accumulates, it can accelerate oxidative damage sustained by the tissues, including mutations that lead to malignancy. In addition to the endogenous antioxidant defense systems, there are supplemental antioxidant compounds found in our diet that can similarly react with and neutralize ROS and RNS. Diet plays a major role in the prevention of oxidative stress accumulation. Many nutrients found in a healthy diet are antioxidants that help offset the imbalance caused by oxidative stress. These nutrients or compounds include retinol (vitamin A), carotenoids, flavonoids, resveratrol, and vitamins C and E. There are other compounds (e.g., vitamin D) that do not react with ROS/RNS directly but function by boosting the endogenous antioxidant defense systems in the tissues. With the accumulation of oxidative metabolites being common causes for many diseases, including prostate cancer, it is logical for science to investigate the ability of antioxidants to contribute to regulation and control of oxidative stress and the efficacy of such treatments. Recently there has been a major focus on preventative research, particularly on the role that dietary antioxidants play in helping maintain oxidative homeostasis. In this chapter we will discuss the recent research that has accumulated on the influence of dietary antioxidants on prostate cancer.

Cancer
http://dx.doi.org/10.1016/B978-0-12-405205-5.00017-9

183

© 2014 Elsevier Inc. All rights reserved.

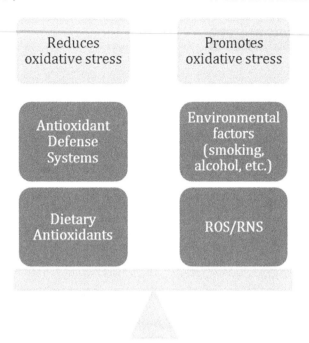

FIGURE 17.1 Oxidative balance in the body. Reactive oxygen species (ROS) and reactive nitrogen species (RNS) are by-products of aerobic respiration. Our bodies counteract the production of ROS/RNS with endogenous antioxidant defense systems. Environmental factors (see figure) or a diet high in fat can increase the oxidative stress in our bodies and cause an imbalance. Dietary antioxidants found in many fruits, vegetables, and other natural foods are important in maintaining oxidative balance in the body.

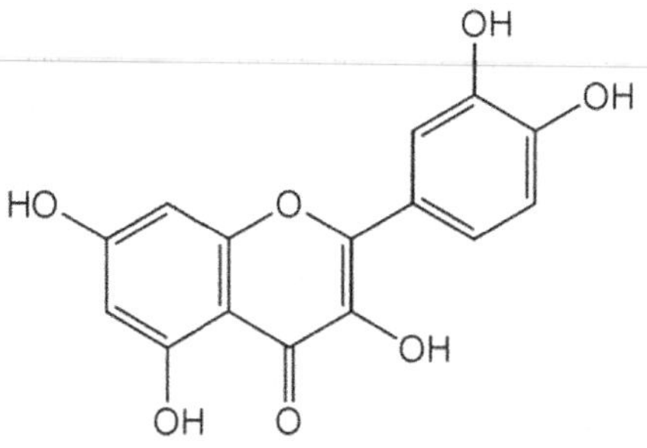

FIGURE 17.2 Chemical structure of quercitin. Quercitin is a polyphenol that belongs to the class of molecules known as flavonoids. Quercitin has several oxygen containing functional groups and is conjugated throughout the entire molecule. *In vitro* studies have shown that quercitin functions to inhibit the androgen receptor in LNCaP cells.

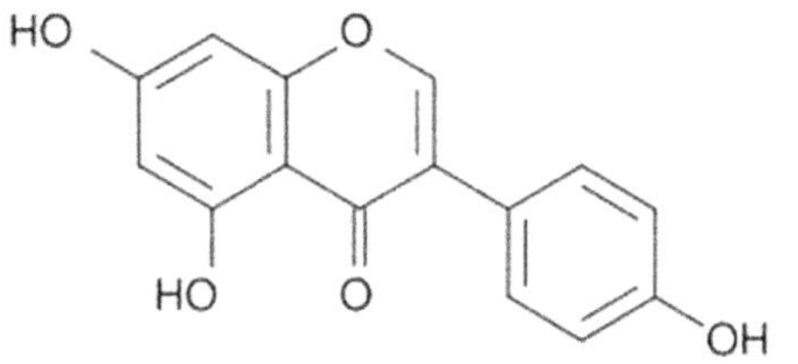

FIGURE 17.3 Chemical structure of genistein. Genistein is a polyphenol flavonoid that is found in soy. Epidemiological studies have shown that the consumption of genistein and other flavonoids found in soy is strongly correlated with decreased incidence of prostate cancer. One proposed mechanism of action for genistein is that it protects against age-related degradation of the antioxidant defense systems in the body. This effect prolongs the protection from antioxidant defense systems and could delay the development and onset of prostate cancer.

DIETARY ANTIOXIDANTS

Flavonoids

Flavonoids are polyphenolic compounds that are found in a variety of dietary sources such as fruits and berries, nuts, seeds, and herbs. Pomegranate juice, soy, and green tea are commonly consumed dietary sources that are high in flavonoid compounds. These foods have been the subject of several research studies done on flavonoids and their impact on prostate cancer. Pomegranate juice contains a great variety of flavonoids. One of the most studied flavonoids in pomegranate fruit is quercetin, which has been shown to be an androgen receptor inhibitor in LNCaP cells (Figure 17.2).[4] The androgen receptor is involved in the development and progression of prostate cancer.[5] Soy products contain a group of flavonoids called isoflavones. One isoflavone found in soy products, Genistein, has been shown to counteract oxidative DNA damage by increasing transcription of genes involved in antioxidant defense systems (Figure 17.3).[6] It is possible that Genistein and other soy isoflavones can delay the age-related degradation of antioxidant defense systems, which would in turn delay the development and onset of prostate cancer. Green tea contains a group of flavonoids called catechins, which

Thakur et al. demonstrated can cause cell cycle arrest and apoptosis in prostate cancer cell lines.[7]

In vitro evidence of flavonoid anticancer effects is supported by epidemiological evidence. Ganry et al. recently reviewed observational research analyzing daily intake of flavonoids, mostly soy isoflavones, and found that the lowest rates of prostate cancer were among populations with the highest intake of dietary flavonoids.[8] Diets in Eastern Asian countries are high in flavonoids mostly due to consumption of soy products and green tea. Data of flavonoid intake compared with rates of prostate cancer for different countries indicate that Japan and China had the lowest rates of prostate cancer and consumed 100 times more flavonoids than western countries.[9,10] Accordingly, Parkin et al. provided a summary of prostate cancer risk with data from GLOBOCAN, which showed that men in Japan and China have a 60- to 80-fold decrease in prostate cancer risk compared with men in North America where flavonoid intake is low and inconsistent.[11] Interestingly another study compared Japanese men who migrated to America and adopted the western diet with those who remained in Japan. They found a significant increase in prostate cancer risk in those who migrated to America and decreased their flavonoid intake.[12,13]

Dietary sources high in flavonoids include several other classes of polyphenols and antioxidant compounds, so it is difficult to isolate dietary flavonoids for clinical studies. One attempt to clinically confirm these results was done in a Phase II study of men with rising PSA postsurgery or postradiotherapy.[14] After daily consumption of pomegranate juice over a period of a few months, PSA doubling time was significantly delayed.[14] Further research is needed in a clinical setting to confirm data collected in those epidemiological and preclinical studies.

Resveratrol

Resveratrol is a stilbenoid with potential antioxidant properties (Figure 17.4). It is commonly found in White Hellebore plants such as *Veratrum album*, a medicinal plant used since antiquity to induce vomiting. Today it can be easily obtained from red wine, grape skins, blueberries, and other plants. Jang et al. reported in 1997 how resveratrol can inhibit the start of carcinogenesis, stopping cancer progression at its initial stage. Since then, further research has been done trying to measure its effects on other cancer types, including prostate cancer,[15] finding that resveratrol selectively activates signaling-dependent apoptosis in human tumor cells.

One group of researchers[16] also showed how low doses of resveratrol inhibit growth and induce apoptosis in LNCaP cells (a cell line that mimics prostate cancer cells), while at the same time not affecting normal epithelial cells in the prostate.[16] Higher doses of resveratrol also decreased proliferation and increased apoptosis in LNCaP cells. However, these are *in vitro* studies and have yet to be confirmed *in vivo* in humans. Recently, *in vivo* studies with nude mice growing LNCaP cells were also conducted with resveratrol, with promising results such as tumor growth delay and no apparent signs of toxicity.[17]

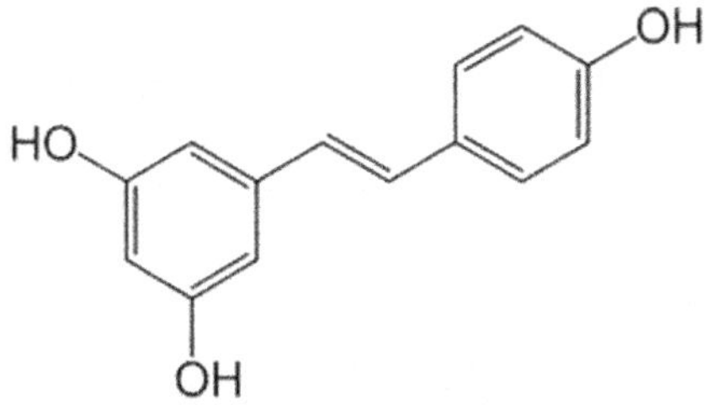

FIGURE 17.4 Chemical structure of resveratrol. Resveratrol is a non-flavonoid polyphenol with three hydroxyl groups attached. The antioxidant properties of resveratrol are due to conjugation throughout the entire molecule. Resveratrol has been shown to inhibit growth and induce apoptosis in prostate cancer cell lines. Common food sources that contain resveratrol are red vine grapes, blueberries, and other plants.

Carotenoids

β-Carotene

Beta-carotene is a conjugated compound that is responsible for the orange pigment in many plants and vegetables. It can be obtained through carrots, sweet potatoes, pumpkins, and leafy green vegetables. α-Tocopherol, a form of vitamin E, enhances β-carotene absorption in the small intestine. β-Carotene can be symmetrically cleaved to form retinol, also known as vitamin A. Retinol is the precursor for compounds important for vision and skin health. Dietary sources of retinol include dairy products and green vegetables. Due to the close metabolic relationship between β-carotene and retinol it is important to consider their physiological effects in tandem.

Two major clinical trials have been reported that investigated β-carotene supplementation in relation to prostate cancer risk. The CARET study was a double-blind placebo controlled trial aimed at evaluating the impact of β-carotene and retinol on lung cancer in heavy smokers and asbestos workers. It had previously been noticed that persons with high β-carotene levels were not as susceptible to lung cancer. A study was then conducted to increase the β-carotene levels in smokers. It was ended 21 months prematurely due to internal analysis showing a significant increase in lung cancer incidence and mortality among participants.[18] Relevant data derived from this study was collected through follow-ups that continued for 11 more years. Neuhouser et al. analyzed the results of the study in relation to risk for prostate cancer, which was a second endpoint of the trial.[19] Treatment groups received daily supplementation of 30 mg β-carotene and 25,000 IU of retinyl palmitate. Multivitamins were allowed to be taken during the study with restrictions to the levels of β-carotene and retinol forms included. Participants that received CARET supplementation and at least one other dietary supplement showed a 52% increase in risk for aggressive prostate cancer compared with the placebo group or the CARET supplementation-only group.[19] A moderately protective effect was observed (35% reduction) in participants who received CARET supplements and took no other dietary supplements.[19] The elevated risk associated with increased supplementation during the active trial was not present in the follow-up study. These results are consistent with the α-Tocopherol β-Carotene (ATBC) study, which showed a 23% increase in risk during the active trial and showed no significant risk in the follow-up study.[20] Both studied populations of heavy smokers. In contrast, the Physician's Health study reported a 32% decrease in prostate cancer risk with supplemental β-carotene in the lowest quartile of serum β-carotene.[21] These results suggest that there may be a protective effect with sufficient serum levels of β-carotene and an adverse effect when β-carotene is above a normal physiological level. Further research is needed to quantify protective physiological levels as well as comparisons between dietary intake of β-carotene and artificial supplements.

Lycopene

Lycopene (also called rhodopurpurin or nonprovitamin A carotenoid) is a powerful antioxidant present in tomatoes and other vegetables and fruits. Unlike most vitamins (e.g., vitamin C), the bioavailability of lycopene increases upon cooking,[22] partially because processing tomato-based foods releases lycopene tightly bound to vegetable fibers, allowing lycopene to be more readily available to be absorbed in the digestive tract. It has long been thought that, due to lycopene's strong antioxidant effects, it can help prevent cellular damage by ROS that could lead to chronic illness or cancer.[23]

Etminan et al. published research on the role of tomato products in preventing prostate cancer and found that increased intake of these foods—cooked or raw—do provide some protection against prostate cancer when tomato product intake is high (10 or more servings per week).[24,25] Other studies, however, failed to find a correlation between lycopene and prostate cancer.[26] A possible explanation has recently been proposed for this inconsistency: Goodman and coworkers reported that different levels of expression of the XRCC1 gene—responsible for aiding in DNA damage repair—affects whether lycopene will decrease prostate cancer risk.[27]

However, Kristal et al. published a report showing that lycopene had no association with prostate cancer risk.[28] In their report, an increase in lycopene serum concentration of 10 μg/dL was associated with a 7% decrease in prostate cancer detection via prostate specific antigen (PSA), but also was correlated with an increase in 8% of cancers detected without a biopsy sample.[28] These conflicting findings led Kristal to suggest additional studies with lycopene before medical authorities can recommend it as a potential anticancer agent.[28]

The mechanism by which lycopene may prevent prostate cancer is still under investigation, but a review article by Johary et al. suggests two mechanisms: nonoxidative and oxidative. In the nonoxidative mechanism, lycopene suppresses carcinogen-phosphorylation of p53 and other tumor suppressor genes, preventing deregulation of these proteins.[29] Additionally, lycopene also helps stop cellular division at the G0-G1 stage of the cell cycle, which could prevent new tumors from becoming aggressive by uncontrolled division. In the oxidative mechanism, lycopene may protect oxidation of important proteins and lipoproteins, thus helping maintain their proper function in the cell. Agarwal and Rao found that men who had low levels of lycopene and high levels of ROS had higher incidence of prostate cancer, hinting at a potential direct benefit of lycopene at preventing prostate cancer in those with high oxidative stress.[25]

Despite these apparent benefits from tomato products, little is known about the precise mechanism by which lycopene works as an anticarcinogen. Before lycopene can be recommended as a cancer protective agent, further studies are needed to ensure that the apparent benefit from lycopene isn't actually due to other nutrients present in foods high in lycopene content.

Vitamin D

Vitamin D and its metabolites are found in several forms in the body. Vitamin D itself does not have antioxidant properties but it acts to boost the glutathione redox system. Calcidiol 25(OH)D and calcitriol 1,25(OH)D are the two principle forms of the vitamin found in the body. Calcitriol is the active form and is also known as vitamin D3. It can be synthesized from calcidiol by a hydroxylase so the physiological effects of each form are tightly coupled.

Li et al. reported in the Physicians Health study that participants who had serum levels of 1,25(OH)D and 25(OH)D below the median were twice as likely to develop prostate cancer as those with both serum levels above the median (odds ratio 2.1).[30] These findings suggest a strong protective effect from vitamin D metabolites in association with prostate cancer. Corder et al. also reported a strong correlation between 1,25(OH)D and 25(OH)D serum levels and prostate cancer risk when frozen serum samples were analyzed (relative risk 0.03 not significant).[31] A second study looking at nonfrozen plasma levels of these two forms of vitamin D and prostate cancer risk also found a nonsignificant inverse correlation.[32] When the data was reviewed they found that there was very little variance across race, family history, concentrations of other metabolites, and several other known risk factors for prostate cancer including season of the year.[32] In an observational study, Ahn et al. found no correlation with vitamin D levels in the serum and prostate cancer risk.[33] The follow-up period for this study was 4 to 5 years. It is possible that this follow-up window was too short to see the long-term results of cancer risk. Other studies conducted found a significant correlation after longer follow-up periods (11–12 years).[34] These findings show that the time interval used to gather follow-up data may be an important variable.

Preclinical data and some clinical data strongly suggest that serum levels of vitamin D metabolites are inversely associated with prostate cancer risk. Other observational studies with large populations contradict these findings.[33] However, as stated earlier, length of follow up may account for the discrepancy in data. Antioxidant interactions are complex and multivariate. Data recorded for vitamin D may be confounded by variance in sun exposure experienced by people in different regions. Further research is needed to validate these findings and to investigate the impact of longer follow-up time and sun exposure.

Vitamin A (Retinol)

Retinol and its derivatives are used in the body to absorb light in the eyes, function as growth factors for epithelial cells, modulate gene function, and possibly as antioxidants. Vitamin A can be found in many different fruits and vegetables such as carrots, broccoli, spinach, tomatoes, cantaloupe, papaya, and mangos. Some other sources include cheddar cheese, eggs, and butter.

Although it has been suggested that diets rich in fruits and vegetables may help prevent or slow down development of prostate cancer because of their high antioxidant content, the actual effect of vegetables in prostate cancer is not known.[35] Jain et al. believe this was caused in part by the inability of previous studies to compare the effects of various antioxidants in tandem with each other.[35] They found that individuals who consumed green vegetables, tomatoes, beans, and cruciferous vegetables (which are rich in antioxidants such as vitamin A, lycopene, and vitamin C) had a small but significantly lower chance of developing prostate cancer.[35] However, when looking only on the effects of lycopene (the main antioxidant found in tomatoes) they could not find significant benefits from increasing its amount in the diet, suggesting that maybe another antioxidant (such as lutein), or a combined effect of several compounds may account for the beneficial effect from tomatoes.[35]

In this study, there were 617 individuals in the treatment population. All 617 had a recent histologically confirmed diagnosis of adenocarcinoma of the prostate. Their average age was 69.8 years old, with an average mass of 79.3 kg, and consumed on average 2,951 calories daily. When Jain et al. divided all 617 prostate cancer cases into four quartiles of food intake (quartile 1 with the lowest food intake, while quartile 4 would have the highest; see Table 17.2), an interesting correlation between increased food intake and higher prostate cancer incidence was observed (Table 17.1).[35] This increase in prostate cancer incidence was surprising because the foods selected for the correlation are considered staples of a healthy diet: vegetables, grains, and fruits. This suggests a possible role between increased caloric consumption with higher prostate cancer incidence.

Increased vitamin A consumption was suggested to have lowered prostate cancer incidence among this same group of individuals, but it is still too early to tell if the correlation is causal or circumstantial. As seen in Table 17.1, higher vegetable intake (including vegetables high in vitamin A) correlated with higher prostate cancer cases, which could mean vitamin A is not as important in protecting oneself from prostate cancer as is consuming fewer calories and eating a more balanced diet.[35] More research is needed to shed more light on the exact role

TABLE 17.1 Prostate Cancer Incidence Based on Increasing Amounts of Food Intake

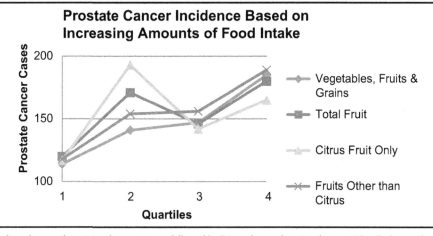

This table groups the 617 confirmed cases of prostate adenocarcinoma followed by Jain and coworkers into four quartiles. Each quartile represents a particular amount of food consumed (listed in Table 17.2). The data suggests that increased food consumption correlated with higher prostate cancer incidence. *Data obtained from Jain et al. (1999).*

TABLE 17.2 Data Showing the Amounts of Food Consumed in Each Quartile for Every Food Category

Food Type	Quartile 1	Quartile 2	Quartile 3	Quartile 4
Vegetables + Fruits + Grains (no beans)	<825 g/day	826–1098 g/day	1099–1403 g/day	>1404 g/day
Total fruit	<183.3 g/day	183.3–342.7 g/day	342.7–514.4 g/day	>514.4 g/day
Citrus fruit only	<34.3 g/day	34.3–145.3 g/day	145.4–265.0 g/day	>265 g/day
Fruit other than citrus	<101.3 g/day	101.2–195.1 g/day	195.2–315.8 g/day	>315.8 g/day

vitamin A plays in preventing and protecting against prostate cancer.

Vitamin E

Vitamin E is a collection of eight fat-soluble tocopherol and tocotrienol molecules that are commonly found in corn and soybean oils, margarine, and other seeds. Its active form, α-tocopherol, is very common and is used by the body to remove ROS originated from fat breakup. Its role as an antioxidant has therefore attracted much attention in the research community for its potential anticancerous properties.

In 1994, a clinical trial called α-Tocopherol β-Carotene (ATBC) attempted to measure the effects of β-carotene and vitamin E in preventing lung cancer. The trial revealed an increase in lung cancer risk in those taking beta carotene supplements, while vitamin E did not increase such risk in similar groups. In fact, the ATBC trial found that patients taking vitamin E supplements had less risk of prostate cancer compared to control groups.[20] This finding motivated another clinical trial (SELECT) that measured the effect of vitamin E and selenium on prostate cancer risks over the course of 7 years among individuals with low or nonexistent prostate cancer markers (males at least 50 years of age, with prostate-specific antigen concentration of 4.0 ng/mL and a digital rectal exam not suspicious for prostate cancer).[36]

Based on the results from the ATBC study, researchers expected a prostate cancer diagnosis decrease between the vitamin E treatment groups compared to those taking placebo supplements. Surprisingly, the opposite occurred: a small but statistical increase in prostate cancer diagnosis among vitamin E-only treatment groups was noticed, while vitamin E + selenium, selenium, or placebo only had no such increase. This has left many scientists wondering about the actual role of vitamin E in prostate cancer prevention. Klein et al. reported in their review that, because the population in the ATBC study was mainly smoking patients, vitamin E may help prevent prostate cancer in smokers only, due to the protective qualities of vitamin E in high oxidative stress environments characteristically present in smokers.[37] The effect may not be seen in nonsmokers because the high oxidative stress isn't present, so any free radical that appears in the nonsmoking population is taken care of by the body's natural antioxidants. The vitamin E excess could therefore be interfering with other cellular pathways that enhance prostate cancer risk ever so slightly. This finding is consistent with the Prostate, Lung, Colorectal, and Ovarian (PLCO) study, which found a statistically significant decrease in prostate cancer aggressiveness among smokers if they took vitamin E supplements.[38]

Another possible explanation for these conflicting effects of vitamin E in prostate cancer risk is that vitamin E may affect the expression of genes involved in protection against antioxidative damage. It is known that mice expressing a defective or null NKX3.1 gene (which is responsible for prostate differentiation during embryogenesis and for protecting the cell against oxidative damage) had their prostate cancer risks increased when treated with antioxidant supplements, including vitamin E.[39] This gene is now being sequenced and analyzed in biospecimens of the SELECT study, in an attempt to correlate prostate cancer risk with a defect in their NKX3.1 gene.

Despite all the research conducted so far with vitamin E and prostate cancer, no definitive answers can, as of yet, be given pertaining to vitamin E's precise role in prostate cancer. However, it appears that vitamin E uptake in those with high risk for prostate cancer (such as smokers) may provide some welcome protection. Further research is required to give clearer answers as to the precise role of vitamin E. The best protection against prostate cancer is to eat a balanced diet, avoid activities that raise ROS in the body such as smoking, and participate in regular exercise.

Vitamin C

Vitamin C (also called ascorbic acid) is an essential nutrient present in most fruits and vegetables, and frequently added to a variety of food products. It is an antioxidant that has long been suspected to help prevent a variety of ailments, from a simple cold to advanced cancer. However, the health benefits of vitamin C in cancer remain unclear, and are still under investigation.

A publication by Fukumura et al. studied the effects of vitamin C on ROS in the presence of chemotherapy agents, finding that vitamin C did maintain its antioxidative properties even under high chemotherapy agent concentrations.[40] In fact, the study found that some chemotherapy medicines (like vinblastine) induced the production of ROS within the tumor, and the ROS produced were successfully neutralized by vitamin C. Although this would lead us to believe vitamin C is a protective agent for prostate cancer, Jain's 1999 study found increased levels of prostate cancer incidence among participants who ate large quantities of fruit, including fruits rich in vitamin C like oranges and other citrus fruits.[35] The explanation for this seemingly conflicting observation has yet to be determined, but it is possible that whatever benefits vitamin C brings in fighting prostate cancer are superseded by other compounds found in the fruits or by an increase in caloric intake.

Recently, vitamin C has also been found to regulate the hypoxia-inducible factor 1 (HIF-1), a transcription factor that regulates glycolysis and angiogenesis. When cancer cells are in a hypoxic environment, HIF-1 is activated and induces angiogenesis toward the tumor to increase oxygen and nutrient availability to the cancerous cells.[41] Currie and colleagues found that human umbilical endothelial cells treated with vitamin C under hypoxic conditions had lower HIF-1 gene activation than cells with no vitamin C.[41] Cells that were under no stress (and as such had only basal HIF-1 gene expression) had their HIF-1 gene completely shut down when in contact with vitamin C. These findings were further confirmed by Ratcliffe et al., who also showed decreased HIF-1 expression in cancerous cell lines upon contact with vitamin C.[42] Since many chemotherapy agents work by preventing angiogenesis (such as Avastar, Nexavar, and Sutent), vitamin C may be a helpful adjuvant to help treat cancer if administered together with these drugs.

As with other antioxidants, the role of vitamin C in prostate cancer prevention is still under investigation and is a hot topic in many scientific discussion forums. There is some evidence pointing toward a beneficial role of vitamin C in prostate cancer prevention at the cellular level, but more studies are needed before its effects in the human body can be accurately determined.

SUMMARY

The imbalance caused by bad diet and unhealthy habits like smoking causes great oxidative stress on the body, which increases prostate cancer risk. This imbalance doesn't appear to be remedied by consuming any one supplement in great amounts (although some supplementation is better than no supplementation at all, if healthy eating habits are difficult to maintain); instead, a balanced diet appears to provide the best source of antioxidants and other nutrients needed to help the body's natural defenses against the damages caused by oxidative stress.

While we cannot categorically list antioxidants with known anti-prostate cancer activity as of yet, the ongoing research from these studies suggests the importance of developing good eating habits early in one's life as an important factor in prostate cancer prevention. These findings are in agreement with the recommendations given by the American Cancer Society, which suggests men ages 50 and up meet with their physician and discuss the pros and cons of testing for prostate cancer.[43] Additionally, maintaining a healthy weight and avoiding tobacco products and alcohol may reduce not only prostate cancer risk, but the risk of developing other cancers as well. It is clear from these studies that smoking and being overweight correlated highly with prostate cancer risk, so improving one's diet and quitting tobacco products is a great start for preventing this, and other types of cancer.

REVIEW OF MAIN POINTS

- Clinical evidence of dietary antioxidants and prostate cancer is inconclusive and needs further research.
- Epidemiological studies show strong correlations between dietary intake of antioxidants through whole foods and decreased prostate cancer risk.
- *In vitro* evidence has shown mechanisms that suggest these antioxidant compounds discussed in this chapter do have tumor suppressive functions.
- Artificial supplementation of antioxidants, in contrast to obtaining antioxidants through a consistent consumption of fruits and vegetables, could be an explanation for the conflicting results seen in clinical evidence.
- Other environmental conditions such as smoking, alcohol intake, and lack of exercise are shown to correlate with increased prostate cancer risk.

References

1. Ferlay J, Shin HR, Bray F, Forman D, Mathers C, Parkin DM. *GLOBOCAN 2008 v2.0, Cancer Incidence and Mortality Worldwide: IARC CancerBase No. 10 [Internet]*. Lyon, France: International Agency for Research on Cancer; 2010. Available from: http://globocan.iarc.fr; [accessed 21 12 12].
2. Gupta-Elera G, Garrett AR, Robinson RA, O'neill KL. The role of oxidative stress in prostate cancer. *Euro J Cancer Prev* 2012;**21**: 155–62.
3. Shan W, Zhong W, Swanlund BS, Oberley TD. Oxidative Stress in Prostate Cancer. In: *Oxidative Stress in Cancer Biology and Therapy*. New York: Springer; 2012. p. 301–31.
4. Xing NZ, Chen Y, Mitchell SH, Young CYF. Quercetin inhibits the expression and function of the androgen receptor in LNCaP prostate cancer cells. *Carcinogenesis* 2001;**22**:409–14.
5. Shiota M, Yokomizo A, Naito S. Oxidative stress and androgen receptor signaling in the development and progression of castration-resistant prostate cancer. *Free Radical Biol Med* 2011;**51**(7): 1320–8.
6. Raschke M, Rowland IR, Magee PJ, Pool-Zobel BL. Genistein protects prostate cells against hydrogen peroxide-induced DNA damage and induces expression of genes involved in the defence against oxidative stress. *Carcinogenesis* 2006;**27**(11):2322–30.
7. Thankur VS, Gupta K, Gupta S. Green tea polyphenols cause cell cycle arrest and apoptosis in prostate cancer cells by suppressing class I histone deacetylases. *Carcinogenesis* 2012;**33**(2):377–84.
8. Ganry O. Phytoestrogens and prostate cancer risk. *Preventative Med* 2005;**41**(1):1–6.
9. Morton MS, Chan PS, Cheng C, Blacklock N, Matos-Ferreira A, Abranches-Monteiro L, et al. Lignans and isoflavonoids in plasma and prostatic fluid in men: Samples from Portugal, Hong Kong, and the United Kingdom. *The Prostate* 1999;**32**(2):122–8.
10. Aldercreutz H, Markkanen H, Watanabe S. Plasma concentrations of phyto-estrogens in Japanese men. *Lancet* 1993;**342**(8881):1209–10.
11. Parkin DM, Bray F, Ferlay J, Pisani P. Global cancer statistics 2002. *CA Cancer J Clin* 2005;**55**:74–108.
12. Lyn-Cook BD, Rogers T, Yan Y, Blann EB, Kadlubar FF, Hammons GJ. Chemopreventive effects of tea extracts and various components on human pancreatic and prostate tumor cells in vitro. *Nutr Cancer* 1999;**35**:80–6.

13. Maskarine G, Singh S, Meng L, Franke AA. Dietary soy intake and urinary isoflavone excretion among women from a multiethnic population. *Cancer Epidemiol Biomark Prev* 1998;7:613–9.

14. Pantuck AJ, Leppert JT, Zomorodian N, Aronson W, Hong J, Barnard RJ, et al. Phase II study of pomegranate juice for men with rising prostate-specific antigen following surgery or radiation for prostate cancer. *Cancer Therapy: Clin* 2006;**12**:4018–26.

15. Jang M, Cai L, Udeani GO, Slowing KV, Thomas CF, Beecher CWW, et al. Cancer chemopreventive activity of resveratrol, a natural product derived from grapes. *Science* 1997;**275**(5297):218–20.

16. Shankar S, Chen Q, Siddiqui I, Sarva K, Srivastava RK. Sensitization or TRAIL-resistant LNCaP cells by resveratrol (3,4′,5 trihydroxystilbene): Molecular mechanisms and therapeutic potential. *J Molecular Signaling* 2007;**2**(7). http://dx.doi.org/10.1186/1750-2187-2-7 published online.

17. Seeni A, Takahashi S, Takeshita K, Tang M, Sugiura S, Sato SY, et al. Suppression of prostate cancer growth by resveratrol in the transgenic rat for adenocarcinoma of prostate (TRAP) model. *Asian Pacific J Cancer Prev* 2008;**9**(1):7–14.

18. Omenn GS, Goodman GE, Thornquist MD, Balmes J, Cullen MR, Glass A, et al. Effects of a combination of beta carotene and vitamin A on lung cancer and cardiovascular disease. *New England J Med* 1996;**334**(18):1150–5.

19. Neuhouser ML, Barnett MJ, Kristal AR, Ambrosone CB, Thornquist M, Goodman GG. Dietary supplement use and prostate cancer risk in the carotene and retinol efficacy trial. *Cancer Epidemiol Biomark Prev* 2009;**18**(8):2202–6.

20. Heinonen OP, Koss L, Albanes D, Taylor PR, Hartman AM, Edwards BK, et al. Prostate cancer and supplementation with α-tocopherol and β-carotene: incidence and mortality in a controlled trial. *J National Cancer Inst* 1998;**90**(6):440–6.

21. Cook NR, Stampfer MJ, Ma J, Manson JE, Sacks FM, Buring JE, et al. β-carotene supplementation for patients with low baseline levels and decreased risks of total and prostate carcinoma. *Cancer* 1999;**86**(9):1783–92.

22. Agarwal A, Shen H, Agarwal S, Rao AV. Lycopene content of tomato products: Its stability, bioavailability and in vivo antioxidant properties. *J Med Food* 2001;**4**(1):9–15.

23. Sahin K, Sahin N, Kucuk O. Lycopene and chemotherapy toxicity. *Nutr Cancer* 2010;**62**(27):988–95.

24. Etminan M, Takkouche B, Caamaño-Isorna F. The role of tomato products and lycopene in the prevention of prostate cancer: A meta-analysis of observational studies. *Cancer Epidemiol Biomark Prev* 2004;**13**:340–5.

25. Agarwal S, Rao AV. Tomato lycopene and its role in human health and chronic diseases. *Can Assoc J* 2000;**163**(6):739–44.

26. Giovannucci E. A review of epidemiologic studies of tomato, lycopene, and prostate cancer. *Exp Biol Med* 2002;**227**(10):852–9.

27. Goodman M, Bostick RM, Ward KC, Terry PD, Van Gils CH, Taylor JA, et al. Lycopene intake and prostate cancer risk: effect modification by plasma antioxidants and the XRCC1 genotype. *Nutr Cancer* 2006;**55**(1):13–20.

28. Kristal AR, Till C, Platz EA, Song X, King IB, Neuhouser ML, et al. Serum lycopene concentration and prostate cancer risk: Results from the prostate cancer prevention trial. *Cancer Epidemiol Biomark Prev* 2011;**20**(4):638–46.

29. Johary A, Jain V, Misra S. Role of lycopene in the prevention of cancer. International Journal of Nutrition, Pharmacology. *Neurological Diseases* 2012;**2**(3):167–70.

30. Li H, Stampfer MJ, Hollis JB, Mucci LA, Gaziano JM, Hunter D, et al. A prospective study of plasma vitamin D metabolites, vitamin D receptor polymorphisms, and prostate cancer. *PLOS Medicine* 2007;**4**(3):e103. http://dx.doi.org/10.1371/journal.pmed.0040103.

31. Corder EH, Guess HA, Hulka BS, Friedman GD, Sadler M, Vollmer RT, et al. Vitamin D and prostate cancer: A prediagnostic study with stored sera. *Cancer Epidemiol Biomark Prev* 1993;**2**:467–72.

32. Platz EA, Leitzmann MF, Hollis BW, Willett WC, Giovannucci E. Plasma 1,25-dihydroxy- and 25-hydroxyvitamin D subsequent risk of prostate cancer. *Cancer Causes & Control* 2004;**15**(3):255–65.

33. Ahn J, Peters U, Albanes D, Purdue MP, Abnet CC, Chatterjee N, et al. Serum vitamin D concentration and prostate cancer risk: A nested case-control study. *J National Cancer Inst* 2008;**100**(11):796–804.

34. Garland CF, Gorham ED, Mohr SB, Garland FC. Vitamin D for cancer prevention: Global perspective. *Ann Epidemiol* 2009;**19**(7):468–83.

35. Jain MG, Hislop GT, Howe GR, Ghadirian P. Plant foods, antioxidants, and prostate cancer risk: findings from case-control studies in Canada. *Nutr Cancer* 1999;**34**(2):173–84.

36. Lippman SM, Klein EA, Goodman PJ, Lucia MS, Thompson IM, Ford LG, et al. Effect of selenium and vitamin E on risk of prostate cancer and other cancers. *J Am Med Assoc* 2009;**301**(1):39–51.

37. Klein EA, Thompson IM, Tangen CM, Crowley JJ, Lucia MS, Goodman PJ, et al. Vitamin E and the Risk of Prostate Cancer: The Selenium and Vitamin E Cancer Prevention Trial (SELECT). *J Am Med Assoc* 2011;**306**(14):1549–56.

38. Gohagan JK, Prorok PC, Hayes RB, Kramer BS, PLCO Project Team. The Prostate, Lung, Colorectal and Ovarian (PLCO) Cancer Screening Trial of the National Cancer Institute: History, organization, and status. *Cont Clin Trials* 2000;**21**(6):251S–72S.

39. Bhatia-Gaur R, Donjacour AA, Sciavolino PJ, Kim M, Desai N, Young P, et al. Roles for Nkx3.1 in prostate development and cancer. *Genes Dev* 1999;**13**(8):966–77.

40. Fukumura H, Sato M, Kezuka K, Sato I, Feng X, Okumura S, et al. Effect of ascorbic acid on reactive oxygen species production in chemotherapy and hyperthermia in prostate cancer cells. *J Physiol Sci* 2012;**62**(3):251–7.

41. Vissers MC, Gunningham SP, Morrison MJ, Dachs GU, Currie MJ. Modulation of hypox inducible factor 1 alpha in cultured primary cells by intracellular ascorbate. *Free Radic Biol Med* 2007;**42**(6):765–72.

42. Knowles HJ, Raval RR, Harris AL, Ratcliffe PJ. Effect of ascorbate on the activity of hypoxia inducible factor in cancer cells. *Cancer Res* 2003;**63**(8):1764–8.

43. American Cancer Society. Prostate cancer: Early detection. *Am Cancer Soc* 2012.

18

Curcumin Analogs, Oxidative Stress, and Prostate Cancer

Alexandra M. Fajardo

University of South Alabama Mitchell Cancer Institute, Mobile, AL, USA

Marco Bisoffi

Chapman University Schmid College of Science and Technology, Biological Sciences, Biochemistry and Molecular Biology, Orange, CA, USA and University of New Mexico Health Sciences Center, Department of Biochemistry and Molecular Biology, School of Medicine Albuquerque, NM, USA

List of Abbreviations

AhR Arylhydrocarbon receptor
AIF Apoptosis inducing factor
ANT Adenine nucleotide translaocase
AP-1 Activated protein 1
AR Androgen receptor
ARE Antioxidant response element
ARNT AhR nuclear translocator
CRPC Castration resistant prostate cancer
CYP Cytochrome P450
Δψ$_m$ Mitochondrial membrane potential
DHT Dihydrotestosterone
Endo G Endonuclease G
ERK Extracellular regulated kinase
GCL Glutamate cysteine ligase
GP Glutathione peroxidase
GR Glutathione reductase
GSH Reduced glutathione
GS-SG Oxidized glutathione
GST Glutathione *S*-transferases
H$_2$O$_2$ Hydrogen peroxide
HO• Hydroxyl radical
HO-1 Heme oxygenase-1
IM Mitochondrial inner membrane
Keap1 Kelch-like ECH-associated protein 1
mPTPC Mitochondrial permeability transition pore complex
NAD(P)H Nicotinamide adenine dinucleotide (phosphate) H
NCI National Cancer Institute
NFκB Nuclear factor kappa B
NQO1 NAD(P)H:quinone oxidoreductase-1
Nrf2 Nuclear factor erythroid 2 related factor 2
O$_2$•$^-$ Superoxide radical
OH Hydroxyl
OM Mitochondrial outer membrane
PIN Prostatic intraepithelial neoplasia
PKC Protein kinase C
ROS Reactive oxygen species

SAR Structure activity relationship
SH Sulfhydryl
SOD Superoxide dismutase
TCDD 2,3,7,8-tetrachlorodibenzo-p-dioxin
TR Thioredoxin reductase
TRAMP Transgenic adenocarcinoma of the mouse prostate
VDAC Voltage dependent ion channel

INTRODUCTION

This chapter covers three main subjects:

- The plant natural product curcumin (diferuloylmethane) and its naturally occurring curcuminoids are present as major polyphenolic constituents in the popular spice turmeric, which is gained from the roots of the plant *Curcuma longa (Linn.)* and other species that are members of the ginger family *Zingiberaceae*, a native plant throughout Southeast Asia.[1] Curcumin, curcuminoids, and chemical analogs of curcumin, which were synthesized by combinatorial chemistry based on the structural components of the original chemical structure, featuring a "combichemical" design,[2-4] will be discussed. Throughout this chapter, the word "analog" will be used instead of "derivative"; although the two are largely overlapping in meaning, "analog" better accommodates the *de novo* synthesis of new structures.
- The effect of natural curcuminoids and chemical analogs of curcumin on the redox status of human cells, taking into consideration the dual capacity to

© 2014 Elsevier Inc. All rights reserved.

act as both anti- and pro-oxidant agents, depending on the chemical nature, dose, and the cellular and molecular context in which they are investigated. While such a dual role, often dependent of their concentration (hormetics), has long been reported for many agents, it has only been relatively recently emphasized for curcumin and related curcuminoids.[5]

- The role of oxidative stress in prostate cancer as it pertains to its contribution to the malignant process underlying the transformation of normal epithelial cells into premalignant precursor of prostate intraepithelial neoplasia (PIN) and the further progression into invasive and metastatic cells. This chapter will focus mainly on reactive oxygen species and will not cover reactive nitrogen species that can lead to nitrosative stress.[6,7]

These individual subjects represent rather large bodies of scientific research reflected in a high number of published studies. Their full coverage would thus exceed the scope of this chapter and the reader is referred to excellent previously published reports reviewing these topics.[1,3-7] However, the combination of the three main subjects, which is the effect of naturally occurring and synthetic agents with curcuminoid structure on the redox status and on the cellular response of prostate cancer cells, is to date a remarkably unexplored field of biomedical investigation. In fact, much of the knowledge about how curcumin and its analogs influence the redox balance in prostate cancer cells is implied, often from data generated in other cell systems, rather than proven.

The scope of this chapter is to explore future avenues and possibilities of utilizing the oxidant properties of curcuminoid agents in the prevention of and fight against the second most lethal malignancy in men in Western countries, prostate cancer.[8-10] This is achieved by exploring the available published information, specific for prostate cancer, on the effects of curcuminoids on components of the cellular machinery responsible for the balanced redox status. To achieve this, the chapter will:

- Briefly introduce the subject of oxidative stress, in particular its factors and pathways as they pertain to prostate cancer, and review the chemistry of curcumin and its analogs, as well as explore the mechanisms of action underlying their anti- and pro-oxidant activities
- Discuss the current knowledge of the effect of curcuminoid agents in prostate cancer cells relative to the mechanisms of action
- Summarize the current status of knowledge of curcuminoid agents and their ability to influence the redox status in prostate cancer cells, and their possible clinical use in the future

PROSTATE CANCER AND OXIDATIVE STRESS

Prostate Cancer—A Brief Introduction

Prostate cancer, or prostatic adenocarcinoma, is a hormonally driven and dependent malignancy of the glandular structures primarily within the peripheral zone of the prostate, which is an organ that contributes to the optimal composition of the ejaculate fluid.[11] Prostate cancer can develop through well-defined precursors including PIN to become a highly invasive and metastatic disease. In addition, prostate cancer cells initially thrive on the relatively high concentrations of androgen (testosterone, dihydrotestosterone [DHT]) produced by the testicles and adrenal glands. This constitutes the basis of androgen ablation therapy, which aims at reducing the androgen concentration systemically and locally and/or blocking the androgen receptor (AR); the latter is a pro-proliferative signaling factor in prostate cancer cells and its blockage leads to cell death.[12] Unfortunately, progression of prostate cancer cells to a more aggressive and metastatic phenotype is possible through a number of pathways, including the androgen-independent activation of the AR through gene amplification and mutation as well as intratumoral synthesis of testosterone.[12,13] This constitutes castration-resistant prostate cancer (CRPC), which often displays simultaneous metastatic capabilities with a propensity for the bone microenvironment causing much morbidity.[14] At this stage, therapeutic options are limited and the disease has become incurable. Of the approximately 900,000 new cases diagnosed annually worldwide, about a third develop this advanced phenotype and results in the estimated annual mortality of 260,000 cases, making this malignancy the sixth most common cause of death from cancer in men.[9]

Prostate Cancer and Oxidative Stress—Possible Factors

Denham Harman introduced his "free radical theory of aging" in the 1950s, according to which the aging process is based on the cumulative production of free radicals over time that affect cellular lifespan by damaging essential biomolecules including DNA, lipids, and proteins.[15] Since that time, the theory was expanded to age-related diseases, including prostate cancer, which tends to develop primarily in men over the age of 55.[6,7] Based on epidemiological and experimental data using cell models and human tissues, it is now widely accepted that prostatic tissue, as it ages, changes from a neutral oxidant to a pro-oxidant state, and that components of the diet, including those that affect cell oxidant state play an important role in influencing this process (Figure 18.1).

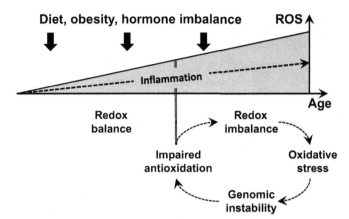

FIGURE 18.1 **Oxidative Stress in Prostate Cancer.** The level of reactive oxygen species (ROS) and inflammation increase with age, which is influenced by diet, obesity, and hormone imbalance. Redox balance changes to redox imbalance, which induces a vicious cycle of oxidative stress leading to genomic instability, impaired antioxidant mechanisms, and further redox imbalance.

Because of the effect of free radicals on pathways that are essential for cellular transformation, it is conceivable that the change toward a pro-oxidant state is linked to the onset of cancer. Many etiologic factors potentially responsible for increasing oxidative stress in prostatic tissues are under active investigation, including diet, obesity, hormone imbalance, and chronic inflammation.[16,17] Once cellular transformation has been initiated, persistent oxidative stress continues to support cancer progression through acquired genomic mutations that lead to the failure of antioxidant defense mechanisms, thereby constituting a vicious cycle of pro-oxidant activity [6,7,18] (Figure 18.1).

Reactive Oxygen Species—A Paradox in (Prostate) Cancer

Aerobic cells have been referred to as "the antioxidant machine" in Nick Lane's book, *Oxygen, the Molecule That Made the World*.[19] Although oxygen can inflict potential harm on cells, it is absolutely necessary for aerobic life. Oxidative stress is induced by deviations from the tightly controlled redox balance, as defined by the levels of reactive oxygen species (ROS) and enzymatic antioxidant activities.[20] These players will be introduced in the next sections in the context of curcuminoids and prostate cancer.

While research on cellular redox balance was initiated with a focus on the deleterious effects of ROS on cellular components and ultimately cell viability, subsequent studies have revealed that ROS are important second messengers that support signal transduction pathways implicated in normal cell function. ROS have thus been called "a double-edged sword."[18,21,22] This concept of a double-edged sword is of particular interest when

considering a change in redox balance as a therapeutic intervention against cancer, and leading some authors to include Shakespearean *innuendos* in the title of their reports, such as: "Cancer cell killing *via* ROS—To increase or decrease, that is the question."[22] This question is justified and applies perfectly to the subject of this chapter, which is the use of curcuminoids against prostate cancer, because curcumin and its analogs can be potent anti- as well as pro-oxidant agents. How does this dual capacity affect their use as potential anticancer agents? To answer this question, the widely accepted "ROS threshold concept" (Figure 18.2) should be considered.[18,21,22]

As outlined earlier, prostate cancer cells are characterized by higher ROS levels when compared to their normal counterparts.[6,7] According to the ROS threshold concept, cancer cells are adapted to a higher level of ROS, which they exploit for elevated survival and proliferation. However, cancer cells are also more sensitive to any further elevation in ROS levels and more prone to cell death because of it. It is conceivable then that antioxidant intervention (e.g., by curcuminoid agents) would decrease ROS levels in normal cells, which would compensate by down-regulating scavenging agents such as glutathione, thereby maintaining redox balance, while cancer cells would be inhibited by the diminished ROS levels. In contrast, normal cells would compensate a pro-oxidant intervention and increased ROS levels by upregulating the scavenging defense system to maintain redox balance, while cancer cells would be pushed toward cell death due to overwhelming the antioxidant system (Figure 18.2). Whether this is a viable antiprostate cancer scheme for the use of bifunctional oxidants such as the xenobiotic curcuminoid agents remains to be shown.

The next section will briefly review the dual function of curcumin and its analogs with respect to the chemistry that underlies their anti- and pro-oxidant activities.

CURCUMIN, CURCUMINOIDS, AND CURCUMIN ANALOGS

Chemistry and Biochemistry of Curcumin, Curcuminoids, and Curcumin Analogs

Curcumin (1*E*,6*E*)-1,7-bis(4-hydroxy-3-methoxyphenyl)-1,6-heptadiene-3,5-dione), or diferuloylmethane, is the principal curcuminoid component of the widely used spice turmeric, which is isolated from the roots of the plant *Curcuma longa*, a member of the ginger family *Zingiberaceae*, native throughout Southeast Asia.[1] The curcuminoids are natural phenols, which exist in several tautomeric forms, including a keto and an energetically more stable enol form[2,3] (Figure 18.3). Curcumin incorporates several functional groups: The aromatic phenol ring systems are connected by

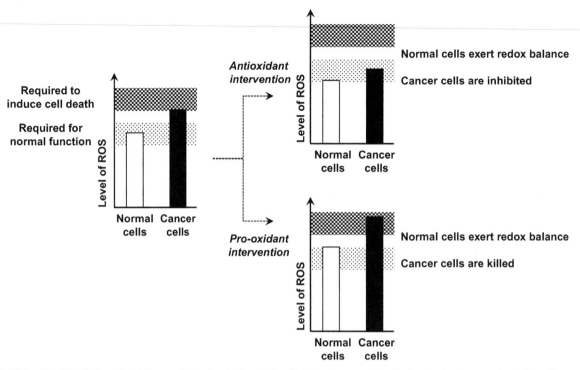

FIGURE 18.2   The "ROS Threshold Concept" for the Anti- and Pro-Oxidant Intervention Redox-Active Agents, Including Curcuminoids.
Bars represent the levels of ROS in normal (white) and cancer (black) cells. The light and dark shadowed areas represent ROS levels required for
normal function and for the induction of cell death, respectively. Antioxidant intervention would inhibit cancer cells, while pro-oxidant intervention would kill cancer cells. Normal cells would compensate both situations and maintain redox balance.

two α,β-unsaturated carbonyl (1,3-ketoenolyl) groups.
These groups are good Michael acceptors and readily
undergo nucleophilic addition by forming irreversible
adducts with the sulfhydryl (SH) group of multiple
target molecules such as glutathione.[23] The diketones
form stable enols and are readily deprotonated to form
enolates.[2-4]

The term "curcuminoid" relates to organic compounds with chemical moieties that are either identical
or similar to those found in curcumin itself. However,
"curcuminoid" is not a very well-defined term, making it difficult to identify the exact rules for inclusion
or exclusion of compounds into this group. In taking a
stringent view, this group includes other components
of turmeric, such as the di-, tetra-, and hexahydrocur-
cumins, the demethoxy- and bisdemethoxycurcumins,
the curcuminols, and cyclocurcumin (Figure 18.3). When
taking a less stringent view, a number of naturally occurring curcuminoids can be included, such as gingerol,
capsaicin, and caffeic acid.[3]

The chemical structure of curcumin is being extensively exploited by combinatorial chemistry and structure activity relationship (SAR) studies due to its relative
simplicity, its well-known bioactivity, and its safe pharmacological profile, including in prostate cancer model
systems.[1,3,8] A major driver of these efforts is its limited
bioavailability, which has greatly barred its rapid development into a clinically used preventive or therapeutic

agent. Curcumin is a hydrophobic compound and
although its physicochemical parameters are not unfa-
vorable for drug development purposes, it has been
shown to be poorly absorbed.[24] Oral and intraperitoneal
administration indicate that serum concentrations are
particularly low and often do not reach more than 0.1%
of intake, followed by rapid clearance in the feces and
in bile/urine,[24] and metabolization to the less bioactive
curcumin glucoronide, curcumin sulphate, and ferulic
acid.[3,24]

There are many reports on the generation of chemical libraries using curcumin as a starting structure.[2-4]
Although detailed information on their bioavailability is largely missing, whenever a more potent effect is
observed, it could be due to enhanced absorption and
distribution, longer retention, and/or decreased systemic secretion. Human data pertaining to the bioavailability of curcumin analogs is currently unavailable,
since these analogs are not yet in clinical trials.

To be able to understand the oxidant actions of agents
with curcuminoid character with respect to prostate cancer, it is necessary to explore the several mechanisms
of action by which curcuminoids can affect the cellular
redox status. Interestingly, curcuminoids exert both anti-
and pro-oxidant effects. This is not entirely surprising,
as an initial pro-oxidant effect by curcumin can lead
to a secondary antioxidant response. Hence, curcumin
has recently been discussed as a potential hormetic

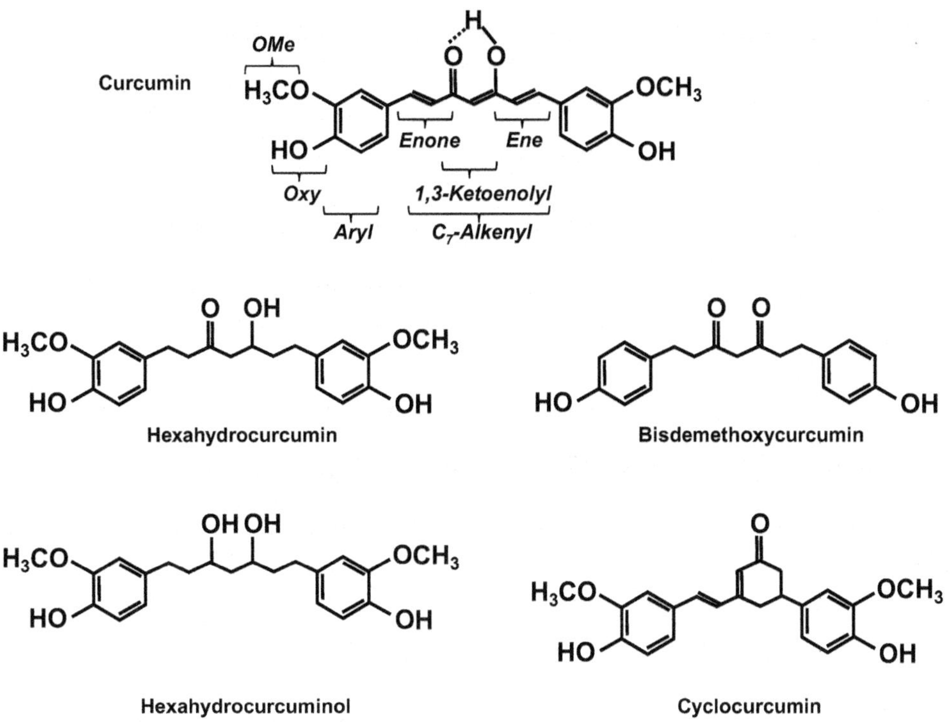

FIGURE 18.3 **The Chemical Structure and Individual Moieties of the Natural Product Curcumin.** All moieties are targets of combinatorial chemistry, including the methoxy, oxy, aryl, enone, ene, 1,3-ketoenolyl, and the C_7-alkenyl groups. Some of the naturally occurring curcuminoids isolated from the popular spice turmeric from the roots of the plant *Curcuma longa (Linn.)* and other species that are members of the ginger family *Zingiberaceae* are shown.

compound.[5] While these effects have not all been shown in prostate cancer cells, they can be reasonably implied by extension, as the targets described in other systems are also present in cells of prostatic origin.

Antioxidant versus Pro-Oxidant Activities of Curcumin, Curcuminoids, and Curcumin Analogs

Antioxidant Activities—Mechanisms

Curcumin, curcuminoids, and the chemical structures described in its various analogs have been recognized as antioxidant agents. This capability is thought to be at the very heart of the chemopreventive activity of curcumin, as postulated from epidemiological data in population studies.[1] In this capacity, it is assumed that long-term exposure to dietary curcumin would counteract the age-dependent increase of oxidative stress in prostate tissue (Figure 18.1).

Different mechanisms of action responsible for the antioxidant activities of curcumin and agents with curcuminoid structure have been identified and further investigated. Generally speaking, there are two major mechanisms, nonenzymatic (direct) and enzymatic (indirect; Figure 18.4). The nonenzymatic mechanisms are due to the scavenging capability of these molecules. Accordingly, curcuminoids have been shown to scavenge free radicals and ROS, including hydroxyl radicals (HO•), superoxide radicals (O_2•[-]), singlet oxygen, and peroxyl radicals. This reaction could occur after the following scheme:

(i) R-OO• + curcumin ⟷ ROOH + curcumin•
(reversible step)
(ii) Curcumin• + X• → nonradical curcuminoid products (irreversible step)

where R-OO• is the radical, ROOH is the oxidized reactant, and X• is another radical species (after[25]).

General conclusions from SAR studies can be made for the structural requirements of curcumin to act as a direct antioxidant. The hydrogen bonding interaction between the phenolic hydroxyl and the ortho-positioned methoxy groups in curcumin markedly influences the O-H bond energy and H-atom abstraction by free radicals. Changes in the composition and distribution of these moieties affect scavenging potential, as shown for bisdemethoxycurcumin and demethoxycurcumin.[26] In general, the phenolic hydroxyl groups are needed for antioxidant activity and the presence of more than one of these groups, as for example in the curcumin derivative bis(3,4-dihydroxycinnamoyl)-methane, confers enhanced activity.[27] The position of the hydroxyl groups also affects the antioxidant potential, with position 2 yielding enhanced antioxidant activity (e.g., in bis(2-hydroxycinnamoyl)-methane), and similar findings were reported for curcumin analogs with shortened (C-5) carbon linkers, with additional hydroxyl substituents on the phenyl rings, and in structures carrying 3-alkoxy-4-hydroxyphenyl units.[28] The β-diketone moiety, which is part of the Michael acceptor reactivity of curcuminoids (see later), was also shown to be involved in the antioxidant effect (e.g., in dimethyltetrahydrocurcumin).[29] In addition, the presence of an alkoxy group

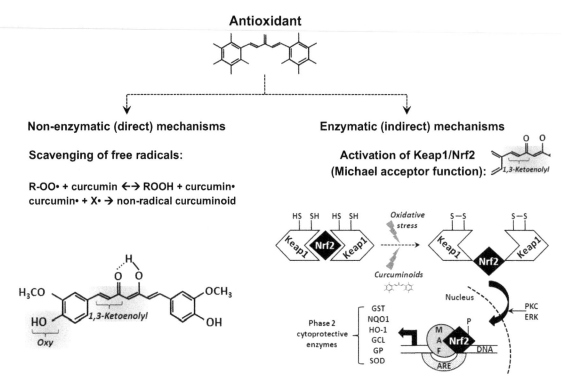

FIGURE 18.4 Possible Mechanisms of Antioxidant Activities of Curcuminoids. Left: Nonenzymatic (direct) mechanisms: Scavenging abilities of free radicals, e.g., superoxide radical $O_2\bullet^-$ and the hydroxyl radical HO•. R-OO• = the radical, ROOH = the oxidized reactant, X• = another radical species. The responsible structural moieties (hydroxyl and α,β-unsaturated carbonyl [1,3-ketoenolyl] groups) are shaded. Right: Enzymatic (indirect) mechanisms: Induction of phase 2 cytoprotective enzymes, e.g., glutathione transferases (GST), NAD(P)H:quinone oxidoreductase-1 (NQO1), heme oxygenase-1 (HO-1), glutamate cysteine ligase (GCL), glutathione peroxidase (GP), and superoxide dismutase (SOD) via Michael acceptor (1,3-ketoenolyl) activation and nuclear translocation of the redox sensor nuclear factor erythroid 2 related factor 2/Kelch-like ECH-associated protein 1 (Nrf2)/Keap1). Bolting flash indicates point of attacks by curcuminoids. SH and S-S = reduced and oxidized sulhydryl groups, respectively; ARE = antioxidant response elements; PKC = protein kinase C; ERK = extracellular regulated kinase.

in the ortho position relative to the hydroxyl group has been reported to potentiate the antioxidant activity.[30,31]

Curcumin and analogs with curcuminoid character may also act as antioxidants by enzymatic (indirect) mechanisms. The best studied is the effect on the nuclear factor erythroid 2 related factor 2/Kelch-like ECH-associated protein 1 (Nrf2)/Keap1) pathway, which regulates the expression of downstream phase 2 cytoprotective (detoxification) proteins by binding to antioxidant response elements (ARE) in their promoters (Figure 18.4). In this scheme, Keap1 acts as a major sensor for the redox status within the cell through its cysteine residues. When these residues are in their reduced form, Keap1 is bound to the transcription factor Nrf2 in the cytoplasm, thereby sequestering it and inhibiting its nuclear translocation. When the cellular redox balance shifts toward a more pro-oxidative state, Keap1 cysteine residues are oxidized, which results in the dissociation of the two proteins, allowing Nrf2 to translocate into the nucleus. Upon dimerization with a number of different cotranscription factors, including Maf proteins, the resultant Nrf2 heterodimers bind to the ARE regulatory region of phase 2 cytoprotective genes, including glutathione S-transferase (GST), NAD(P)H:quinone

oxidoreductase-1 (NQO1), heme oxygenase 1 (HO-1), glutamate cysteine ligase (GCL), glutathione peroxidase (GP), and superoxide dismutase (SOD)[32] (Figure 18.4).

A xenobiotic agent with curcuminoid structure could induce the Nrf2/Keap1 pathway *via* both its anti-, as well as its pro-oxidant capability. The antioxidant mechanism is based on the Michael acceptor capability, as shown in excellent works by Albena Dinkova-Kostova and Paul Talalay.[33,34] Accordingly, the α,β-unsaturated 1,3-diketone moiety of curcumin is a very attractive site for nucleophilic attacks by sulfhydryl groups. Keap1 carries a number of cysteine residues that are prone to be directly oxidized by ROS or by xenobiotics with Michael acceptor capability. It is thus conceivable to assume that the α,β-unsaturated 1,3-diketone moiety of curcumin and its analogs are able to oxidize Keap1, disrupt its interaction with Nrf2, and trigger its transcriptional activity for phase 2 antioxidant genes (Figure 18.4). Validity for this concept was provided by a comprehensive SAR study of curcuminoids, as shown by the induction of the activities of the Nrf2 downstream targets NQO1 and GST.[35] This study indicated that hydroxyl groups at position 2 in the phenolic rings greatly enhanced the capability of curcuminoid structures containing α,β-unsaturated

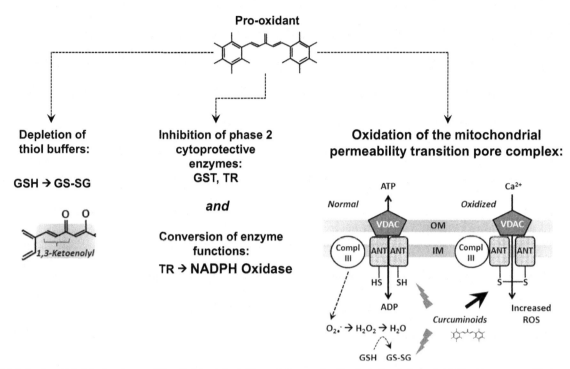

FIGURE 18.5 Possible Mechanisms of Pro-Oxidant Activities of Curcuminoids. Left: Depletion of thiol buffers, e.g., oxidation of reduced glutathione (GSH) to oxidized glutathione (GS-SG) via Michael acceptor (1,3-ketoenolyl) action. Middle: Inhibition of phase 2 cytoprotective enzymes, e.g., glutathione transferase (GST) and thioredoxin reductase (TR) via unknown mechanisms; and conversion of TR to NADPH oxidase by alkylation of TR cysteines and selenocyteines. Right: Oxidation of the mitochondrial permeability transition pore complex. Bolting flash indicates point of attacks by curcuminoids. ANT = adenine nucleotide translocase; VDAC = voltage dependent ion channel; OM and IM = mitochondrial outer and inner membrane, respectively; Compl III = complex III; SH and S-S = reduced and oxidized sulfhydryl groups, respectively.

carbonyl groups with sulfhydryl groups to induce phase 2 cytoprotective proteins.[35] Nrf2 activation also involves its phosphorylation by several different kinases, for example protein kinase C (PKC) and extracellular signal-regulated kinase (ERK; Figure 18.4). These kinases have been shown to be redox sensitive and directly activated by high levels of ROS.[36,37]

Pro-oxidant Activities—Mechanisms

The same redox capacity of curcumin and agents with curcuminoid moieties discussed earlier under the anti-oxidant effects can be responsible for strong pro-oxidant activities (Figure 18.5).

Agents with curcuminoid character that feature α, β-unsaturated 1,3-diketone moieties can be strong Michael acceptors. They react with nucleophilic centers, which are characteristic for reduced glutathione and thioredoxin, two of the most prominent members of the cellular reservoir of natural thiols. It is thus conceivable that curcumin and agents with curcumin character rapidly deplete these thiols and shift the redox balance toward a pro-oxidant state. An additional effect of agents able to bind glutathione is its cellular extrusion without affecting the redox status of the cell. This has been termed "nonoxidative loss of glutathione" and was shown for several cytotoxic and chemotherapeutic agents.[38]

An example of the diverse and often contradictory actions of curcumin and its analogs is the inhibitory actions on antioxidant phase 2 cytoprotective enzymes that are often induced by these agents (see earlier). This is the case for the thioredoxin system, such as the inhibition of thioredoxin reductase. This enzyme is crucial in the replenishing of reduced thioredoxin as an important cellular thiol buffer. Accordingly, curcumin and several curcumin analogs featuring furan moieties have been shown to convert thioredoxin reductase from an antioxidant to a pro-oxidant enzyme, changing it to a NADPH oxidase that transfers electrons from NADPH to oxygen, by alkylating cysteines and seleno-cyteines in the catalytically active sites of the enzyme.[39] Independent of these actions of curcuminoids on thio-redoxin reductase activity, curcumin can also significantly down-regulate thioredoxin reductase expression by as yet unknown mechanisms.[40] Similarly, curcumin and several structurally different curcumin analogs are able to inhibit GST expression in the low micromolar range, which is one of the key enzymes in the replenishing function of glutathione as an important cellular thiol buffer. These studies were conducted in different cell systems and indicated that GST inhibition may involve the upstream inhibition of the binding of transcription factors activated protein 1 (AP-1) and nuclear factor kappa B (NFκB) to recognition sites located on the GSTP1-1 gene promoter.[41]

Curcumin and several of its analogs can exert their cell death via mitochondrial processes in conjunction with ROS production.[8,10] This is not surprising, as the mitochondria are the primary source of cellular ROS production. However, the exact mechanism of action of xenobiotic agents with curcuminoid character within the mitochondria remains largely unexplained and may differ greatly in different cell systems. An intriguing possibility is that they affect the mitochondrial permeability transition pore complex (mPTPC). The mPTPC is a multicomponent channel spanning the inner and outer mitochondrial membrane. Essential components are the adenine nucleotide translocase (ANT) located in the inner membrane and the voltage-dependent ion channel (VDAC) located in the outer membrane.[42] The mPTPC is responsible for the mitochondrial ADP import and ATP export, and for the exchange of metabolites between the mitochondrial matrix and the cytosol. The mPTPC component ANT is also a sensor for ROS. It carries redox-sensitive cysteine residues that are normally reduced. In this state, ANT binds nucleotides and function as an ATP/ADP translocase. When these cysteine residues are oxidized, ANT undergoes a conformational change resulting in its inability to bind nucleotides and allowing calcium (Ca^{2+}) to flux into the mitochondria where it induces additional pore opening.[42] This mitochondrial permeability transition leads to changes in the mitochondrial membrane potential ($\Delta\psi_m$), outer membrane permeabilization, release of apoptotic proteins, including cytochrome C, apoptosis-inducing factor (AIF), and endonuclease G (Endo G), and ultimately cell death. Intramitochondrial ROS levels are kept in balance by the cellular redox systems. Accordingly, the superoxide radical $O_2{}^{\bullet-}$ produced by complex III of the electron transport chain, is converted to hydrogen peroxide (H_2O_2) by SOD and further neutralized by catalase or GP to water thereby oxidizing glutathione. The cysteine residues of ANT are kept reduced by the thioredoxin/thioredoxin reductase and/or the glutathione/glutathione reductase systems.

Several mechanisms by which curcumin and curcumin analogs could interfere with the normal function of the mPTPC can be hypothesized (Figure 18.5). First, the Michael acceptor reactivity could deplete the natural thiol pool such as glutathione and thioredoxin, leading to elevated ROS, oxidized ANT, pore opening, changes in the membrane potential $\Delta\psi_m$, and membrane permeabilization. Second, it is not inconceivable that curcuminoids could directly interfere with ANT to oxidize the redox-sensitive cysteine residues. This seems to be in agreement with earlier studies that suggested that curcumin affects the mPTPC by the involvement of "oxidation of membrane thiol functions."[43] Third, curcumin

and fluorinated derivatives have been shown to reduce iron (Fe^{3+}) to Fe^{2+}, which converts H_2O_2 to hydroxyl anion OH^- and hydroxyl radical $HO\bullet$ in a Fenton reaction, leading to the oxidation of ANT.[30,44] Remarkably, in this pro-oxidant scheme, curcumin acts as an antioxidant, which represents a perfect example of the intricate interdependence of the two systems.

Most of these antioxidant and pro-oxidant mechanisms were not elucidated in prostate cancer cells. However, it is conceivable that they would be at work in cells of prostatic or uroepithelial provenance. The next section will elucidate the current knowledge on the oxidant actions of curcuminoid agents in cells of prostatic origin.

THE POTENTIAL OF CURCUMIN, CURCUMINOIDS, AND CURCUMIN ANALOGS AS OXIDANT AGENTS IN PROSTATE CANCER

Molecular Targets of Curcumin, Curcuminoids, and Curcumin Analogs in Prostate Cancer

Prostate cancer is characterized by a myriad of molecular changes, some or even the majority of which may well be just consequential bystanders.[11,45] However, despite this heterogeneous background of deregulated molecular pathways, some of them have been reported to be strongly causative for the development and progression of disease, driving the development from normal epithelial cells to the well-characterized precursor PIN, and on to invasive and metastasizing carcinoma. In fact, some pathways have been recognized and proven to be necessary for the survival of prostate cancer cells; these constitute "oncogenic addiction."[12] An imbalance in the pathways responsible for the maintenance of the cellular redox status may fuel the development of such oncogenic addiction.

Backed by extensive epidemiological and population data, curcumin is widely accepted to be a dietary supplement with chemopreventive efficacy against including prostate cancer.[8] In a substantial number of in vitro and in vivo experimental studies, curcumin and its various analogs have proven to be strongly pleiotropic agents with a plethora of reported targets of importance to the biology of prostate cancer cells.[8] In fact, these targets include oncogenic pathways known to be essential for cell survival, proliferation, and resistance to therapeutic intervention. Interference with these pathways has been shown to lead to cell death of prostate cancer cells in vitro and to the inhibition of tumor formation in vivo.[10] This section will focus on the cellular and molecular targets that are directly related to the redox machinery of the cell. There is direct evidence that curcumin affects the transcriptional regulation of genes involved in oxidative stress in prostate cancer cells.[46] However, although curcumin and agents with curcuminoid character are proven effectors of the cellular redox status, this particular aspect of their bioactivity remains to date remarkably unexplored in the specific context of prostate cancer. The current information regarding the anti- and pro-oxidant actions of curcuminoids as they relate to cells of prostatic origin will be reviewed. In the cases where the exact mechanisms are not known, the mechanisms outlined in the previous section can be implied.

Curcumin, Curcuminoids, and Curcumin Analogs as Antioxidants in Prostate Cancer Cells

The published literature is rich in reports on the use of natural products with curcuminoid character with efficacies against prostate cancer cells ([8] and references therein). It is also well established that curcumin and by extrapolation, curcumin analogs, are potent antioxidants ([47]and references therein). However, the term antioxidant is often used as an accompanying adjective when the effect of curcumin and curcuminoids are experimentally investigated. In many of these studies antioxidant mechanisms are assumed as opposed to proven by experimentation. Surprisingly, there are relatively few studies focused on prostate cancer that include analyses specific to the antioxidant capabilities of curcuminoid compounds and to the components of the cellular redox machinery. In addition, clinical trials using the pharmacologically safe natural product curcumin typically are not designed to specifically address its antioxidant capacities, as they tend to focus on more clinical end-points.[48] However, the antioxidant mechanisms outlined in the previous section (Figure 18.4) can be reasonably implied for cells of prostatic origin. An example for this is the ability of curcumin to inhibit the pro-proliferative effects of H_2O_2 in LNCaP cells due to its free radical scavenging activity.[49]

One of the strongest effects of curcumin and its analogs with respect to their antioxidant capability in cells of the prostate is the induction of phase 2 cytoprotective genes (see earlier). In this context, curcumin was shown to reactivate the master regulator of cellular antioxidant defense systems Nrf2 by restoring its expression *via* demethylation of CpG islands in its promoter, therefore acting as DNA hypomethylation agent. Nrf2 was previously shown to be epigenetically silenced during the progression of prostate tumorigenesis in transgenic adenocarcinoma of the mouse prostate (TRAMP) mice.[50]

Curcumin and Curcumin Analogs as Pro-Oxidants in Prostate Cancer

It is remarkable that most scientific reports on the oxidant effect of curcumin and its various analogues in prostate cancer cells related to their pro-oxidant, as opposed to their antioxidant capacity. This is surprising in the light of the fact that this polyphenol natural product has been and continues to be a parade example for antioxidant activity. In fact, curcumin and its naturally occurring and synthetic analogs are being increasingly recognized as pro-oxidant agents based on the mechanisms discussed in the previous section.

Adams and coworkers characterized a synthetic fluorinated curcumin analog, 3,5-Bis-(2-fluorobenzylidene)-4-piperidone (termed EF24; Figure 18.6) with respect to its capability to induce cell cycle arrest and apoptosis by means of a redox-dependent mechanism in DU-145 human prostate cancer cells.[51] EF24 was previously identified as a curcumin analog with higher antitumorigenic potential in the 60 cancer cell line test panel provided by the National Cancer Institute (NCI).[52] EF24 caused G2/M arrest followed by the induction of apoptosis, as shown by caspase-3 activation, phosphatidylserine externalization, and an increased number of cells with a sub-G1 DNA fraction. In addition, EF24 induced a depolarization of the mitochondrial membrane potential, indicating the involvement of mitochondrial deregulation and subsequent activation of apoptotic pathways. EF24, as well as an additional derivative (termed EF31) featuring nitrogen substitutions for the *ortho*-positioned fluorine groups, was later shown to act similarly in breast cancer cells when precombined with glutathione.[52] With respect to the mechanism(s) of action of these curcumin analogs in relation to oxidative stress, they were shown to induce ROS by depleting the pool of natural thiol buffers, that is, reduced glutathione (GSH) and thioredoxin via their Michael acceptor reactivity toward the sulfhydryl groups of GSH and thioredoxin.[51]

The capability of curcumin to act as a pro-oxidant agent in prostate cancer cells has been confirmed in different studies utilizing different cell models and schemes. For example, ROS induction by curcumin followed by changes in the mitochondrial membrane potential and cell death was shown in the widely used human prostate cancer cell line LNCaP.[53] Formulations of curcumin complexations with palladium and bipyridines induced growth inhibition and cell death in various prostate cancer cells, including LNCaP, DU-145, and PC-3, through an ROS dependent mechanism involving complex 1 and mitochondrial membrane depolarization.[54] This study also revealed another potential mechanism of curcuminoids, the down-regulation of the important phase 2 cytoprotective enzyme GSTp1[54], which led to a shift toward a pro-oxidant cellular status.

An important consequence of ROS induction by curcumin has been shown to be the selective down-regulation of proteins that are implicated in prostate tumorigenesis and cancer progression. A prominent example for this action was shown in the context of

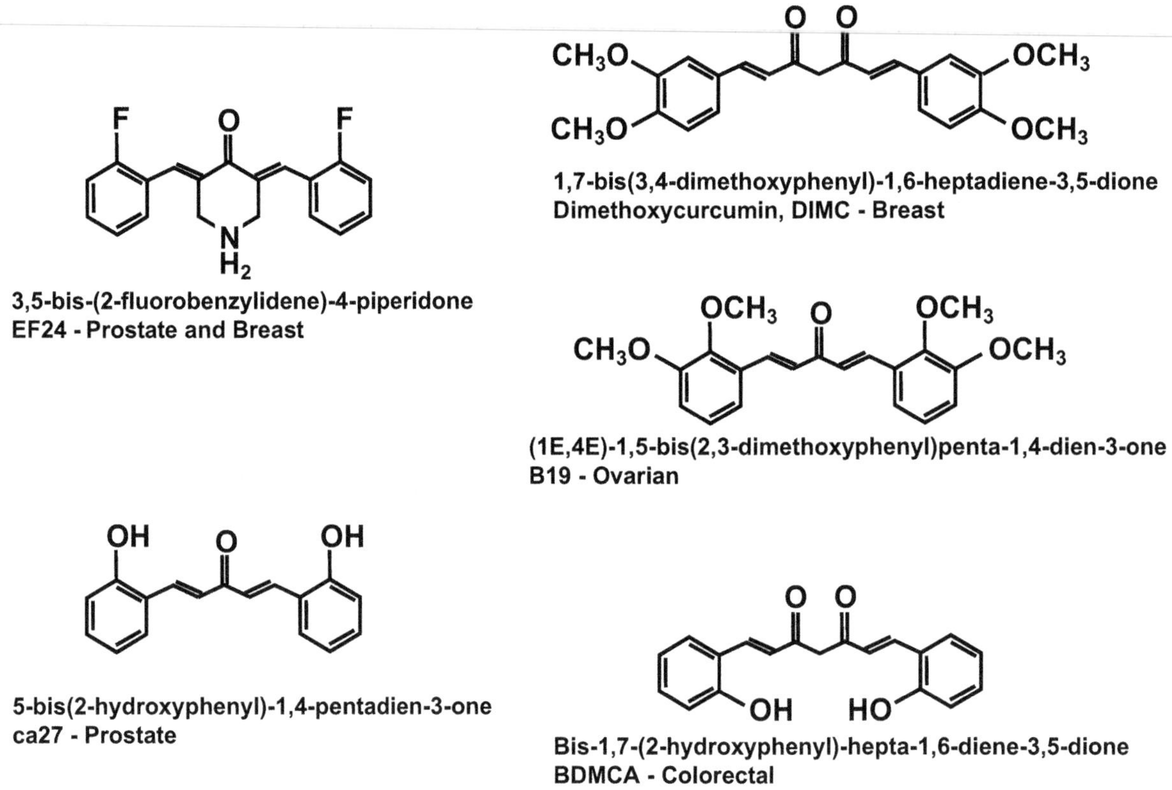

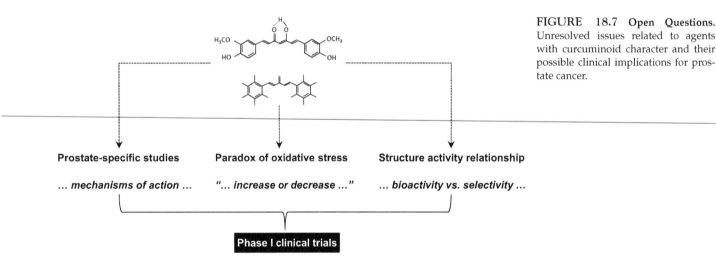

FIGURE 18.6 **Synthetic Curcumin Analogs That Exert Effects Via a Proven Mechanism of Action That Includes Oxidant Activities.** EF24 and ca27 are to date the only two examples of synthetic curcuminoids that exert cytotoxic effects in prostate cancer cells *via* pro-oxidation.[51,56] EF24, DIMC, and B19 are pro-oxidant curcumin analogs in breast and ovarian cancer cells;[52,58,59]. BDMCA is an antioxidant curcumin analog in colorectal cancer cells.[60]

FIGURE 18.7 Open Questions. Unresolved issues related to agents with curcuminoid character and their possible clinical implications for prostate cancer.

the induction of cytochrome P450 enzymes by the toxic environmental contaminant 2,3,7,8-tetrachlorodibenzo-p-dioxin (TCDD). TCDD mediated P450 induction occurs through the activation of the arylhydrocarbon receptor (AhR) in the cytoplasm, which translocates to the nucleus, dimerizes with the AhR nuclear translocator (ARNT), and drives the expression of CYP1A1 and 1B1. TCDD is known to mediate cell transformation and

carcinogenesis via the generation of genotoxic metabolites. Accordingly, curcumin was shown to down-regulate nuclear AhR and ARNT in normal human embryonic kidney cells and normal prostate cells and to inhibit malignant cell transformation. Of interest to this chapter, these actions were dependent on the concomitant induction of ROS leading to oxidative stress via an unidentified mechanism.[55] Another protein target of importance to prostate cancer development and progression is the androgen receptor (AR), which is the major target of the mainstay therapeutic intervention of androgen ablation.[11-13] The expression of the AR is retained and often increased during cancer progression. Therefore, AR protein down-regulation could be an effective therapeutic approach against prostate cancer. Fajardo and coworkers showed that the curcumin analog 5-Bis(2-hydroxyphenyl)-1,4-pentadien-3-one (termed ca27; Figure 18.6) down-regulates AR expression at low micromolar concentrations in several prostate cancer cell lines including the androgen-dependent LNCaP and LAPC4 and the androgen ablati on resistant C4-2 cells.[56] Pertinent to the link to oxidative stress, ca27-mediated AR down-regulation was rapidly induced via the generation of ROS, as the antioxidant N-acetyl cysteine was able to inhibit this effect. Interestingly, the prostate cancer cells exerted an antioxidant response to ca27 by inducing the expression of the cellular redox sensor Nrf2 and the phase 2 cytoprotective enzymes NQO1 and aldoketoreductase 1C1. The latter shows that pro-oxidant effects and antioxidant responses are intricately linked in the cellular redox status of a cell.[56] The effect of curcumin analogs on the expression of the AR has been extensively covered by other authors, notably by researchers associated with Kuo-Hsiung Lee.[57] While these authors presented a number of curcumin analogs with this interesting capacity, the role of oxidative stress was not specifically addressed.

SUMMARY POINTS

- The development and progression of prostate cancer is characterized by increasing oxidative stress due to elevated levels of free radicals and other reactive oxygen species.
- Both the antioxidant and the pro-oxidant capacity of curcumin and curcumin analogs can be exploited as therapeutic strategies against prostate cancer.
- More studies are necessary to determine the oxidant mechanisms of action of curcuminoid agents specifically in cells of prostate origin, in order to pursue their possible clinical implications for prostate cancer (Figure 18.7).
- Structural modification by combinatorial chemistry approaches, as featured in de novo synthesized chemical analogs, will in the future improve the bioavailability of agents with curcuminoid character (Figure 18.7).
- Structure analysis relationship studies of curcumin analogs are necessary to identify agents with a pharmacologically safe profile that discriminates between prostate cancer and normal cells, in order to facilitate the development of phase I clinical trials (Figure 18.7).

References

1. Gupta SC, Patchva S, Koh W, Aggarwal BB. Discovery of curcumin, a component of golden spice, and its miraculous biological activities. Clin Exp Pharmacol Physiol 2012;39(3):283–99.
2. Agrawal DK, Mishra PK. Curcumin and its analogues: Potential anticancer agents. Med Res Rev 2010;30(5):818–60.
3. Anand P, Thomas SG, Kunnumakkara AB, Sundaram C, Harikumar KB, Sung B, et al. Biological activities of curcumin and its analogues (Congeners) made by man and Mother Nature. Biochem Pharmacol 2008;76(11):1590–611.
4. Mosley CA, Liotta DC, Snyder JP. Highly active anticancer curcumin analogues. Adv Exp Med Biol 2007;595:77–103.
5. Speciale A, Chirafisi J, Saija A, Cimino F. Nutritional antioxidants and adaptive cell responses: An update. Curr Mol Med 2011;11(9):770–89.
6. Khandrika L, Kumar B, Koul S, Maroni P, Koul HK. Oxidative stress in prostate cancer. Cancer Lett 2009;282(2):125–36.
7. Gupta-Elera G, Garrett AR, Robison RA, O'Neill KL. The role of oxidative stress in prostate cancer. Eur J Cancer Prev 2012;21(2):155–62.
8. Aggarwal BB. Prostate cancer and curcumin: Add spice to your life. Cancer Biol Therapy 2008;7(9):1436–40.
9. Jemal A, Bray F, Center MM, Ferlay J, Ward E, Forman D. Global cancer statistics. CA Cancer J Clin 2011;61(2):69–90.
10. Khan N, Adhami VM, Mukhtar H. Apoptosis by dietary agents for prevention and treatment of prostate cancer. Endocr Relat Cancer 2010;17(1):R39–52.
11. Grossfeld GD, Carroll PR. Prostate Cancer. Hamilton, London: BC Decker Inc; 2001.
12. Knudsen KE, Scher HI. Starving the addiction: New opportunities for durable suppression of AR signaling in prostate cancer. Clin Cancer Res 2009;15(15):4792–8.
13. Amaral TM, Macedo D, Fernandes I, Costa L. Castration-resistant prostate cancer: Mechanisms, targets, and treatment. Prostate Cancer 2012;2012:327253.
14. Jin JK, Dayyani F, Gallick GE. Steps in prostate cancer progression that lead to bone metastasis. Int J Cancer 2011;128(11):2545–61.
15. Harman D. Aging: A theory based on free radical and radiation chemistry. J Gerontol 1956;11(3):298–300.
16. Fleshner NE, Klotz LH. Diet, androgens, oxidative stress and prostate cancer susceptibility. Cancer Metastasis Rev 1998;17(4):325–30.
17. Omabe M, Ezeani M. Infection, inflammation and prostate carcinogenesis. Infect Genet Evol 2011;11(6):1195–8.
18. Trachootham D, Alexandre J, Huang P. Targeting cancer cells by ROS-mediated mechanisms: A radical therapeutic approach? Nature Rev Drug Discov 2009;8(7):579–91.
19. Lane N. Oxygen, the Molecule That Made the World. Oxford, UK: Oxford University Press; 2002.
20. Sies H. Oxidative stress: Oxidants and antioxidants. Exp Physiol 1997;82(2):291–5.
21. Acharya A, Das I, Chandhok D, Saha T. Redox regulation in cancer: A double-edged sword with therapeutic potential. Oxidat Med Cell Longevity 2010;3(1):23–34.
22. Wang J, Yi J. Cancer cell killing via ROS: To increase or decrease, that is the question. Cancer Biol Therapy 2008;7(12):1875–84.
23. Awasthi S, Pandya U, Singhal SS, Lin JT, Thiviyanathan V, Seifert Jr WE, et al. Curcumin-glutathione interactions and the role of human glutathione S-transferase P1-1. Chem Biol Int 2000;128(1):19–38.
24. Anand P, Kunnumakkara AB, Newman RA, Aggarwal BB. Bioavailability of curcumin: Problems and promises. Molecular Pharm 2007;4(6):807–18.
25. Itokawa H, Shi Q, Akiyama T, Morris-Natschke SL, Lee KH. Recent advances in the investigation of curcuminoids. Chinese Med 2008;3:11.

26. Ahsan H, Parveen N, Khan NU, Hadi SM. Pro-oxidant, anti-oxidant and cleavage activities on DNA of curcumin and its derivatives demethoxycurcumin and bisdemethoxycurcumin. *Chem Biol Int* 1999;**121**(2):161–75.

27. Priyadarsini KI, Maity DK, Naik GH, Kumar MS, Unnikrishnan MK, Satav JG, et al. Role of phenolic O-H and methylene hydrogen on the free radical reactions and antioxidant activity of curcumin. *Free Radic Biol Med* 2003;**35**(5):475–84.

28. Weber WM, Hunsaker LA, Abcouwer SF, Deck LM, Vander Jagt DL. Anti-oxidant activities of curcumin and related enones. *Bioorg Med Chem* 2005;**13**(11):3811–20.

29. Sugiyama Y, Kawakishi S, Osawa T. Involvement of the beta-diketone moiety in the antioxidative mechanism of tetrahydrocurcumin. *Biochem Pharm* 1996;**52**(4):519–25.

30. Ligeret H, Barthelemy S, Zini R, Tillement JP, Labidalle S, Morin D. Effects of curcumin and curcumin derivatives on mitochondrial permeability transition pore. *Free Radic Biol Med* 2004;**36**(7):919–29.

31. Venkatesan P, Rao MN. Structure-activity relationships for the inhibition of lipid peroxidation and the scavenging of free radicals by synthetic symmetrical curcumin analogues. *J Pharm Pharmacol* 2000;**52**(9):1123–8.

32. Surh YJ, Kundu JK, Na HK. Nrf2 as a master redox switch in turning on the cellular signaling involved in the induction of cytoprotective genes by some chemopreventive phytochemicals. *Planta Med* 2008;**74**(13):1526–39.

33. Dinkova-Kostova AT, Cheah J, Samouilov A, Zweier JL, Bozak RE, Hicks RJ, et al. Phenolic Michael reaction acceptors: Combined direct and indirect antioxidant defenses against electrophiles and oxidants. *Med Chem* 2007;**3**(3):261–8.

34. Dinkova-Kostova AT, Talalay P. Direct and indirect antioxidant properties of inducers of cytoprotective proteins. *Mol Nutr Food Res* 2008;**52**(Suppl. 1):S128–38.

35. Dinkova-Kostova AT, Massiah MA, Bozak RE, Hicks RJ, Talalay P. Potency of Michael reaction acceptors as inducers of enzymes that protect against carcinogenesis depends on their reactivity with sulfhydryl groups. *Pro Natl Acad Sci U S A* 2001;**98**(6):3404–9.

36. Chen CA, Chen TS, Chen HC. Extracellular signal-regulated kinase plays a proapoptotic role in podocytes after reactive oxygen species treatment and inhibition of integrin-extracellular matrix interaction. *Exp Biol Med* 2012;**237**(7):777–83.

37. Konishi H, Tanaka M, Takemura Y, Matsuzaki H, Ono Y, Kikkawa U, et al. Activation of protein kinase C by tyrosine phosphorylation in response to H2O2. *Pro Natl Acad Sci U S A* 1997;**94**(21):11233–7.

38. Ghibelli L, Coppola S, Rotilio G, Lafavia E, Maresca V, Ciriolo MR. Non-oxidative loss of glutathione in apoptosis via GSH extrusion. *Biochem Biophys Res Commun* 1995;**216**(1):313–20.

39. Qiu X, Liu Z, Shao WY, Liu X, Jing DP, Yu YJ, et al. Synthesis and evaluation of curcumin analogues as potential thioredoxin reductase inhibitors. *Bioorg Med Chem* 2008;**16**(17):8035–41.

40. Cai W, Zhang B, Duan D, Wu J, Fang J. Curcumin targeting the thioredoxin system elevates oxidative stress in HeLa cells. *Toxicol Appl Pharmacol* 2012;**262**(3):341–8.

41. Appiah-Opong R, Commandeur JN, Istyastono E, Bogaards JJ, Vermeulen NP. Inhibition of human glutathione S-transferases by curcumin and analogues. *Xenobiotica* 2009;**39**(4):302–11.

42. Fruehauf JP, Meyskens Jr FL. Reactive oxygen species: A breath of life or death? *Clin Cancer Res* 2007;**13**(3):789–94.

43. Morin D, Barthelemy S, Zini R, Labidalle S, Tillement JP. Curcumin induces the mitochondrial permeability transition pore mediated by membrane protein thiol oxidation. *FEBS Lett* 2001;**495**(1-2):131–6.

44. Ligeret H, Barthelemy S, Bouchard Doulakas G, Carrupt PA, Tillement JP, Labidalle S, et al. Fluoride curcumin derivatives: New mitochondrial uncoupling agents. *FEBS Lett* 2004;**569**(1-3):37–42.

45. Mackinnon AC, Yan BC, Joseph LJ, Al-Ahmadie HA. Molecular biology underlying the clinical heterogeneity of prostate cancer: an update. *Arch Pathol Lab Med* 2009;**133**(7):1033–40.

46. Thangapazham RL, Shaheduzzaman S, Kim KH, Passi N, Tadese A, Vahey M, et al. Androgen responsive and refractory prostate cancer cells exhibit distinct curcumin regulated transcriptome. *Cancer Biol Ther* 2008;**7**(9):1427–35.

47. Ak T, Gulcin I. Antioxidant and radical scavenging properties of curcumin. *Chem Biol Int* 2008;**174**(1):27–37.

48. Gupta SC, Patchva S, Aggarwal BB. Therapeutic Roles of Curcumin: Lessons Learned from Clinical Trials. *AAPS J* 2012.

49. Polytarchou C, Hatziapostolou M, Papadimitriou E. Hydrogen peroxide stimulates proliferation and migration of human prostate cancer cells through activation of activator protein-1 and up-regulation of the heparin affin regulatory peptide gene. *J Biol Chem* 2005;**280**(49):40428–35.

50. Khor TO, Huang Y, Wu TY, Shu L, Lee J, Kong AN. Pharmacodynamics of curcumin as DNA hypomethylation agent in restoring the expression of Nrf2 via promoter CpGs demethylation. *Biochem Pharmacol* 2011;**82**(9):1073–8.

51. Adams BK, Cai J, Armstrong J, Herold M, Lu YJ, Sun A, et al. EF24, a novel synthetic curcumin analog, induces apoptosis in cancer cells via a redox-dependent mechanism. *Anti-cancer drugs* 2005;**16**(3):263–75.

52. Sun A, Lu YJ, Hu H, Shoji M, Liotta DC, Snyder JP. Curcumin analog cytotoxicity against breast cancer cells: Exploitation of a redox-dependent mechanism. *Bioorg Med Chem Lett* 2009;**19**(23):6627–31.

53. Shankar S, Srivastava RK. Involvement of Bcl-2 family members, phosphatidylinositol 3′-kinase/AKT and mitochondrial p53 in curcumin (diferulolylmethane)-induced apoptosis in prostate cancer. *Int J Oncol* 2007;**30**(4):905–18.

54. Valentini A, Conforti F, Crispini A, De Martino A, Condello R, Stellitano C, et al. Synthesis, oxidant properties, and antitumoral effects of a heteroleptic palladium(II) complex of curcumin on human prostate cancer cells. *J Med Chem* 2009;**52**(2):484–91.

55. Choi H, Chun YS, Shin YJ, Ye SK, Kim MS, Park JW. Curcumin attenuates cytochrome P450 induction in response to 2,3,7,8-tetrachlorodibenzo-p-dioxin by ROS-dependently degrading AhR and ARNT. *Cancer sci* 2008;**99**(12):2518–24.

56. Fajardo AM, MacKenzie DA, Ji M, Deck LM, Vander Jagt DL, Thompson TA, et al. The curcumin analog ca27 down-regulates androgen receptor through an oxidative stress mediated mechanism in human prostate cancer cells. *Prostate* 2012;**72**(6):612–25.

57. Shi Q, Shih CC, Lee KH. Novel anti-prostate cancer curcumin analogues that enhance androgen receptor degradation activity. *Anti Cancer Agents Med Chem* 2009;**9**(8):904–12.

58. Zhang X, Zhang HQ, Zhu GH, Wang YH, Yu XC, Zhu XB, et al. A novel mono-carbonyl analogue of curcumin induces apoptosis in ovarian carcinoma cells via endoplasmic reticulum stress and reactive oxygen species production. *Mol Med Rep* 2012;**5**(3):739–44.

59. Kunwar A, Jayakumar S, Srivastava AK, Priyadarsini KI. Dimethoxycurcumin-induced cell death in human breast carcinoma MCF7 cells: evidence for pro-oxidant activity, mitochondrial dysfunction, and apoptosis. *Arch Toxicol* 2012;**86**(4):603–14.

60. Devasena T, Menon VP, Rajasekharan KN. Prevention of 1,2-dimethylhydrazine-induced circulatory oxidative stress by bis-1,7-(2-hydroxyphenyl)-hepta-1,6-diene-3,5-dione during colon carcinogenesis. *Pharmacol Rep* 2006;**58**(2):229–35.

Oxidative Stress and Inflammatory Factors in Lung Cancer: Role of n-3 PUFAs

Concetta Finocchiaro, Maurizio Fadda
Department of Clinical Nutrition, San Giovanni Battista Hospital, Turin, Italy

Marina Schena
Department of Oncology - San Giovanni Battista Hospital, Turin, Italy

Maria G. Catalano
Department of Medical Sciences, University of Turin, Turin, Italy

Marina Maggiora, Rosa A. Canuto, Giuliana Muzio
Department of Clinical and Biological Sciences, University of Turin, Turin, Italy

List of Abbreviations

PUFAs polyunsaturated fatty acids
EPA eicosapentaenoic acid
DHA docosahexaenoic acid
ARA arachidonic acid
ROS reactive oxygen species
PPARs peroxisome proliferator activated receptors
CRP C-reactive protein

INTRODUCTION

Reactive oxygen species (ROS) are generated in response to both endogenous and exogenous stimuli,[1] and are involved in cancer generation and therapy.[2–9] Lung cancer is caused by tobacco smoking or inhalation of inorganic substances, such as arsenic, asbestos, and radon, which are nontobacco carcinogens closely associated with lung cancer. These agents can cause genetic and epigenetic alterations in normal cells, determining their transformation into tumor cells, acting directly on genes and pathways involved in lung cancer development. Moreover, these agents also exert their carcinogenesis effects through oxidative stress.[10]

Chronic and cumulative oxidative stress induces deleterious modifications to a variety of macromolecular components, such as DNA, lipids, and proteins. The association between reactive oxygen species (ROS), carcinogenesis, and the progression of lung cancer has been widely demonstrated. The high percentage of oxidants in cigarette smoke contributes to smoking-associated carcinogenesis.[11] Asbestos fibers alter the DNA of lung cells, increasing cell proliferation[12] by changing the redox state of the cells. Several mechanisms are responsible for the development of oxidative stress in cancer patients: the altered energy metabolism caused by the impossibility of normal nutrition in patients with anorexia, nausea and vomiting reduces the availability of glucose, proteins, and vitamins, leading to increased free-radical production.[13–15] Further, chronic nonspecific activation of the immune system, with excessive production of inflammatory cytokines, is responsible for increased production of ROS.[16]

The presence of a chronic inflammatory microenvironment, which is associated with oxidative stress, is also important in the carcinogenesis process. Chronic oxidative stress can cause a continuous local inflammatory

Cancer
http://dx.doi.org/10.1016/B978-0-12-405205-5.00019-2

© 2014 Elsevier Inc. All rights reserved.

response, which is thought to induce tissue destruction; chronic inflammation is considered to be a preneoplastic state that induces gene mutations, inhibits apoptosis, and stimulates angiogenesis and cell proliferation. Chronic lung inflammation, induced by cigarette smoking, together with oxidative stress and cell alterations such as increased cell proliferation, angiogenesis, and the inhibition of apoptosis, all have an influence on lung tumor growth.[17]

Conversely, some antineoplastic agents, such as radiation and chemotherapeutics, and in particular alkylating agents and cisplatin, are thought to exert their antineoplastic effect by inducing oxidative stress: they are considered to be significant inducers of cell damage via ROS-mediated toxicity.[2,18]

One important capability possessed by ROS is that of killing noncycling cells, inducing damage in macromolecules and mitochondria, and as a consequence causing apoptosis; conversely, cytotoxic drugs, whose action is not mediated by ROS, exert their effect especially on cycling cells. Within a tumor, there are both cycling and noncycling cells. Blocking cell cycle stops the invasiveness, an important stage in metastasis formation. Further, increased ROS production may overcome the resistance of cancer cells to antineoplastic agents, such as oxaliplatin or doxorubicin.[4]

Doxorubicin is frequently used in treating solid tumors, and induces cell-cycle arrest and apoptosis by inducing DNA damage. Anthracyclines are known for their complex cytotoxic mechanism involving inhibition of topoisomerase II, RNA polymerase, and cytochrome c oxidase, intercalation into DNA, chelation of iron, and also generation of ROS.[5] More recently, it has also been suggested that doxorubicin may increase intracellular levels of the lipid messengers ceramide and diacylglycerol, and decrease intracellular glutathione.[6,7] It has also been suggested that the killing capacity of ROS is strongly potentiated by hyperthermia; treatment of tumor cells with a combination of hyperthermia, hydrogen peroxide, and standard chemotherapeutic agents determined the best outcome.[8]

Endogenous enzymatic antioxidant defenses (superoxide dismutase, glutathione peroxidase, and catalase) can counterbalance oxidative damage by chelating superoxide and various peroxides.[18] Thus the use of antioxidants in combination therapy may have an adverse effect on those anticancer drugs that act on tumor cells by increasing ROS levels to induce cell death. The production of ROS, whether induced by chemotherapeutic agents or by inhibiting antioxidant enzymes,[8] can also cause cell injury by stimulating lipid peroxidation. Patients with acute myeloid leukemia receiving cytarabine and daunorubicin as chemotherapy regimen showed increased plasma levels of malondialdehyde during the 14-day postchemotherapy period, and decreased erythrocyte superoxide dismutase and glutathione peroxidase activity, both of these being antioxidant enzymes.[9]

Increasing the substrate available for lipid peroxidation in tumor cells may be an important goal in enhancing the effect of antineoplastic agents. Recent studies have suggested that n-3 polyunsaturated fatty acids (PUFAs) may enhance oxidative stress in cancer cells and, at the same time, may decrease the inflammatory response during radiotherapy and chemotherapy. Long-chain n-3 PUFAs are easily oxidized, and the peroxidation products are considered crucial to explain the cytotoxicity of these fatty acids against cancer cells, as well as their ability to inhibit cancer-cell growth. The use of n-3 PUFAs as adjuvant of chemotherapy may thus enable two goals to be achieved: to decrease cancer growth, and to reduce the pro-inflammatory process, which is involved in cancer progression and cachexia.

GENERAL INFORMATION CONCERNING EPA AND DHA

Eicosapentaenoic acid (EPA) and docosaexaenoic acid (DHA) are n-3 PUFAs present in large amounts in fish; they can be synthesized in humans from vegetables containing α linolenic acid (Figure 19.1).

Effect of EPA and DHA on Oxidative Stress and Cell Growth

DHA has been shown to decrease human A549 lung adenocarcinoma cell proliferation, and to induce lipid peroxidation to extents dependent on the concentration used (Figure 19.2, panels A and B). The mechanism underlying these effects involves changes in the fatty acid composition of biomembranes, induction of lipid peroxidation, and induction of the nuclear receptors peroxisome proliferator-activated receptors (PPARs).[19] DHA has also been shown to increase the inhibiting effect of ionizing radiation on the growth of A549 cells, by increasing oxidative stress: a significant decrease in cell proliferation and colony formation, and a significant increase in lipid peroxidation and apoptotic cell death, have been found to occur. Moreover, DHA supplementation of the diet of athymic mice transplanted with A549 xenograft produced increased DHA incorporation into normal lung tissue and lung cancer, the DHA increase being more marked in lung cancer than in normal lung tissue. Because lung cancer tends to be difficult to treat, a DHA-enriched diet may be considered a promising strategy to enhance the effect of treating lung cancer with ionizing radiation, by increasing oxidative stress and cancer cell death.[20]

Addition of vitamin E, an antioxidant, completely reversed cell proliferation and lipid peroxidation, when

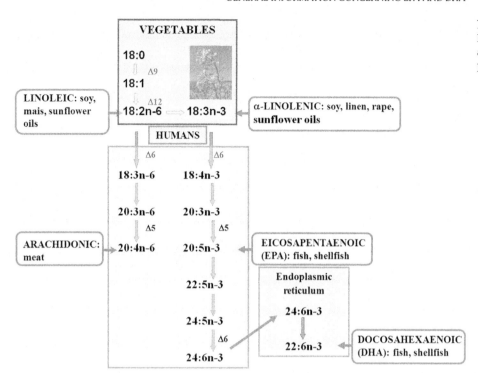

FIGURE 19.1 Synthesis of PUFAs in human beings from linoleic and α-linolenic acids taken with vegetables. (See the color plate.)

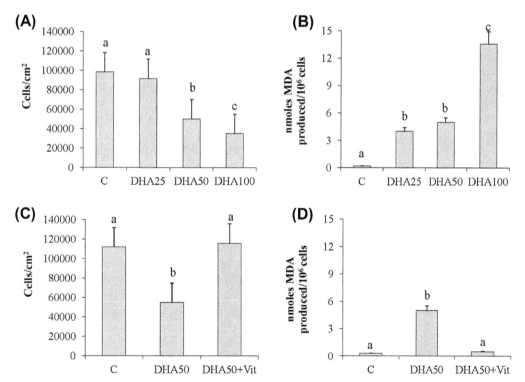

FIGURE 19.2 Cell proliferation and lipid peroxidation in human lung adenocarcinoma A549 cells treated with docosahexaenoic acid (DHA; panels A and B) and with DHA+vitamin E (panels C and D). Cell proliferation is expressed as number of cells/cm². Data are means ± S.D. of five experiments. Lipid peroxidation is expressed as nmoles of malondialdehyde (MDA) produced per 10^6 cells and data are means ± S.D. of five experiments. Means with different letters are significantly different from one another ($p < 0.05$) as determined by variance analysis followed by the Newman–Keuls test. Abbreviations: C, control cells; DHA25, 25μM docosahexaenoic acid; DHA50, 50μM docosahexaenoic acid; DHA100, 100μM docosahexaenoic acid; DHA50+Vit, 50μM docosahexaenoic acid + 10 μM Vitamin E.

DHA was added to A549 cells alone (Figure 19.2, panels C and D) or in combination with ionizing radiation.[19,20] The contemporary administration of EPA and DHA with the diet, in athymic nude mice subcutaneously transplanted with A427 lung adenocarcinoma cells, inhibited tumor growth and increased the level of lipid peroxidation products, in comparison with nude mice untreated with n-3 fatty acids. Again, growth inhibition and elevated lipid peroxidation in the n-3 diet group were counteracted by dietary vitamin E supplementation.[21]

These results were confirmed (unpublished data) by experiments carried out *in vitro* in lung A427 cells, which showed that administration of EPA plus DHA reduced tumor growth, inducing lipid peroxidation and PPARs.

Effect of EPA and DHA on Cancer Cachexia

The two compounds may be considered a potential natural support for cachexia treatment, having been shown to possess anti-inflammatory properties, downregulating both pro-inflammatory cytokine production and the acute-phase protein response in cancer patients.[22-27] Table 19.1 summarizes the anti-inflammatory effects of n-3 PUFAs, and Figure 19.3 shows the mechanisms involved in these effects. They occur through three different signaling transduction pathways: n-3 PUFAs inhibit NFkB transcription activity by acting on IkB, or through the activation of PPARs, leading to physical interaction of PPARs with NFkB subunits and consequent NFkB activity inhibition; moreover, n-3 PUFAs also inhibit arachidonic acid (ARA) metabolism, competing for cyclooxygenase and lypooxygenase.

One common thread that is responsible for cachexia is the increased production of IL-1, IL-2, IL-6, and TNF; these pro-inflammatory cytokines are recognized to play a central role in the pathogenesis of cancer-related cachexia.[27,28] Cachexia is an altered metabolic state characterized by anorexia, weight loss, asthenia, anemia, and alterations in carbohydrate, lipid, and protein metabolism.[18,29] This syndrome is the biggest cause of morbidity and mortality in patients with advanced cancer.[30] As a result of cachexia, increased energy expenditure and reduced dietary intake of nutrients combine to cause a negative energy balance and weight loss.

The interaction of cytokines with peptides/neuropeptides and neurotransmitters has been studied.[31] Since n-3 fatty acids, such as DHA and EPA, have well-established beneficial effects on human health, owing to the anti-inflammatory effects described earlier,[25,26] fish-oil supplementation has been proposed for the treatment of the cancer cachexia syndrome. It is emphasized[33,34-36] that conventional nutritional support can only partially stop lean mass reduction in cancer patients: it is only possible to increase muscle mass by resolving the metabolic alterations. Nutritional support is often wasted due to

TABLE 19.1   Anti-Inflammatory Effect of n-3 PUFAs

↓ eicosanoids from arachidonic acid
↓ endocannabinoids containing arachidonic acid
↑ endocannabinoids containing EPA or DHA
↑ resolvins and protectins
↓ cell adhesion molecules expression and leucocyte-endothelium interaction
↓ leucocyte chemotaxis
↓ inflammatory cytokine production
↓ reactivity of T lymphocytes

the hyper-metabolic state of the inflammatory pattern, and for this reason, nutritional therapy in cancer patients should aim to reduce the inflammatory state; it is thus important to use natural substances possessing both nutritional and anticachectic properties;[37,38] n-3 PUFAs might have this effect[22-24,38,39] and thus n-3 PUFA administration has been tested in patients with advanced lung cancer, to explore these compounds' ability to improve patients' nutritional status and reduce the inflammatory pattern. Some studies[22-24,38,39] only explored the administration of EPA, while others used fish oil or both PUFAs (EPA plus DHA). In particular, EPA was shown not only to have anti-inflammatory properties[22-24,40,41] but also to inhibit activation of the ubiquitin proteasome pathway, which is involved in atrophy of skeletal muscle in animal models.[30]

A 2006 study[42] compared administration of a dose of 2 or 4 g of EPA to placebo, in 518 cancer patients (gastrointestinal and lung), over an 8-week period. The results indicated that the 4 g dose provided no benefit, but that 2 g EPA per day conferred a potentially clinically relevant treatment effect. Many recent studies have addressed this subject, but without reaching any conclusions concerning survival improvement or weight increase. The Cochrane analysis, published in 2009,[43] concluded that there was insufficient evidence to draw any conclusions about EPA supplementation in cancer patients with cachexia.

In a trial in which fish oil or EPA plus DHA (the two PUFAs present in fish oil) were administered, it was shown that the two PUFAs, but not EPA alone, suppress inflammatory cytokines and reduce weight loss in patients with advanced cancer.[28,31,33,44-46] A recent study[47] published the results of a randomized, case-controlled, double-blind trial of 40 patients with stage III NSCLC, who received chemotherapy and radiotherapy, together with either supplements containing 2 g of EPA or isocaloric control supplements. After five weeks of treatment, it was found that levels of inflammatory markers had decreased during chemotherapy and that IL-6 production was lower in the intervention group than in

MECHANISMS OF ANTI-INFLAMMATORY
EFFECT OF n-3 PUFAs

FIGURE 19.3 Mechanisms underlying the anti-inflammatory effect of n-3 PUFAs.

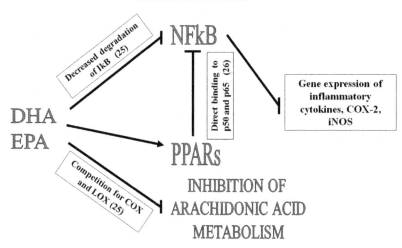

the control group. The same researchers[48] investigated the effects of an oral nutritional supplement containing EPA plus DHA, evidencing that patients receiving n-3 PUFAs showed higher Karnofsky performance status and higher physical activity than the control group.

A recent study,[49] investigating the effect of EPA (510 mg) plus DHA (340 mg) versus placebo on inflammatory condition and nutritional status in patients with advanced inoperable non-small-cell lung cancer subjected to chemotherapy, examined weight gain or loss and trends in C-reactive protein (CRP), IL-6, and PGE2 production after 66 days of administration. The n-3 group showed a mean weight gain of 3.4 kg, which was statistically significant versus the placebo group, in which weight remained similar to the starting value. CPR and IL-6 decreased in a statistically significant manner versus controls at the end of the experimental time (Figure 19.4, panels A and B). The n-3 group showed a significant decrease in the pro-inflammatory PGE2 in during treatment, but no variation occurred in the placebo group (Figure 19.4, panel C). The percentage content of EPA and DHA evaluated in total lipids from plasma (Figure 19.5, panels A and B) increased significantly in the n-3 group compared to the placebo group; the percentage plasma content of ARA (Figure 19.5, panel C) increased significantly in the n-3 group between T_0 and end of administration. These fatty acids tended toward plasma levels found in healthy persons, since they are normally decreased in lung-cancer patients (Table 19.2). The decrease in PGE2 is probably the cause of the increase in ARA. These results were confirmed in *in vitro* experiments on a model of cachexia. Enrichment of A427 cells with EPA+DHA decreased not only growth but also IL-6 and PGE2 release. When used to grow murine muscle C2C12 cells, medium taken from A427 cells with reduced content of pro-inflammatory factors enabled C2C12 cells to differentiate. No myosin or myotubes

were observed in C2C12 cells grown in medium conditioned without EPA + DHA (unpublished data).

CONCLUSIONS

Experimental data from studies in which EPA, DHA, or both together were administered in lung cancer cells *in vitro*, combined with those reported in studies administering these fatty acids in mice transplanted with lung cancer, support the view that n-3 PUFAs have an anticancer effect, and that they can increase the benefit of chemotherapeutic agents. However, data reported in the literature are not all in agreement concerning the effect of n-3 PUFAs in patients with lung cancer.

The Cochrane review[43] on EPA, used alone in treating cancer cachexia, neither confirmed nor rejected its use in clinical practice; the systematic review suggests that there is little evidence of harm deriving from the use of EPA. A comparison among the various studies is difficult, because of methodological differences (lung, pancreatic, or gastrointestinal cancer).[30,42,44–46] Moreover, failure to reveal effects might be attributed to sample heterogeneity, reduction of patient number at the end of studies, lack of patient compliance, or subclinical toxicity of high doses of EPA.[50] Other studies[30,33,40,41] that have examined the effect of fish oil in cachectic patients have also expressed the opinion that valid conclusions are difficult to draw, for several reasons (short duration of trial, poor tolerability of supplementation, inability of patients to complete the study).

The administration of EPA + DHA in patients with advanced cancer reduced weight loss and postsurgical morbidity, improved appetite, improved quality of life, and led to a progressive decrease in inflammatory parameters during chemotherapy, evidencing an anti-inflammatory action of n-3 PUFAs.[35,44,45,51] Although

FIGURE 19.4 Comparison of changes in CRP, IL-6 and PGE2 plasma content in placebo and n-3 groups. Data are expressed as means ± S.D. and are the plasma content of CRP (panel A), IL-6 (panel B), and PGE2 (panel C). * "paired t test" $p < 0.05$ T_1, T_2 versus T_0; § "unpaired t test" $p < 0.05$ n-3 group versus placebo group Placebo, group of patients with lung cancer untreated with EPA+DHA; n-3 group, group of patients with lung cancer treated with EPA+DHA.

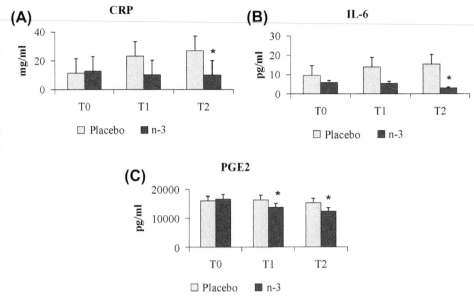

FIGURE 19.5 Changes in EPA, DHA, and ARA in total lipid from plasma in placebo and n-3 groups. Data are expressed as means ± S.D. and are the percentage content of EPA (panel A), of DHA (panel B), and of ARA (panel C). *"paired t test" $p < 0.05$ T_1, T_2 versus T_0; § "unpaired t test" $p < 0.05$ n-3 group versus placebo group Placebo, group of patients with lung cancer untreated with EPA+DHA; n-3 group, group of patients with lung cancer treated with EPA+DHA.

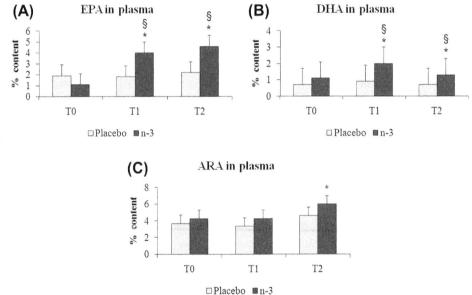

TABLE 19.2 Percentage Content of PUFAs in Total Lipid Extracted from Plasma

Fatty Acids	Healthy Group	Placebo + n-3 Group (T0)
ARA (20:4)	6.9 ± 2.2	3.8 ± 2.0 §
EPA (20:5)	1.4 ± 1.0	1.5 ± 1.3
DHA (22:6)	3.2 ± 1.8	0.9 ± 0.8 §

Data are expressed as means ± S.D. and are the percentage content of ARA, EPA and DHA.
Placebo, patients with lung cancer untreated with EPA+DHA; n-3 group, patients with lung cancer treated with EPA+DHA; T0, the beginning of clinical study.
§ "unpaired t test" p<0.05 placebo + n-3 groups versus healthy group.

numerous studies have addressed this subject and there is great interest in scientific circles concerning n-3 fatty acids, there is at present no agreement that might justify application of the results to cancer patients; it appears probable that it is better to use both n-3 fatty acids, EPA and DHA, than EPA alone.

SUMMARY POINTS

1. Oxidative stress causes inhibition of cell proliferation and death in lung cancer cells.

2. n-3 fatty acids induce oxidative stress.

3. Inflammatory factors are involved in cancer cachexia.

4. n-3 fatty acids reduce inflammatory indexes.

5. n-3 fatty acids may play an important role in the prevention and treatment of anorexia and cachexia.

Acknowledgments

This work was supported by grants from Compagnia San Paolo and from Regione Piemonte.
All authors have read and approved the final manuscript. None of the authors had any conflict of interests.

References

1. Ziech D, Franco R, Georgkalikas AG. The role of reactive oxygen species and oxidative stress in environmental carcinogenesis and biomarker development. *Chem Biol Interact* 2010;**188**:334–9.

2. De Bont R, Labereke NV. Endogenous DNA damage in human: A review of quantitative data. *Mutagenesis* 2004;**19**:169–85.

3. Weijl NI, Cleton FJ, Osanto S. Free radicals and antioxidants in chemotherapy-induced toxicity. *Cancer Tret Rev* 1997;**23**:209–40.

4. Laurent A, Nicco C, Chereau C, Goulvestre C, Alexandre J, Alves A, et al. Controlling tumor growth by modulating endogenous production of reactive oxygen species. *Cancer Res* 2005;**65**:948–56.

5. Müller I, Niethammer D, Bruchelt G. Anthracycline-derived chemotherapeutics in apoptosis and free radical cytotoxicity. *Int J Mol Med* 1998;**1**:491–4.

6. Martínez R, Navarro R, Lacort M, Ruiz-Sanz JI, Ruiz-Larrea MB. Doxorubicin induces ceramide and diacylglycerol accumulation in rat hepatocytes through independent routes. *Toxicol Lett* 2009;**190**:86–90.

7. Huwiler A, Kotelevets N, Xin C, Pastukhov O, Pfeilschifter J, Zangemeister-Wittke U. Loss of sphingosine kinase-1 in carcinoma cells increases formation of reactive oxygen species and sensitivity to doxorubicin-induced DNA damage. *Br J Pharmacol* 2011;**162**:532–43.

8. Lehmann K, Rickenbacher A, Jang JH, Oberkofler CE, Vonlanthen R, von Boehmer L, et al. New insight into hyperthermic intraperitoneal chemotherapy: induction of oxidative stress dramatically enhanced tumor killing in *in vitro* and *in vivo* models. *Ann Surg* 2012;**256**:730–7.

9. Esfahani A, Ghoreishi Z, Nikanfar A, Sanaat Z, Ghorbanihaghjo A. Influence of chemotherapy on the lipid peroxidation and antioxidant status in patients with acute myeloid leukemia. *Acta Med Iran* 2012;**50**:454–8.

10. Hubaux R, Becker-Santos DD, Enfield KS, Lam S, Lam WL, Martinez VD. Arsenic, asbestos and radon: Emerging players in lung tumorigenesis. *Environ Health* 2012;**11**:89.

11. Prior WA. Cigarette smoke radicals and the role of free radicals in chemical carcinogenity. *En Health Perspect* 1997;**105**:875–82.

12. Mossman BT, Gee JB. Asbestosis-related disease. *N Engl J Med* 1998;**320**:1721–30.

13. Hardman WE. Omega-3 fatty acids to augment cancer therapy. international research conference of food. *Nutr Cancer* 2002:3508S–11S.

14. Burns CP, Halabi S, Clamon G. Phase II study of high-dose fish oil capsules for patients with cancer-related Cachexia. *Cancer* 2004;**101**:370–8.

15. Larsson SC, Kumlin M, Ingelman-Sundberg M. Dietary long-chain n-3 fatty acids for prevention of cancer: a review of potential mechanisms. *Am J Clin Nutr* 2004;**79**:935–45.

16. Mantovani G, Macciò A, Lai P, Massa E, Ghiani M, Santona MC. Cytokine activity in cancer-related anorexia/cachexia: role of megestrol acetate and medroxyprogesterone acetate. *Semin Oncol* 1998;**25**:45–52.

17. Milara J, Cortijo J. Tobacco, inflammation, and respiratory tract cancer. *Curr Pharm Des* 2012;**18**:3901–38.

18. Lowenfels AB, Maisononneuve P, Cavallini G. International pancreatitis study group. Pancreatitis and risk of pancreatic cancer. *N Engl J Med* 1993;**328**:1433–7.

19. Trombetta A, Maggiora M, Martinasso G, Cotogni P, Canuto RA, Muzio G. Arachidonic and docosahexaenoic acids reduce the growth of A549 human lung-tumor cells increasing lipid peroxidation and PPARs. *Chem Biol Interact* 2007;**165**:239–50.

20. Kikawa KD, Herrick JS, Tateo RE, Mouradian M, Tay JS, Pardini RS. Induced oxidative stress and cell death in the A549 lung adenocarcinoma cell line by ionizing radiation is enhanced by supplementation with docosahexaenoic acid. *Nutr Cancer* 2010;**62**:1017–24.

21. Maehle L, Lystad E, Eilertsen E, Einarsdottír E, Høstmark AT, Haugen A. Growth of human lung adenocarcinoma in nude mice is influenced by various types of dietary fat and vitamin E. *Anticancer Res* 1999;**19**:1649–55.

22. Calder PC. Omega-3 polyunsaturated fatty acids and inflammatory processes: Nutrition or pharmacology? *Br J Clin Pharmacol* 2013;**75**:645–62.

23. Calder PC. Immunomodulation by omega-3 fatty acids. *Prostaglandins Leukot Essent Fatty Acids* 2007;**77**:327–35.

24. Jho DH, Cole SM, Lee EM. Role of omega-3 fatty acid supplementation in inflammation and malignancy. *Integr Cancer Ther* 2004;**3**:98–111.

25. Wall R, Ross RP, Fitzgerald GF, Stanton C. Fatty acids from fish: The anti-inflammatory potential of long-chain omega-3 fatty acids. *Nutr Rev* 2010;**68**:280–9.

26. Muzio G, Trombetta A, Maggiora M, Martinasso G, Vasiliou V, Lassen N, et al. Arachidonic acid suppresses growth of human lung tumor A549 cells through down-regulation of ALDH3A1 expression. *Free Radic Biol Med* 2006;**40**:1929–38.

27. Martin F, Santolaria F, Batista N. Cytokine levels (IL-6 and INF-gamma), acute phase response and nutritional status as prognostic factors in lung cancer. *Cytokine* 1999;**11**:80–6.

28. Simons JP, Schols AM, Buurman WA, Wouters EF. Weight loss and low body cell mass in males with lung cancer: Relationship with systemic inflammation, acute-phase response, resting energy expenditure, and catabolic and anabolic hormones. *Clin Sci (Lond)* 1999;**97**:123–5.

29. Evans WJ, Morley JE, Argiles J. Cachexia: A new definition. *Clin Nutr* 2008;**27**:203–9.

30. Fearon KC, Von Meyenfeldt MF, Moses AGW. Effect of a protein and energy dense n-3 fatty acid enriched oral supplement on loss of weight and lean tissue in cancer cachexia: A randomised double blind trial. *Gut* 2003;**52**:1479–86.

31. Laviano A, Inui A, Marks DL, Meguid MM, Pichard C, Rossi Fanelli F, et al. Neural control of the anorexia-cachexia syndrome. *Am J Physiol Endocrinol Metab* 2008;**295**:E1000–8.

32. Im DS. Omega-3 fatty acids in anti-inflammation (pro-resolution) and GPCRs. *Prog Lipid Res* 2012;**51**:232–7.

33. Bruera E, Strasser F, Palmer JL. Effect of fish oil on appetite and other symptoms in patient with advanced cancer and anorexia/cachexia: a double-blind, placebo-controlled study. *J Clin Oncol* 2003;**21**:129–34.

34. Ravasco P, Monteiro-Grillo I, Marques VP. Impact of nutrition on outcome: A prospective randomized controlled trial in patients with head and neck cancer undergoing radiotherapy. *Head Neck* 2005;**27**:659–68.

35. Ross PJ, Ashley S, Norton A. Do patients with weight loss have a worse outcome when undergoing chemotherapy for lung cancer? *Br J Cancer* 2004;**90**:1905–11.

36. Jatoi A. W-3 fatty acid supplementation for cancer-associated weight loss. *Nutr Clin Pract* 2005;**20**:394–9.

37. Calder PC. Polyunsaturated fatty acids and inflammatory processes: New twists in an old tale. *Biochimie* 2009;**91**:791–5.

38. Moses AW, Slater C, Preston T. Reduced total energy expenditure and physical activity in cachectic patients with pancreatic cancer can be modulated by an energy and protein dense oral supplementation enriched with n-3 fatty acids. *Br J Cancer* 2004;**90**:996–1002.

39. Brown T, Zelnik D, Dobs A. Fish oil supplementation in the treatment of cachexia in pancreatic cancer patients. *Int J Gastrointest Cancer* 2003;**34**:143–50.

40. Guarcello M, Riso S, Buosi R. EPA-enriched oral nutritional support in patients with lung cancer: Effects on nutritional status and quality of life. *Nutr Ther Metab* 2007;**25**:25–30.

41. Elia M, Van Bokhorst-de van der Schueren MA, Garvey J. Enteral (oral or tube administration) nutritional support and eicosapentanoic acid in patients with cancer: A systematic review. *Int J Oncol* 2006;**28**:5–23.

42. Fearon KC, Barber MD, Moses AG. Double-Blind, Placebo-Controlled, Randomized Study of Eicosapentaenoic Acid Diester in Patients with Cancer Cachexia. *J Clin Oncol* 2006;**24**:3401–7.

43. Dewey A, Baughan C, Dean TP. Eicosapentaenoic acid (EPA, an omega-3 fatty acid from fish oils) for the treatment of cancer cachexia. *The Cochrane Library* 2009(Issue. 1).

44. Barber MD, Ross JA, Voss AC. The effect of an oral nutritional supplement enriched with fish oil on weight-loss in patients with pancreatic cancer. *Br J Cancer* 1999;**81**:80–6.

45. Barber MD, Mc Millan DC, Preston T. Metabolic response to feeding in weight-losing pancreatic cancer patients and its modulation by a fish-oil-enriched nutritional supplement. *Clin Sci (Lond)* 2000;**98**:389–99.

46. Wigmore SJ, Barber MD, Ross JA. Effect of oral eicosapentaenoic acid on weight loss in patients with pancreatic cancer. *Nutr Cancer* 2000;**36**:177–84.

47. Van der Meij B, Langius J, Smit E. Oral Nutritional Supplements Containing (n-3) Polyunsaturated Fatty Acids Affect the Nutritional Status of Patients with Stage III Non-Small Cell Lung Cancer during Multimodality Treatment. *J Nutr* 2010;**140**: 1774–80.

48. van der Meij BS, Langius JA, Spreeuwenberg MD, Slootmaker SM, Paul MA, Smit EF, et al. Oral nutritional supplements containing n-3 polyunsaturated fatty acids affect quality of life and functional status in lung cancer patients during multimodality treatment: An RCT. *Eur J Clin Nutr* 2012;**66**:399–404.

49. Finocchiaro C, Segre O, Fadda M, Monge T, Scigliano M, Schena M, et al. Effect of n-3 fatty acids on patients with advanced lung cancer: A double-blind, placebo-controlled study. *Br J Nutr* 2012;**108**:327–33.

50. McLean CH, Newberry SJ, Mojica WA. Effects of Omega-3 fatty acids on cancer risk, a systematic review. *JAMA* 2006;**295**: 403–15.

51. Colomer R, Moreno-Nogueira J, Garcia-Luna P. n-3 Fatty acids, cancer and cachexia: A systematic review on the literature. *Br J Nutr* 2007;**97**:823–31.

Antioxidative Stress Actions of Cocoa in Colonic Cancer

Sonia Ramos, Luis Goya, Maria Angeles Martín

Department of Metabolism and Nutrition, Instituto de Ciencia y Tecnologia de Alimentos y Nutricion (ICTAN-CSIC), Ciudad Universitaria, Madrid, Spain

List of Abbreviations

ACF aberrant crypt foci
AKT protein kinase B
AOM azoxymethane
AP-1 activator protein-1
ARE antioxidant response element
CAT catalase
CDK cyclins-dependent kinase
COX cyclooxygenase
CRC colorectal cancer
DOC deoxycholic
EC (-)-epicatechin
ERK extracellular regulated kinase
γ-GCS gamma-glutamyl cysteine synthase
GPx glutathione peroxidase
GR glutathione reductase
GSH glutathione
GST glutathione-Stranferases
HO-1 heme oxygenase-1
IL interleukin
iNOS inducible nitric oxide synthase
JNK c-Jun N-terminal kinase
Keap1 Kelch-like ECH associating protein-1
LDH lactate dehydrogenase
MAPK mitogen-activated protein kinase
NF-κB nuclear factor kappa B
Nrf2 nuclear-factor-E2-related factor 2
PB2 procyanidin B2
PGE2 prostaglandins E2
PI3K PI-3-kinase, phosphatidylinositol-3-kinase
PPAR poly-(ADP-ribose) polymerase
ROS reactive oxygen species
SOD superoxide dismutase
t-BOOH tert-butylhydroperoxide
TNFα tumor necrosis factor α

INTRODUCTION

Colorectal cancer (CRC) is one of the major causes of cancer-related mortality in most of the developed world.[1] Environmental factors, including dietary and lifestyle, play a crucial role in their etiology even though it is also attributable to inherited and acquired genetic alterations.[2] Cancer is a multistage process conventionally defined by the initiation, promotion, and progression stages. In particular, development of CRC typically follows several consecutive steps from normal epithelial cells via aberrant crypts and progressive adenoma stages to carcinomas *in situ* and then metastasis. Along this process, oxidative stress has the potential to affect a large array of carcinogenic pathways involved in proliferation of initiated cells and enhanced malignant transformation.[3] In fact, the gastrointestinal tract, especially the colon, is constantly exposed to reactive oxygen species (ROS), generated during normal cellular metabolism and pathological processes.[4] ROS overproduction may provoke structure and function damages in colonic cells and induce somatic mutations and neoplastic transformation.[3] Because of this, the suppression of oxidative stress by natural antioxidant compounds has gained interest as an effective approach in CRC prevention. Chemoprevention, defined as the use of natural or synthetic compounds to prevent, block, or reverse the development of cancers seems to be an attractive option in this field and the possible impact of several nutritional agents with antioxidant and anti-inflammatory properties has been intensively studied in recent years.[5]

Accordingly, cocoa and its naturally flavonoid compounds have shown a potential ability to act as highly effective antioxidant and chemopreventive agents.[6] Flavanols are polyphenolic compounds extensively

© 2014 Elsevier Inc. All rights reserved.

found in vegetables, fruits, and plant-derived beverages that present a potent antioxidant activity.[7] Cocoa, the dried and fermented seeds derived from *Theobroma cacao*, has the highest flavanol content of all foods on a per-weight basis and is a significant contributor to the total dietary intake of flavonoids.[8] Actually, for many individuals, cocoa products constitute a larger proportion of the diet than foodstuffs containing bioactive compounds with similar properties such as green tea, wine, or soy beans.[9] Cocoa flavanols are powerful antioxidant agents acting directly as ROS scavengers, metal ion chelators, and free radical reaction terminators and indirectly by stimulating phase II detoxifying and antioxidant defense enzymes.[10] Additionally, polyphenolic compounds can exhibit other anticarcinogenic properties independently of their conventional antioxidant activity.[11] Based on these findings, cocoa polyphenols could be considered as promising candidates for colon cancer chemoprevention.

Nevertheless, health effects derived from cocoa flavonoids depend on their bioavailability (absorption, distribution, metabolism, and elimination), a factor that is also influenced by their chemical structure.[12] In this regard, cocoa contains high amounts of flavanols (-)-epicatechin (EC), (+)-catechin and their dimers procyanidins B2 (PB2) and B1 (Figure 20.1), although other polyphenols such as quercetin, isoquercitrin (quercetin 3-O-glucoside), quercetin 3-O-arabinose, hyperoside (quercetin 3-O-galactoside), naringenin, luteolin, and apigenin have also been found in minor quantities.[13] Interestingly, as compared to other flavonoid-containing foodstuffs, cocoa products exhibit a high concentration of procyanidins that are poorly absorbed in the intestine and consequently their beneficial effects would be restricted to the gastrointestinal tract where they may have an important antioxidant and anticarcinogenetic local function.[14]

In general, the evidence for chemoprevention by any bioactive substance is achieved from a combination of epidemiological, animal, and basic mechanistic studies. In view of that, the mode of action of cocoa and its flavanols has been recently investigated, especially in cell culture systems. However, it remains to be demonstrated whether these mechanisms are involved in cancer prevention in humans. In this chapter we review the different *in vitro* studies that have identified the potential targets and mechanisms whereby cocoa and its polyphenolic compounds could interfere with colonic cancer cells. Then we show the potential antioxidant and chemopreventive activity of cocoa in an animal model of colon cancer. Finally, some evidence from human studies is illustrated.

CHEMOPREVENTIVE MECHANISM OF COCOA POLYPHENOLS IN CULTURED COLON CANCER CELLS

In the recent past years, cocoa and their polyphenolic compounds have been widely studied for their actions against colon cancer cells and related molecular

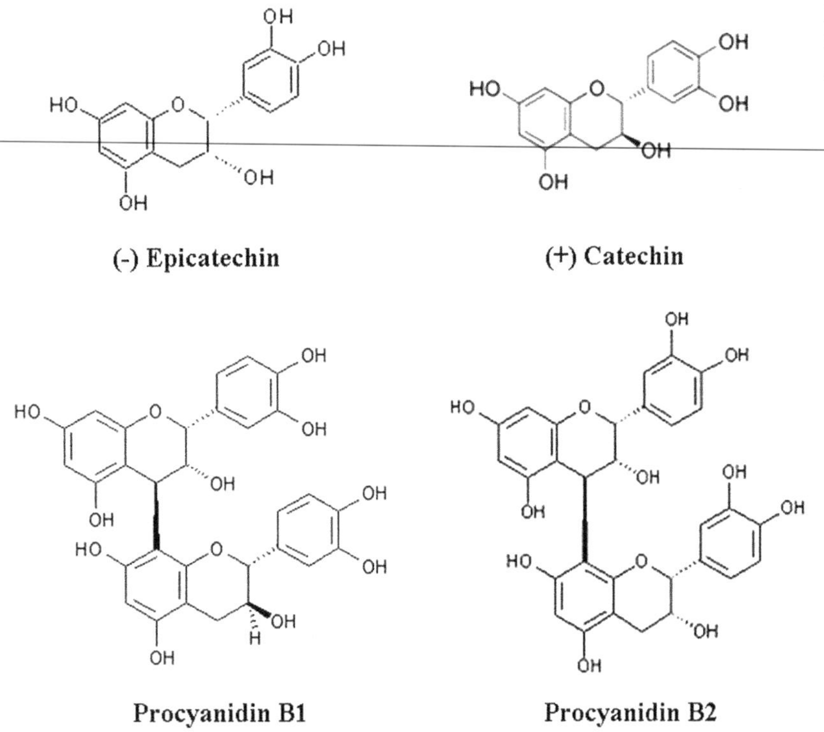

FIGURE 20.1 **Main Flavonoids Present in Cocoa.** Chemical structures of (-)-epicatechin and (+)-catechin and their respective dimers procyanidins B2 and B1.

(-) Epicatechin

(+) Catechin

Procyanidin B1

Procyanidin B2

TABLE 20.1 Effects of Cocoa and Cocoa Polyphenols on Colonic Cancer Cultured Cells.[a]

	Polyphenol	Result	Reference
Antioxidant	Cocoa	↓acrylamide-induced GSH depletion, ↓ ROS generation, ↑ γ-GCS, ↑ GST	[17]
	Hexamer procyanidins	↓ DOC-induced cytotoxicity, ↓ oxidant generation, ↓ NADPH oxidase, ↓ Ca^{2+}	[21,22]
	Procyanidin B2	↓ acrylamide-induced GSH depletion, ↓ ROS generation, ↑ γ-GCS, ↑ GST	[17]
		↑ GPx, GST and GR and Nrf2 translocation, ↓ t-BOOH-induced ROS production and LDH	[18,20]
	Epicatechin	= GPx, GST and GR and Nrf2 translocation, ↓ t-BOOH-induced ROS production and LDH	[18]
	Catechin	↓ lipid peroxidation, ↓ ROS formation, ↑ GPx, ↑ GR, ↑ Nrf2, ↑ HO-1	[19]
Cell cycle	Polymer procyanidins	G2/M arrest, ↓ ornithine decarboxyalse, ↓ S-adenosylmethionine decarboxylase	[24]
	Epicatechin	S arrest	[25]
Apoptosis	Hexamer procyanidins	↓ DOC-induced caspase-3, ↓ PPAR cleavage	[21]
	Procyanidin B2	↓ t-BOOH-induced caspase-3	[18]
	Epicatechin	↓ t-BOOH-induced caspase-3	[18]
Proliferation/survival	Cocoa	↓ acrylamide-induced p-JNK	[20]
	Hexamer procyanidins	↓ DOC-induced AKT, ERK, p38 and AP-1	[21]
	Procyanidin B2	↑ ERK, ↑p38	[20]
		= ↑ proliferation, = ↑ p-AKT = ↑ ERK	[26]
	Epicatechin	= ↑ proliferation, = p-AKT, = p-ERK	[26]
Anti-inflammatory	Cacao	↓ PGE2, ↓ IL-8, = NF-κB, ↑ COX-1	[28]
		↓ TNF-induced IL-8, COX-2, iNOS and NFκB activation	[29]
	Hexamer procyanidins	↓ TNF-induced NF-κB activation and iNOS	[30]

[a] The arrows indicate an increase (↑) or decrease (↓) in the levels or activity of the different analyzed parameters. In certain cases, opposing results have been obtained since the studies were carried out in different colonic cell types and/or the final effects may depend on the dose and time of treatment with the phenolic compound.

mechanisms (Table 20.1). All these studies have shown that the pathways responsible for the potential chemopreventive activity of cocoa and its flavonoids are mainly related to their antioxidant and anti-inflammatory properties and their ability to inhibit proliferation and to induce apoptotic cell death (Figure 20.2).

Antioxidant Effects

Aerobic organisms cannot avoid free radical and reactive oxygen species (ROS) generation. Overproduction of ROS may lead to the formation of highly reactive oxidation products, activation of carcinogens, formation of oxidized DNA bases, and DNA strand breaks. These alterations might cause errors during DNA replication and genetic alterations, increase transformation frequencies, modulate transcription of redox-regulated proteins,

ultimately leading to enhanced cell proliferation and tumor promotion/progression.[11] In a physiological situation, cells maintain the balance between generation and neutralization of ROS through the enzymatic and nonenzymatic defenses, such as glutathione (GSH), catalase (CAT), glutathione peroxidase (GPx), glutathione reductase (GR), superoxide dismutase (SOD), nitric-oxide synthase, lipooxygenase, xanthine oxidase, and so on. However, when the cellular balance is altered and cellular defenses overwhelmed, cells can be damaged, as just mentioned.

Cocoa and its flavonoids exert strong antioxidant effects. Thus, cocoa possesses a potent antioxidant capacity as compared with other foods or products, such as teas and red wine, and this property has been related to its flavonoid content.[15] Interestingly, the antioxidant properties of cocoa and its flavonoids are partly

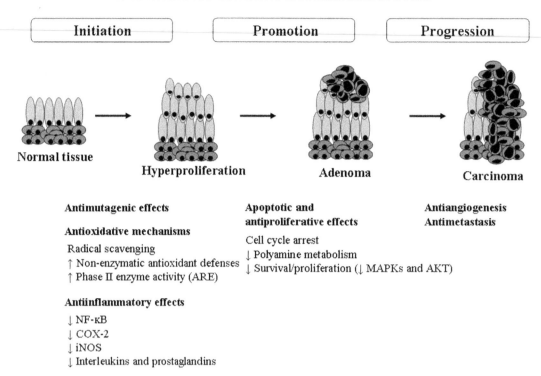

FIGURE 20.2   **Mechanisms Involved in the Potential Chemopreventive Effects of Cocoa and Its Flavonoids against Colorectal Cancer.** The arrows indicate an increase (↑) or decrease (↓) in the levels or activity of the different analyzed parameters.

based on their structural characteristics, including the hydroxylation of the basic flavan-ring system, especially 3',4'-dihydroxylation of the B-ring (catechol structure), the oligomer chain length, and the stereochemical features of the molecule.[16] These structural characteristics of flavanols represent the molecular basis for their hydrogen donating (radical-scavenging) properties and their metal-chelating antioxidant properties. In addition, cocoa and its flavonoids can prevent the DNA-damage caused by free radicals or carcinogenic agents acting through the modulation of enzymes related to oxidative stress (CAT, GR, GPx, SOD, etc.) and the alteration of the procarcinogenic metabolism by inhibiting phase-I drug-metabolizing enzymes (cytochrome P450) or activating phase II conjugating-enzymes (glucuronidation, sulfation, acetylation, methylation, and conjugation).

Protective Effects

Prevention of ROS generation and the preservation of the cellular antioxidant defenses seem to represent important chemopreventive mechanisms of natural polyphenols.[11] In this line, intestinal Caco-2 cells pretreated with a cocoa phenolic extract or with the pure cocoa flavanols EC and PB2 at physiological concentrations (i.e., 10 µg/mL for cocoa phenolic extract and 10 µM for EC and PB2, respectively) for 20 hours counteracted acrylamide-induced cytotoxicity (5 mM for 24 h) by inhibiting GSH consumption and ROS generation.[17] Both cocoa phenolic extract

and PB2 almost completely blocked the decrease of GSH induced by acrylamide and totally abrogated the subsequently increased ROS generation, whereas these effects were only partially restored with EC. This result suggests that the minor effect exerted by EC could be partially ascribed to the fact that EC mainly acted as scavenger of free radicals. However, similar to what was reported for other polyphenols and antioxidants,[11] PB2 and a cocoa phenolic extract could protect cell constituents not only by neutralizing several types of radicals but also by upregulating antioxidant defenses as well as by interacting with signaling pathways involved in cell survival. Thus, PB2 and the cocoa phenolic extract increased the levels of gamma-glutamyl cysteine synthase (γ-GCS) and glutathione-S-transferase (GST) in the mentioned experimental conditions in Caco-2 cells.[17]

Pure PB2 and EC (1–10 µM) decreased ROS production but did not affect GSH content in Caco-2 cells, and PB2 (1–10 µM) evoked a substantial increase in GPx, GR, and GST activity after 20 h of incubation.[18] Thus, pretreatment of Caco2 cells with EC and PB2 for 20 h before the oxidative insult induced by the potent pro-oxidant *tert*-butylhydroperoxide (*t*-BOOH at 400 µM) attenuated or blunted ROS production, respectively. In addition EC and PB2 protected cells from necrosis, as lactate dehydrogenase (LDH) leakage decreased after 1 and up to 6 h of incubation with the pro-oxidant, respectively.[18] All together this suggests that at least two mechanisms could be involved in the protection of Caco2 cells afforded by flavanols: (1) the inherent antioxidant capacity to quench ROS and (2) the improvement of the endogenous antioxidant defenses.

Effects on Phase I and II Enzymes

Enzymes of the phase I of drug metabolism (cytochromes P450) transform xenobiotics by adding functional groups that render these compounds more water soluble. Phase I functionalization may be required to efficiently detoxify carcinogens.[11] Phase II enzymes such as GST and sulfotransferases conjugate transformed phase I metabolites and xenobiotics to endogenous ligands like GSH, glucuronic, acetic, or sulfuric acid, and enhance excretion and detoxification in the form of these conjugates.[11] Therefore, reduction of elevated phase I enzyme activities to physiological levels and enhanced excretion of carcinogens via upregulation of phase II enzymes are considered a strategy in chemoprevention. In addition, the transcription factor NF-E2-related factor-2 (Nrf2) and the Kelch-like ECH-associated protein 1 (Keap1) are considered chemopreventive targets because both proteins participate in the regulation of the antioxidant response element (ARE). Thus, the modification of the protein Keap1 can lead to the accumulation of Nrf2 in the nucleus and the subsequent ARE activation.[11]

Cocoa and its phenolic compounds also exert their protective effect toward oxidative stress through the modulation of phase I and II enzyme activities and Nrf2. Accordingly, catechin (100 μM) increased the expression of Nrf2 and heme oxygenase-1 (HO-1) in a time-dependent manner in intestinal Int-407 cells.[19] PB2 (1-10 μM) alone also evoked a dose-dependent increase in GPx, GR, and GST after incubating Caco-2 cells for 20 h, which could be related to an improved cell response to an oxidative challenge.[18] Hence, cells treated with 10 μM PB2 for 20 h, and then submitted to an oxidative stress induced by t-BOOH (400 μM, for 1 or 6 h, respectively) showed a reduced ROS production, restricted activation of caspase 3, and higher viability than cells plainly submitted to the stressor.[18]

Furthermore, PB2 (10 μM, for 20 h) showed a protective effect against the oxidative injury induced by 400 μM t-BOOH in Caco-2 cells through the upregulation of the expression and activity of GST P1 via a mechanism that involved extracellular regulated kinase (ERK) and p38 mitogen-activated protein kinase (MAPK) activation and Nrf2 translocation.[20] Thus, PB2 treatment increased the protein levels of Nrf2 in the nucleus at 3 h, peaked at 6 h, and continued elevated up to 20 h of treatment. Accordingly, this procyanidin significantly enhanced the mRNA levels and activity of GST P1 at 4 to 20 h of incubation, which was accompanied by an increment in the levels of protein expression at 8 and 20 h.[20]

Similarly, cocoa procyanidins protected Caco-2 cells from the loss of integrity induced by a lipophilic oxidant.[21,22] Interestingly, a hexameric procyanidin fraction isolated from cocoa (2.5–20 μM) interacted with the Caco-2 cell membranes preferentially at the water–lipid interface without affecting their integrity after 30 min of incubation. Moreover, the hexameric procyanidin fraction inhibited the deoxycholic (DOC)-induced cytotoxicity and partly prevented the generation of oxidants following NADPH oxidase inhibition, as well as DOC-triggered increase in cellular calcium.[21,22] The limited effects on LDH release observed after 6 h of incubation for lower molecular weight procyanidins (i.e., monomer-tetramer) stress the relevance of the membrane-related effects of larger procyanidins. These differential actions have been explained as a compromise between the incorporation of the compounds into the cells that decreases as procyanidin oligomerization increases and the adsorption to the cell surface that increases as procyanidin oligomerization increases.

Effects on Apoptosis and Proliferation

Apoptosis and proliferation in cells are also modulated by ROS generation.[11,23] Indeed, suppression of cell proliferation, as well as induction of differentiation and apoptosis, are important approaches in cancer chemoprevention.

Cell Cycle

Deregulated cell cycle and resistance to apoptosis are hallmarks of cancer.[11] Cell cycle control is a highly regulated process that involves the modulation of different cell cycle regulatory proteins, such as cyclins, cyclin-dependent kinases (CDKs), CDK inhibitors, and so on.[11] ROS generation could induce an alteration of cell cycle-specific proteins that can affect and/or block the continuous proliferation of cancer cells.

Procyanidin-enriched extracts and procyanidins inhibited Caco-2 cell growth.[24] After 48 h of incubation, procyanidin extracts, with a flavanol and procyanidin content of 501 mg/g, caused only 25% growth inhibition, whereas the procyanidin-enriched extracts (flavanol and procyanidin content: 941 mg/g), induced a 75% growth inhibition. On the contrary, cocoa powder samples, which consisted of a flavanol and procyanidin content of 141 mg/g, showed no growth inhibitory effects in Caco-2 cells. Moreover, 50 μg/mL procyanidin-enriched extracts blocked the cell cycle at G2/M phase, without inducing apoptosis, and decreased the polyamine metabolism by inhibiting the ornithine decarboxylase and S-adenosylmethionine decarboxylase activities, which has been partly related to the accumulation of cells at the G2/M phase.[24] Similarly, incubation of LoVo cancer cells with 690 and 1380 μg/mL EC for 24 h induced S phase arrest in the cell cycle progression but it did not induce apoptosis.[25] Importantly, lower concentrations of EC seemed to promote slight proliferation of LoVo cells.

Apoptosis

ROS generation has been described as a critical upstream activator of the development of apoptosis.[23] At the molecular level, the existence of two mechanisms for the activation of programmed cell death has been widely reported: (1) the extrinsic pathway, which is mediated by death receptors, and (2) the intrinsic mechanism (mitochondria-mediated)[23] that is regulated by pro- and antiapoptotic proteins of the Bcl-2 family.[23] Both cascades converge in a common executor mechanism involving DNA endonucleases that activate proteases (caspases) and lead to cellular death.[11,23]

Treatment of cells with natural antioxidants prevents the cytotoxicity induced by oxidative stress inducers through the ability of these compounds to restrain the increase in ROS levels and the subsequent activation of caspase-3, which leads to apoptosis induction.[23] Consistent with this, 20 h of incubation with EC or PB2 (10 μM) effectively reduced the apoptotic effects induced by t-BOOH (400 μM for 4 h) in Caco-2 cells.[18] Similarly, pretreatment with 10 μM hexameric procyanidins for 30 min also delayed the DOC-induced Caco-2 cell apoptosis, as restrained caspase-3 activation, given that poly-(ADP-ribose) polymerase (PPAR) cleavage was observed only after 6 h incubation.[21]

Proliferation/Survival

Phosphatidyl-inositol-3-kinase (PI3K)/protein kinase B (AKT), growth factor receptors/Ras/MAPKs, and nuclear factor kappa B (NF-κB), which also importantly contribute to the inflammatory process (see later), constitute the most important signaling pathways regulating cell proliferation and survival.[11]

Cocoa phenolic compounds can interact with signaling proteins and modulate their activity. In this line, pretreatment with 10 μM hexameric procyanidins (30 min) prevented oncogenic events initiated by DOC through the interaction with Caco-2 cell membranes, and inhibited the DOC-promoted activation of AKT, ERK, and p38, as well as the downstream transcription factor activator protein-1 (AP-1).[21] Interestingly, PB2 and EC (10–50 μM) did not have an obvious effect on Caco-2 and SW480 colon carcinoma cells after 24 h of incubation. However, PB2 promoted cell growth in SW480 cells by increasing p-AKT and p-ERK levels[26] and activated Nrf2 translocation and increased both GST P1 protein and activity, via ERK and p38 pathways, which were also essential routes for the cytoprotective effect exerted by the flavanol in Caco-2 cells.[20] This different response depending on the distinct chemical structure of the compound and the different degree of cell differentiation highlights the importance of an integrated approach to study the biological effects of phytochemicals.[26]

Anti-inflammatory Effects

Manifestation of oxidative stress by infections, immune diseases, and chronic inflammation has been associated with carcinogenesis.[11] Thus, chronic inflammation is a risk for colorectal cancer.[11,27] Most inflammation-associated colorectal cancers are characterized by the activation of the transcription factor NF-κB and inflammatory mediators such as tumor necrosis factor (TNFα), cyclooxygenase-(COX)-2, and such, since all these proteins also are related to cell proliferation, antiapoptotic activity, angiogenesis, and metastasis.[11]

Cocoa extract inhibited the inflammatory mediator prostaglandins E2 (PGE2) in human intestinal Caco-2 cells.[28] Thus, cells incubated with a polyphenolic extract of cocoa (equivalent to 50 μM of gallic acid) for 4 h and stimulated with interleukin-(IL)-1β for 24 or 48 h showed a decrease in PGE2 synthesis, whereas IL-8 secretion and NF-κB activity remained at high levels.[28] Surprisingly, in the absence of pro-inflammatory stimulus, the cocoa polyphenolic extract induced a basal PGE2 synthesis in Caco-2 cells after 24 h of incubation. This effect has been associated with an induction of COX-1, which seems to be implicated in maintaining the mucosal integrity.[28]

More recently, pretreatment with a cocoa phenolic extract, at a physiological concentration (10 μg/mL) for 20 h, reduced the increase in inflammatory markers such as IL-8 secretion, COX-2, and inducible nitric oxide synthase (iNOS) expression induced by the pro-inflammatory agent TNFα (40 ng/mL for 24 h) in Caco-2 cells.[29] In this work, cocoa phenolic extract selectively decreased both phosphorylated levels of c-Jun N-terminal kinase and nuclear translocation of NK-κB induced by TNFα, indicating that this pathway could be an important mechanism contributing to the reduction of intestinal inflammation.

EC and procyanidins can inhibit NF-κB at different levels in the activation pathway. A decrease in cell oxidants that are involved in NF-κB activation is a potential mechanism of modulation by these compounds. Thus, incubation of Caco-2 cells for 30 min with 2.5 to 20 μM hexameric procyanidins prior to treatment with 10 ng/mL TNFα for a further 5 to 30 min inhibited the TNFα-induced NF-κB activation (inhibitor of κB phosphorylation and degradation, p50 and RelA nuclear translocation, and NF-κB–DNA binding), iNOS expression, and cell oxidant increase.[30] These effects have been suggested to occur because hexameric procyanidins can inhibit NF-κB activation by interacting with the plasma membrane of intestinal cells, and through these interactions preferentially inhibit the binding of TNFα to its receptor and the subsequent NF-κB activation.[30]

CHEMOPREVENTIVE MECHANISM OF COCOA IN AN *IN VIVO* MODEL OF COLON CANCER

Studies with colonic cell culture models have clearly demonstrated the antioxidant and chemopreventive abilities of cocoa and its flavonoids, but only experimental models for colorectal cancer could offer the opportunity to assess the contribution of this natural dietary compound to the potential prevention of CRC. To this end, carcinogen-induced rodent models have been shown to mimic many features of human nonfamilial colorectal cancer (non-genetic-based),[31] which is the most frequent and occurs sporadically. The induction of colon tumors is achieved by the administration of carcinogens such as nitrosamines, heterocyclic amines, aromatic amines, 1,2-dimethylhydrazine, and azoxymethane (AOM). Among these, the AOM model has been extensively used to examine the chemopreventive effect of numerous compounds on CRC.[32] Administration of AOM to rodents induces the development of colonic preneoplastic lesions (aberrant crypt foci, ACF) that may progress into cancer with time.[33] ACF represent the earliest identifiable intermediate precancerous lesions during colon carcinogenesis in both laboratory animals and humans[34] and can be identified microscopically on the surface of the colon mucosa after methylene blue staining (Figure 20.3). These cryptal lesions often appear within 2 to 3 weeks of carcinogen treatment and can be distinguished by their increased size, thicker epithelial lining, enlarged pericryptal zone, and slit-like luminal opening. In this line, increased number and multiplicity of ACF are associated with an increased risk for the development of CRC.[35]

Contrary to the strong evidence for the antioxidant and cancer preventive activity of cocoa and its components in cultured cells, there is only one recent study in rats that has demonstrated the potential chemopreventive ability of cocoa on colon carcinogenesis.[36] In this study, the well-defined AOM colon cancer model was used. Male Wistar rats were fed a cocoa-enriched diet (12%) starting two weeks before the carcinogenic induction (AOM, 20 mg/kg bw) and throughout the experimental period (8 weeks). As expected, all the rats that were injected with AOM developed aberrant crypts (100% incidence). Nevertheless, the cocoa-enriched diet significantly reduced the AOM-induced ACF formation and especially those ACF with a larger number of crypts (≥ 4 crypts), which exhibit a higher tendency to progress into malignancy (Table 20.2). Therefore, this study showed for the first time that a cocoa-enriched diet was able to suppress the early phase of chemically induced colon carcinogenesis.

The most relevant *in vivo* mechanisms involved in the chemopreventive effects elicited by cocoa are briefly described next. These mechanisms included the prevention of oxidative stress, cell proliferation and cell inflammation, and the ability of a cocoa diet to induce cell apoptosis (Figure 20.4).

Cocoa-Prevented AOM-Induced Oxidative Stress in Colon Tissues

The suppressive effect of cocoa on AOM-induced preneoplastic lesions has been associated with its antioxidative properties. AOM is metabolized in the liver to a methyl-free radical, which in turn generates hydroxyl radical or hydrogen peroxide capable of oxidized DNA, RNA, lipids, or protein of colonic epithelial cells.[37] As a consequence, the levels of lipid and protein peroxidation, indicative of oxidative injury, increased in the colon of animals treated with AOM.[38] However, in animals fed a cocoa diet the increased levels of protein and lipid oxidative damage induced by AOM were strongly prevented,

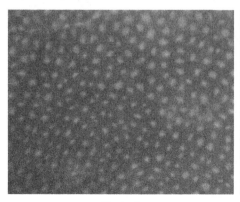

Control

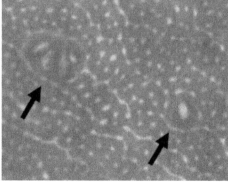

AOM

FIGURE 20.3 Mucosal Surface of Colon from Control and Azoxymethane (AOM)-Injected Rats. Mucosae of colon were stained with methylene blue and observed under light microscope (40× magnification). The presence of aberrant crypt foci (ACF) is indicated by arrows.

TABLE 20.2 Effect of Dietary Cocoa on Aberrant Crypt Foci (ACF) Formation in Azoxymethane (AOM)-Treated Rats.

	ACF Formation	ACF/cm^2	Crypts Multiplicity of ACF			
			1Crypt	2Crypts	3Crypts	≥4Crypts
Control	0/8	0	0	0	0	0
Cocoa	0/8	0	0	0	0	0
Control +AOM	12/12	16.1 ± 6.2[a]	6.3 ± 1.9[a]	5.8 ± 2.8[a]	3.0 ± 1.4[a]	0,87 ± 0,23[a]
Cocoa + AOM	12/12	8.8 ± 2.5[b]	3.5 ± 1.0[b]	3.7 ± 1.5[a]	1.3 ± 0.6[b]	0,14 ± 0,05[b]

Values are means ± SD. Means in a column without a common letter differ, $P<0.05$

demonstrating that cocoa possesses a potent antioxidative effect *in vivo* on the stressed colonic tissue. In particular, cocoa feeding was able to avoid oxidative stress by reverting to control values the diminished levels of GSH and the activities of GPx, GR, and GST provoked by the toxicant. Since GSH and antioxidant enzymes participate in the detoxification of xenobiotics, carcinogens, free radicals, and peroxides,[39] it can be suggested that cocoa could prevent AOM-induced ACF formation by reinforcing the endogenous defense capacity in colon tissues to counteract carcinogen-induced toxicity. Consequently, the increased cellular defense in the colon of cocoa-fed animals treated with AOM seems to be an effective strategy to protect against carcinogen-induced toxicity and largely accounts for the chemoprotective activity of cocoa.

Cocoa-Prevented Cell Proliferation in AOM Treated Animals

Besides inducing oxidative damage and genomic instability, ROS can specifically activate certain redox-sensitive signaling pathways and contribute to CRC initiation/promotion through the regulation of cellular proliferation and survival.[40] Among these, PI3K/AKT and ERK/MAPKs are within the most important pathways activated in response to oxidative stress, and play important roles in the carcinogenesis of many types of cancers including colon cancer.[3] Accordingly, AOM treatment clearly elevated the proliferative activity of the colonic mucosa and this increase was accompanied by the phosphorylation of AKT and ERKs and the overexpression of cyclin D1, a preneoplastic marker involved in cell cycle progression. However, cocoa intake prevented all these processes induced by AOM, suggesting that cocoa, by its ability to restrain oxidative stress, could also inhibit the consequent activation of signaling pathways involved in proliferation and thereby the progression of preneoplasia in the colonic epithelial cells. Supporting this, a recent *in vitro* study has showed that flavonoids such as luteolin and quercetin have antiproliferative and

proapoptotic effects in human CRC cells through the regulation of the ERK/MAPK and the PI3K pathways.[41]

Cocoa-Prevented AOM-Induced Inflammation in Colon Tissues

In recent years, considerable evidence has demonstrated that ROS are also involved in the link between chronic inflammation and cancer.[3] Redox status has an impact on the transcription factor NF-κB, which regulates the expression of the pro-inflammatory enzymes COX-2 and iNOS. Both enzymes are implicated in chronic inflammation, causing a microenvironment that contributes to the development of preneoplastic lesions in the colon carcinogenesis.[42] In fact, iNOS and COX-2 have been found to be increased in human CRC and in AOM-induced rat colon carcinogenesis.[43] In this study, the cocoa-rich diet was able to suppress the intestinal inflammation induced by AOM through the inhibition of NF-κB signaling and the down-regulation of the pro-inflammatory enzyme expressions of COX-2 and iNOS.[29] iNOS expression is frequently observed in dysplastic, but not hyperplastic, ACF indicating that iNOS plays an important role in the early stages of tumor formation.[44] On the other hand, tumorigenic mechanisms of COX-2 include inhibition of apoptosis via increased Bcl-2 and activation of proliferation via MAPK or PI3K/AKT signaling pathways.[45] Thus, the effect of a cocoa-rich diet preventing iNOS and COX-2 expression induced by AOM seems to be related to the inhibition of ACF formation observed in the AOM group treated with cocoa.

Cocoa-Induced Apoptosis in AOM-Treated Animals

During the promotion/progression phase of carcinogenesis, apoptosis is the main biological event involved in the removal of the initiated/mutated colonic epithelial cells.[3] In fact, many natural dietary compounds have been shown to suppress ACF formation by increasing

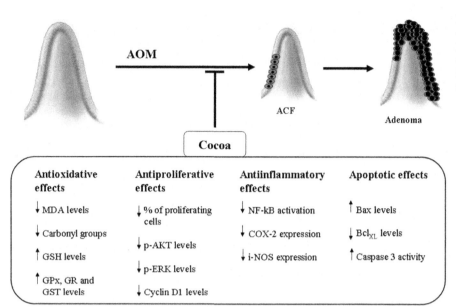

FIGURE 20.4 Mechanisms Involved in the Inhibition of Azoxymethane (AOM)-Induced Aberrant Crypt Foci Formation by a Cocoa-Enriched Diet. The arrows indicate an increase (↑) or decrease (↓) in the levels or activity of the different analyzed parameters.

apoptosis.[46] Accordingly, cocoa supplementation clearly induced apoptosis in the colon tissue of AOM-treated rats. Indeed, cocoa was able to modulate the expression of pro- and antiapoptotic proteins (Bax and Bcl-x$_L$, respectively) and to provoke caspase-3 activation, suggesting that cocoa induces apoptosis with the participation of the mitochondrial pathway. These data are in agreement with other results illustrating the apoptotic effect as the major mechanism for chemoprevention of different polyphenolic plant constituents.[11] Consequently, the pro-apoptotic *in vivo* effect of cocoa seems to be a complementary mechanism both to reduce preneoplastic lesions induced by AOM and to prevent promotion/progression of carcinogenesis, thus playing an important role in its anticarcinogenic potential.

HUMAN STUDIES

Cell culture and animal studies have indicated cancer preventive efficiency of cocoa polyphenols; however, whether these natural compounds could exert chemopreventive properties in patients remains to be elucidated. Epidemiologic studies of cocoa intake and colon cancer risk are few, and those assessing mortality by cancer provide only weak support for a benefit of cocoa. Consequently, confirmation of colon cancer preventive efficacy in humans requires large and long-lasting controlled clinical trials.

Epidemiologic Studies

The most prominent case in support of the cancer preventive effect of cocoa in humans is the Kuna tribe in Panama. The Kuna living in the San Blas district of Panama drink a flavanol-rich cocoa as their main beverage, contributing more than 900 mg/day and thus probably have the most flavonoid-rich diet of any population.[47] When death certificates were examined to compare cause-specific death rates from year 2000 to 2004 in mainland Panama and the San Blas islands where Kuna live, the rate of cardiovascular disease and cancer among island-dwelling Kuna was much lower than in mainland Panama. In support of the anticancer effect, data from the Iowa Women's Study established an inverse epidemiological relation between catechin consumption and the incidence of rectal cancer in postmenopausal women.[48]

On the contrary, a French study showed no significant association between a high chocolate dietary pattern and any stage of colorectal disease ranging from polyps, to adenomas, and colorectal cancer.[49] Similarly, an adenoma study in North Carolina also failed to observe a significantly lower prevalence of adenomatous polyps and colorectal cancer with chocolate consumption.[50] In addition, in a case-control study in Burgundy (France), chocolate was identified as a risk factor for colorectal cancer.[51]

Intervention Studies

There are properly no human intervention studies attempting to show a correlation between cocoa intake and cancer prevention, but a few human intervention trials indicate that cocoa favorably affects intermediary factors in cancer progression.[6] In this regard, several recent studies have focused on the modulation of antioxidant and anti-inflammatory status by consumption of cocoa products. In a study by Spadafranca et al.[52] dark chocolate consumption significantly improved DNA resistance to oxidative stress. In this study healthy subjects were assigned to a daily intake of 45 g of dark chocolate or white chocolate for 14 days and oxidative damage to mononuclear blood cells DNA was reduced in the dark chocolate group 2 h after consumption; 22 h later the effect disappeared. Similarly, cocoa consumption reduced NFκB activation in peripheral blood mononuclear cells in healthy voluntaries,[53] but biomarkers of inflammation, including IL-6, were unaffected in patients at high risk of cardiovascular disease consuming cocoa powder.[54]

Although the results from the limited epidemiological and human trials that are available are not conclusive, cocoa has consistently demonstrated an ability to increase serum antioxidant status, therefore theoretically reducing cancer risk. Nevertheless, more well-designed epidemiological and intervention studies are needed to demonstrate the potential CRC preventive activities of cocoa in humans.

SUMMARY POINTS

- Potential cellular mechanisms of action for cocoa and its main phenolic compounds include the modulation of the redox status and multiple key elements in signal transduction pathways related to cell proliferation, differentiation, apoptosis, and inflammation.
- Studies performed in animals have demonstrated that cocoa and its main phenolic components would prevent and/or slow down the initiation-promotion of colon cancer.
- Human intervention studies have reported favorable changes in biomarkers for antioxidant status.
- Daily consumption of small amounts of flavanols and procyanidins from cocoa or chocolate, in conjunction with usual dietary intake of flavonoids, would constitute a natural approach to potentially prevent colon cancer with minimal toxicity.
- Caution is mandatory when attempting to extrapolate the *in vivo* cellular observations to *in vivo* animal tumor models and, most importantly, to humans.
- Cocoa products deserve further investigations since molecular mechanisms of action are not completely elucidated.

References

1. Siegel R, Naishadham D, Jemal A. Cancer statistics. *CA Cancer J Clin* 2012;**62**:10–29.
2. Benson AB. Epidemiology, disease progression, and economic burden of colorectal cancer. *J Manag Care Pharm* 2007;**13**:S5–18 Care Pharm.
3. Reuter S, Gupta SC, Chaturvedi MM, Aggarwal BB. Oxidative stress, inflammation, and cancer: how are they linked? *Free Radic Biol Med* 2010;**49**:1603–16.
4. Klaunig JE, Kamendulis LM. The role of oxidative stress in carcinogenesis. *Annu Rev Pharmacol Toxicol* 2004;**44**:239–67.
5. Pan M-H, Lai Ch-S, Wu J-Ch, Ho Ch-T. Molecular mechanisms for chemoprevention of colorectal cancer by natural dietary compounds. *Mol Nutr Food Res* 2011;**55**:32–45.
6. Maskarinec G. Cancer protective properties of cocoa: a review of the epidemiologic evidence. *Nutr Cancer* 2009;**61**:573–9.
7. Scalbert A, Williamson G. Dietary intake and bioavailability of polyphenols. *J Nutr* 2000;**130**:2073–85.
8. Lamuela-Raventós RM, Romero-Pérez AI, Andrés-Lacueva C, Tornero A. Health effects of cocoa flavonoids. *Food Sci Tech Int* 2005;**11**:159–76.
9. Cooper KA, Donovan JL, Waterhouse AL, Williamson G. Cocoa and health: a decade of research. *Br J Nutr* 2008;**99**:1–11.
10. Vinson JA, Proch J, Bose P, Muchler S, Taffera P, Shuta D, et al. Chocolate is a powerful ex vivo and in vivo antioxidant, an anti-atherosclerotic agent in an animal model, and a significant contributor to antioxidants in the European and American Diets. *J Agric Food Chem* 2006;**54**:8071–6.
11. Ramos S. Cancer chemoprevention and chemotherapy: dietary polyphenols and signalling pathways. *Mol Nutr Food Res* 2008;**52**:507–26.
12. Manach C, Williamson G, Morand C, Scalbert A, Rémésy C. Bioavailability and bioefficacy of polyphenols in humans. I. Review of 97 bioavailability studies. *Am J Clin Nutr* 2005;**81**:230S–42S.
13. Sánchez-Rabaneda F, Jáuregui O, Casals I, Andrés-Lacueva C, Izquierdo-Pulido M, Lamuela-Raventós RM. Liquid chromatographic/electrospray ionization tandem mass spectrometric study of the phenolic composition of cocoa (*Theobroma cacao*). *J Mass Spectrom* 2003;**38**:35–42.
14. Ramiro-Puig E, Castell M. Cocoa: antioxidant and immunomodulator. *Br J Nutr* 2009;**101**:931–40.
15. Lee KW, Kim YJ, Lee HJ, Lee CY. Cocoa has more phenolic phytochemicals and a higher antioxidant capacity than teas and red wine. *J Agric Food Chem* 2003;**51**:7292–5.
16. Andújar I, Recio MC, Giner RM, Ríos JL. Cocoa polyphenols and their potential benefits for human health. *Oxidative Med Cell Longevity* 2012. http://dx.doi.org/10.1155/2012/906252.
17. Rodríguez-Ramiro I, Ramos S, Bravo L, Goya L, Martín MA. Procyanidin B2 and a cocoa polyphenolic extract inhibit acrylamide-induced apoptosis in human Caco-2 cells by preventing oxidative stress and activation of JNK pathway. *J Nutr Biochem* 2011;**22**:1186–94.
18. Rodríguez-Ramiro I, Martín MA, Ramos S, Bravo L, Goya L. Comparative effects of dietary flavanols on antioxidant defences and their response to oxidant-induced stress on Caco2 cells. *Eur J Nutr* 2011;**50**:313–22.
19. Cheng YT, Wu CH, Ho CY, Yen GC. Catechin protects against ketoprofen-induced oxidative damage of the gastric mucosa by up-regulating Nrf2 in vitro and in vivo. *J Nutr Biochem* 2013;**24**: 475–83. http://dx.doi.org/10.1016/j.jnutbio.2012.01.010.
20. Rodríguez-Ramiro I, Ramos S, Bravo L, Goya L, Martín MA. Procyanidin B2 induces Nrf2 translocation and glutathione S-transferase P1 expression via ERKs and p38-MAPK pathways and protect human colonic cells against oxidative stress. *Eur J Nutr* 2011;**51**:881–92.
21. Da Silva M, Jaggers GK, Verstraeten SV, Erlejman AG, Fraga CG, Oteiza PI. Large procyanidins prevent bile-acid-induced oxidant production and membrane-initiated ERK1/2, p38, and Akt activation in Caco-2 cells. *Free Rad Biol Med* 2012;**52**:151–9.
22. Erlejman AG, Fraga CG, Oteiza PI. Procyanidins protect Caco-2 cells from bile acid- and oxidant-induced damage. *Free Rad Biol Med* 2006;**41**:1247–56.
23. Ramos S. Effects of dietary flavonoids on apoptotic pathways related to cancer chemoprevention. *J Nutr Biochem* 2007;**18**:427–42.
24. Carnésecchi S, Schneider Y, Lazarus SA, Coehlo D, Gossé F, Raul F. Flavanols and procyanidins of cocoa and chocolate inhibit growth

and polyamine biosynthesis of human colonic cancer cells. *Cancer Lett* 2002;**175**:147–55.

25. Tan X, Hu D, Li S, Han Y, Zhang Y, Zhou D. Differences of four catechins in cell cycle arrest and induction of apoptosis in LoVo cells. *Cancer Lett* 2000;**158**:1–6.

26. Ramos S, Rodríguez-Ramiro I, Martín MA, Goya L, Bravo L. Dietary flavanols exert different effects on antioxidant defenses and apoptosis/proliferation in Caco-2 and SW480 colon cancer cells. *Toxicol In Vitro* 2011;**25**:1771–81.

27. Pan MH, Lai CS, Ho CT. Anti-inflammatory activity of natural dietary flavonoids. *Food Funct* 2010;**1**:15–31.

28. Romier-Crouzet B, Van De Walle J, During A, Joly A, Rousseau C, Henry O, et al. Inhibition of inflammatory mediators by polyphenolic plant extracts in human intestinal Caco-2 cells. *Food Chem Toxicol* 2009;**47**:1221–30.

29. Rodríguez-Ramiro I, Ramos S, López-Oliva E, Agis-Torres A, Bravo L, Goya L, et al. Cocoa polyphenols prevent inflammation in the colon of 2 azoxymethane-treated rats and in TNF-a-stimulated Caco-2 cells. *Br J Nutr* 2013;**110**:206–15. http://dx.doi.org/10.1017/S0007114512004862 2012.

30. Erlejman AG, Jaggers G, Fraga CG, Oteiza PI. TNFα-induced NF-κB activation and cell oxidant production are modulated by hexameric procyanidins in Caco-2 cells. *Arch Biochem Biophys* 2008;**476**:186–95.

31. Rosenberg DW, Giardina C, Tanaka T. Mouse models for the study of colon carcinogenesis. *Carcinogenesis* 2009;**30**:183–96.

32. Reddy BS. Studies with the azoxymethane-rat preclinical model for assessing colon tumor development and chemoprevention. *Environ Mol Mutagen* 2004;**44**:26–35.

33. Pritchard CC, Grady WM. Colorectal cancer molecular biology moves into clinical practice. *Gut* 2011;**60**:116–29.

34. Raju J. Azoxymethane-induced rat aberrant crypt foci: relevance in studying chemoprevention of colon cancer. *World J Gastroenterol* 2008;**14**:6632–5.

35. Kim J, Ng J, Arozulllah A, Ewing R, Llor X, Carroll RE, et al. Aberrant crypt focus size predicts distal polyp histopathology. *Cancer Epidemiol Biomark Prev* 2008;**17**:1155–62.

36. Rodríguez-Ramiro I, Ramos S, López-Oliva E, Agis-Torres A, Gómez-Juaristi M, Mateos R, et al. Cocoa-rich diet prevents azoxymethane-induced colonic preneoplastic lesions in rats by restraining oxidative stress and cell proliferation and inducing apoptosis. *Mol Nutr Food Res* 2011;**55**:1895–9.

37. Chen J, Huang X-F. The signal pathways in azoxymethane-induced colon cancer and preventive implications. *Cancer Biol Therap* 2009;**8**:1313–7.

38. Ashokkumar P, Sudhandiran G. Protective role of luteolin on the status of lipid peroxidation and antioxidant defense against azoxymethane-induced experimental colon carcinogenesis. *Biomed Pharmacoth* 2008;**62**:590–7.

39. Masella R, Di Benedetto R, Varì R, Filesi C, Giovannini C. Novel mechanisms of natural antioxidant compounds in biological systems: involvement of glutathione and glutathione-related enzymes. *J Nutr Biochem* 2005;**16**:577–86.

40. Storz P. Reactive oxygen species in tumor progression. *Front Biosci* 2005;**10**:1881–96.

41. Xavier CP, Lima CF, Preto A, Seruca R, Fernandes-Ferreira M, Pereira-Wilson C. Luteolin, quercetin and ursolic acid are potent inhibitors of proliferation and inducers of apoptosis in both KRAS and BRAF mutated human colorectal cancer cells. *Cancer Lett* 2009;**281**:162–70.

42. Itzkowitz SH, Yio X. Colorectal cancer in inflammatory bowel disease: The role of inflammation. *Am J Physiol Gastrointest Liver Physiol* 2004;**287**:G7–17.

43. Half E, Arber N. Colon cancer: Preventive agents and the present status of chemoprevention. *Expert Opin Pharmacother* 2009;**10**:211–9.

44. Romier B, Van De Walle J, During A, Larondelle Y, Schneider YJ. Modulation of signalling nuclear factor-kB activation pathway by polyphenols in human intestinal Caco-2 cells. *Br J Nutr* 2008;**100**:542–51.

45. Surh YJ, Chun KS, Cha HH, Han SS, Keum YS, Park KK, et al. Molecular mechanisms underlying chemopreventive activities of anti-inflammatory phytochemicals: down-regulation of COX-2 and iNOS through suppression of NF-kappa B activation. *Mutat Res* 2001;**481**:243–68.

46. Pan M-H, Lai CS, Wu J-Ch, Ho Ch-T. Molecular mechanisms for chemoprevention of colorectal cancer by natural dietary compounds. *Mol Nutr Food Res* 2011;**55**:32–45.

47. Bayard V, Chamorro F, Motta J, Hollenberg NK. Does flavanol intake influence mortality from nitric oxide-dependent processes? Ischemic heart disease, stroke, diabetes mellitus, and cancer in panama. *Int J Med Sci* 2007;**4**:53–8.

48. Arts IC, Hollman PC, Bueno de Mesquita HB, Feskens EJ, Kromhout D. Dietary catechins and epithelial cancer incidence: the Zutphen elderly study. *Int J Cancer* 2001;**92**:298–302.

49. Rouillier P, Senesse P, Cottet V, Valléau A, Faivre J, Boutron-Ruault MC. Dietary patterns and the adenomacarcinoma sequence of colorectal cancer. *Eur J Nutr* 2005;**44**:311–8.

50. McKelvey W, Greenland S, Sandler RS. A second look at the relation between colorectal adenomas and consumption of foods containing partially hydrogenated oils. *Epidemiology* 2000;**11**:469–73.

51. Boutron-Ruault MC, Senesse P, Faivre J, Chatelain N, Belghiti C, Méance S. Foods as risk factors for colorectal cancer: a case-control study in Burgundy (France). *Eur J Cancer Prev* 1999;**8**:229–35.

52. Spadafranca A, Martinez Conesa C, Sirini S, Testolin G. Effect of dark chocolate on plasma epicatechin levels, DNA resistance to oxidative stress and total antioxidant activity in healthy subjects. *Br J Nutr* 2010;**103**:1008–114.

53. Vázquez-Agell M, Urpí-Sarda M, Sacanella E, Camino-López S, Chiva-Blanch G, Llorente-Cortés V, et al. *Cocoa consumption reduces NF-κB activation in peripheral blood mononuclear cells in humans. Nutr Metab Cardiovasc Dis* 2013;**23**:257–63. http://dx.doi.org/10.1016/j.numecd.2011.03.015.

54. Monagas M, Khan N, Andrés-Lacueva C, Casas R, Urpí-Sarda M, Llorach R, et al. Effect of cocoa powder on the modulation of inflammatory biomarkers in patients at high risk of cardiovascular disease. *Am J Clin Nutr* 2009;**90**:1144–50.

Green Tea Polyphenols and Reduction of Oxidative Stress in Liver Cancer

Yasutaka Baba, Sadao Hayashi, Nanako Tosuji, Shunro Sonoda, Masayuki Nakajo

Department of Radiology, Graduate School of Medical and Dental Sciences, Kagoshima University, Kagoshima-city, Kagoshima, Japan

List of Abbreviations

GTP green tea polyphenol
d-ROMs derivatives of reactive oxygen metabolites
BAP biological antioxidative potential
HAI hepatic arterial infusion chemotherapy
HCC hepatocellular carcinoma

INTRODUCTION

Hepatocellular carcinoma (HCC) is the sixth-most common cancer and third-most common cause of cancer death in the world, according to a worldwide survey.[1] Epidemiological surveys have revealed that carcinogenesis in HCC is highly associated with the burden of chronic hepatitis virus infection.[2-4] In addition, nonalcoholic steatohepatitis (NASH) has recently been associated with the occurrence of HCC.[5]

Green tea is a traditional Asian beverage ingested for health, relaxation, and maintenance of physical condition. Recent studies have revealed that green tea polyphenols (GTPs; catechins) are active moieties with antioxidant and apoptotic effects on cancer cells, and thus may be useful for cancer prevention and treatment[6-13] (Figures 21.1–21.6).

Previous reports have shown that antioxidant agents reduce oxidative stress and increase antioxidative effects of the human body. According to the Cochrane Reviews,[10] GTPs are presumably effective for cancer prevention with the intake of 3 to 5 cups of green tea a day (providing a minimum of 250 mg of catechins per day), although the authors imply that this recommendation does not hinge on a representative number of randomized controlled trials.

In nutritional research, Kim et al.[14] reported that oxidative stress levels are lower in vegetarians than in omnivores by measuring diacron-reactive oxygen metabolites as a marker of oxidative stress.

OXIDATIVE STRESS ASSOCIATED WITH HCCS

The burden of chronic viral hepatitis is reportedly associated with the carcinogenic process of HCC[2-4] together with NASH[5,15] accounting for most cases of liver cirrhosis and consequent development of HCCs. Increased oxidative stress and decreased antioxidant levels in the body are thought to be key to hepatocellular carcinogenesis.[16,17] In models of hepatitis B or C infection, oxidative stress generates HCC by inducing iron overload and transforming growth factor (TGF)-β-mediated fibrosis, or by suppression of hepcidin leading to stepwise carcinogenesis of HCC.[18-23]

Our study suggested the potential for GTPs to reduce oxidative stress and increase the antioxidative potential of HCC patients.[24] In the clinical setting, we have many opportunities to provide hepatic arterial infusion chemotherapy (HAI) for patients with inoperable HCC.[24]

In the process of treatment, we encountered hepatic steatohepatitis due to direct arterial distribution of chemotherapeutic drugs or systemically toxic symptoms such as nausea, myelosuppression, and liver dysfunction. To overcome these adverse effects, we looked for "key drugs" to reduce side effects likely induced by oxidative stress. Our colleagues[25,26] reported *in vivo* and *in vitro* studies of GTPs that may be feasible for use

Cancer
http://dx.doi.org/10.1016/B978-0-12-405205-5.00021-0

© 2014 Elsevier Inc. All rights reserved.

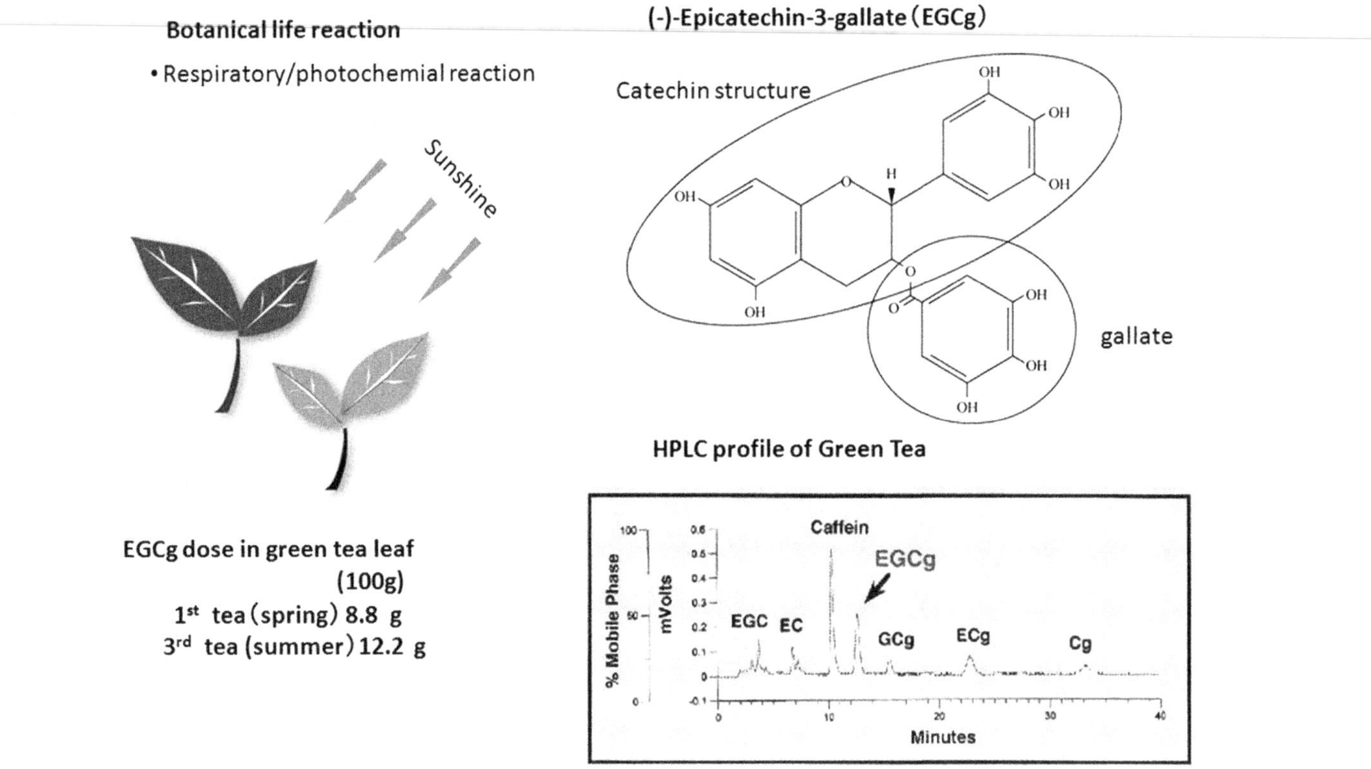

Botanical life reaction
• Respiratory/photochemial reaction
Sunshine
EGCg dose in green tea leaf (100g)
1st tea (spring) 8.8 g
3rd tea (summer) 12.2 g
(-)-Epicatechin-3-gallate (EGCg)
Catechin structure
gallate
HPLC profile of Green Tea
Caffein
EGCg
EGC
EC
GCg
ECg
Cg
% Mobile Phase
mVolts
Minutes

FIGURE 21.1 Polyphenol elements of green tea. Note: EGC, (-)-epigallocatechin; EC, (-)-epicatechin; GCg, (-)-gallocatechin-3-gallate; ECg(-), epicatechin-3-gallate; Cg, catechin gallate.

Human peripheral monocyte culture (+Staphylococcal Exotoxin E (SEE)
Unpublished data (Sonoda S, Fukumoto R, Espey M.)
Put Green Tea polyphenol
HPBL+SEE+GT(1,2,3,4)
+ SEE
- SEE
Green Tea Polyphenol
No
Yes
+SEE
- SEE
Superoxide dose
Response time

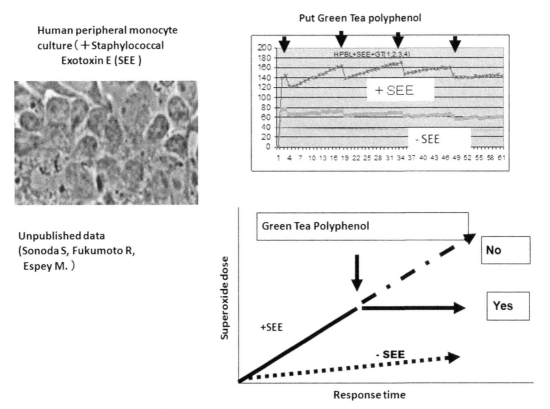

FIGURE 21.2 Antioxidant effects of green tea polyphenols.

in the treatment of solid cancer by reducing oxidative stress and increasing antioxidant levels in cancer patients. Based on these previous studies, we decided to use GTP tablets as a possible "key drug" for HCC patients receiving HAI.

Nineteen patients with inoperable HCC were allocated into two groups (receiving 3 or 6 tablets/day) and followed for 12 months to observe changes in derivatives of reactive oxygen metabolites (d-ROM) and biological antioxidative potential (BAP), as well as clinical laboratory data.

On the assumption that oxidative stress is continuously produced during HAI treatment, we checked and followed levels of oxidative stress (d-ROM) and antioxidant (BAP) using an automated reading system (FRAS4 Free Radical Analytical System, SEAC s.r.l., Florence, Italy) in reference to clinical laboratory data. We found significant reductions in d-ROM and unchanged levels of BAP in the group receiving 6 tablets/day and unchanged levels of d-ROM and decreased levels of BAP in the group receiving 3 tablets/day after 12 months (Tables 21.1 and 21.2). Total serum bilirubin (T-bil) levels were significantly increased at 3 and 9 months and red blood cell (RBC) counts were decreased at 6 months in the group receiving 6 tablets/day. These results suggest that increased T-bil may be associated with a higher turnover of GTP-bound RBCs and enhanced oxidant-scavenging capacity in the circulation, as postulated by other investigators.[27] If this hypothesis is correct, reductions in d-ROMs may coincide with increased RBC counts and decreased total bilirubin levels. One green tea tablet contains 79 mg of epigallocatechin-3-gallate (EGCG), so six tablets contain 474 mg of polyphenols, which should represent a sufficient amount for clinical antioxidant efficacy. However, our results hinge on a single-arm, prospective, randomized study, not a randomized control study. To address these limitations, we have initiated a fully randomized control trial studying two newly enrolled patient groups: one taking GTP tablets and the other taking placebo tablets. These groups are to be followed up using the same protocol applied in our previous study.

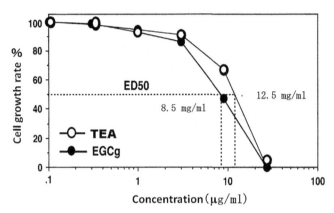

FIGURE 21.3 Growth rate restriction of ATL cells induced by green tea polyphenols.

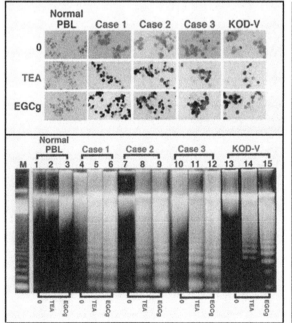
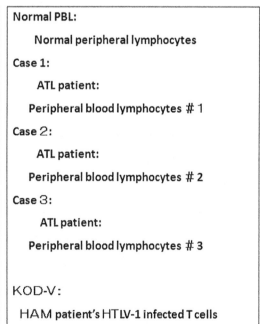

FIGURE 21.4 Apoptosis induction in ATL cells and HTLV-1-infected cells by green tea. (See the color plate.)

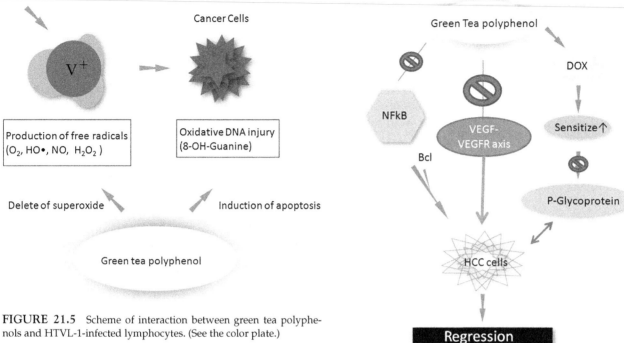

FIGURE 21.5 Scheme of interaction between green tea polyphenols and HTVL-1-infected lymphocytes. (See the color plate.)

FIGURE 21.6 Scheme of interaction between green tea polyphenols and hepatocellular carcinoma (HCC). Note: NF-κB, nuclear factor κB; Bcl, B-cell lymphoma; VEGF, vascular endothelial growth factor; DOX, doxorubicin; ⊘, blockade. (See the color plate.)

TABLE 21.1   d-ROM Value Reduction by Green Tea Polyphenol (GTP) Intake in 3 and 6 Tablet Groups

	Pre	3 Months	6 Months	9 Months	12 Months	P Value
	n = 9	n = 9	n = 9	n = 6	n = 5	
6 Tablet group	430.5(74.6)	401.5(64.4)	376.7(60.0)[a]	348.8(59.19)[b]	365.4(57.7)[c]	[a]P = 0.0463, [b]P = 0.0436
	n = 10	n = 10	n = 10	n = 7	n = 6	
3 Tablet group	391.8(87.4)	397.9(130.2)	410.5(151.1)	373.8(90.9)	377.3(100.3)	NS

Note: All numbers are the average d-ROM readings and the standard deviation (SD). NS, not statistically significant.
d-ROMs, derivatives of reactive oxygen metabolites; GTP, green tea polyphenol. The units for d-ROM are Carratelli units (Carr units).
[a]*Significant difference in d-ROM readings taken before and 6 months after the intake of 6 GTP tablets.*
[b]*Significant difference in d-ROM readings taken before and 9 months after the intake of 6 GTP tablets.*
[c]*Not yet analyzed due to the small number of subjects (n = 5). (Reproduced from reference 24 with the permission of the publisher.)*

TABLE 21.2   BAP Value Change by Green Tea Polyphenol (GTP) Intake in 3 and 6 Tablet Groups

	Pre	3 Months	6 Months	9 Months	12 Months	P Value
	n = 9	n = 9	n = 9	n = 6	n = 5	
6 Tablet group	2653.1(368.8)	2540.1(175.2)	2482.3(211.6)	2483.0(148.7)	2627.6(297.2) [a]	NS
	n = 10	n = 10	n = 10	n = 7	n = 6	
3 Tablet group	2694.8(230.0)	2526(160.5)	2639.6(165.9)	2652.8(263.1)	2590.6(158.9)[b]	[b] P = 0.0078

Note: All numbers are the average BAP readings and the standard deviation (SD).
[a]*Not yet analyzed due to the small number of patients (n = 5).*
[b]*Significant difference in BAP readings taken before and 12 months after the intake of 3 GTP tablets. NS, not statistically significant; BAP, biological antioxidative potential; GTP, green tea polyphenol. The units for BAP are micromole/l. (Reproduced from reference 24 with the permission of the publisher.)*

ANTITUMOR EFFECTS OF GTPS

Nishikawa et al.[28] reported that EGCG induces apoptosis in human lens epithelial (HLE) HCC cell lines both *in vivo* and *in vitro*. They speculated that the mechanism of apoptosis hinges on the down-regulation of B cell lymphoma (Bcl)-2a and Bcl-xl through the inactivation of nuclear factor (NF)-κB. Shirakami et al.[29] reported that EGCG inhibits the growth of HuH7 HCC cells by activating the vascular endothelial growth factor (VEGF)-tyrosine kinase VGEF receptor (VEGFR) axis. They concluded that EGCG has inhibiting effects on tumor growth rates via the VEGF-VEGFR axis and potential anticancer effects on HCC cell lines.[29] Although both studies only monitored levels *in vitro* and *in vivo*, we think that GTPs including EGCG might have impacts on anticancer efficacy in patients with HCC.

COMBINED EFFECTS OF GREEN TEA AND ANTICANCER DRUGS

Liang et al. reported that low (nontoxic) doses of GTPs could sensitize doxorubicin (DOX)-resistant HCC cells and render them to be susceptible to therapeutic doses of DOX. Such sensitizing effects of GTPs may involve the suppression of multiple drug resistance receptor 1 (MDR1) or by enhanced accumulation of intracellular DOX, eventually inhibiting P-glycoprotein (P-gp).[7] Taking these findings together, GTPs have a dual function, exerting anticancer effects via apoptosis and reversing anticancer drug resistance by blocking P-gp.

METABOLISM OF GTPS IN THE HUMAN BODY

Most GTPs take on conjugated forms in the blood or urine after ingestion of green tea.[30] Measuring the total amount of GTPs in the blood and urine is thus difficult. Although the content of GTPs in blood varies among tested samples,[30,31] the maximum plasma concentration (Cmax) of nonconjugated tea polyphenols (EGCG, epigallocatechin (EGC), epicatechin (EC)) is detectable at around two hours after ingestion. When the ingested dose of GTPs was doubled, Cmax increased 2.7- to 3.4-fold.[30,31] However, if the ingested dose of GTPs was increased beyond three-fold, no further increase in Cmax was identified. This saturation phenomenon has not yet been fully explained, but suggests that multiple molecular interactions between GTPs and mucosal receptors in the digestive tract may be involved. Meanwhile, GTPs are largely excreted into urine within 8 to 9 hours and completely excreted within 24 hours after ingestion.[30,31] Kuzuhara

et al.[32] reported that EGCG has the potential to bind both DNA and RNA, preventing injury from oxidative stress.

PREVIOUS REPORTS OF GTPS AND HTLV-1-ASSOCIATED T-CELL LEUKEMIA (ADULT T-CELL LEUKEMIA)

Prior to our clinical study on GTPs for patients with HCC, our colleagues reported that GTPs inhibited *in vitro* proliferation of human T-cell lymphotropic virus (HTLV)-infected lymphocytes and adult T-cell leukemia cells through antioxidant effects and apoptotic cell death by suppressing HTLV-pX gene expression and inducing apoptotic responses.[25] Meanwhile, Sonoda et al.[26] revealed that daily intake of green tea tablets (700 mg of polyphenols) for 5 months diminished the HTLV-1 provirus load in the peripheral blood lymphocytes of HTLV-1-infected patients. These studies prompted us to expand our clinical research into GTPS for HCC associated with HBV/HCV.

CONTROVERSIAL RESULTS CONCERNING THE EFFICACY OF GTPS

In contrast to our results for antioxidative effects of GTPs, some reports have found that GTPs, particularly EGCG, may induce oxidative damage to cellular DNA.[33–35] The major cause of oxidative damage may be pro-oxidant effects in which EGCG may induce oxidative DNA damage in the presence of metal ions (Cu^{2+} or Fe^{2+}).[34] Our results for measurements of d-ROM and BAP could not reveal whether GTPs with EGCG comprising the major component is associated with the development of carcinogenesis. However, further investigation is needed to verify the exact role of GTPs in HCCs and whether an endogenous antioxidant system could be generated by the consumption of green tea.[36]

SUMMARY POINTS

- Hepatocellular carcinoma (HCC) is the sixth most common cancer and third most common cause of cancer death according to a nationwide survey.
- Increased oxidative stress and decreased antioxidant state in a body may be key to the occurrence of HCC.
- Green tea is a traditional Asian beverage taken for good health, relaxation, and maintenance of physical strength. Recent studies have revealed that green tea catechins (polyphenols) are active moieties with antioxidant and apoptotic effects on cancer cells, and these pharmacological effectors are thus potentially useful for cancer prevention and treatment.

- We described our initial results for the efficacy of green tea polyphenols in reducing oxidative stress and antioxidant potential in patients with HCC receiving arterial infusion chemotherapy.
- Clarification of the exact roles of GTPs in oxidative stress and antioxidant potential is needed among patients with HCC using further fully randomized controlled trial studies.

Acknowledgement

We wish to thank Mrs. Yukiko Baba for creating the graphics for the figures.

References

1. Parkin DM, Bray F, Ferlay J, Pisani P. Global cancer statistics, 2002. *CA Cancer J Clin* 2005;**55**(2):74–108.
2. El Khoury AC, Klimack WK, Wallace C, Razavi H. Economic burden of hepatitis C-associated diseases in the United States. *J Viral Hepat* 2012;**19**(3):153–60.
3. El-Serag HB. Epidemiology of viral hepatitis and hepatocellular carcinoma. *Gastroenterology* 2012;**142**(6):1264–73.
4. Sarkar M, Stewart S, Yu A, Chen MS, Nguyen TT, Khalili M. Hepatocellular carcinoma screening practices and impact on survival among hepatitis B-infected Asian Americans. *J Viral Hepat* 2012;**19**(8):594–600.
5. Starley BQ, Calcagno CJ, Harrison SA. Nonalcoholic fatty liver disease and hepatocellular carcinoma: A weighty connection. *Hepatology* 2010;**51**(5):1820–32.
6. Ellinger S, Muller N, Stehle P, Ulrich-Merzenich G. Consumption of green tea or green tea products: is there an evidence for antioxidant effects from controlled interventional studies? *Phytomedicine* 2011;**18**(11):903–15.
7. Liang G, Tang A, Lin X, Li L, Zhang S, Huang Z, et al. Green tea catechins augment the antitumor activity of doxorubicin in an in vivo mouse model for chemoresistant liver cancer. *Int J Oncol* 2010;**37**(1):111–23.
8. Park IJ, Lee YK, Hwang JT, Kwon DY, Ha J, Park OJ. Green tea catechin controls apoptosis in colon cancer cells by attenuation of H2O2-stimulated COX-2 expression via the AMPK signaling pathway at low-dose H2O2. *Ann N Y Acad Sci* 2009;**1171**:538–44.
9. Lee AS, Jung YJ, Kim DH, Lee TH, Kang KP, Lee S, et al. Epigallocatechin-3-O-gallate decreases tumor necrosis factor-alpha-induced fractalkine expression in endothelial cells by suppressing NF-kappaB. *Cell Physiol Biochem* 2009;**24**(5-6):503–10.
10. Boehm K, Borrelli F, Ernst E, Habacher G, Hung SK, Milazzo S, et al. Green tea (*Camellia sinensis*) for the prevention of cancer. *Cochrane Database Syst Rev* 2009;(3) CD005004.
11. Yang CS, Lambert JD, Ju J, Lu G, Sang S. Tea and cancer prevention: Molecular mechanisms and human relevance. *Toxicol Appl Pharmacol* 2007;**224**(3):265–73.
12. Landis-Piwowar KR, Huo C, Chen D, Milacic V, Shi G, Chan TH, et al. A novel prodrug of the green tea polyphenol (-)-epigallocatechin-3-gallate as a potential anticancer agent. *Cancer Res* 2007;**67**(9):4303–10.
13. Suganuma M, Kurusu M, Suzuki K, Tasaki E, Fujiki H. Green tea polyphenol stimulates cancer preventive effects of celecoxib in human lung cancer cells by upregulation of GADD153 gene. *Int J Cancer* 2006;**119**(1):33–40.
14. Kim MK, Cho SW, Park YK. Long-term vegetarians have low oxidative stress, body fat, and cholesterol levels. *Nutr Res Pract* 2012;**6**(2):155–61.
15. Hashimoto E, Tokushige K. Hepatocellular carcinoma in non-alcoholic steatohepatitis: Growing evidence of an epidemic? *Hepatol Res* 2011;**42**(1):1–14.
16. Maki A, Kono H, Gupta M, Asakawa M, Suzuki T, Matsuda M, et al. Predictive power of biomarkers of oxidative stress and inflammation in patients with hepatitis C virus-associated hepatocellular carcinoma. *Ann Surg Oncol* 2007;**14**(3):1182–90.
17. Choi J. Oxidative stress, endogenous antioxidants, alcohol, and hepatitis C: Pathogenic interactions and therapeutic considerations. *Free Radic Biol Med* 2012;**52**(7):1135–50.
18. Machida K, Cheng KT, Lai CK, Jeng KS, Sung VM, Lai MM. Hepatitis C virus triggers mitochondrial permeability transition with production of reactive oxygen species, leading to DNA damage and STAT3 activation. *J Virol* 2006;**80**(14):7199–207.
19. Machida K, Cheng KT, Sung VM, Lee KJ, Levine AM, Lai MM. Hepatitis C virus infection activates the immunologic (type II) isoform of nitric oxide synthase and thereby enhances DNA damage and mutations of cellular genes. *J Virol* 2004;**78**(16): 8835–43.
20. Machida K, Cheng KT, Sung VM, Shimodaira S, Lindsay KL, Levine AM, et al. Hepatitis C virus induces a mutator phenotype: Enhanced mutations of immunoglobulin and protooncogenes. *Proc Natl Acad Sci U S A* 2004;**101**(12):4262–7.
21. de Mochel NS, Seronello S, Wang SH, Ito C, Zheng JX, Liang TJ, et al. Hepatocyte NAD(P)H oxidases as an endogenous source of reactive oxygen species during hepatitis C virus infection. *Hepatology* 2010;**52**(1):47–59.
22. Teixeira R, Marcos LA, Friedman SL. Immunopathogenesis of hepatitis C virus infection and hepatic fibrosis: New insights into antifibrotic therapy in chronic hepatitis C. *Hepatol Res* 2007;**37**(8): 579–95.
23. Liu RM, Gaston Pravia KA. Oxidative stress and glutathione in TGF-beta-mediated fibrogenesis. *Free Radic Biol Med* 2009;**48**(1): 1–15.
24. Baba Y, Sonoda J, Hayashi S, Tosuji N, Sonoda S, Makisumi K, et al. Reduction of oxidative stress in liver cancer patients by oral green tea polyphenol tablets during hepatic arterial infusion chemotherapy. *Exp Ther Med* 2012;**4**:452–8.
25. Li HC, Yashiki S, Sonoda J, Lou H, Ghosh SK, Byrnes JJ, et al. Green tea polyphenols induce apoptosis in vitro in peripheral blood T lymphocytes of adult T-cell leukemia patients. *Jpn J Cancer Res* 2000;**91**(1):34–40.
26. Sonoda J, Koriyama C, Yamamoto S, Kozako T, Li HC, Lema C, et al. HTLV-1 provirus load in peripheral blood lymphocytes of HTLV-1 carriers is diminished by green tea drinking. *Cancer Sci* 2004;**95**(7):596–601.
27. Kimura M, Umegaki K, Kasuya Y, Sugisawa A, Higuchi M. The relation between single/double or repeated tea catechin ingestions and plasma antioxidant activity in humans. *Eur J Clin Nutr* 2002;**56**(12):1186–93.
28. Nishikawa T, Nakajima T, Moriguchi M, Jo M, Sekoguchi S, Ishii M, et al. A green tea polyphenol, epigalocatechin-3-gallate, induces apoptosis of human hepatocellular carcinoma, possibly through inhibition of Bcl-2 family proteins. *J Hepatol* 2006;**44**(6):1074–82.
29. Shirakami Y, Shimizu M, Adachi S, Sakai H, Nakagawa T, Yasuda Y, et al. (-)-Epigallocatechin gallate suppresses the growth of human hepatocellular carcinoma cells by inhibiting activation of the vascular endothelial growth factor-vascular endothelial growth factor receptor axis. *Cancer Sci* 2009;**100**(10):1957–62.
30. Lee MJ, Wang ZY, Li H, Chen L, Sun Y, Gobbo S, et al. Analysis of plasma and urinary tea polyphenols in human subjects. *Cancer Epidemiol Biomark Prev* 1995;**4**(4):393–9.
31. Yang CS, Chen L, Lee MJ, Balentine D, Kuo MC, Schantz SP. Blood and urine levels of tea catechins after ingestion of different amounts of green tea by human volunteers. *Cancer Epidemiol Biomark Prev* 1998;**7**(4):351–4.

32. Kuzuhara T, Tanabe A, Sei Y, Yamaguchi K, Suganuma M, Fujiki H. Synergistic effects of multiple treatments, and both DNA and RNA direct bindings on, green tea catechins. *Mol Carcinog* 2007;**46**(8):640–5.

33. Oikawa S, Furukawa A, Asada H, Hirakawa K, Kawanishi S. Catechins induce oxidative damage to cellular and isolated DNA through the generation of reactive oxygen species. *Free Radic Res* 2003;**37**(8):881–90.

34. Furukawa A, Oikawa S, Murata M, Hiraku Y, Kawanishi S. (-)-Epigallocatechin gallate causes oxidative damage to isolated and cellular DNA. *Biochem Pharmacol* 2003;**66**(9):1769–78.

35. Babich H, Schuck AG, Weisburg JH, Zuckerbraun HL. Research strategies in the study of the pro-oxidant nature of polyphenol nutraceuticals. *J Toxicol* 2011 2011:467305. Epub 2011 Jun 26.

36. Lambert JD, Elias RJ. The antioxidant and pro-oxidant activities of green tea polyphenols: A role in cancer prevention. *Arch Biochem Biophys* 2010;**501**(1):65–72.

Quercetin's Potential to Prevent and Inhibit Oxidative Stress-Induced Liver Cancer

Ming-Ta Sung

Department of Medical Research and Education, Taipei Veterans General Hospital, Taipei, Taiwan

Yin-Chiu Chen

Institute of Environmental and Occupational Health Sciences, School of Medicine, National Yang-Ming University, Taipei, Taiwan

Chin-Wen Chi

Department of Medical Research and Education, Taipei Veterans General Hospital and Institute of Pharmacology, School of Medicine, National Yang-Ming University, Taipei, Taiwan

List of Abbreviations

GSH Glutathione
HBV hepatitis B virus
HBx hepatitis B virus X protein
HCC hepatocellular carcinoma
HCV hepatitis C virus
H_2O_2 hydrogen peroxide
HSC hepatic stellate cells
MnSOD manganese superoxide dismutase
NF-κB nuclear factor-κB
8-OHdG 8-hydroxydeoxyguanosine
ROS reactive oxygen species
SOD superoxide dismutase
Trx thioredoxin

INTRODUCTION

Hepatocellular carcinoma (HCC) is the third leading cause of death in cancer worldwide and is the most common type of liver cancer. Other types of liver cancer include cholangiocarcinoma, angiosarcoma, and hepatoblastoma, which represent 15% of liver cancers. More than 80% of HCC occur in the developing countries such as Southeast Asia and Far East Asia.[1] Chronic hepatitis, fibrosis, and cirrhosis are all liver diseases with higher risk to become HCC. The main etiology for HCC is represented by persistent Hepatitis B Virus (HBV) and Hepatitis C Virus (HCV) infections; thus countries with higher HBV

or HCV carriers usually accompany with high incidence of HCC. The chronic HBV carriers have about 100 times higher possibility than uninfected individuals to develop liver cancer. Additionally, alcohol abuse, cigarette smoking, aflatoxin B1 intake, and obesity are the other risk factors correlated with the development of HCC.[2]

Currently, surgical resection is the most effective treatment for HCC, but this approach has limitations for patients with large (> 5 cm) or multiple tumors.[3] Liver transplantation is another option for treatment of patients with HCC, but limited numbers of donors have restricted extensive application of this therapy. Moreover, because of the lack of specific markers for early detection, HCC is usually diagnosed at late stage. Thus, only a small percentage of patients are suitable for partial hepatectomy or liver transplantation.

Although transcatheter arterial chemoembolization (TACE), percutaneous ethanol injection (PEI), and radiofrequency thermal ablation (RFA) are the other options for nonsurgical therapies, HCC patients receiving these treatments still had higher recurrence rates and lower survival rates than HCC patients treated with surgery.[3] Multiple mechanisms such as drug resistance account for the low response rate in chemotherapy, so understanding the mechanisms of HCC development is urgently needed to develop more effective strategies for HCC prevention and treatment.

Cancer
http://dx.doi.org/10.1016/B978-0-12-405205-5.00022-2

© 2014 Elsevier Inc. All rights reserved.

Interestingly, ample evidence has showed that increased reactive oxygen species (ROS) and decreased antioxidants were found in various types of liver disease. Moreover, many studies have focused on the significance of oxidative stress in liver cancer, and have indicated that ROS plays important roles in the development and progression of liver cancer.[4] Indeed, excessive ROS production causes increased levels of oxidative stress that is considered to contribute to a variety of human diseases, including neurodegenerative diseases, aging, and cancer.[5] Thus, in this chapter we focus on the roles of ROS in the development of HCC.

ROLE OF OXIDATIVE STRESS IN THE DEVELOPMENT OF LIVER CANCER

ROS and Oxidative Stress

ROS, acting as an intracellular signaling molecule, plays a role in several physiologic processes, such as proliferation, cell-cycle regulation, apoptosis, signaling transduction, Ca^{2+} accumulation, and cytokine gene expression, etc. The most important ROS, including hydrogen peroxide (H_2O_2), superoxide anion ($O2^{-}\cdot$), and the highly reactive hydroxyl radical ($\cdot OH$), are the by-products of metabolic reactions generated by all aerobic cells (Figure 22.1). Under normal physiologic conditions, cells maintain redox homeostasis and/or a balance between ROS and antioxidants. ROS is normally regulated at relatively low steady-state levels, thus temporary elevated ROS will be scavenged immediately by endogenous antioxidative enzyme systems.

The oxidative stress condition plays a key role in the initiation and development of cancer by modifying cellular processes, such as oxidative DNA damage and lipid

peroxidation product, caused primarily by hydroxyl radicals.[6] Abnormal production of ROS together with deficient antioxidants may cause redox imbalance and oxidative stress to the cells, further resulting in abnormal signal transduction, irreversible damage responses, and other pathological conditions. These ROS mediated or activated signal pathways, including mitogen-activated protein kinase (MAPK), nuclear factor-κB (NF-κB), phosphatidylinositol 3-kinase (PI3K), p53, and β-catenin/Wnt, are involved in the development and progression of HCC.[7] Moreover, initiation of cancer in humans by inflammation-generated oxidative stress is further supported by the presence and accumulation of pro-mutagenic DNA adduct 8-hydroxydeoxyguanosine (8-OHdG) in adjacent nontumor liver tissue as compared to HCC.[8] The increased level of ROS is associated with the increase of the expression of inducible manganese superoxide dismutase (MnSOD). Therefore, relatively high values of serum MnSOD levels and activity have been observed in patients with chronic hepatitis, liver cirrhosis, and HCC as compared to healthy subjects.[9]

Mechanisms of ROS-Induced Oxidative Stress

ROS are highly reactive molecules, and robust and sustained ROS can induce oxidative stress and direct damage to multiple cellular components like DNA, lipids, and proteins (Table 22.1). Thus, persistent oxidative stress ultimately may lead to uncontrollable cell growth and malignant tumor formation.

Oxidative damage can induce the formation of DNA adducts and chromosomal double-stranded breaks, resulting in gene mutagenesis, genomic instability, aneuploidy, and amplification of DNA. More than 100 types of oxidative DNA adducts, including 8-OHdG, have been identified. Moreover, oxidative DNA damages such as 8-OHdG accumulation lead to cell-cycle arrest, which inhibits cell growth and induces apoptosis.[10] ROS-induced lipid peroxidation produces various products such as malondialdehyde (MDA), and may

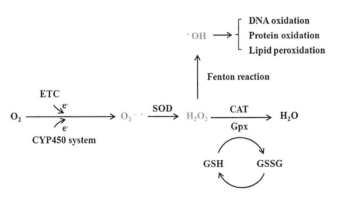

FIGURE 22.1   Intercellular ROS System. The most important ROS include superoxide anion (O2$^{-}\cdot$), hydrogen peroxide (H_2O_2), and highly reactive hydroxyl radical ($\cdot OH$). Endogenous antioxidant enzymes are superoxide dismutase (SOD), catalase (CAT), and glutathione peroxidase (Gpx). Electrons (e^{-}) are produced by cytochrome P450 (CYP450) system and leakage of electron transport chain (ETC). GSH and GSSG are reduced glutathione and oxidized glutathione.

TABLE 22.1   ROS Induced Oxidative Damage to DNA, Protein, and Lipid

Target	Oxidative Damages
DNA	Loss of genomic integrity by deoxyribose oxidation, strand breakage, removal of nucleotides, and modification of bases.[10]
Protein	Alternation of cellular signal pathways by site-specific amino acid modification, fragmentation of the peptide chain, altered electric charge, enzyme inactivation, and increased susceptibility of proteins to proteolysis.[11]
Lipid	Increase in membrane fluidity and permeability by oxidation of polyunsaturated fatty acids.[11]

react with DNA to form DNA adducts. Indeed, increased plasma MDA concentrations found in many diseases such as chronic myeloid leukemia has been considered as a biomarker for evaluating oxidative damage and disease progression.[11]

The effect of oxidative stress on protein is the conformational changes of redox-related proteins, including kinases, phosphatases, and transcription factors, which cause the changes of enzymatic activities or protein-protein interactions.[12] For example, thioredoxin (Trx)-ASK1 complex regulates MAPK activity in response to ROS (Figure 22.2a).[13] Reduced Trx, an intracellular ASK1 inhibitor, directly interacts to the ASK1 and inhibits the activation of apoptosis-regulating kinase ASK1. After ROS treatment, Trx is oxidized and dissociates from ASK1, thereby resulting in activation of ASK1. Thus, activated ASK1 can trigger its downstream mitogen activated protein signaling pathway, activate JNK, and promote cell apoptosis.

Another example is the Nrf2-Keap1 complex, which regulates the expression of many cytoprotective genes in response to ROS (Figure 22.2b).[14] Keap1, a negative regulator of cytoplasmic Nrf2, controls the subcellular distribution of Nrf2. After ROS treatment, oxidative stress disrupts Keap1-Nrf2 interactions by oxidative reaction, thereby resulting in release of Nrf2, and subsequently translocates into the nucleus. The, translocated Nrf2 dimerized with MAF can bind to antioxidant response element (ARE) to promote various antioxidative genes expression.

Several studies have suggested the important connection between NF-κB, inflammation, and cancers.[15] NF-κB has been found to be constitutively activated in various cancers, such as HCC. Transcription factor NF-κB regulated by ROS is an oxidative stress-inducible antiapoptotic transcription factor (Figure 22.2c). Activation of NF-κB is mediated by the degradation of its inhibitor IκB, resulting in translocation of NF-κB from the cytoplasm into the nucleus and regulation of gene expression. Moreover, it has been demonstrated that suppression of NF-κB activity inhibits the development of liver cancer in Mdr2−/− knockout mice.[16] These results suggest that NF-κB activated by ROS plays an essential role in promoting inflammation-associated cancer.

Oxidative Stress and Liver Cancer

Mitochondria, using electron transport chain and ATP synthase to generate ATP, are the major producers of ROS in hepatocytes. During ATP synthesis by aerobic respiration, molecular oxygen is converted to superoxide and other ROS metabolites. Another main source of ROS production in hepatocytes is the cytochrome P450 (CYP450) system, which is associated with detoxifying toxic substances such as drug and alcohol in the liver.

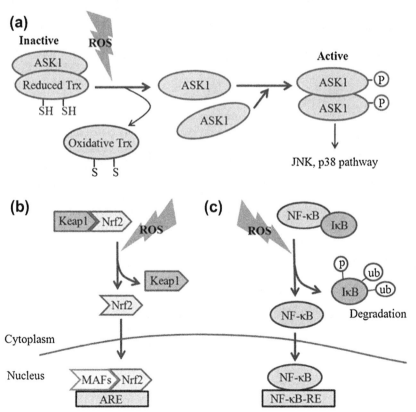

FIGURE 22.2 Regulation of Protein Activation by ROS. (a) Reduced thioredoxin (Trx), an intracellular apoptosis signal-regulating kinase 1 (ASK1) inhibitor, directly interacts with ASK1 and inhibits the activation of ASK1. After ROS treatment, Trx is oxidized and dissociates from ASK1, thereby resulting in activation of ASK1 and its downstream JNK and p38 pathways. (b) Keap1, a negative regulator of cytoplasmic Nrf2, controls the subcellular distribution of Nrf2. After ROS-induced oxidative stress disrupts Keap1-Nrf2 interactions, it results in release of Nrf2 and subsequently translocates into the nucleus to bind MAFs. After heterodimerization with the MAFs, the MAFs-Nrf2 binds to the antioxidant response element (ARE). (c) Upon ROS insult, activation of NF-κB is mediated by the degradation of its inhibitor IκB, resulting in translocation of NF-κB into the nucleus and binding to response element (NF-κB-RE).

In addition to hepatocytes, ROS is also produced by Kupffer and inflammatory cells in the liver.

ROS has been considered a key player in the development and progression of many pathological conditions, including virus-induced hepatitis, fibrosis, and liver cancer (e.g., HCC; Figure 22.3).[17] For example, fatty liver disease caused by obesity, excessive alcohol consumption, or hepatitis virus infection is the most common type of liver disease. However, persistent production of ROS leads to membrane lipid peroxidation and ATP depletion, and thereby promotes fatty liver progress to steatohepatitis, fibrosis, cirrhosis, or liver cancer.[18]

Hepatic stellate cells (HSC), involved in synthesis of extracellular matrix and collagen, play a central role in the pathogenesis of liver fibrosis and liver cancer.[19] Liver injury and inflammation generated ROS induces HSC proliferation and activation, which promotes transdifferentiation from quiescent HSC into collagen-producing cells, resulting in increased collagen synthesis.[20] It has also been found that MMP2 is necessary for ROS-induced proliferation of HSC, and ERK1/2 and PI3K signaling pathways are involved in ROS-induced MMP2 expression in HSC. Moreover, platelet-derived growth factor (PDGF) and transforming growth factor-β (TGF-β) released from immune cells are two characterized cytokines involved in HSC activation. PDGF acts as the mediator to stimulate HSC proliferation and leads to increases in collagen deposition and fibrosis, which is markedly suppressed by treatment with the antioxidants.[21] TGF-β induces ROS production, decreases the concentration of glutathione (GSH), and stimulates fibrogenic response of HSC.[22] ROS producers Kupffer cells and damaged hepatocytes are also associated with HSC activation. Therefore, chronic liver inflammation persistently generates excessive ROS and induces higher levels of oxidative stress, ultimately resulting in liver fibrosis, cirrhosis, or liver cancer.

Hepatitis virus infection induced liver inflammation is another source of oxidative stress to activate HSC and contribute to hepatic fibrogenesis and liver cancer. Experiments using HBV gene-carrying transgenic mice or HBV DNA-transfected HepG2 hepatoma cells clearly showed that HBV can induce oxidative stress and inhibit tumor suppressor genes such as GSTP1.[23] The level of peroxide, an indicator of oxidative stress, is significantly higher in patients with chronic hepatitis, suggesting that oxidative stress plays an important role in viral hepatitis. Moreover, oxidative DNA damage is also found in HBV or HCV patients. Human hepatitis B virus X protein (HBx) encoded by HBV is found to be associated with an increase of ROS during chronic HBV infection. HBx directly interacts with voltage-dependent anion channel of mitochondria and alters the mitochondrial membrane potential, resulting in increases of the intracellular ROS level.[24] Moreover, HBx induces oxidative stress via calcium signaling and constitutively activates transcription factors STAT-3 and NF-κB.[25] HBx also induces lipid peroxidation through inhibiting the expression of selenoprotein P, resulting in upregulation of tumor necrosis factor-α expression in human HepG2 hepatoma cells.[26] Moreover, DNA adduct 8-OHdG, the marker of DNA oxidative damage, is found to be increased in patients with chronic hepatitis, liver cirrhosis, and hepatocellular carcinoma as compared to controls.[27] Studies using HCV core protein-expressing transgenic mice found that increased accumulation of ROS and an unbalanced oxidant/antioxidant state in the liver is associated with the development of HCC.[28]

Excessive alcohol use has been linked to the progression of liver disease from fatty liver to alcoholic hepatitis and cirrhosis, which may ultimately cause liver cancer.[29] Alcohol consumption induces ROS production, suppresses cellular antioxidant levels, and increases

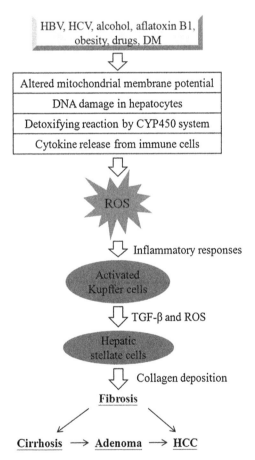

FIGURE 22.3   Role of ROS in the Development of Liver Diseases. HBV, HCV, alcohol, aflatoxin B, obesity, and diabetes mellitus (DM) are risk factors correlated with the development of liver diseases. ROS induced inflammatory responses activate Kupffer cells, the macrophages in the liver. Activated Kupffer cells secrete ROS and TGF-β to activate hepatic stellate cells (HSC), leading to collagen deposition in liver and resulting in fibrosis. Long-term effects are cirrhosis, then producing adenoma, and eventually generating hepatocellular carcinoma (HCC).

oxidative stress in many tissues, especially the liver, the major site for ethanol metabolism.[30] Alcohol is enzymatically converted into many metabolites, including ROS and acetaldehyde by alcohol dehydrogenase and CYP2E1. Acetaldehyde produced by the oxidation of alcohol promotes cell death by inducing lipid peroxidation and by increasing free radical-mediated toxicity. Acetaldehyde also stimulates HSC proliferation and collagen synthesis. Kupffer cells are the resident macrophages of the liver that elicit inflammatory and innate immune reaction in response to alcohol-induced liver injury and contribute to activation of HSCs.[31] Chronic ethanol consumption suppresses hepatic mitochondrial energy metabolism, leading to lower ATP production and redox shift. Recent studies also found that heavy alcohol consumption significantly increased the risk of developing liver cancer in cirrhotic patients with HBV infection.[32] In brief, excessive hepatic oxidative stress induced by HBV, HCV, alcohol, aflatoxin B, or obesity indeed plays a key role in the development of liver cancer.

ANTIOXIDANTS AS THE OXIDATIVE STRESS SCAVENGER

Endogenous Antioxidants

Although robust ROS is toxic and harmful to humans, our body still has developed several antioxidative systems to neutralize and eliminate it. The endogenous antioxidants that directly target to ROS to protect cells from oxidative damage can be divided into enzymatic and nonenzymatic antioxidants. These antioxidative enzymes are superoxide dismutase (SOD), catalase (CAT), glutathione peroxidase (Gpx), and Trx.[33] SOD catalyzes the superoxide anion into oxygen and hydrogen peroxide. Glutathione peroxidase, an important water-soluble antioxidant, can directly neutralize cellular ROS such as lipid peroxides. Catalase, found in peroxisomes in most eukaryotic cells, catalyzes the conversion of hydrogen peroxide to water and oxygen. For optimum catalytic activity of these antioxidative enzymes, trace minerals such as selenium, iron, copper, zinc, and manganese are needed and used as cofactors. The nonenzymatic antioxidants include glutathione, lipophilic α-tocopherol (vitamin E) and ubiquinol-10.

Antioxidants act as scavengers that directly remove oxygen-derived ROS. During this reaction, the antioxidants can transfer an electron or donate a proton to ROS. Consequently, ROS is neutralized to form a relatively stable product; meanwhile, the antioxidant itself becomes oxidized.

Antioxidants in the Food and Their Protective Effect against Oxidative Stress

As described earlier, oxidative stress plays a key role in liver cancer caused by viral infection or alcohol abuse. Thus, several strategies have been used to prevent ROS induced oxidative damage. In addition to endogenous antioxidants, natural antioxidants such as curcumin, silymarin, green tea, vitamins E, and quercetin available in food are beneficial in reducing ROS production in the liver (Table 22.2). Also, deep seawater from 200 m below the surface of the sea has high levels of minerals, such as magnesium (Mg), calcium (Ca), and potassium (K). Deep seawater drinking has been found to have antioxidant activity to maintain higher liver glutathione and has the potential to prevent high-fat diet induced hepatosteatosis and hepatic inflammation.[34] Whether drinking water with high levels of Ca or Mg will reduce serum cholesterol and oxidative stress in the liver awaits further investigation.

Curcumin, the main active compound of turmeric, has been used widely to treat inflammatory disorders, including sprain and arthritis.[35] In addition to anti-inflammatory activity, curcumin also exhibits its antioxidative activity to scavenge the hydroxyl radical. Indeed, curcumin reduces the risk of DNA damage by protecting DNA from oxidative stress. Curcumin exerts beneficial effects on the inhibition of thioacetamide-induced hepatic cirrhosis in animal models, which is mainly due to its anti-inflammatory activities.[36] Curcumin can regulate acute/chronic inflammation cascades and prevent

TABLE 22.2 List of Protective Effects of Antioxidants on Oxidative Stress-Induced Liver Injury

Antioxidant	Injury Type	Protective Effects	Reference
Curcumin	Ethanol-induced oxidative damage	↓MDA; ↑HO-1; ↑GSH	[37]
Silymarin	N-nitrosodiethylamine-induced HCC	↓MDA; ↑GSH; ↑Activities of antioxidant enzymes	[40]
Green tea	Diet-induced steatohepatitis	↓TNF-α; ↓NF-κB activity; ↑GSH	[41]
Vitamin E	Alcohol-induced oxidative damage	↓MDA; ↓NF-κB activity; ↑SOD	[44]
Quercetin	Ethanol-induced oxidative damage	↓MDA; ↓GSH; ↑CAT activity	[47]

alcohol-induced liver injury through reducing the activation of NF-κB or inducing the expression of heme oxygenase-1.[37] Curcumin upregulates matrix metalloproteinase expression via PPARγ in HSC, thereby inhibiting extracellular matrix formation and development of fibrosis. Curcumin exerts antiviral activity to suppress the production of HBV particles by upregulating p53 expression, which is the reason why curcumin has been used to treat HBV-induced liver disease in Asia.[38]

Silymarin, the purified extract from milk thistle, has shown protective effects against different types of acute and chronic liver diseases. In view of the hepatoprotective effects silymarin has been used for treatment in clinical liver diseases, such as viral hepatitis, toxic hepatitis, liver cirrhosis, and alcoholic liver damage.[39] Moreover, silymarin can stabilize the liver cell membranes and exert antioxidative activity to defend against oxidative attack, promote liver regeneration, reduce inflammation, inhibit liver fibrogenesis, and prevent liver cancer.[40] The anti-inflammatory and anticancer effects of silymarin are mediated through suppressing TNF-induced activation of NF-κB, which is induced by various carcinogens and inflammatory agents.

Green tea containing catechins is made from leaf extract of the plant *Camellia sinensis*. It has antioxidative activity to scavenge superoxide, singlet oxygen, and hydrogen peroxide. It can also enhance the expression of other antioxidants and detoxifying enzymes to prevent DNA from oxidative damage. Similar to other flavonoids, green tea can inhibit chronic inflammation and carcinogenesis through suppressing NF-κB activation.[41] Treatment with natural green tea extract also inhibits HBV production *in vitro*.[42]

Vitamin E, a fat-soluble antioxidant, can reduce the production of ROS and prevent membrane fatty acids from lipid peroxidation. In patients with alcoholic liver disease it was found that serum vitamin E concentration is significantly decreased with a concomitant increased level of lipid peroxidation, indicating that antioxidant status is negatively correlated with the level of oxidative damage.[43] Therefore, maintenance of normal concentrations of vitamin E seems to be essential to prevent oxidant-induced lipid peroxidation. In the mice model, alcohol treatment leads to a decrease in SOD, and an increase in lipid peroxidation and apoptosis; however, vitamin E supplementation recovered redox status, and reduced oxidative stress and apoptosis.[44] These results together suggest that vitamin E can be used as a potential therapeutic agent against alcohol-induced liver injury.

Quercetin, belonging to the flavonoid family, has antioxidative activity. It is ubiquitously found in vegetables and fruits such as apples, onions, nuts, and berries. It has been reported that quercetin is the major flavonoid found in foods and the average intake of quercetin is between 6 and 31 mg per day.[45] Quercetin was absorbed in the gastrointestinal tract after food intake and digestion.

In addition to antioxidative ability that scavenges highly reactive ROS, quercetin is also involved in many physiological and pathological processes. For example, quercetin can prevent low-density lipoprotein oxidation, inhibit platelet aggregation, and promote cardiovascular smooth muscle relaxation. Quercetin also protects against many diseases such as osteoporosis, heart diseases, aging, and cancer.

In vitro antioxidative effects of quercetin on the hepatocytes have been studied. Quercetin is effective in chelating iron ions and then scavenges free radicals. By scavenging superoxide, quercetin also reduces liver ischemia-reperfusion injury. Quercetin can induce expression and activation of heme oxygenase-1 via MAPK/Nrf2 pathways and then prevent alcohol-induced oxidative stress in human hepatocytes;[46] quercetin pretreatment also protects rats' liver from alcohol-induced oxidative damage.[47] Regarding the antioxidative activity of quercetin, it has been reported that quercetin treatment can increase glutamylcysteine synthetase activity leading to a dose-dependent increase of GSH biosynthesis in human hepatoma HepG2 cells.[48] Organic hydroperoxides, such as *tert*-butyl hydroperoxide (*t*-BHP), have been shown to cause oxidative DNA damage in human hepatoma HepG2, and can be counteracted by natural dietary antioxidant quercetin (5 µM).[49] This protective effect of quercetin is caused by reducing oxidative stress, increasing the capacity of DNA repair and modulating cell antioxidative responses. In addition, quercetin (1 µM) can induce the expression of antioxidative peroxiredoxins, including peroxiredoxins 3 and 5, to protect cells against oxidative stress-induced damage by activating the Nrf2/NRF1 transcription pathway.[50] By inhibiting NF-κB activation and COX-2 levels, low-dose quercetin shows its anti-inflammatory property to attenuate TNF-α-induced inflammation in HepG2 hepatoma cells.[51] Besides an antioxidative effect, quercetin also has an antiproliferation effect on hepatoma cells, suggesting an anticancer activity and chemopreventive potential of quercetin.

Cytotoxic Effect of Quercetin on HCC

Quercetin has been shown to inhibit the growth of various human cancer cell lines, including leukemia, ovarian cancer, colorectal cancer, breast cancer, and prostate cancer, suggesting that quercetin may have anticancer potential for cancer treatment.[52] Quercetin represses cell growth and induces apoptosis of MiaPaCa-2 pancreatic cancer cells through inhibiting the activity of the epidermal growth factor receptor tyrosine kinase.[53] Quercetin-induced cytotoxic effect on pancreatic cancer cells is caused by depolarizing mitochondria, releasing mitochondrial cytochrome c, activating caspase-3, and thereby inducing apoptosis.[54] Moreover, quercetin also shows its anticancer effect on HeLa cells, which

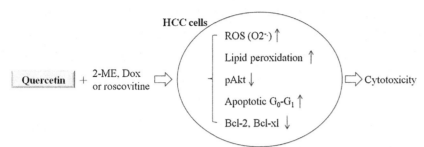

FIGURE 22.4 Cytotoxic Effects of Quercetin on Hepatocellular Carcinoma (HCC) Cells. High-dose quercetin combined with 2-methoxyestradiol (2-ME), doxorubicin (Dox), or rescovitine show cytotoxic potential to inhibit growth of HCC cells by triggering oxidative damages and apoptotic signaling transduction.

is mediated by AMPK-mediated down-regulation of heat shock protein HSP70 and activation of caspase-3.[55]

Several studies have shown the anticancer effect of quercetin on hepatoma cells (Figure 22.4). Previous study showed that high-dose 50 µM quercetin inhibits the growth of ML-3 murine hepatoma cells in a dose-dependent manner.[56] In cell-cycle analysis, quercetin-treated cells were arrested at G_0-G_1 phases. *In vivo* animal study also showed that quercetin treatment of tumor-bearing mice resulted in a significant increase in survival. On the other hand, in 40 µM quercetin treated-HepG2 and HA22T/VGH human hepatoma cells, the levels of apoptotic sub-G_0/G_1 phase, annexin V, lipid peroxidation and ROS were also increased as compared to untreated hepatoma cells.[57] This quercetin-induced cytotoxic effect on human hepatoma cells is dose- and time-dependent.

Interestingly, quercetin can improve the apoptotic effect of chemotherapeutic drugs arsenic trioxide and paclitaxel in leukemia and hepatoma cells, respectively. Combination treatment with 40 µM quercetin and 2-methoxyestradiol induced a synergistic cytotoxic effect in HA22T/VGH hepatoma cells as compared to the cells treated with quercetin or 2-methoxyestradiol alone (Figure 22.5).[58] Combined treatment of quercetin and 2-methoxyestradiol also increased the level of apoptotic sub-G_0/G_1 phase and annexin V (+) population in HA22T/VGH hepatoma cells. Moreover, the amount of superoxide radicals, the expression of MnSOD, and the activity of SOD were increased but mitochondrial membrane potential was decreased after combination treatment with quercetin and 2-methoxyestradiol in HA22T/VGH hepatoma cells. Recently, it has been found that 20 µM quercetin can enhance the effect of doxorubicin-mediated apoptosis in hepatoma cells by increasing p53 activity and down-regulating Bcl-XL expression; meanwhile, quercetin also protects normal liver cells from doxorubicin-mediated cytotoxicity.[59] It was also found that this combined treatment of quercetin and doxorubicin significantly suppressed liver tumor growth in a mice xenograft model. Additionally, combination treatment of cell cycle inhibitor roscovitine with quercetin shows enhanced cytotoxic effect in HepG2 and Hep3B hepatoma cells through reducing the expression of pAkt, Bcl-2, and proactive forms of caspase 9 and caspase 3

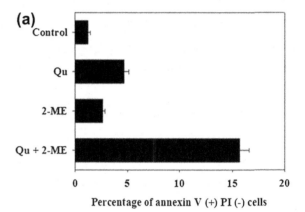

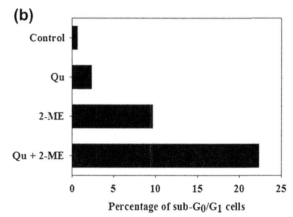

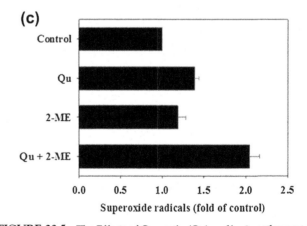

FIGURE 22.5 The Effects of Quercetin (Qu) and/or 2-methoxyestradiol (2-ME) on HA22T/VGH Hepatoma Cells. Cells were treated with 40 µM Qu, 5 µM 2-ME alone, or Qu in combination with 2-ME for 72 hr. Dimethyl sulfoxide was used as vehicle control. (a) The level of annexin V binding, (b) percentage of sub-G_0/G_1 cells, and (c) production of superoxide radicals were analyzed by flow cytometry. PI: propidium iodide.

resulting in apoptosis.[60] These results indicated that quercetin not only has a cytotoxic effect on cancer cells but also has potential to treat HCC when used in combination with paclitaxel, 2-methoxyestradiol, or doxorubicin.

SUMMARY POINTS

- ROS-induced oxidative stress has been considered as a key player in the development and progression of various liver diseases, including viral hepatitis, fibrosis, cirrhosis, and liver cancer.
- In addition to endogenous antioxidants, quercetin available in food is a natural antioxidant to protect humans against oxidative stress.
- The protective effect of quercetin is caused by reducing ROS production, increasing the capacity of DNA repair, and promoting cell antioxidative responses.
- Quercetin has been shown to inhibit the growth of various human cancer cell lines, including leukemia, ovarian cancer, colorectal cancer, breast cancer, prostate cancer, and liver cancer (e.g., hepatocellular carcinoma), suggesting that quercetin may have anticancer potential for cancer treatment.
- Combination treatment with quercetin and chemotherapeutic drug doxorubicin induced a synergistic cytotoxic effect in liver cancer cells.
- Quercetin not only shows its antioxidative effect to prevent oxidative damage-induced liver injury but also has a potential anticancer effect for liver cancer treatment.

References

1. Yang JD, Roberts LR. Hepatocellular carcinoma: A global view. *Nat Rev Gastroenterol Hepatol* 2010;**7**:448–58.
2. Farazi PA, DePinho RA. Hepatocellular carcinoma pathogenesis: From genes to environment. *Nat Rev Cancer* 2006;**6**:674–87.
3. Tang ZY. Hepatocellular carcinoma – cause, treatment and metastasis. *World J Gastroenterol* 2001;**7**:445–54.
4. Ha HL, Shin HJ, Feitelson MA, Yu DY. Oxidative stress and antioxidants in hepatic pathogenesis. *World J Gastroenterol* 2010;**16**:6035–43.
5. Finkel T. Oxidant signals and oxidative stress. *Curr Opin Cell Biol* 2003;**15**:247–54.
6. Moller P, Wallin H. Adduct formation, mutagenesis and nucleotide excision repair of DNA damage produced by reactive oxygen species and lipid peroxidation product. *Mutat Res* 1998;**410**:271–90.
7. Marra M, Sordelli IM, Lombardi A, Lamberti M, Tarantino L, Giudice A, et al. Molecular targets and oxidative stress biomarkers in hepatocellular carcinoma: An overview. *J Transl Med* 2011;**9**:171.
8. Jungst C, Cheng B, Gehrke R, Schmitz V, Nischalke HD, Ramakers J, et al. Oxidative damage is increased in human liver tissue adjacent to hepatocellular carcinoma. *Hepatology* 2004;**39**:1663–72.
9. Clemente C, Elba S, Buongiorno G, Guerra V, D'Attoma B, Orlando A, et al. Manganese superoxide dismutase activity and incidence of hepatocellular carcinoma in patients with Child-Pugh class A liver cirrhosis: A 7-year follow-up study. *Liver Int* 2007;**27**:791–7.
10. Marnett LJ. Oxyradicals and DNA damage. *Carcinogenesis* 2000;**21**:361–70.
11. Ahmad R, Tripathi AK, Tripathi P, Singh S, Singh R, Singh RK. Malondialdehyde and protein carbonyl as biomarkers for oxidative stress and disease progression in patients with chronic myeloid leukemia. *In Vivo* 2008;**22**:525–8.
12. Veal EA, Day AM, Morgan BA. Hydrogen peroxide sensing and signaling. *Mol Cell* 2007;**26**:1–14.
13. Nadeau PJ, Charette SJ, Toledano MB, Landry J. Disulfide Bond-mediated multimerization of Ask1 and its reduction by thioredoxin-1 regulate H(2)O(2)-induced c-Jun NH(2)-terminal kinase activation and apoptosis. *Mol Biol Cell* 2007;**18**:3903–13.
14. Taguchi K, Motohashi H, Yamamoto M. Molecular mechanisms of the Keap1-Nrf2 pathway in stress response and cancer evolution. *Genes Cells* 2011;**16**:123–40.
15. Karin M. NF-kappaB as a critical link between inflammation and cancer. *Cold Spring Harb Perspect Biol* 2009;**1** a000141.
16. Pikarsky E, Porat RM, Stein I, Abramovitch R, Amit S, Kasem S, et al. NF-kappaB functions as a tumour promoter in inflammation-associated cancer. *Nature* 2004;**431**:461–6.
17. Klaunig JE, Kamendulis LM. The role of oxidative stress in carcinogenesis. *Annu Rev Pharmacol Toxicol* 2004;**44**:239–67.
18. Amann T, Bataille F, Spruss T, Muhlbauer M, Gabele E, Scholmerich J, et al. Activated hepatic stellate cells promote tumorigenicity of hepatocellular carcinoma. *Cancer Sci* 2009;**100**:646–53.
19. Moreira RK. Hepatic stellate cells and liver fibrosis. *Arch Pathol Lab Med* 2007;**131**:1728–34.
20. Galli A, Svegliati-Baroni G, Ceni E, Milani S, Ridolfi F, Salzano R, et al. Oxidative stress stimulates proliferation and invasiveness of hepatic stellate cells via a MMP2-mediated mechanism. *Hepatology* 2005;**41**:1074–84.
21. Adachi T, Togashi H, Suzuki A, Kasai S, Ito J, Sugahara K, et al. NAD(P)H oxidase plays a crucial role in PDGF-induced proliferation of hepatic stellate cells. *Hepatology* 2005;**41**:1272–81.
22. Liu RM, Gaston Pravia KA. Oxidative stress and glutathione in TGF-beta-mediated fibrogenesis. *Free Radic Biol Med* 2010;**48**:1–15.
23. Niu D, Zhang J, Ren Y, Feng H, Chen WN. HBx genotype D represses GSTP1 expression and increases the oxidative level and apoptosis in HepG2 cells. *Mol Oncol* 2009;**3**:67–76.
24. Lim W, Kwon SH, Cho H, Kim S, Lee S, Ryu WS. HBx targeting to mitochondria and ROS generation are necessary but insufficient for HBV-induced cyclooxygenase-2 expression. *J Mol Med (Berl)* 2010;**88**:359–69.
25. Waris G, Huh KW, Siddiqui A. Mitochondrially associated hepatitis B virus X protein constitutively activates transcription factors STAT-3 and NF-kappa B via oxidative stress. *Mol Cell Biol* 2001;**21**:7721–30.
26. Yi YS, Park SG, Byeon SM, Kwon YG, Jung G. Hepatitis B virus X protein induces TNF-alpha expression via down-regulation of selenoprotein P in human hepatoma cell line, HepG2. *Biochim Biophys Acta* 2003;**1638**:249–56.
27. Shimoda R, Nagashima M, Sakamoto M, Yamaguchi N, Hirohashi S, Yokota J, et al. Increased formation of oxidative DNA damage, 8-hydroxydeoxyguanosine, in human livers with chronic hepatitis. *Cancer Res* 1994;**54**:3171–2.
28. Moriya K, Nakagawa K, Santa T, Shintani Y, Fujie H, Miyoshi H, et al. Oxidative stress in the absence of inflammation in a mouse model for hepatitis C virus-associated hepatocarcinogenesis. *Cancer Res* 2001;**61**:4365–70.
29. Ronis MJ, Korourian S, Blackburn ML, Badeaux J, Badger TM. The role of ethanol metabolism in development of alcoholic steatohepatitis in the rat. *Alcohol* 2010;**44**:157–69.
30. Zhu H, Jia Z, Misra H, Li YR. Oxidative stress and redox signaling mechanisms of alcoholic liver disease: updated experimental and clinical evidence. *J Dig Dis* 2012;**13**:133–42.

31. Cubero FJ, Urtasun R, Nieto N. Alcohol and liver fibrosis. *Semin Liver Dis* 2009;**29**:211–21.

32. Lin CW, Lin CC, Mo LR, Chang CY, Perng DS, Hsu CC, et al. Heavy alcohol consumption increases the incidence of hepatocellular carcinoma in hepatitis B virus-relatedcirrhosis. *J Hepatol* 2013;**58**:730–5.

33. Diplock AT. Defense against reactive oxygen species. *Free Radic Res* 1998;**29**:463–7.

34. Chen IS, Chang YY, Hsu CL, Lin HW, Chang MH, Chen JW, et al. Alleviative effects of deep-seawater drinking water on hepatic lipid accumulation and oxidation induced by a high-fat diet. *J Chin Med Assoc* 2013;**76**:95–101.

35. Goel A, Kunnumakkara AB, Aggarwal BB. Curcumin as Curecumin: from kitchen to clinic. *Biochem Pharmacol* 2008;**75**:787–809.

36. Bruck R, Ashkenazi M, Weiss S, Goldiner I, Shapiro H, Aeed H, et al. Prevention of liver cirrhosis in rats by curcumin. *Liver Int* 2007;**27**:373–83.

37. Bao W, Li K, Rong S, Yao P, Hao L, Ying C, et al. Curcumin alleviates ethanol-induced hepatocytes oxidative damage involving heme oxygenase-1 induction. *J Ethnopharmacol* 2010;**128**:549–53.

38. Kim HJ, Yoo HS, Kim JC, Park CS, Choi MS, Kim M, et al. Antiviral effect of *Curcuma longa* Linn extract against hepatitis B virus replication. *J Ethnopharmacol* 2009;**124**:189–96.

39. Pradhan SC, Girish C. Hepatoprotective herbal drug, silymarin from experimental pharmacology to clinical medicine. *Indian J Med Res* 2006;**124**:491–504.

40. Ramakrishnan G, Raghavendran HR, Vinodhkumar R, Devaki T. Suppression of N-nitrosodiethylamine induced hepatocarcinogenesis by silymarin in rats. *Chem Biol Interact* 2006;**161**:104–14.

41. Park HJ, Lee JY, Chung MY, Park YK, Bower AM, Koo SI, et al. Green tea extract suppresses NFkappaB activation and inflammatory responses in diet-induced obese rats with nonalcoholic steatohepatitis. *J Nutr* 2012;**142**:57–63.

42. Xu J, Wang J, Deng F, Hu Z, Wang H. Green tea extract and its major component epigallocatechin gallate inhibits hepatitis B virus in vitro. *Antiviral Res* 2008;**78**:242–9.

43. Masalkar PD, Abhang SA. Oxidative stress and antioxidant status in patients with alcoholic liver disease. *Clin Chim Acta* 2005;**355**:61–5.

44. Kaur J, Shalini S, Bansal MP. Influence of vitamin E on alcohol-induced changes in antioxidant defenses in mice liver. *Toxicol Mech Methods* 2010;**20**:82–9.

45. Hertog MG, Kromhout D, Aravanis C, Blackburn H, Buzina R, Fidanza F, et al. Flavonoid intake and long-term risk of coronary heart disease and cancer in the seven countries study. *Arch Intern Med* 1995;**155**:381–6.

46. Yao P, Nussler A, Liu L, Hao L, Song F, Schirmeier A, et al. Quercetin protects human hepatocytes from ethanol-derived oxidative stress by inducing heme oxygenase-1 via the MAPK/Nrf2 pathways. *J Hepatol* 2007;**47**:253–61.

47. Liu S, Hou W, Yao P, Zhang B, Sun S, Nussler AK, et al. Quercetin protects against ethanol-induced oxidative damage in rat primary hepatocytes. *Toxicol In Vitro* 2010;**24**:516–22.

48. Scharf G, Prustomersky S, Knasmuller S, Schulte-Hermann R, Huber WW. Enhancement of glutathione and g-glutamylcysteine synthetase, the rate limiting enzyme of glutathione synthesis, by chemoprotective plant-derived food and beverage components in the human hepatoma cell line HepG2. *Nutr Cancer* 2003;**45**:74–83.

49. Alia M, Ramos S, Mateos R, Granado-Serrano AB, Bravo L, Goya L. Quercetin protects human hepatoma HepG2 against oxidative stress induced by tert-butyl hydroperoxide. *Toxicol Appl Pharmacol* 2006;**212**:110–8.

50. Miyamoto N, Izumi H, Miyamoto R, Kondo H, Tawara A, Sasaguri Y, et al. Quercetin induces the expression of peroxiredoxins 3 and 5 via the Nrf2/NRF1 transcription pathway. *Invest Ophthalmol Vis Sci* 2011;**52**:1055–63.

51. Granado-Serrano AB, Martin MA, Bravo L, Goya L, Ramos S. Quercetin attenuates TNF-induced inflammation in hepatic cells by inhibiting the NF-kappaB pathway. *Nutr Cancer* 2012;**64**:588–98.

52. Yang CS, Landau JM, Huang MT, Newmark HL. Inhibition of carcinogenesis by dietary polyphenolic compounds. *Annu Rev Nutr* 2001;**21**:381–406.

53. Lee LT, Huang YT, Hwang JJ, Lee PP, Ke FC, Nair MP, et al. Blockade of the epidermal growth factor receptor tyrosine kinase activity by quercetin and luteolin leads to growth inhibition and apoptosis of pancreatic tumor cells. *Anticancer Res* 2002;**22**:1615–27.

54. Mouria M, Gukovskaya AS, Jung Y, Buechler P, Hines OJ, Reber HA, et al. Food-derived polyphenols inhibit pancreatic cancer growth through mitochondrial cytochrome C release and apoptosis. *Int J Cancer* 2002;**98**:761–9.

55. Jung JH, Lee JO, Kim JH, Lee SK, You GY, Park SH, et al. Quercetin suppresses HeLa cell viability via AMPK-induced HSP70 and EGFR down-regulation. *J Cell Physiol* 2010;**223**:408–14.

56. Chi CW, Chang YF, Ou YR, Hsieh CC, Lui WY, Peng FK, et al. Effect of quercetin on the in vitro and in vivo growth of mouse hepatoma cells. *Oncol Rep* 1997;**4**:1021–4.

57. Chang YF, Chi CW, Wang JJ. Reactive oxygen species production is involved in quercetin-induced apoptosis in human hepatoma cells. *Nutr Cancer* 2006;**55**:201–9.

58. Chang YF, Hsu YC, Hung HF, Lee HJ, Lui WY, Chi CW, et al. Quercetin induces oxidative stress and potentiates the apoptotic action of 2-methoxyestradiol in human hepatoma cells. *Nutr Cancer* 2009;**61**:735–45.

59. Wang G, Zhang J, Liu L, Sharma S, Dong Q. Quercetin Potentiates Doxorubicin Mediated Antitumor Effects against Liver Cancer through p53/Bcl-xl. *PLoS One* 2012;**7**:e51764.

60. Sharma A, Bhat MK. Enhancement of carboplatin- and quercetin-induced cell death by roscovitine is Akt dependent and p53 independent in hepatoma cells. *Integr Cancer Ther* 2011;**10**:NP4–14.

Capsaicin Mediated Oxidative Stress in Pancreatic Cancer

Palika Datta, Kartick C. Pramanik

Department of Biomedical Sciences and Cancer Biology Center, Texas Tech University of Health Sciences Center, Amarillo, TX, USA

Sudhir Mehrotra

Department of Biochemistry, Lucknow University, Lucknow, India

Sanjay K. Srivastava

Department of Biomedical Sciences and Cancer Biology Center, Texas Tech University of Health Sciences Center, Amarillo, TX, USA

List of Abbreviations

ASK-1 Apoptosis signal-regulating kinase-1
Bax Bcl2 associated X protein
Bcl2 B-cell lymphoma 2
DNA Deoxyribonucleic acid
ERK Extracellular signal-regulated kinase
ETC Electron transport chain
FADH Flavin adenine dinucleotide
GSH Glutathione
GSSG Glutathione disulfide
JNK c-Jun NH_2-terminal kinase
MAPK Mitogen activated protein kinase
NADH Nicotinamide adenine dinucleotide
PanIN Pancreatic Intraepithelial neoplasia
PARP Poly ADP ribose polymerase
PKC Protein kinase C
RNS Reactive nitrogen species
ROS Reactive oxygen species
SOD Superoxide dismutase
Thr Threonine
Trx Thioredoxin
UV Ultraviolet

INTRODUCTION

Reactive Oxygen Species (ROS) is a phrase used to describe a number of reactive molecules and free radicals derived from molecular oxygen. These molecules produced as by-products during the mitochondrial electron transport of aerobic respiration or by oxido-reductase enzymes have the potential to cause a number of deleterious events. Mitochondrion is the major source of ROS, which are generated during respiration and involved in maintaining the intracellular redox state. ROS are considered to be important signaling molecules that play a role in gene expression, cell growth, and survival as well as oxygen sensing in various cell types.[1,2] While ROS have a physiological role through mediating several signal transduction pathways, high levels of ROS can cause oxidative stress, which can subsequently result in cell death by influencing apoptotic pathways.[3] Oxidative stress is an expression used to describe various harmful processes resulting from an imbalance between the excessive formation of ROS and/or RNS and limited antioxidant defenses. The increased oxidative stress is associated with various cancer models including pancreatic cancer.

Pancreatic cancer is one of the most challenging of human malignancies. Of the several types, exocrine pancreatic tumors (i.e., adenocarcinoma) account for 95% of all pancreatic cancers and arise from cells lining the pancreatic duct. It is the fourth most common cause of cancer-related deaths in the United States with an overall 5-year survival rate of less than 5%. Cancer of

Cancer
http://dx.doi.org/10.1016/B978-0-12-405205-5.00023-4

© 2014 Elsevier Inc. All rights reserved.

pancreas grows slowly, taking years and even decades to develop. It is one of the most lethal cancers, as the signs and symptoms of the disease show only once a more advanced stage of cancer is reached. Genetic mutations in K-Ras have been seen to be responsible for 95% of pancreatic cancer, indicating their role in molecular pathogenesis.[4,5] Local oxidative stress from chronic pancreatitis, extracellular matrix proteins, or Kras mutations causes an increase in ROS production, which in turn influences downstream propagation of signaling pathways.

Limited efficacy of conventional chemotherapy is the cause of the need for new therapeutic approaches for the treatment of pancreatic cancer. Epidemiological data suggest that diets rich in fruits, vegetables, and certain spices may prevent some types of cancer including pancreatic cancer.[6–8] Flavonoids are common in plant-based foods but are found in highest concentrations in onions, apples, berries, kale, and broccoli, and have been seen to reduce the risk of developing pancreatic cancer.[9] Among the various spices, capsaicin, an active component in chili peppers, is a homovanillic acid derivative (N-vanillyl-8-methyl-383 nonenamide) and is found to have cancer chemopreventive properties. The role of capsaicin mediated oxidative stress and apoptosis in pancreatic cancer cells *in vitro* and *in vivo* is discussed in this chapter.

APOPTOTIC PATHWAYS

Apoptosis is a highly regulated form of cell death and is regulated by a series of biochemical pathways. There are two major pathways involved: the extrinsic pathway, which is receptor mediated, and the intrinsic pathway, which is mitochondria mediated. The intrinsic pathway can be activated by stress, UV radiation, or by drugs that disturb the redox balance within the cell. Mitochondria play a crucial role in the intrinsic pathway of apoptosis. Permeabilization of the outer membrane occurs in response to various stimuli and is regulated by many proteins including the Bcl2 family. Proapoptotic proteins include Bax, Bak, Bim, and Bid whereas antiapoptotic proteins such as Bcl-2 and Bcl-xL constitute the Bcl2 family.[10] Dysregulation of these pathways leads to the progression of cancer. Recent studies suggest that capsaicin has a strong evidence of having potential as a chemopreventive agent against various types of cancer.[11–14] A study has shown antiproliferative properties of capsaicin by inducing apoptosis in pancreatic cancer cells AsPC-1 and BxPC-3.[15] Cell death was found to be associated with generation of reactive oxygen species, which occurred due to disruption of mitochondrial membrane potential (Table 23.1). Capsaicin was able to upregulate the proapoptotic proteins such

TABLE 23.1 Capsaicin-Mediated Oxidative Stress in Pancreatic Cancer

	Cell Organelles	Effects
Oxidative stress	Mitochondria	Oxidation of cardiolipin resulting in the disruption of mitochondrial membrane potential Activation of Bax (proapoptotic protein) Release of cytochrome C into cytosol Inhibition of complex I and III of ETC complex
	Cytosol	Phosphorylation of JNK Oxidation of antioxidant enzyme Glutathione and inhibition of SOD and catalase Oxidation of cysteine residues in Trx Dissociation of Trx-Ask1 complex Activation of Bim and Bax Down-regulation of survivin expression

as Bax, which in turn caused the release of cytochrome c leading to the activation of caspase cascade (Table 23.2).

Another signal transduction pathway being affected by capsaicin was the activation of the JNK pathway, which supposedly caused apoptosis in the pancreatic cancer cell lines. This observation was proved using JNK inhibitors, which reversed the effects of capsaicin-induced apoptosis in AsPC-1 and BxPC-3 cells.[15] It is known that JNK is regulated by oxidative stress[16] and causes apoptosis in prostate cancer cells.[17] Pancreatic cancer cells, when treated with antioxidants to block ROS, offered significant protection against capsaicin-induced mitochondrial membrane potential disruption and apoptosis. Above all, capsaicin treatment did not cause any significant ROS generation or apoptosis in normal human pancreatic ductal epithelial (HPDE-6) cells, suggesting that effects of capsaicin were specific to cancer cells. This indicated that capsaicin was able to cause apoptosis in pancreatic cancer cells through generation of ROS and JNK phosphorylation (Table 23.1), thus leading to the activation of the mitochondrial intrinsic death pathway.[15]

ROS AND ETC COMPLEX IN THE MITOCHONDRIA

Reactive oxygen species (ROS) are chemically reactive molecules and mostly formed during respiration in the mitochondria.[18,19] The electron transport chain (ETC) complexes in the mitochondria have been recognized as the main source of ROS. During the breakdown of glucose into acetyl CoA, NADH, and FADH the reducing equivalents are released and are utilized by the electron

TABLE 23.2 Molecular Targets of Capsaicin

Targets	Cell Lines	Modulation	References
Cardiolipin	BxPC-3 and AsPC-1	Oxidation	15,28
Cytochrome C (cytosol)	BxPC-3 and AsPC-1	Increase	15,28
Bax	BxPC-3 and AsPC-1	Increase	15
AIF	BxPC-3 and AsPC-1	Increase	15
Bcl-2	BxPC-3 and AsPC-1	Decrease	15
Survivin	BxPC-3 and AsPC-1	Decrease	15,28,46
Caspase -3	BxPC-3 and AsPC-1	Increase	15,28,46
Caspase- 9	BxPC-3 and AsPC-1	Increase	15,28,46
PARP	BxPC-3 and AsPC-1	Increase	15,28,46
Glutathione	BxPC-3 and AsPC-1	Oxidation	15,28
SOD	BxPC-3 and AsPC-1	Decrease	28
Catalase	BxPC-3 and AsPC-1	Decrease	28
p-JNK	BxPC-3 and AsPC-1	Increase	15
ETC complex I	BxPC-3 and AsPC-1	Inhibition	28
ETC complex III	BxPC-3 and AsPC-1	Inhibition	28
Thioredoxin (Trx)	BxPC-3 and AsPC-1	Decrease	46
p-ASK1 (Thr845)	BxPC-3 and AsPC-1	Increase	46
p-MKK4 (Thr 261)	BxPC-3 and AsPC-1	Increase	46
p-MKK7 (Ser 271/Thr275)	BxPC-3 and AsPC-1	Increase	46

transport chain. In the mitochondrial electron transport chain, electrons move from an electron donor (NADH or QH_2) to a terminal electron acceptor (O_2) via a series of redox reactions, which comprises Complex I (NADH dehydrogenase), Complex II (succinate dehydrogenase), Complex III (ubiquinol cytochrome c reductase), and Complex IV (cytochrome-c oxidase). These reactions are coupled to the creation of a proton gradient across the mitochondrial inner membrane. Thus the formation and metabolism of ROS in and by the mitochondria can make an impact on the cytosolic redox state. If the cell is subjected to external stress such as UV light or chemical poison there is an excessive production of ROS, which results in the production of oxidative stress that can cause damage to several mitochondrial components including proteins, lipids, and DNA.

Several studies have targeted the ETC complex for induction of apoptosis and cancer treatment.[20–27] A recent study has shown the involvement of mitochondrial ETC Complex I and III in capsaicin-induced apoptosis.[28] The increased generation of ROS was due to capsaicin-induced inhibition of Complex-I and III activities mediated in the pancreatic cancer cells (Table 23.2). Further it was seen that capsaicin was not able to mediate any effect in BxPC-3 rho cells (which lack mitochondrial DNA and hence normal oxidative phosphorylation gets inactive). Antioxidants catalase or EUK-134 were able to block capsaicin-mediated ROS generation and apoptosis. These results indicate that capsaicin was responsible for generation of ROS and that ROS were involved in inducing apoptosis in pancreatic cancer cells. Therefore it is now evident that ROS plays a fundamental role in cell death signaling

pathways, and intracellular accumulation of ROS can lead to cell death.[18,19,29]

Oxidative stress induced by ROS is characterized by an imbalance between oxidant-producing systems and antioxidant defense mechanisms, which can trigger cell damage by oxidizing macromolecular structures (lipids, proteins, and DNA) and modifying their biological functions, which ultimately causes cell death. Thus, depending on their levels in the cell, ROS can either be beneficial or harmful. Redox control systems have direct interactions with components of the apoptotic signaling pathways. The generation of ROS by a cascade of reactions is efficiently blocked by various endogenous antioxidants to overcome their potentially injurious actions.[19] The two most important antioxidant systems are those of thioredoxin and glutathione (GSH). GSSG is formed upon GSH oxidation and the level of oxidative stress in the cell is measured by the ratio of GSH/GSSG.[30] Capsaicin was able to increase the levels of the oxidized (GSSG) form of glutathione in pancreatic cancer cells. In a time dependent study it was seen that capsaicin steadily caused an increase in the GSSG levels and reduced GSH levels.[28] Another important antioxidant enzyme is superoxide dismutase (SOD), which scavenges the superoxide radicals generated by ETC complex in the mitochondria. Furthermore glutathione peroxidase enzyme activity was also inhibited by capsaicin in pancreatic cancer cells. These observations were further confirmed when ectopic expression of antioxidant enzyme catalase blocked the generation of ROS and apoptosis induced by capsaicin in BxPC-3 cells.[28] Similar results were observed *in vivo* in the xenograft models in the athymic nude mice. Capsaicin treatment increased the levels of GSSG and reduced the activities of antioxidant enzymes in the tumors of capsaicin-treated mice. These results revealed that capsaicin treatment disrupts the cellular redox equilibrium resulting in increased oxidative stress in both *in vitro* and *in vivo* models.

THIOREDOXIN AND ASK-1 SIGNALING PATHWAY

Although ROS have been considered only to damage cells, there is evidence that shows that oxidative stress and ROS inducible stress evokes many intracellular events such as proliferation, gene activation, cell-cycle arrest, and apoptosis.[31] Thioredoxin (Trx) is another major redox system in the cell, which plays an important role in cell growth and apoptosis.[32] Trx is a small ubiquitous protein of 12 kDa. Its active site contains two cysteine residues at positions 32 and 35 that undergo reversible oxidation to form a disulfide bond during the transfer of reducing equivalents to a disulfide substrate.[33] Trx protects the cell against many forms of induced oxidative stress.[34–37] It is also known that the association of Trx with apoptosis signal-regulating kinase-1 (ASK-1) causes an inhibitory effect on ASK-1 regulated functions.[38,39] Apoptosis signal-regulating kinase-1 is a MAPK kinase kinase (MPAKKK) and it activates several MAPK kinases such as MKK 3, 4, 6, and 7, which in turn activates c-Jun N-terminal kinase (JNK) and the p38 MAPK pathways.[40] These kinases further phosphorylate and activate specific proapoptotic proteins such as Bim[41,42] and Bax[43] resulting in apoptosis. ASK-1 is functionally active when it is phosphorylated at Thr845 whereas phosphorylation of ASK-1 at Ser-83 by AKT/ PKC inhibits its activity.[44] ASK-1 is negatively regulated when associated with Trx under reducing environment. In the presence of oxidative stress, Trx gets oxidized and then dissociates from ASK-1-Trx complex resulting in the activation of ASK1 and apoptosis (Table 23.1). Trx has been shown to be overexpressed in many tumors including pancreatic cancer.[45]

In a recent study it was demonstrated that capsaicin targets Trx-ASK-1 signaling in pancreatic cancer cells to induce apoptosis.[46] Capsaicin treatment increased the phosphorylation of ASK-1 at Thr 845 in AsPC-1 and BxPC-3 pancreatic cancer cells whereas expression of survivin was significantly reduced. On the other hand, cleavage of caspase-9, caspase-3, and poly (ADP-ribose) polymerase (PARP) was increased. In a time-dependent study, down-regulation of Trx coincided with phosphorylation of ASK1 by capsaicin treatment indicating that inhibition of Trx correlates well with ASK1 activation. Activation of ASK1 was abrogated in the presence of antioxidants, indicating that capsaicin-mediated ROS generation was involved in ASK1 dependent pathway

leading to apoptosis (Table 23.2). Transient overexpression of ASK-1 increased apoptosis in the cancer cells whereas Trx overexpression blocked capsaicin-induced ROS generation and apoptosis. Thus, capsaicin-mediated oxidative stress causes dissociation of the Trx-ASK-1 complex, ASK-1 gets phosphorylated at Thr845 and its activation leads to apoptosis.

Xenograft tumor models showed substantial reduced tumor growth in athymic nude mice when fed capsaicin (5mg/kg) by oral gavage as compared to control mice. Tumor growth in the capsaicin treated group was reduced by 73% with no noticeable side effects. Therefore in the *in vivo* studies too, capsaicin was able to reduce levels of Trx in the tumors from capsaicin treated mice and also showed activation of ASK-1 and increased the expression of apoptotic proteins.[46] Furthermore luciferase transfected PanC-1 pancreatic cancer cells were orthotopically implanted in the pancreas of athymic nude mice followed by oral administration of 5 mg capsaicin/kg. This treatment showed drastic suppression of the growth of orthotopic pancreatic tumors. The average wet weight of tumor from capsaicin treated mice was about 80% less as compared to control mice.

SUPPRESSION OF PANCREATITIS BY CAPSAICIN

Earlier studies have shown that patients with chronic pancreatitis have a higher risk of developing pancreatic cancer.[47] Capsaicin has also shown to have chemopreventive effects against pancreatitis and pancreatic intraepithelial neoplasia (PanIN) lesion formation in the pancreas of a unique genetically engineered mice model LSL-Kras^{G12D}/Pdx-Cre.[48] These mice developed chronic pancreatitis and PanIN lesions after a single dose of

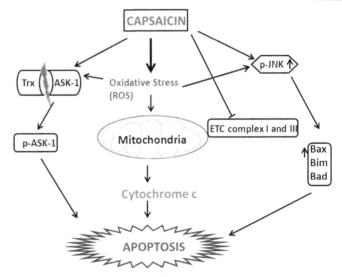

FIGURE 23.1 Mechanism of capsaicin in pancreatic cancer cells.

caerulin and mimic the progressive form of human disease. Mice were fed a daily diet containing 10 or 20 p.p.m capsaicin for 8 weeks. It was observed that capsaicin in the diet significantly reduced the severity of chronic pancreatitis and PanIN lesions. Further analysis of the tumors from these mice revealed that capsaicin significantly reduced the phosphorylation of ERK, c-JUN, and Hedgehog/GLI1 activation in capsaicin-treated mice as compared to control mice. Hence, capsaicin was quite effective in inhibiting pancreatic carcinogenesis. However, it still remains to be seen whether capsaicin was able to induce oxidative stress related activation of signaling pathways in this model.

SUMMARY POINTS

- Capsaicin, a constituent of chili pepper, has been shown to mediate mitochondrial dependent apoptosis through generation of ROS.
- Capsaicin has also shown to target mitochondrial electron transport chain to generate ROS, activate JNK, and disrupt Trx-ASK-1 interaction to induce mitochondrial apoptosis in pancreatic cancer cells (Figure 23.1).
- Administration of 5mg/kg capsaicin by oral gavage or providing 10 p.p.m capsaicin in diet not only blocked chronic pancreatitis and early carcinogenesis but also suppressed the growth of implanted pancreatic tumors substantially.
- This provides a foundation for the development of capsaicin as a chemopreventive agent for further clinical investigation.

Acknowledgments

This work was supported in part by R01 grants CA129038 and CA106953 (to Sanjay K. Srivastava) awarded by The National Cancer Institute, NIH, USA.

References

1. Brown DI, Griendling KK, Nox proteins in signal transduction. *Free Radic Biol Med* 2009;**47**(9):1239–53.
2. Gillespie MN, Pastukh V, Ruchko MV. Oxidative DNA modifications in hypoxic signaling. *Ann New York Acad Sci* 2009;**1177**:140–50.
3. Andersen JK. Oxidative stress in neurodegeneration: Cause or consequence? *Nat Med* 2004;**10**(Suppl):S18–25.
4. Motojima K, Urano T, Nagata Y, Shiku H, Tsurifune T, Kanematsu T. Detection of point mutations in the Kirsten-ras oncogene provides evidence for the multicentricity of pancreatic carcinoma. *Ann Surgery* 1993;**217**(2):138–43.
5. Rivera JA, Rall CJ, Graeme-Cook F, Fernandez-del Castillo C, Shu P, Lakey N, et al. Analysis of K-ras oncogene mutations in chronic pancreatitis with ductal hyperplasia. *Surgery* 1997;**121**(1):42–9.
6. Bhutani M, Pathak AK, Nair AS, Kunnumakkara AB, Guha S, Sethi G, et al. Capsaicin is a novel blocker of constitutive and interleukin-6-inducible STAT3 activation. *Clin Cancer Res* 2007;**13**(10):3024–32.
7. Block G, Patterson B, Subar A. Fruit, vegetables, and cancer prevention: A review of the epidemiological evidence. *Nutr Cancer* 1992;**18**(1):1–29.
8. Satyanarayana MN. Capsaicin and gastric ulcers. *Crit Rev Food Sci Nutr* 2006;**46**(4):275–328.
9. Zhang R, Loganathan S, Humphreys I, Srivastava SK. Benzyl isothiocyanate-induced DNA damage causes G2/M cell cycle arrest and apoptosis in human pancreatic cancer cells. *J Nutr* 2006;**136**(11):2728–34.
10. Cory S, Huang DC, Adams JM. The Bcl-2 family: Roles in cell survival and oncogenesis. *Oncogene* 2003;**22**(53):8590–607.
11. Modly CE, Das M, Don PS, Marcelo CL, Mukhtar H, Bickers DR. Capsaicin as an in vitro inhibitor of benzo(a)pyrene metabolism and its DNA binding in human and murine keratinocytes. *Drug Metab Disposit* 1986;**14**(4):413–6.
12. Yoshitani SI, Tanaka T, Kohno H, Takashima S. Chemoprevention of azoxymethane-induced rat colon carcinogenesis by dietary capsaicin and rotenone. *Int J Oncol* 2001;**19**(5):929–39.
13. Surh YJ, Lee SS. Capsaicin, a double-edged sword: Toxicity, metabolism, and chemopreventive potential. *Life Sci* 1995;**56**(22):1845–55.
14. Teel RW. Effects of capsaicin on rat liver S9-mediated metabolism and DNA binding of aflatoxin. *Nutr Cancer* 1991;**15**(1):27–32.
15. Zhang R, Humphreys I, Sahu RP, Shi Y, Srivastava SK. In vitro and in vivo induction of apoptosis by capsaicin in pancreatic cancer cells is mediated through ROS generation and mitochondrial death pathway. *Apoptosis* 2008;**13**(12):1465–78.
16. Shen HM, Liu ZG. JNK signaling pathway is a key modulator in cell death mediated by reactive oxygen and nitrogen species. *Free Radic Biol Med* 2006;**40**(6):928–39.
17. Sanchez AM, Sanchez MG, Malagarie-Cazenave S, Olea N, Diaz-Laviada I. Induction of apoptosis in prostate tumor PC-3 cells and inhibition of xenograft prostate tumor growth by the vanilloid capsaicin. *Apoptosis* 2006;**11**(1):89–99.
18. Orrenius S, Gogvadze V, Zhivotovsky B. Mitochondrial oxidative stress: Implications for cell death. *Ann Rev Pharmacol Toxicol* 2007;**47**:143–83.
19. Fleury C, Mignotte B, Vayssiere JL. Mitochondrial reactive oxygen species in cell death signaling. *Biochimie* 2002;**84**(2–3):131–41.
20. Wolvetang EJ, Johnson KL, Krauer K, Ralph SJ, Linnane AW. Mitochondrial respiratory chain inhibitors induce apoptosis. *FEBS Lett* 1994;**339**(1–2):40–4.
21. Isenberg JS, Klaunig JE. Role of the mitochondrial membrane permeability transition (MPT) in rotenone-induced apoptosis in liver cells. *Toxicol Sci* 2000;**53**(2):340–51.
22. Chung WG, Miranda CL, Maier CS. Epigallocatechin gallate (EGCG) potentiates the cytotoxicity of rotenone in neuroblastoma SH-SY5Y cells. *Brain Res* 2007;**1176**:133–42.
23. Tada-Oikawa S, Hiraku Y, Kawanishi M, Kawanishi S. Mechanism for generation of hydrogen peroxide and change of mitochondrial membrane potential during rotenone-induced apoptosis. *Life Sci* 2003;**73**(25):3277–88.
24. Neuzil J, Weber T, Schroder A, Lu M, Ostermann G, Gellert N, et al. Induction of cancer cell apoptosis by alpha-tocopheryl succinate: Molecular pathways and structural requirements. *FASEB J* 2001;**15**(2):403–15.
25. Dong LF, Low P, Dyason JC, Wang XF, Prochazka L, Witting PK, et al. Alpha-tocopheryl succinate induces apoptosis by targeting ubiquinone-binding sites in mitochondrial respiratory complex II. *Oncogene* 2008;**27**(31):4324–35.
26. Rohlena J, Dong LF, Neuzil J. Targeting the mitochondrial electron transport chain complexes for the induction of apoptosis and cancer treatment. *Curr Pharm Biotechnol* 2013;**14**(3):377–89.
27. Kang J, Pervaiz S. Mitochondria: redox metabolism and dysfunction. *Biochem Res Int* 2012 Epub 2012 Apr 24.

28. Pramanik KC, Boreddy SR, Srivastava SK. Role of mitochondrial electron transport chain complexes in capsaicin mediated oxidative stress leading to apoptosis in pancreatic cancer cells. *PloS ONE* 2011;**6**(5):e20151.

29. Carmody RJ, McGowan AJ, Cotter TG. Reactive oxygen species as mediators of photoreceptor apoptosis in vitro. *Exp Cell Res* 1999;**248**(2):520–30.

30. Trachootham D, Zhou Y, Zhang H, Demizu Y, Chen Z, Pelicano H, et al. Selective killing of oncogenically transformed cells through a ROS-mediated mechanism by beta-phenylethyl isothiocyanate. *Cancer Cell.* 2006;**10**(3):241–52.

31. Nakamura H, Nakamura K, Yodoi J. Redox regulation of cellular activation. *Ann Rev Immunol* 1997;**15**:351–69.

32. Powis G, Mustacich D, Coon A. The role of the redox protein thioredoxin in cell growth and cancer. *Free Radic Biol Med* 2000;**29**(3–4):312–22.

33. Holmgren A. Thioredoxin. *Ann Rev Biochem* 1985;**54**:237–71.

34. Nakamura H, Matsuda M, Furuke K, Kitaoka Y, Iwata S, Toda K, et al. Adult T cell leukemia-derived factor/human thioredoxin protects endothelial F-2 cell injury caused by activated neutrophils or hydrogen peroxide. *Immunol Lett* 1994;**42**(1–2):75–80.

35. Sasada T, Iwata S, Sato N, Kitaoka Y, Hirota K, Nakamura K, et al. Redox control of resistance to cis-diamminedichloroplatinum (II) (CDDP): Protective effect of human thioredoxin against CDDP-induced cytotoxicity. *J Clin Investigat* 1996;**97**(10):2268–76.

36. Takagi Y, Mitsui A, Nishiyama A, Nozaki K, Sono H, Gon Y, et al. Overexpression of thioredoxin in transgenic mice attenuates focal ischemic brain damage. *Proc Natl Acad Sci United States Am* 1999;**96**(7):4131–6.

37. Yokomise H, Fukuse T, Hirata T, Ohkubo K, Go T, Muro K, et al. Effect of recombinant human adult T cell leukemia-derived factor on rat lung reperfusion injury. *Respiration* 1994;**61**(2):99–104.

38. Liu Y, Min W. Thioredoxin promotes ASK1 ubiquitination and degradation to inhibit ASK1-mediated apoptosis in a redox activity-independent manner. *Circul Res* 2002;**90**(12):1259–66.

39. Saitoh M, Nishitoh H, Fujii M, Takeda K, Tobiume K, Sawada Y, et al. Mammalian thioredoxin is a direct inhibitor of apoptosis signal-regulating kinase (ASK) 1. *EMBO J* 1998;**17**(9):2596–606.

40. Ichijo H, Nishida E, Irie K, ten Dijke P, Saitoh M, Moriguchi T, et al. Induction of apoptosis by ASK1, a mammalian MAPKKK that activates SAPK/JNK and p38 signaling pathways. *Science* 1997;**275**(5296):90–4.

41. Cai B, Chang SH, Becker EB, Bonni A, Xia Z. p38 MAP kinase mediates apoptosis through phosphorylation of BimEL at Ser-65. *J Biol Chem* 2006;**281**(35):25215–22.

42. Lei K, Davis RJ. JNK phosphorylation of Bim-related members of the Bcl2 family induces Bax-dependent apoptosis. *Proc Natl Acad Sci United States Am* 2003;**100**(5):2432–7.

43. Kim BJ, Ryu SW, Song BJ. JNK- and p38 kinase-mediated phosphorylation of Bax leads to its activation and mitochondrial translocation and to apoptosis of human hepatoma HepG2 cells. *J Biol Chem* 2006;**281**(30):21256–65.

44. Kim AH, Khursigara G, Sun X, Franke TF, Chao MV. Akt phosphorylates and negatively regulates apoptosis signal-regulating kinase 1. *Molecular Cell Biol* 2001;**21**(3):893–901.

45. Ohashi S, Nishio A, Nakamura H, Asada M, Tamaki H, Kawasaki K, et al. Overexpression of redox-active protein thioredoxin-1 prevents development of chronic pancreatitis in mice. *Antioxidants & Redox Signaling* 2006;**8**(9–10):1835–45.

46. Pramanik KC, Srivastava SK. Apoptosis signal-regulating kinase 1-thioredoxin complex dissociation by capsaicin causes pancreatic tumor growth suppression by inducing apoptosis. *Antioxidants & Redox Signaling* 2012;**17**(10):1417–32.

47. Boreddy SR, Sahu RP, Srivastava SK. Benzyl isothiocyanate suppresses pancreatic tumor angiogenesis and invasion by inhibiting HIF-alpha/VEGF/Rho-GTPases: pivotal role of STAT-3. *PloS ONE* 2011;**6**(10):e25799.

48. Bai H, Li H, Zhang W, Matkowskyj KA, Liao J, Srivastava SK, et al. Inhibition of chronic pancreatitis and pancreatic intraepithelial neoplasia (PanIN) by capsaicin in LSL-KrasG12D/Pdx1-Cre mice. *Carcinogenesis* 2011;**32**(11):1689–96.

Tocotrienols in Pancreatic Cancer Treatment and Prevention

Kanishka Chakraborty

Department of Internal Medicine, Division of Oncology, James H. Quillen College of Medicine, East Tennessee State University, Johnson City, TN, USA

Victoria Palau Ramsauer

Department of Pharmaceutical Sciences, Bill Gatton College of Pharmacy, East Tennessee State University, Johnson City, TN, USA

William Stone

Department of Pediatrics, James H. Quillen College of Medicine, East Tennessee State University, Johnson City, TN, USA

Koyamangalath Krishnan

Department of Internal Medicine, Division of Oncology, James H. Quillen College of Medicine, East Tennessee State University, Johnson City, TN, USA

List of Abbreviations

GT3 gamma tocotrienol
DT3 delta tocotrienol

INTRODUCTION

Oxidative stress is a documented factor in the mechanisms conducive to the development and progression of inflammation and cancer. Thus nutrients with antioxidant properties have long been explored in research, for the prevention and treatment of both chronic diseases associated with inflammatory processes (atherosclerosis, hypertension, diabetes) and cancer. Among these nutrients, vitamin E holds promise for use in clinical practice along with vitamin D and selenium. Vitamin E was first found in green leafy vegetables,[1] and its chemical structure was first described in 1938.[2] There are two main forms of vitamin E, tocopherols and tocotrienols, which exist as eight different naturally occurring isoforms.[3]

SOURCE OF TOCOTRIENOLS

Tocotrienols are found in annatto, palm oil, rice bran, coconut oil, and barley (Table 24.1). The tocotrienol content in reference to tocopherols is 100% in annatto,[4] and almost 70% in palm oil.[5] Tocotrienols are isoprenoids with an unsaturated phytyl chain.[6] This unique structural configuration allows tocotrienols to cross saturated fatty layers of tissue more readily, thus likely favoring greater anticancer effects[7] (Figure 24.1). The concentration of tocotrienols in human plasma and tissues is lower than tocopherols due to faster degradation. However, tocotrienols accumulate within cells at significantly higher concentrations as compared to tocopherols.[8,9] The bioavailability of tocotrienols is dependent on multiple factors including shorter half-life, the presence of tocopherols, food processing, and nutritional status. The ability to achieve a better uniform distribution in the phospholipid bilayer of the plasma membrane makes tocotrienols better antioxidants than tocopherols.[2]

© 2014 Elsevier Inc. All rights reserved.

TABLE 24.1   Sources of Tocotrienols

Source of Tocotrienols			
Amaranth	Grape Seed	Palm Fruit	Rye
Annato Seed	Hazelnut	Poppy	Safflower
Barley	Maize	Pumpkin Seed	Walnut
Flax Seed	Oat	Rice Bran	Wheat

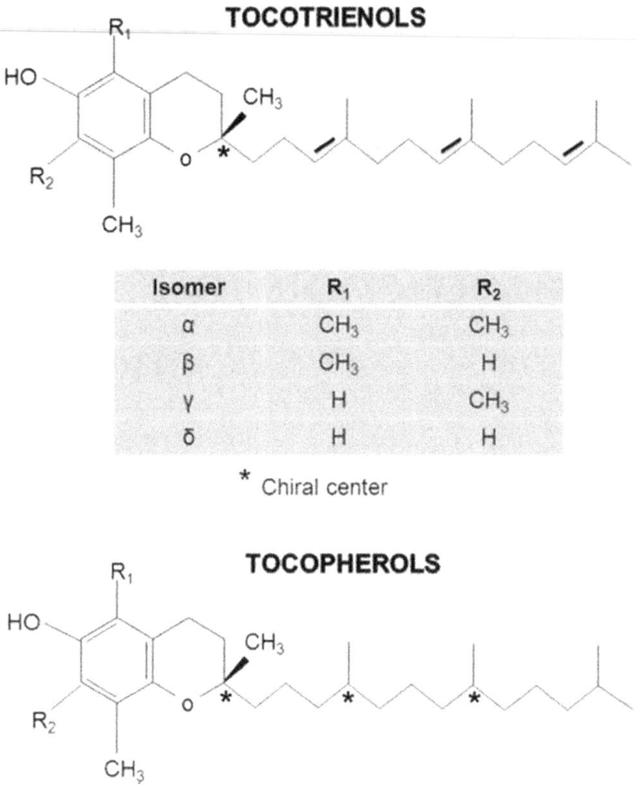

USE OF TOCOTRIENOLS IN CANCER

Tocotrienols exert an antiproliferative activity against malignant cells but have no deleterious effect on normal cells (Figure 24.2). Despite these properties, only 1% of vitamin E research has focused on tocotrienols. Recently, several studies have brought to the forefront the superior anticancer properties of tocotrienols.[9] These studies have demonstrated the antiproliferative and proapoptotic activity of tocotrienols in diverse cancer cell lines, including breast, pancreas, prostate, colorectal, cervical, and skin. Tocotrienols efficacy rests on its pleiotropic effects on cancer pathways including HMG CoA reductase, EGFR, ErbB2, AKT, ERK, p-GSK, c-JUN, caspases, NF-kappa B, and Bax/Bcl2. A novel evolving concept is the role of tocotrienols in causing differential distribution of ceramide in pancreatic and breast cancer cell lines. Thus, the antineoplastic properties of tocotrienols result from a combination of factors including antioxidant properties, upregulation of apoptotic pathways and down-regulation of both proliferative pathways, and crucial cell surface receptors.[9–11]

OXIDATIVE STRESS IN CANCER AND THE PREVENTIVE ROLE OF TOCOTRIENOLS

Chronic inflammation has long been linked to the development of cancer. The processes of tumor proliferation, metastasis, immortality, and chemoresistance all can be associated with chronic inflammation.[12] The antioxidant properties of tocotrienols may have a potent role in counteracting these processes. In the past, a significant part of vitamin E research was focused on its role as an antioxidant in prevention of cardiovascular disease and dyslipidemia. However, a better understanding of cancer pathobiology has helped to shed light on the detrimental role of oxidative stress in causing cellular damage, thus leading to alteration of function of nucleic acids, proteins, and lipids, and eventually initiating carcinogenesis.[13] Free radicals and nonradicals can both cause oxidative stress, and reactive oxygen species can be endogenous or exogenous. Exogenous or environmental oxidants are a major concern in carcinogenesis. They

TOCOTRIENOLS

Isomer	R₁	R₂
α	CH₃	CH₃
β	CH₃	H
γ	H	CH₃
δ	H	H

* Chiral center

TOCOPHEROLS

FIGURE 24.1   Isoforms of Vitamin E. Tocotrienols and tocopherols and the substituents that define their isoforms. Tocotrienols have a characteristic farnesylated tail, which may provide superior anticancer properties as compared to tocopherols. (Wong RS, Radhakrishnan AK. Tocotrienol research: past into present. Nutr Rev. 2012 Sep;70(9):483-90. http://dx.doi.org/10.1111/j.1753-4887.2012.00512.x. Epub 2012 Aug 17.Review. PubMed PMID: 22946849).

include cigarette smoking, ozone exposure, hyperoxia, ionizing radiation, and heavy metal ions.[14–19] There are two counteracting mechanisms of antioxidation in our bodies; those that involve enzymatic antioxidants, such as superoxide dismutase, catalase, glutathione peroxidase, thioredoxin, and glutathione transferase, as well as nonenzymatic antioxidants, such as vitamin C, vitamin A, vitamin E, glutathione, and beta carotene.[13]

Oxidative stress can cause a multitude of cellular damages including changes in DNA sequence and structure, alteration in lipid and protein functions, upregulation of stress-induced transcription factors and formation of pro-inflammatory and anti-inflammatory cytokines. Reactive oxygen species can lead to DNA base degradation, DNA breaks, purine and pyrimidine alteration, mutations, deletions, and cross-linking with proteins. All these processes alone or in combination can lead to carcinogenesis.[13] For example, ionizing radiation can cause double-strand DNA breaks and impair cell survival.[20] Oxidative stress can disintegrate the bilayer arrangement of membrane lipids disrupting normal membrane bound receptors and enzyme functions, thus increasing tissue

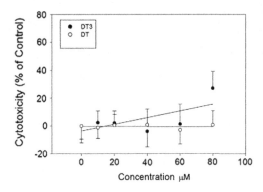 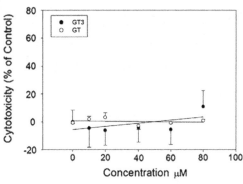

FIGURE 24.2 Neither tocotrienols nor tocopherols have a cytotoxic effect on normal cells. Nontumorigenic pancreatic duct ephithelial HPDE-E6E7 cells were treated with δ-tocotrienol (DT3), δ-tocopherol (DT), γ-tocotrienol (GT3) and γ-tocopherol (GT). Viable cells were measured by the MTT assay and results are shown as percentage cytotoxicity of the respective control. Values shown are the means ±SEM obtained from three independent experiments.

permeability.[21] Similarly, peptide chains and amino acids also become susceptible to proteolytic degradation, secondary to oxidative stress. Most importantly, changes in cysteine and methionine residues lead to conformational alterations in proteins. Under oxidative stress, disruption of glutathione balance (GSH/GSSG ratio) can lead to activation of AP-1, NF-kappa B, and hypoxia inducible factor (Figure 24.3). Reactive oxygen species may also cause activation of transcription factors by mediating signal transduction through epidermal growth factor receptors, endothelial growth factor receptors, platelet-derived growth factors, and serine/threonine kinases. Oxidants can also cause alterations in proliferation, differentiation, and apoptosis markers like JNK and p38.[18,22,23] Furthermore, activation of NF-kappa B by reactive oxygen species can lead to upregulation of several immune response mediating genes including IL-1b, IL-6, tumor necrosis factor-a, IL-8, and several adhesion molecules.[24]

Tocotrienols play an important role in counteracting cellular inflammatory responses secondary to oxidative stress, thus exerting an anticancer property. Tocotrienols, especially gamma tocotrienol, mediate the function of NF-kappa B, STAT3 (signal transduction and activators), and COX-2.[12] Specifically, tocotrienols can cause suppression of NF-kappa B activation in different cell lines like human lung adenocarcinoma (H1299), human breast cancer (MCF7), multiple myeloma (U266), and pancreas (MiaPaCa-2 and BxPc-3).[25, 26] Gamma tocotrienol also inhibits STAT3 activation. Constitutive activation of STAT3 is a common phenomenon in chronic lymphocytic leukemia, multiple myeloma, acute myeloid leukemia, and gastric cancer.[27] Overexpression of COX-2 is now known to be associated with cancer and many other diseases. Tocotrienols can inactivate not only COX-2 but also many other pro-inflammatory cytokines including TNF-alpha, IL-6, IL-4, and IL-8.[28,29] Wu et al. showed these effects of tocotrienols by using palm tocotrienol rich fraction (TRF) in lipopolysaccharide-induced human monocytic (THP-1) cells.[28]

EFFECT OF TOCOTRIENOLS IN MEDIATING CELLULAR PATHWAYS

Besides their role as antioxidant and anti-inflammatory agents, tocotrienols also mediate multiple cell cycle pathways. All these anti-neoplastic roles of tocotrienols have been demonstrated by various research groups.[9,27,30,31]: promotion of death receptors (DRs) like CD95, TNF receptor, and TNF-related apoptosis-inducing ligand (TRAIL) receptor; induction of apoptosis either through activation of caspase-8 and -3 or through caspase-9; overexpression of nuclear factor (Erythroid derived 2)-like 2; suppression of hypoxia inducible factors (HIFs); down-regulation of PI3K and ErbB2, and prevention of VEGF stimulated angiogenesis. The ability of tocotrienols to selectively inhibit the HMG CoA reductase pathway through posttranslational degradation of HMG CoA reductase, inhibition of the PI3 kinase/Akt, ERK pathways through down-regulation of Her2/ErbB2 receptors, as well as the suppression of the activity of transcription factor NF kappa B, could be the bases for some of these properties (Figure 24.4).

TOCOTRIENOLS IN PANCREATIC CANCER

Pancreatic cancer is the fourth leading cause of cancer death in the United States (36). The overall prognosis, even with locally advanced disease, continues to be very grim. The average age of diagnosis is around 60 to 70 years, and for advanced disease 5-year survival is less than 5%. For patients with metastatic disease, even with the best available treatment option, average survival is less than 12 months. The biggest challenge results from significant treatment-related side effects, thus compromising quality of life, even when survival is minimally prolonged. For that reason, a better understanding of carcinogenesis at the molecular level, in order to find novel targeted therapeutics with minimal toxicity, is quintessential to

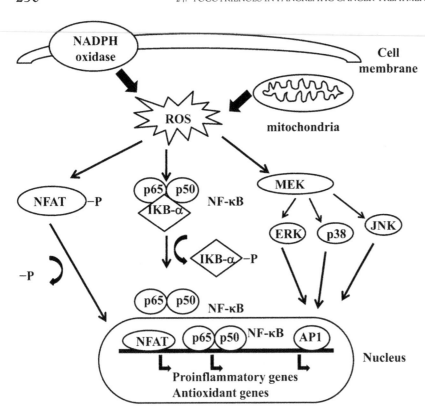

FIGURE 24.3 Effects of oxidative stress on signal transduction in the cell. (Birben E, Sahiner UM, Sackesen C, Erzurum S, Kalayci O. Oxidative stress and antioxidant defense. World Allergy Organ J. 2012 Jan;5(1):9-19.   http://dx.doi.org/10.1097/WOX.0b 013e3182439613. Epub 2012 Jan 13. PubMed PMID: 23268465; PubMed Central PMCID: PMC3488923).

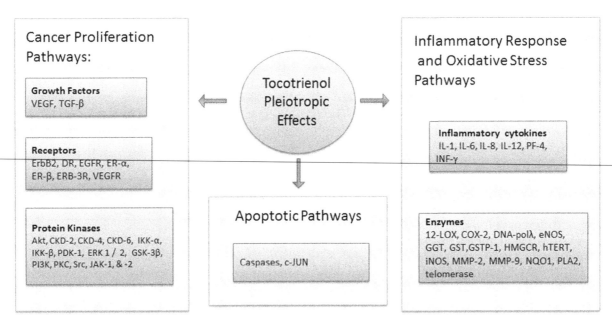

FIGURE 24.4 Tocotrienol Pleiotropic Effects. Tocotrienols affect proliferative and apoptotic signaling cascades, as well as cellular pathways involved in antioxidant and anti-inflammatory processes in the cell.

changing disease prognosis. As there are no feasible modes of surveillance or screening techniques available, it is even more relevant to explore natural agents capable of exerting pleiotropic effects on cellular pathways, with hope to find a prevention strategy, such as the use of folic acid during pregnancy to prevent neural tube defect in newborns.

Somatic mutation of the K-ras gene at codon 12 occurs in over 95% of pancreatic ductal adenocarcinoma.[32,33] Likely, the K-ras mutation is an early event to transform preinvasive pancreatic lesion into pancreatic adenocarcinoma.[34,35] These precursor lesions are known as PanINs (pancreatic intraepithelial neoplasia). There are three grades of PanIN lesions. PanIN-1 is commonly found in the epithelial lining of the pancreatic ducts of half of all elderly persons. As few as 1 in 500 PanIN-1 lesions will progress to a malignant lesion, unlike PanIN-2 and PanIN-3, which have a higher risk of progressing to invasive cancer. Once these preinvasive lesions are transformed to being invasive, it is uniformly fatal. These PanIN lesions have increased rates of the K-Ras oncogenic mutation in addition to alterations in other proteins such as p16INK4a and p53.[36] Constitutively activated K-Ras can trigger signaling via ERK/mitogen-activated protein (MAP) kinase and phosphatidylinositol 3-kinase (PI3-kinase)/AKT (Figure 24.5). Drugs targeting these signaling pathways are the main focus of research work because of their potential therapeutic value.[37]

As detailed earlier in this discussion, the antitumor activity of tocotrienols may be related to their antioxidant potency, ability to induce cell cycle arrest or apoptosis, ability to inhibit HMG-CoA reductase[25,38], prevent angiogenesis, and down-regulate transcription factor NF-κB,[25] which is a target of the PI3-kinase/AKT pathway in pancreatic cancer cells.[39] Ras inhibitors have the potential of being effective chemotherapeutic agents as proliferation of the majority of pancreatic cancer cells is regulated through an aberrant oncogenic Ras signaling pathway and its downstream effectors.

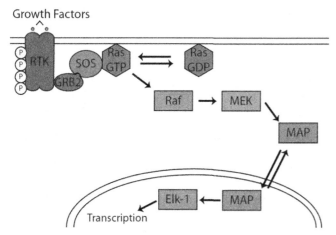

FIGURE 24.5 Cell proliferation pathway in pancreatic cancer. The Ras-Raf-Erk pathway under the control of activated growth factor receptors. (Steven Kennedy, Hannah Berrett and Robert J. Sheaff (2012). Disruption of Cell Cycle Machinery in Pancreatic Cancer, Pancreatic Cancer - Molecular Mechanism and Targets, Prof. Sanjay Srivastava (Ed.), ISBN: 978-953-51-0410-0, InTech, Available from: http://www.intechopen.com/books/pancreatic-cancermolecular-mechanism-and-targets/disruption-of-cell-cycle-machinery-in-pancreatic-cancer).

IN VITRO DATA IN PANCREATIC CANCER

The unique structure of tocotrienols has raised significant interest to explore its impact on tumorigenesis of pancreatic cancer. Tocotrienols have a farnesylated tail[40] that could selectively down-regulate 3-hydroxy-3-methylglutaryl-coenzyme A reductase, a necessary component in the biosynthesis of farnesyl pyrophosphate, which is essential in the activation of Ras protein by protein farnesyl transferase. Some preclinical data with inhibitors of farnesyl transferase have shown their effect only on H-ras- and N-ras-transformed cells, but not on K-ras,[41-43] which is the commonly mutated gene in pancreatic cancer. A recent study has shown that tocotrienols can modulate Ras signaling pathways, regardless of the status of Ras, in cells with wild-type Ras as well as in cells with mutated Ras.[9] Tocotrienols also down-regulate ERK phosphorylation and protein levels, and have the opposite effect on c-Jun. Tocotrienols are also able to down-regulate the activation of AKT, but not the protein levels. Receptor tyrosine kinase ErbB2 initiates signaling pathways via tyrosine activation on its cytoplasmic domain and subsequent interactions with docking proteins.[44,45] In normal tissues as well as in cancers of the breast, uterus, pancreas, urinary bladder, lung, gallbladder, stomach, and head and neck these interactions play a major role. Tocotrienols affect receptor tyrosine kinase ErbB2, thus down-regulating both MAPK and AKT pathways.[9] Tocotrienols (γ and δ) also show selective cytotoxicity to pancreatic cancer cells as compared to normal pancreatic cells.

The antitumor activity of tocotrienols is exerted through the modulation of the intracellular Ras signaling via the Raf/MEK/ERK pathway in K-ras mutated pancreatic cancer cell lines. The ERK pathway is involved in cellular proliferation, differentiation, and survival. ERK phosphorylation promotes cell proliferation in pancreatic cancer cells and prevents apoptosis; conversely, inhibition of the ERK pathway promotes apoptosis via activation of caspase-6, -8, and -9.[46] Research by our group has shown that tocotrienols (GT3 and DT3) cause a reduction in total ERK and phosphorylated ERK, as well as affecting survival/apoptotic pathways to induce growth arrest.

As important downstream effectors of Ras protein, AKT regulates cell survival/apoptosis, proliferation, angiogenesis, protein synthesis, and drug resistance. K-ras mutations that lock pancreatic cancer cells in a constitutively active state occur early and in about 75 to 90% of pancreatic adenocarcinomas. The PI3-kinase AKT pathway is activated in about 60% of pancreatic cancers.[47] The presence of phosphorylated AKT leads to the gene expression of several downstream molecules that include transcription factors, mediators of survival, and cell cycle regulators such as mTOR and GSK-3β.[48,49] AKT overexpression has been reported in pancreatic cancers as having a major role in cell proliferation and survival.[50] Tocotrienols cause a reduction in AKT phosphorylation at serine 473 at both low and high concentrations, thus inducing cell growth inhibition and apoptosis. This effect is mediated by cell signal control upstream of AKT without change in total AKT protein level.[9]

ErbB2 mediated downstream activation of ERK and PI3-kinase/AKT pathways are related to tumor cell proliferation and metastatic potential. Phosphorylation at Tyr 1248/1253 of ErbB2 activates this pathway and inhibition of this residue leads to significantly diminished transforming potential of ErbB2.[51] Tocotrienols cause down-regulation of the phosphorylated form (pY1248 ErbB2) and ErbB2 at the messenger level. These effects of tocotrienols on tumor cell proliferation are evident in both a time- and concentration-dependent manner.[9] A mild effect was observed on the expression levels of ErbB3, but not on the activation levels of the receptor. These preclinical studies were conducted in MIA PaCa-2 cells, which are known to express ErbB2. All these facts further consolidate a promise of tocotrienols as novel agents in treating and preventing pancreatic malignancies irrespective of ErbB2 status using a mechanism different from those of known agents such as trastuzumab.

Tocotrienols also cause a strong increase in the level of expression of c-Jun, as well as in the phosphorylation level of the protein. c-Jun activation increases the stability of the molecule and mediates apoptosis.[52] Recent studies have shown inhibition of the expression of p-mTOR, p-70 S6 kinase, and p-GSK-3β by tocotrienols. The PI3-kinase/AKT signaling pathway phosphorylates mTOR, which plays a key role in cell growth and homeostasis. Tocotrienols inhibit mTOR phosphorylation at serine 2448 in MIA PaCa-2 cells, resulting in reduced translation and cell growth. GSK-3β is also an important component of the PI3-kinase/AKT survival pathway. GSK-3 is a serine–threonine kinase that phosphorylates and inactivates glycogen synthase.[53] It can be inactivated by AKT-mediated phosphorylation of serine 9. Gamma tocotrienol (GT3) is able to reverse this phosphorylation and consequently inactivate GSK-3 by inhibiting AKT phosphorylation.

As a member of the Forkhead family of transcription factors, FOXO3a acts as a tumor suppressor by promoting cell cycle arrest and apoptosis. FOXO3a is a key target of the phosphatidylinositol 3-kinase/AKT survival pathway. Phosphorylated AKT inhibits FOXO3a by phosphorylation, thus sequestering it in the cytoplasm.[54] ERK also shows similar effect on FOXO3a. Upon dephosphorylation of both AKT and ERK, FOXO3a is rapidly dephosphorylated by PP2A, resulting in nuclear translocation and transcriptional activity conducive to apoptosis or cell cycle arrest.[55] Tocotrienols inhibit both the ERK and the AKT pathway leading to dephosphorylation of FOXO3a and export to the nucleus.

Tocotrienols (GT3>DT3) show antitumor activity at various levels of cell signaling pathways including down-regulating cell-surface receptor ErbB2; reducing phosphorylation of ERK, AKT, and mTOR; reversing phosphorylation of GSK-3β; causing upregulation and nuclear export of FOXO3a; and upregulating c-Jun and also by direct cellular cytotoxicity.

Multiple research studies[26,56] have demonstrated tocotrienols exert a multipronged effect on pancreatic cancer cell proliferation by means of their direct effects on HMG-CoA reductase, ErbB2, AKT, ERK, NF-κB, FOXO3a, c-Jun, p-GSK-3β, mTOR, and RSK or through the modulation of upstream or downstream signaling cascades. As ErbB2 is upstream of NF-κB, down-regulation of ErbB2 may have a role in inhibiting the NF-κB signaling pathway. Most recently the beneficial effect of tocotrienols has also been proven in animal models.[57] More preclinical data is needed and eventually clinical trials to consider the use of tocotrienols for prevention or as an adjunct to chemotherapy agents to meet challenges of pancreatic cancer.

IN VIVO DATA IN PANCREATIC CANCER

The data showing the beneficial roles of tocotrienols in prevention/treatment of pancreatic cancer is quite impressive. Based on this information more researchers are now exploring the use of tocotrienols (gamma and delta) in animal models and human subjects in the setting of pancreatic cancer. As a first step, before its use in human subjects, researchers have shown that oral administration of delta tocotrienols in mice has led to the presence of a bioactive level in pancreatic tissue, raising the possibility of using it in clinical trial.[58] Subsequent use of delta tocotrienol in genetically (LSL-Kras$^{G12D/+}$; Pdx-1-Cre) mutated mouse model showed increased median survival, complete disappearance of invasive components of pancreatic cancer, reduction of proliferation of mouse pancreatic intraepithelial neoplasm, as well as upregulation of apoptotic pathways and down-regulation of proliferation pathways in delta tocotrienol treated mice compared to vehicle treated and no treatment groups.[57] Most interestingly, a couple of phase I clinical trials are ongoing to assess (1) safety and tolerability of delta tocotrienols in healthy subjects[59] and (2) dose escalation study among subjects with resectable pancreatic cancer to assess safety, pharmacokinetics, and pharmacodynamics before proceeding with phase II study.[60] Eventually a phase III clinical trial will be needed to bring beneficial use of tocotrienols from bench research to clinical practice.

CONCLUSION AND FUTURE DIRECTION

The effects of tocotrienols in pancreatic cancer cell lines and in early experiments in animal models are encouraging. More work on animal and genetic models of pancreatic cancer is needed to gather more data to support these initial observations.

SUMMARY POINTS

- Oxidative stress is a documented factor in the pathogenesis of inflammation and cancer.
- Out of many nutrients with antioxidant properties, vitamin E holds promise for use in clinical practice.
- Tocotrienols, one of the lesser explored but more potent vitamin E derivatives, exert their antiproliferative activity against malignant cells but not on normal cells.
- Tocotrienols have shown antiproliferative and proapoptotic activity in diverse cancer cell lines including breast, pancreas, prostate, colorectal, cervical, and skin.
- Tocotrienols play an important role in counteracting cellular inflammatory response secondary to oxidative stress thus exerting an anticancer property.
- Tocotrienols mediate function of NF-kappa B, STAT3 (signal transduction and activators), and COX-2.
- Tocotrienols show pleiotropic effects on cancer pathways including HMG CoA reductase, EGFR, ErbB2, AKT, ERK, p-GSK, c-JUN, caspase, NF-kappa B, and Bax/Bcl2.
- Tocotrienols show anticancer effects by combination of factors including antioxidant properties, upregulation of apoptotic pathways and down-regulation of both proliferative pathways and cell surface receptors.
- Additional preclinical data and eventually clinical trials will be needed to consider the use of tocotrienols for prevention or as an adjunct to chemotherapy agents in the management of pancreatic cancer.

References

1. Evans HM, Bishop KS. On the Existence of a Hitherto Unrecognized Dietary Factor Essential for Reproduction. *Science* 1922;**56**(1458):650–1.
2. Packer L, Weber SU, Rimbach G. Molecular aspects of alpha-tocotrienol antioxidant action and cell signalling. *J Nutr* 2001;**131**(2):369S–73S.
3. Sen CK, Khanna S, Roy S. Tocotrienols: Vitamin E beyond tocopherols. *Life Sci* 2006;**78**(18):2088–98.
4. Frega N, Mozzon M, Bocci F. Identification and Estimation of Tocotrienols in the Annatto Lipid Fraction by Gas Chromatography–Mass Spectrometry. *J Am Oil Chem Soc* 1998;**75**(12):1723–7.

5. Sundram K, Sambanthamurthi R, Tan YA. Palm fruit chemistry and nutrition. *Asia Pacific J Clin Nutr* 2003;**12**(3):355–62.
6. He L, Mo H, Hadisusilo S, Qureshi AA, Elson CE. Isoprenoids suppress the growth of murine B16 melanomas in vitro and in vivo. *J Nutr* 1997;**127**(5):668–74.
7. Das S, Lekli I, Das M, et al. Cardioprotection with palm oil tocotrienols: comparison of different isomers. *Am J Physiol Heart Circ Physiol* 2008;**294**(2):H970–8.
8. Hayes KC, Pronczuk A, Liang JS. Differences in the plasma transport and tissue concentrations of tocopherols and tocotrienols: observations in humans and hamsters. *Proc Soc Exp Biol Med Soc Exp Biol Med* 1993;**202**(3):353–9.
9. Shin-Kang S, Ramsauer VP, Lightner J, et al. Tocotrienols inhibit AKT and ERK activation and suppress pancreatic cancer cell proliferation by suppressing the ErbB2 pathway. *Free Radical Biol Med* 2011;**51**(6):1164–74.
10. Yap WN, Chang PN, Han HY, et al. Gamma-tocotrienol suppresses prostate cancer cell proliferation and invasion through multiple-signalling pathways. *Br J Cancer* 2008;**99**(11):1832–41.
11. Kannappan R, Gupta SC, Kim JH, Aggarwal BB. Tocotrienols fight cancer by targeting multiple cell signaling pathways. *Genes Nutr* 2012;**7**(1):43–52.
12. Nesaretnam K, Meganathan P. Tocotrienols: inflammation and cancer. *Ann New York Acad Sci* 2011;**1229**:18–22.
13. Birben E, Sahiner UM, Sackesen C, Erzurum S, Kalayci O. Oxidative stress and antioxidant defense. *World Allergy Organizat J* 2012;**5**(1):9–19.
14. Denissenko MF, Pao A, Tang M, Pfeifer GP. Preferential formation of benzo[a]pyrene adducts at lung cancer mutational hotspots in P53. *Science* 1996;**274**(5286):430–2.
15. Peluso M, Munnia A, Hoek G, et al. DNA adducts and lung cancer risk: A prospective study. *Cancer Res* 2005;**65**(17):8042–8.
16. Moen I, Oyan AM, Kalland KH, et al. Hyperoxic treatment induces mesenchymal-to-epithelial transition in a rat adenocarcinoma model. *PloS One* 2009;**4**(7):e6381.
17. Glaviano A, Nayak V, Cabuy E, et al. Effects of hTERT on metal ion-induced genomic instability. *Oncogene* 2006;**25**(24):3424–35.
18. Kambach DM, Sodi VL, Lelkes PI, Azizkhan-Clifford J, Reginato MJ. ErbB2, FoxM1 and 14-3-3zeta prime breast cancer cells for invasion in response to ionizing radiation. *Oncogene* 2013.
19. Stohs SJ, Bagchi D. Oxidative mechanisms in the toxicity of metal ions. *Free Radic Biol Med* 1995;**18**(2):321–36.
20. Vilenchik MM, Knudson AG. Endogenous DNA double-strand breaks: Production, fidelity of repair, and induction of cancer. *Proc Natl Acad Sci United States Am* 2003;**100**(22):12871–6.
21. Girotti AW. Mechanisms of lipid peroxidation. *J Free Radic Biol Med* 1985;**1**(2):87–95.
22. Neufeld G, Cohen T, Gengrinovitch S, Poltorak Z. Vascular endothelial growth factor (VEGF) and its receptors. *FASEB J* 1999;**13**(1):9–22.
23. Salmeen A, Park BO, Meyer T. The NADPH oxidases NOX4 and DUOX2 regulate cell cycle entry via a p53-dependent pathway. *Oncogene* 2010;**29**(31):4473–84.
24. Sarkar D, Lebedeva IV, Emdad L, Kang DC, Baldwin Jr AS, Fisher PB. Human polynucleotide phosphorylase (hPNPaseold-35): A potential link between aging and inflammation. *Cancer Res* 2004;**64**(20):7473–8.
25. Ahn KS, Sethi G, Krishnan K, Aggarwal BB. Gamma-tocotrienol inhibits nuclear factor-kappaB signaling pathway through inhibition of receptor-interacting protein and TAK1 leading to suppression of antiapoptotic gene products and potentiation of apoptosis. *J Biol Chem* 2007;**282**(1):809–20.
26. Kunnumakkara AB, Sung B, Ravindran J, et al. {Gamma}-tocotrienol inhibits pancreatic tumors and sensitizes them to gemcitabine treatment by modulating the inflammatory microenvironment. *Cancer Res* 2010;**70**(21):8695–705.

27. Kannappan R, Yadav VR, Aggarwal BB. gamma-Tocotrienol but not gamma-tocopherol blocks STAT3 cell signaling pathway through induction of protein-tyrosine phosphatase SHP-1 and sensitizes tumor cells to chemotherapeutic agents. *J Biol Chem* 2010;**285**(43):33520–8.

28. Wu SJ, Liu PL, Ng LT. Tocotrienol-rich fraction of palm oil exhibits anti-inflammatory property by suppressing the expression of inflammatory mediators in human monocytic cells. *Molecular Nutr Food Res* 2008;**52**(8):921–9.

29. Yam ML, Abdul Hafid SR, Cheng HM, Nesaretnam K. Tocotrienols suppress proinflammatory markers and cyclooxygenase-2 expression in RAW264.7 macrophages. *Lipids* 2009;**44**(9):787–97.

30. Takahashi K, Loo G. Disruption of mitochondria during tocotrienol-induced apoptosis in MDA-MB-231 human breast cancer cells. *Biochem Pharmacol* 2004;**67**(2):315–24.

31. Hsieh TC, Elangovan S, Wu JM. Differential suppression of proliferation in MCF-7 and MDA-MB-231 breast cancer cells exposed to alpha-, gamma- and delta-tocotrienols is accompanied by altered expression of oxidative stress modulatory enzymes. *Anticancer Res* 2010;**30**(10):4169–76.

32. Smit VT, Boot AJ, Smits AM, Fleuren GJ, Cornelisse CJ, Bos JL. KRAS codon 12 mutations occur very frequently in pancreatic adenocarcinomas. *Nucleic Acids Res* 1988;**16**(16):7773–82.

33. Hong SM, Vincent A, Kanda M, et al. Genome-wide somatic copy number alterations in low-grade PanINs and IPMNs from individuals with a family history of pancreatic cancer. *Clin Cancer Res* 2012;**18**(16):4303–12.

34. Hingorani SR, Petricoin EF, Maitra A, et al. Preinvasive and invasive ductal pancreatic cancer and its early detection in the mouse. *Cancer cell* 2003;**4**(6):437–50.

35. Grippo PJ, Nowlin PS, Demeure MJ, Longnecker DS, Sandgren EP. Preinvasive pancreatic neoplasia of ductal phenotype induced by acinar cell targeting of mutant Kras in transgenic mice. *Cancer Res* 2003;**63**(9):2016–9.

36. Baumgart M, Werther M, Bockholt A, et al. Genomic instability at both the base pair level and the chromosomal level is detectable in earliest PanIN lesions in tissues of chronic pancreatitis. *Pancreas* 2010;**39**(7):1093–103.

37. Maurer G, Tarkowski B, Baccarini M. Raf kinases in cancer-roles and therapeutic opportunities. *Oncogene* 2011;**30**(32):3477–88.

38. Parker RA, Pearce BC, Clark RW, Gordon DA, Wright JJ. Tocotrienols regulate cholesterol production in mammalian cells by post-transcriptional suppression of 3-hydroxy-3-methylglutaryl-coenzyme A reductase. *J Biol Chem* 1993;**268**(15):11230–8.

39. Asano T, Yao Y, Zhu J, Li D, Abbruzzese JL, Reddy SA. The PI 3-kinase/Akt signaling pathway is activated due to aberrant Pten expression and targets transcription factors NF-kappaB and c-Myc in pancreatic cancer cells. *Oncogene* 2004;**23**(53):8571–80.

40. Suzuki YJ, Tsuchiya M, Wassall SR, et al. Structural and dynamic membrane properties of alpha-tocopherol and alpha-tocotrienol: Implication to the molecular mechanism of their antioxidant potency. *Biochemistry* 1993;**32**(40):10692–9.

41. Song SY, Meszoely IM, Coffey RJ, Pietenpol JA, Leach SD. K-Ras-independent effects of the farnesyl transferase inhibitor L-744,832 on cyclin B1/Cdc2 kinase activity, G2/M cell cycle progression and apoptosis in human pancreatic ductal adenocarcinoma cells. *Neoplasia* 2000;**2**(3):261–72.

42. Zhang FL, Kirschmeier P, Carr D, et al. Characterization of Ha-ras, N-ras, Ki-Ras4A, and Ki-Ras4B as in vitro substrates for farnesyl protein transferase and geranylgeranyl protein transferase type I. *J Biol Chem* 1997;**272**(15):10232–9.

43. Whyte DB, Kirschmeier P, Hockenberry TN, et al. K- and N-Ras are geranylgeranylated in cells treated with farnesyl protein transferase inhibitors. *J Biol Chem* 1997;**272**(22):14459–64.

44. Klapper LN, Kirschbaum MH, Sela M, Yarden Y. Biochemical and clinical implications of the ErbB/HER signaling network of growth factor receptors. *Adv Cancer Res* 2000;**77**:25–79.

45. Hynes NE, Lane HA. ERBB receptors and cancer: The complexity of targeted inhibitors. *Nature Rev Cancer* 2005;**5**(5):341–54.

46. Boucher MJ, Morisset J, Vachon PH, Reed JC, Laine J, Rivard N. MEK/ERK signaling pathway regulates the expression of Bcl-2, Bcl-X(L), and Mcl-1 and promotes survival of human pancreatic cancer cells. *J Cell Biochem* 2000;**79**(3):355–69.

47. Schlieman MG, Fahy BN, Ramsamooj R, Beckett L, Bold RJ. Incidence, mechanism and prognostic value of activated AKT in pancreas cancer. *Br J Cancer* 2003;**89**(11):2110–5.

48. Hanada M, Feng J, Hemmings BA. Structure, regulation and function of PKB/AKT—A major therapeutic target. *Biochimica et Biophysica Acta* 2004;**1697**(1-2):3–16.

49. Franke TF. PI3K/Akt: Getting it right matters. *Oncogene* 2008;**27**(50):6473–88.

50. Semba S, Moriya T, Kimura W, Yamakawa M. Phosphorylated Akt/PKB controls cell growth and apoptosis in intraductal papillary-mucinous tumor and invasive ductal adenocarcinoma of the pancreas. *Pancreas* 2003;**26**(3):250–7.

51. Holbro T, Hynes NE. ErbB receptors: Directing key signaling networks throughout life. *Ann Rev Pharmacol Toxicol* 2004;**44**:195–217.

52. Behrens A, Sibilia M, Wagner EF. Amino-terminal phosphorylation of c-Jun regulates stress-induced apoptosis and cellular proliferation. *Nature Genetics* 1999;**21**(3):326–9.

53. Cross DA, Alessi DR, Cohen P, Andjelkovich M, Hemmings BA. Inhibition of glycogen synthase kinase-3 by insulin mediated by protein kinase B. *Nature* 1995;**378**(6559):785–9.

54. Brunet A, Park J, Tran H, Hu LS, Hemmings BA, Greenberg ME. Protein kinase SGK mediates survival signals by phosphorylating the forkhead transcription factor FKHRL1 (FOXO3a). *Molecular Cell Biol* 2001;**21**(3):952–65.

55. Yang JY, Zong CS, Xia W, et al. ERK promotes tumorigenesis by inhibiting FOXO3a via MDM2-mediated degradation. *Nature Cell Biol* 2008;**10**(2):138–48.

56. Hussein D, Mo H. d-delta-Tocotrienol-mediated suppression of the proliferation of human PANC-1, MIA PaCa-2, and BxPC-3 pancreatic carcinoma cells. *Pancreas* 2009;**38**(4):e124–36.

57. Husain K, Centeno BA, Chen DT, et al. Prolonged survival and delayed progression of pancreatic intraepithelial neoplasia in LSL-KrasG12D/+;Pdx-1-Cre mice by vitamin E delta-tocotrienol. *Carcinogenesis* 2013.

58. Husain K, Francois RA, Hutchinson SZ, et al. Vitamin E delta-tocotrienol levels in tumor and pancreatic tissue of mice after oral administration. *Pharmacology* 2009;**83**(3):157–63.

59. *Institute HLMCCaR. Vitamin E δ-Tocotrienol Multiple Dose in Healthy Subjects.* October 7, 2011. http://clinicaltrials.gov/ct2/show/NCT01450046?term=tocotrienol+AND+Moffitt&rank=1 [accessed 03.10.13].

60. *Institute HLMCCaR. Vitamin E δ-Tocotrienol Administered to Subjects With Resectable Pancreatic Exocrine Neoplasia.* http://clinicaltrials.gov/show/NCT00985777 [September 25, 2009].

Fern Extract, Oxidative Stress, and Skin Cancer

Concepción Parrado

Department of Histology and Pathology, Faculty of Medicine, University of Málaga, Spain

Angeles Juarranz

Biology Department, Sciences School, Universidad Autónoma de Madrid, Madrid, Spain

Yolanda Gilaberte

Department of Dermatology, San Jorge Hospital, Huesca, Spain

Neena Philips

School of Natural Sciences, University College, Fairleigh Dickinson University, Teaneck, NJ, USA

Salvador Gonzalez

Dermatology Service, Memorial Sloan-Kettering Cancer Center, New York, NY, USA

List of Abbreviations

8-OH-dG 8-Hydroxy-2′-deoxyguanosine
AP-1 Activator protein 1
CIE Commission Internationale de L' Éclairage (CIE)
COLIα**1** Collagen type I
CD Common deletion
COX-2 Cyclooxygenase-2
CPDs Ciclobutane pyrimidine dimers
DC Dendritic cells
ECM Extracellular matrix
eLC Epidermal Langerhans cells
GP Glutathione peroxidase
GST Glutathione S-transferase
HPLC High-performance liquid chromatography
IL Interleukine
iNOS Inducible nitric oxide synthase
MMP Matrix metalloprotease
NB-UVB Narrow-band UVB
NF-κβ Nuclear factor kappa beta
NIEHS National Institute of Environmental Health Sciences
NO Nitric oxide
NOS Nitric oxide synthase
PG Prostaglandin
PL Polypodium leucotomos
PMLE Polymorphic light eruption
PUVA Psoralens + UVA
ROS Reactive oxygen species

SOD Superoxide dismutase
SSR Simulator simulating radiation
TGF- β Transforming growth factor beta
Th1 T helper 1 lymphocyte
TIMP Inhibitor of metalloproteinase
TNF-α Tumor necrosis factor-alpha
UCA Urocanic acid
UV Ultraviolet light

INTRODUCTION

Skin cancer accounts for at least 40% of all human malignancies. Ultraviolet (UV) radiation is a major causal agent in most skin cancers.[1] This is due to the deleterious effects of UV on the skin including inflammation, oxidative stress, DNA damage, deregulation of cellular signaling pathways, and immunosuppression, which are some of the molecular hallmarks that lead to skin cancer.[2,3]

In addition to direct ionizing damage to the cells and molecules of the skin, UV radiation also generates reactive oxygen species (ROS), which exert a multitude of biological effects. Some are beneficial and physiological, but their inappropriate generation may lead to

Cancer
http://dx.doi.org/10.1016/B978-0-12-405205-5.00025-8

© 2014 Elsevier Inc. All rights reserved.

damaging alterations, participating in the pathogenesis of increasing number of diseases. It is also well documented that ROS can stimulate cell division and enhance tumor growth.[4]

The use of botanic supplements endowed with substantial antioxidant activities has generated wide interest to decrease the risk of skin disease induced by UV radiation-dependent oxidative stress. Oral sun screening agents can provide a degree of protection by blocking, at least partially, the abnormal generation of ROS that underlies most of these diseases. In this context, a hydrophilic extract obtained from the aerial parts of the fern *Polypodium leucotomos* (PL, Fernblock®) exhibits strong photoprotective properties following its topical or oral administration. *In vitro* and *in vivo* studies have shown that PL scavenges superoxide anions, hydroxyl radicals, and singlet oxygen.[5–7] It also inhibits lipid peroxidation and increases the survival of human keratinocytes exposed to UV radiation.[5–8]

In addition, preclinical and clinical research has shown that PL modulates the immune/inflammatory responses and prevents UV radiation-induced skin cancer in several animal models. The molecular basis of these effects likely includes a marked reduction of ROS-mediated DNA damage.[9–12]

P. leucotomos is readily absorbed and metabolized cellularly.[7] These effects suggest that PL bears *in vivo* potential as an anti-skin-aging and anticancer agent.[10,13–16]

In this chapter, we will address the composition of *P. leucotomos* extract and its antioxidant properties. To substantiate the molecular basis of the effect of oral PL in preventing the adverse effects of UV radiation on skin, we will summarize what is known of the molecular mechanisms and targets of PL in preventing photodamage and skin cancer. Briefly, these include (1) signaling pathways that induce or regulate ROS generation and/or effects; (2) pathways of DNA damage and DNA repair; (3) role of ROS in inflammation; (4) immune evasion due to oxidative stress; (5) UV-induced tumor progression, and (6) PL use in other nonskin malignancies.

PHOTOPROTECTIVE AGENTS

In addition to physically blocking UV radiation (e.g., by limiting exposure or wearing appropriate clothes, hats, and sunglasses), photoprotective measures can be divided into sunscreens and other photoprotective agents.

Sunscreens are defined as substances that protect the skin from the harmful effects of solar UV radiation by absorbing, reflecting, scattering, or otherwise deflecting UV photons, avoiding their absorption by the components of the skin. These substances physically shield the skin, preventing photons from reaching the epidermis. They are generally very useful to prevent solar erythema and sunburn caused by high-energy UV photons. Importantly, sunscreens block high-energy UV photons, but not low-energy ones. Some of the long-term effects of UV exposure (e.g., photoaging) are caused by repeated exposure to low-energy photons; therefore sunscreens do not confer appropriate protection against this phenomenon.[7]

On the other hand, nonsunscreen photoprotective agents group a wide array of substances that compensate, ameliorate, repair, or otherwise prevent the short- and long-term effects of UV radiation. These substances are usually antioxidants that, among other effects, prevent the UV exposure-mediated overwhelming of the natural antioxidant systems of the skin that contribute to DNA damage, immunosupression, and photoaging.[7] Since these agents are designed to counter the effect of UV radiation locally as well as systemically, their application is not necessarily limited to the skin, but can be systemic (i.e. oral, intravenous, etc.). Of these, oral treatments are garnering the most attention due to the relative ease of application; however, toxicity must be very carefully assessed since oral uptake ensures direct access of toxic substances to the processing organs of the body (e.g., liver and kidney), increasing the risk in case of poor tolerance or toxicity.

Many botanical and phytochemical extracts are used to treat common ailments because the plants they are obtained from are either edible, or their toxicity and absorption properties are well characterized. An important fraction of these bear powerful antioxidant moieties, hence they are good candidates to counter the increase in systemic oxidative stress caused by UV radiation. It is critical to underline that these compounds are not meant to substitute for the use of sunscreens and physical blockers, but they are conceived and designed to raise the antioxidant threshold of the whole body. Classic antioxidants include many herbal extracts. Among these, PL is an extract from the fern *Polypodium leucotomos*, which has been widely studied and proven useful as a photoprotector, both orally and topically.[7]

In the next sections, its molecular composition and existing data on its effects at the cellular and systemic levels are described.

General Features of the *Polypodium* Genus and *P. leucotomos* Species

The fern genus *Polypodium* (Greek poly = many, pod = feet, due to the rhizomes shaped like small feet) is also known by the name of *Phlebodium*. This genus belongs to the family of Polypodiaceas and encompasses a large number of extensive distributions of species. They mainly grow in dark, damp places in subtropical areas. They are characterized by their high tolerance to environmental stress, mainly to drought, and there are species adapted to high salinity, high temperatures, or frost.[17] To survive in these adverse conditions, these ferns have developed anatomical adaptations (e.g., sugar accumulation to stabilize membranes and proteins) and produce different substances that support their survival against environmental stress. These include phenolic compounds, resins, ecdysones, saponins, and some specific carbohydrates. These plants have developed a diversity of antioxidants that scavenge and remove the ROS, preventing oxidative damage and degradation of the tissue. A major component is polyphenols. A study in the rhizomes of *Polypodium vulgare* reveals that these mainly localize to cytoplasmic vacuoles and cellular membrane compartments (e.g., endoplasmic reticulum), whereas only very low levels are found in the cytoplasm.[17]

Polyphenols comprise the major antioxidant moiety of extracts made of these ferns. *P. leucotomos* is the most common species used to prepare extracts, but some others have also been used with similar purposes and results, likely due to the similarities in terms of production of antioxidant agents and phylogenetic proximity.

P. leucotomos has also been described as *P. aureum* and some of its common names include "calaguala" and "anapsos." It mainly grows at altitudes between 700 and 2500 m in certain regions of Central and South America. *P. leucotomos* was first introduced in Europe in 1788 upon the return of the botanical expedition led by Ruiz and funded by the Spanish crown.[7]

Early Evidence of the Beneficial Effect of the Ferns of the Polypodiacea Family

The first clue of the medicinal properties of the ferns of the family Polypodiaceae goes back to the classic Greek botanist and pharmacist Dioscoride, who attributes some medicinal properties to the roots of these ferns, among them the accelerated healing of skin cracks upon topical application. In some regions of Central and South America, the species *Polypodium leucotomos*, and other species belonging to the same genus, have long been used in traditional medicine to treat various inflammatory disorders and skin tumors.

Empirical studies have demonstrated their effectiveness in the treatment of various diseases of the skin such as vitiligo,[18] psoriasis,[19] and atopic dermatitis,[20] interestingly, most of the disorders in which *Polypodium* exerts a beneficial effect involving inflammation and/or hyperproliferation of skin cells.[21]

Conversely, direct application of the plant has little, if any, significant effect in the treatment of psoriasis and atopic dermatitis. This lack of effect is probably due to the low concentration of the active agents in the whole plant, which has led to preparation of concentrated hydrophilic extracts from the leaves or the rhizomes.[22]

Composition

Although the molecular composition of the extracts of the ferns of the genus Polypodium is not completely characterized, one or several individual constituents seem to be responsible for most of their beneficial effects, particularly those endowed with antioxidant properties. Each extract may contain a specific component, or display characteristic ratios of different components that control their beneficial effects. This depends not only on the species but also the anatomical origin of the biological material used to generate the extract, that is, the aerial parts of the ferns or the rhizomes. In this regard, Vasänge determined the presence of five specific flavonoids in extracts of the *P. decumanun* and *P. triseriale* species, both from Honduras.[41] Other, more recent studies, have addressed the molecular composition of the *P. leucotomos* extract that is commercially available for oral administration. These studies determined that the total phenols content is 250mg/l.

Using High-Performance Liquid Chromatography (HPLC), different phenolic compounds were detected according to their retention time.[23] The most abundant were phenolic acids, specifically caffeic acid, chlorogenic acid, and ferulic acid.[24] To study whether these could be responsible for the beneficial effects of the extract, their absorption rate was studied *in vitro*. These phenolic compounds displayed good absorption through an epithelial monolayer *in vitro*, ranging between 70 and 100%.[24] Absorption of these compounds seemed to depend on a concentration-dependent, saturable active transport mechanism, as absorption was more effective at a dose of 50μM than 200Mm.[24] Importantly, three of the major phenolics of the PL extract (coumaric, ferulic, and vanillic acids) were catabolized by phase I and II enzymes, which resulted in extended persistence of the compounds; conversely, chlorogenic and caffeic acids were unstable and disappeared rapidly.[24]

Nonphenolic compounds were also found, for example, adenosine, which inhibits the protease elastase, that are released by neutrophils as part of the inflammatory cascade.[21]

MOLECULAR, CELLULAR, AND CLINICAL EVIDENCE OF ANTIOXIDANT PROPERTIES OF FERNBLOCK®

Oral intake of PL protects against the effect of UV-induced oxidative stress. Early studies described the antitumoral effect of PL extracts and others from similar ferns.[25] These results were confirmed in a hairless albino mouse model, in which PL inhibited skin tumor formation after UVB irradiation.[10]

The antitumoral properties of PL likely reside in its protective effect against UV-induced DNA damage. PL decreases the UVB-induced formation of thymine dimers,[12] likely through its effect in preventing photodamage to the endogenous mechanisms of DNA repair.[16] Also, PL blocked the effect of UV radiation on the expression of COX-2, which is an inducible enzyme responsible for prostaglandin synthesis and also involved in carcinogenesis.[3] Finally, PL induced activation of the tumor suppressor p53.[16]

Regarding the anti-inflammatory properties of PL, different *Polypodium* extracts were used in folkloric medicine for the treatment of skin inflammatory diseases such as psoriasis, vitiligo, and atopic dermatitis.[18–20] In more controlled scenarios, PL successfully blocked the inflammatory response in the skin of guinea pigs irradiated with UVB radiation or in a small group of human volunteers irradiated with small amounts of UVA, in which total photoprotection was achieved.[5] In a larger sample of human volunteers, PL prevented acute sunburn and reduced the phototoxic effect of exposure to sunlight after oral ingestion of psoralens.[9] This postulates its use as adjuvant in phototherapy, for example, PUVA. PUVA therapy is a very successful treatment for psoriasis and other skin pathologies; however, it is not widely used due to its deleterious side effects, which may include skin cancer.[11] Use of PL as an adjuvant reduced phototoxicity during PUVA therapy.[9,11] Moreover, oral PL inhibited PUVA-induced sunburn and infiltration of neutrophils and mast cells, and inhibited the observed decrease of eLC associated with these treatments.[11] This effect is probably due to decreased apoptosis and DNA damage, and correlates well with reduced skin photodamage, including a marked reduction in sunburn cells and inflammatory infiltrates, decreased levels of UV-induced DNA damage and enhanced epidermal cell proliferation.[11] These data suggest the possible use of PL as an adjuvant in other indications of phototherapy, such as repigmentation of vitiligo vulgaris.[26,27]

In addition to its antioxidant activity, PL bears promise in the treatment and prevention of photoaging due to its effects on extracellular matrix remodeling. PL inhibits several matrix MMPs, by inducing tissue inhibitor of metalloproteinase (TIMP), Transforming growth factor beta (TGF-β), elastin, and different types of collagen,[15] thus promoting regeneration and compensating for the deleterious effects of irradiation that cause photoaging.

At a cellular level, PL prevents lipid peroxidation and membrane damage induced by UV[8] and also blocks UV-mediated disarray of the actin cytoskeleton and loss of adhesive cell–cell and cell–matrix contacts.[28] Finally, PL inhibits fibroblast and keratinocyte cell death induced by UV radiation.[29]

Fernblock® in DNA Photodamage and Repair

It is well known that UV-B radiation induces specific DNA damage by two major mechanisms: one is promoting the appearance of specific DNA byproducts; for example, cyclobutane pyrimidine dimers (CPDs, especially thymine-thymine) and pyrimidine-pyrimidone dimers.[30] These UV lesions are formed through a photochemical reaction, and its efficiency depends on the wavelength of the UV light, following direct UV energy absorption by DNA bases.

Administered in oral form, PL inhibits UV-mediated DNA damage and mutagenesis. It prevented the UV-induced accumulation of cyclobutane pyrimidine dimers in a mouse model[16] as well as in a small-sample clinical study of healthy human volunteers.[12] This may be due to an improved function of the DNA repair systems, perhaps due to decreased oxidative damage.[31]

UV radiation also damages DNA through ROS. These may promote the generation of 8-hydroxy-2′-deoxyguanosine (8-OH-dG),[32] which is a marker of DNA oxidative damage and is mutagenic, favoring GC –> TA mutations.[31]

In vitro and *in vivo* experiments have proven that UV radiation increases the percentage positive (8-OH-dG) cells.[33–35]

The DNA damage caused by UVR is prevented by PL, interfering with the same mechanism mentioned earlier. PL reduced systemic oxidative damage, as shown by the reduction in the percentage of 8-OH-dG-positive cells.[16] In addition, PL decreased the levels of 8-OH-dG even before UV irradiation, supporting the idea that PL raises the systemic antioxidant threshold.[16]

Finally, another small-sample clinical study has revealed that PL decreases UVA-dependent mitochondrial DNA damage as evidenced by decreased common deletions (CD),[36] which are mitochondrial markers of chronic UVA radiation in fibroblasts and keratinocytes (Table 25.1).

Fernblock® Effect on Free Radicals during Inflammation

UV photons induce skin erythema, vasodilatation and elevated blood flow, activation of endothelial cells, and expression of pro-inflammatory molecules, which causes leukocyte infiltration and recruitment.[37,38]

Evidence of the role of PL in preventing UV-mediated inflammation comes from experiments that revealed that PL prevented sunburn and erythema in UV-treated human skin,[5,9,12] and also in PUVA-based therapy,[11] which is often used in the treatment of psoriasis, vitiligo, and other inflammatory skin conditions.[18–20] The molecular basis of its anti-inflammatory properties can be explained in terms of its ability to suppress the

TABLE 25.1 Skin Antioxidative Effects of Fernblock®

Skin UV Induced Effect	Macroscopic/Cellular Events	Reference
Photoaging	Reduction of UVR-induced photoaging	10
Skinfold	Reduction of the UVR-induced skinfold thickness	10
Erythema	Inhibition of UVR-induced erythema	12, 14, 52
Pigmentation	Reduction PUVA-induced hyperpigmentation Increase in repigmentation when NB-UVB phototherapy is combined in vitiligo or PUVA in vitiligo	11 26, 27
Sunburn cells	Inhibition of UVR-induced sunburn cells	11
Epidermal proliferation	Inhibition of UVR-induced epidermal proliferation	12, 13
Dermal collagen fibers	Increase the synthesis of types I, III, and V collagen in nonirradiated fibroblasts	15
Dermal elastic fibers	Reduction of UVR-induced elastosis; PL reinforce the network of dermal elastic fibers	8, 10, 13
Dermal vessel	Inhibition of PUVA-induced vasodilation Inhibits UVR-induced angiogenesis *in vivo*	11 39

TABLE 25.2 Beneficial Effects of Fernblock® in DNA Photodamage

In Vivo Studies	Molecular Mechanism/ Cellular Target	Reference
Human	Inhibition of UV-induced accumulation of CPD	12
Human	Decreases UVA-dependent induced mitochondrial DNA damage	36
Hairless Xpc(+/-) mice	Inhibition of the UV-induced increase 8-hydroxy-2'-deoxyguanosine	16
Hairless Xpc(+/-) mice	Inhibition of UV-induced mutation of DNA	16

expression of pro-inflammatory molecules and markers (e.g., TNF-α and inducible NOS (iNOS),[39,40] among others). This is consistent with its ability to block the transcriptional activation of activator protein 1 (AP-1) and NF-κB induced by UV radiation.[40] Experiments using a solar simulating radiation (SSR) proved this. Pretreatment of human keratinocytes with PL inhibited SSR-mediated increase of tumor necrosis factor TNFα-1 and also abrogated nitric oxide (NO) production. Consistent with this, PL blocked the induction of iNOS elicited by SSR.

We are, thus, sure that PL works by inhibiting the inflammatory effects of UVR. UV radiation includes ROS-induced peroxidation, which damages cellular membranes and induces activation of different isoforms of nitric oxide synthase (NOS). The inflammatory cascade also enables production of several species of ROS, amplifying DNA damage. ROS also upregulate the expression of the cyclooxygenase-2 (COX-2) enzyme, which is implicated in premalignant as well as malignant stages of cancer development.[41]

In this regard, as will be discussed further, the antitumoral effect of PL is also due to suppression of the induction of proinflammatory and growth-promoting genes by down-regulating the activation of the crucial transcription factors NFκB and AP-1 induced by UV radiation.[40] Some of the genes responsive to NFκB are oxidative sources. These include COX-2. COX-2 is one of the most important components of the inflammatory response, and its reaction products are responsible for cytotoxicity in models of inflammation. The mechanisms by which COX-2 produces cell damage involve the biosynthesis of prostaglandins, which generates free oxygen radicals that can result in injury. Importantly, PL inhibits UV-induced expression of COX-2.[16] This is important as COX-2 induces the synthesis of PGI2, which is a potent inducer of vasodilatation and inhibits platelet aggregation. Together, these effects account for decreased leukocyte extravasations and presence of mast cells in the irradiated area.[11,12,16] PL also inhibits apoptosis and cell death,[40,41] thereby preventing apoptosis/necrosis-triggered inflammation (Table 25.2).

Fernblock® Prevents UV Radiation-Mediated Immunosuppression

UV radiation induces immunosuppression, anergy, and immunological tolerance. This is mediated by a marked decrease of the numbers of eLCs, which leads to T helper 1 lymphocyte (Th1) clonal anergy.[42]

The decrease of eLC, which decreases the level of immune surveillance in the skin, is likely related to the role of UV radiation in promoting immune evasion by tumor cells.[43,44] Additional correlative evidence is provided by organ transplant patients undergoing immunosuppressive therapy, who have an elevated risk of developing both nonmelanoma skin cancer and melanoma, especially in cases of history of frequent sun exposure.[45]

PL efficiently blocked epidermal LC depletion upon UV irradiation and prevented the appearance of abnormal dendritic cell (DC) morphologies.[11,12,29] Similar results were obtained using blood dendritic cells (DC) irradiated using a solar simulator. In these experiments, PL inhibited DC apoptosis and promoted secretion of anti-inflammatory cytokines and inhibited expression of pro-inflammatory cytokines (e.g., TNF-α) by irradiated DC.[53] The molecular mechanism of enhanced DC survival seems to implicate inhibition of trans-urocanic acide (UCA) isomerization,[46] blockade of iNOS expression induced by UV radiation, which generates altered nitrogen oxide metabolites that cause immunosuppression by eliminating skin DC and enhancement of endogenous systemic antioxidant systems that lead to decreased oxidized intermediates (e.g., oxidized glutathione;[29] Table 25.3).

Photoimmunosuppression has been related to trans-UCA interaction (Table 25.4).[47] Trans-UCA (trans-UCA;(2E)-3-(1H-imidazol-4-yl)prop-2-enoic acid) is a deamination product of histidine, present at a high concentration in the stratum corneum. Trans-UCA shows a similar UV radiation-absorption spectrum as the UV wavelength dependence for immune suppression and upon exposure to UV radiation, trans-UCA photoisomerizes to its cis-isomer. In addition to alterations in antigen presentation and processing, the immunosuppressive effects of UV radiation are mediated by immunomodulatory molecules, which include both pro-inflammatory and anti-inflammatory molecules such as prostaglandin (PG) E2, TNF-α, and IL-10.[48]

Fernblock®, an Anti-UV-Induced Tumor Progression Agent

Some of these types of oxidative DNA and nucleotide damage are known to be mutagenic.[49] As mentioned previously, in addition to its direct effect on DNA, UV radiation can also induce oxidative stress-mediated mutations in the cellular genome through an indirect, ROS-mediated mechanism. In human skin cancers of sun-exposed areas, mutations induced by the presence of oxidative DNA damage frequently target the ras oncogene and p53 tumor suppressor gene. UVA radiation is also involved in photocarcinogenesis via ROS generation and lipid peroxidation. Persistent oxidative stress in cancer may also cause activation of transcription factors and proto-oncogenes such as c-fos and c-jun as well as

TABLE 25.3 Beneficial Effects of Fernblock® in Inflammation Induced by UV Radiation

In Vitro Studies	*In Vivo* Studies	Molecular Mechanism/Cellular Target	Reference
Human keratinocytes		Inhibition of UVR-induced increase of TNF-α Inhibition of UVR-induced increase of NO and iNOS	40
Human keratinocytes		Inhibition of the IVR-induced activation of NFκB and AP-1	40
	Hairless albino mice (Skh-1)	Reduction UVR-induced mast cell infiltration	39
	Hairless Xpc(+/-) mice	Inhibition of the UVR-induced neutrophil and macrophage infiltration	16
	Hairless Xpc(+/-) mice	Inhibition of the UVR-induced COX-2 expression	16
	Human	Inhibition of PUVA-induced vasodilation	11
	Human	Reduction UVR-induced mast cell infiltration	11 ,12

TABLE 25.4 Beneficial Effects of Fernblock® in Photoimmunosuppression

In Vitro Studies	*In Vivo* Studies	Molecular Mechanism/Cellular Target	Reference
Histochemical		PL interferes with the cis-UCA isomerization implicated in photoimmunosuppression	46
Human keratinocytes		Inhibition of expression of pro-inflammatory cytokines such as TNF-α	40
Immature human DCs		Protects DCs from UV-induced apoptosis	43
Immature human DCs		Induces DCs production of anti-inflammatory cytokines (IL-12)	43
	C57BL/6 mice	Inhibition of local and systemic photoimmunosuppression	51
	Hairless rats	Reduction of glutathione oxidation in blood and epidermis	29
	Hairless rats	Reduction of glutathione oxidation in blood and epidermis	29
	Human	Inhibition of UVR-mediated Langerhans cell depletion.	11, 12

genetic instability that may also contribute to maintaining malignancy.[2]

Several studies have documented the antitumor properties of different fern extracts, including PL.[25] For example, in a hairless albino mice model, PL blocked skin tumor formation and photoaging as a result of exposure to UVB, even after discontinuation of the treatment for 8 weeks.[10] A number of features support the anticarcinogenic capability of PL. In addition to its ROS-blocking antioxidant features, its antimutagenic properties also protect from immortalizing mutations leading to carcinogenesis.[16] As outlined earlier, PL inhibits the increase of COX-2 induced by UV radiation,[16] which is also involved in carcinogenesis.[3]

Also, PL induced activation of the tumor suppressor p53.[16] p53 expression and posttranslational activation (e.g., by phosphorylation) correlated with tumor suppressive activities. p53 decreases oxidative damage to the DNA, which also contributes to its antitumor activity and promotes the clearance of oxidative photoproducts, particularly CPDs, which are the main mutagens implicated in skin cancer. PL enhanced both p53 expression and activation in irradiated Xpc +/-mice.[16] In agreement with increased p53 activity, PL also decreased UV radiation-induced cell proliferation.[12]

Activation of p53 has been shown to decrease the expression of COX-2, thereby reducing the inflammatory response.[16] Levels of phospho-p53ser15, presumably active form of p53, were increased in the skin of PL-treated mice, which inversely correlated with decreased COX-2 levels, suggesting that orally administered PL reduces UV-induced COX-2 levels in mouse skin, at least in part, by activating tumor suppressor protein p53.

Importantly, the antioxidant activity of PL was not lost during digestion, but reached the bloodstream (increased ORAC in plasma). In addition, the observed increase in superoxide dismutase (SOD), glutathione peroxidase (GP), and glutathione transferase (GST) activities cannot be explained in terms of increased expression as erythrocytes are anucleated. Instead, it was postulated that PL may play an allosteric activating role on these enzymes. This is particularly striking for SOD, because in vehicle-treated and irradiated animals, SOD activity decreased, but PL prevented such decrease. PL did not significantly affect their expression, suggesting an activating modulatory effect.

Finally, PL complemented the effect of ascorbate in limiting melanoma cell growth and their ability to remodel the extracellular matrix (ECM), among other effects, by increasing the expression of the metalloprotease inhibitor TIMP-1[15] (see next; Table 25.5).

Fernblock® Prevention of Matrix Remodeling and Other Cellular Effects

PL exhibits a strong antiaging effect. It prevents the morphological effects associated with increased oxidative stress, which include a dramatic disorganization of the microfilaments and loss of cell–matrix and cell–cell anchorage points.[28] Additional antiaging effects of PL include inhibition of the expression and activation of several matrix metalloproteases and increased expression of an endogenous metalloprotease inhibitor, TIMP. Inhibition of MMP-1 expression, at protein or promoter level, by P. leucotomos in keratinocytes and fibroblasts has been recently reported[8]. In fibroblasts, P. leucotomos inhibited the expression of MMP-2 and simultaneously stimulated TIMPs. In melanoma cells, P. leucotomos preferentially inhibited MMP-1, which is consistent with inhibited degradation of interstitial

TABLE 25.5 Fernblock®, an Anti-UV-Radiation-Induced Tumor Progression Agent

In Vitro Studies	In Vivo Studies	Molecular Mechanism/Cellular Target	Reference
Dermal fibroblasts Melanoma cells		Increase the expression of the inhibitor TIMP-1 after UV radiation	15
	Hairless albino mice	Reduction in the number of mice showing skin tumors at 8 weeks after the cessation of chronic UVB exposure	10
	Hairless albino mice	Inhibits angiogenesis	39
	Hairless albino mice	Enhance the antioxidant plasma capacity	13
	Hairless albino mice Hairless Xpc(+/-) mice	Increase the number of p53(+) cells after acute UV radiation	13 16
	Hairless Xpc(+/-) mice	Inhibition of UV-induced DNA mutations	16
	Hairless Xpc(+/-) mice	Inhibition of the UV-induced COX-2 expression	16
	Hairless Xpc(+/-) mice	Reduce the UV-induced number of proliferating epidermal cells	16
	Human	Reduce the UV-induced number of proliferating epidermal cells	12

collagen degradation, and stimulated TIMP-2, implicating inhibition of basement membrane remodeling.

The inhibition of MMP-1 promoter in melanoma cells by *P. leucotomos* was via the AP-1 sequence, suggesting an involvement of the AP-1 transcription factor.[50]

The effect of *P. leucotomos* on MMPs and TIMPs expression in melanoma cells mirrored that of lutein.[50] Regarding a direct effect on the ECM, *P. leucotomos* stimulated deposition of types I, III, and V collagen in nonirradiated fibroblasts, and types I and V collagen in UV-radiated fibroblasts.[15] Its stimulatory effect on types I and V collagen was observed in both UVA- or UVB-irradiated fibroblasts, though UVB radiation decreased the level of stimulation of types I and V expression, and UVA radiation significantly counteracted the stimulation of collagen type I (COLIα1) promoter activity by *P. leucotomos* Conversely, UV radiation counteracted the induction of type III collagen by *P. leucotomos*. This data implies that *P. leucotomos* promotes the assembly of fibrillar collagens in sun-protected skin, and of types I and V collagen in UV radiation-exposed skin. *P. leucotomos* also modulates the expression of cytokines that control ECM remodeling and the biology of the cells implicated in this process. Specifically, PL promotes expression of TGF-β in nonirradiated or UV irradiated fibroblasts, but inhibited it in melanoma cells, which in turn may be responsible for the observed inhibition of MMP-1 expression induced by *P. leucotomos* in these cells. This effect may be responsible for the observed inhibition of angiogenesis *in vivo*.[15] In this regard, the effect of *P. leucotomos* was largely similar to that of ascorbic acid.

Stimulation of collagen expression by *P. leucotomos* was associated with increased TGF-β expression, whereas UVB radiation-mediated inhibition of collagen synthesis correlated with decreased expression of TGF-β. However, UV radiation did not counteract *P. leucotomos* stimulation of TGF-β expression, which suggests that UV and PL regulate TGF-β expression by separate, yet related, pathways. [15]

In summary, *P. leucotomos* demonstrated dual protective effects on the ECM for antiaging properties through its effect on ECM proteolytic enzymes and the stimulation of TIMPs, expression/assembly of structural ECM collagens (types I, III, V), and TGF-β in fibroblasts[50]. *P. leucotomos*'s anticancer effects encompass the inhibition of MMPs and TGF-β and stimulation of TIMPs in melanoma cells (Table 25.6)

POTENTIAL USE OF FERNBLOCK® IN THE TREATMENT OF OTHER PATHOLOGICAL SKIN CONDITIONS

Idiopathic Photodermatosis

These lesions include clinical conditions that emerge from exposure of the skin to normal sunlight. Some examples include polymorphic light eruption (PMLE), solar urticaria, chronic actinic dermatitis, and actinic prurigo.[51,52]

A very recent study has addressed the potential of PL to counter the occurrence of PMLE.[52] Despite the small number of patients and taking into account that this was an open study, its results suggested that PL has a beneficial effect in the treatment of PMLE, which may be extended to other idiopathic photodermatoses.

Vitiligo

One of the most efficient methods to treat vitiligo vulgaris is narrow band (311–312 nm) UVB phototherapy, which stimulates melanocyte reservoirs to counter depigmentation. A double blind, placebo-controlled study has showed that the conjoined use of PL with narrow-band UVB (NB-UVB) increases the repigmentation

TABLE 25.6 Fernblock® Prevents Matrix Remodeling and Other Cellular Effects

In Vitro Studies	Molecular Mechanism/Cellular Target	Reference
Fibroblasts, Keratinocytes	Inhibit UVR-induced MMP-1	8
Fibroblasts	Inhibit UVR-induced MMP-2 Stimulate TIMPs	15
Fibroblast	Increase the synthesis of types I, III, and V collagen in nonirradiated fibroblasts Increase the synthesis of types I and V collagen in UV-radiated fibroblasts	15
Fibroblast	Stimulatory effects on TGF-β expression in nonirradiated or UV-irradiated	15
Melanoma	Inhibit MMP-1 Stimulate TIMP-2 Inhibit TGF-β expression in melanoma cell	15

of the head and neck area of vitiligo patients.[46] Together with its beneficial effect on PUVA-therapy,[11,27] these studies suggest that PL may be a beneficial, general use adjuvant in phototherapy protocols.

FERNBLOCK: A ROAD TO (PRESENT AND FUTURE) PHOTOPROTECTION

Fernblock exhibits a wide array of beneficial effects and displays no significant toxicity or allergenic properties. Its dual route of administration—topical and oral—suggests that it prevents UV-induced damage (taken orally before exposure) and is also useful to protect the skin during exposure. It may also contribute to the healing and regeneration of the skin that is required postexposure. These regenerative properties also underlie its potential as an antiaging and anticancer tool. Most of its beneficial effects are related to its antioxidant and ROS-scavenging capability, but its ability to prevent apoptosis and block the improper ECM rearrangements that occur during oxidative damage suggest that its profile may extend beyond skin care and be useful as a systemic antioxidant tool. Further research will be aimed to study its effect in other parameters related to aging, such as telomere length and telomerase activity.

SUMMARY POINTS

Photoprotective and antitumoral effects of oral systemic administration of an extract of *Polypodium Leucotomos* (Fernblock®) Include the following:

- Prevents sunburn and erythema mediated by UV radiation
- Blocks photoinduction of TNFα, iNOS, AP-1, NF-κB
- Prevents DNA photodamage and favors its repair
- Inhibits UV-induced expression of COX-2
- Prevents cell death and apoptosis
- Blocks t-UCA photoisomerization
- Enhances natural cutaneous and plasmatic antioxidant systems
- Prevents photoinduced depletion of Langerhans cells and preserves their function
- Blocks lipid peroxidation
- Inhibits cytoskeletal disarray and MMP expression and activation

Acknowledgments

Salvador Gonzalez is a consultant for Industrial Farmaceutica Cantabria (IFC). This work has been partially supported by two grants: from the Carlos III Health Institute, Ministry of Science and Innovation, Spain (PS09/01099) and from the Comunidad Autónoma de Madrid (S2010/BMD-2359).The authors want to dedicate this review chapter to the memory of Dr. Vicente Garcia Villarrubia, a very special scientist and teacher, in recognition of his remarkable contribution in the field.

References

1. De Vries E, Arnold M, Altsitsiadis E, Trakatelli M, Hinrichs B, Stockfleth E, et al. Potential impact of interventions resulting in reduced exposure to ultraviolet (UV) radiation (UVA and UVB) on skin cancer incidence in four European countries. 2010-2050. *Br J Dermatol* 2012;**167**:53–62.
2. Nishigori C. Cellular aspects of photocarcinogenesis. *Photochem Photobiol Sci* 2006;**5**:208–14.
3. Rundhaug JE, Pavone A, Kim E, Fischer SM. The effect of cyclooxygenase-2 overexpression on skin carcinogenesis is context dependent. *Mol Carcinog* 2007;**46**:981–92.
4. Nishigori C, Hattori Y, Toyokuni S. Role of reactive oxygen species in skin carcinogenesis. *Antioxid Redox Signal* 2004;**6**:561–70.
5. Gonzalez S, Pathak MA. Inhibition of ultraviolet-induced formation of reactive oxygen species, lipid peroxidation, erythema and skin photosensitization by Polypodium leucotomos. *Photodermatol Photoimmunol Photomed* 1996;**12**:45–56.
6. Gomes AJ, Lunardi CN, Gonzalez S, Tedesco AC. The antioxidant action of Polypodium leucotomos extract and kojic acid: Reactions with reactive oxygen species. *Braz J Med Biol Res* 2001;**34**:1487–94.
7. Gonzalez S, Gilaberte Y, Philips N. Mechanistic insights in the use of a Polypodium leucotomos extract as an oral and topical photoprotective agent. *Photochem Photobiol Sci* 2010;**9**:559–63.
8. Philips N, Smith J, Keller T, Gonzalez S. Predominant effects of Polypodium leucotomos on membrane integrity, lipid peroxidation, and expression of elastin and matrixmetalloproteinase-1 in ultraviolet radiation exposed fibroblasts, and keratinocytes. *J Dermatol Sci* 2003;**32**:1–9.
9. Gonzalez S, Pathak MA, Cuevas J, Villarubia VG, Fitzpatrick TB. Topical or oral administration with an extract of Polypodium leucotomos prevents acute sunburn and psolaren-induced phototoxic reactions as well as depletion of Langerhans cells in human skin. *Photodermatol Photoimmunol Photomed* 1997;**13**:50–60.
10. Alcaraz MV, Pathak MA, Rius F, Kollias N, González S. An extract of Polypodium leucotomos appears to minimize certain photoaging changes in a hairless albino mouse animal model. *Photodermatol Photoimmunol Photomed* 1999;**15**:120–6.
11. Middelkamp-Hup MA, Pathak MA, Parrado C, Garcia-Caballero T, Rius-Diaz F, Fitzpatrick TB, et al. Orally administered Polypodium leucotomos extract decreases psoralen-UVA-induced phototoxicity, pigmentation, and damage of human skin. *J Am Acad Dermatol* 2004;**50**:41–9.
12. Middelkamp-Hup MA, Pathak MA, Parrado C, Goukassian D, Rius-Diaz F, Mihm MC, et al. Oral Polypodium leucotomos extract decreases ultraviolet-induced damage of human skin. *J Am Acad Dermatol* 2004;**51**:910–8.
13. Rodríguez-Yanes E, Juarranz Á, Cuevas J, Gonzalez S, Mallol J. Polypodium leucotomos decreases UV-induced epidermal cell proliferation and enhances p53 expression and plasma antioxidant capacity in hairless mice. *Exp Dermatol* 2012;**21**:638–40.
14. Aguilera P, Carrera C, Puig-Butille JA, Badenas C, Lecha M, González S, et al. Benefits of oral Polypodium Leucotomos extract in MM high-risk patients. *J Eur Acad Dermatol Venereol* 2013;**27**(9):1095–100.
15. Philips N, Conte J, Chen YJ, Natrajan P, Taw M, Keller T, et al. Beneficial regulation of matrixmetalloproteinases and their inhibitors, fibrillar collagens and transforming growth factor-beta by Polypodium leucotomos, directly or in dermal fibroblasts,

ultraviolet radiated fibroblasts, and melanoma cells. *Arch Dermatol Res* 2009;**301**:487–95.

16. Zattra E, Coleman C, Arad S, Helms E, Levine D, Bord E, et al. Oral Polypodium leucotomos decreases UV-induced Cox-2 expression, inflammation, and enhances DNA repair in Xpc ± mice. *Am J Pathol* 2009;**175**:1952–61.

17. Bagniewska-Zadworna A, Zenkteler E, Karolewski P, Zadworny M. Phenolic compound localisation in Polypodium vulgare L. rhizomes after mannitol-induced dehydration and controlled desiccation. *Plant Cell Rep* 2008;**27**:1251–9.

18. Mohammad A. Vitiligo repigmentation with Anapsos (Polypodium leucotomos). *Int J Dermatol* 1989;**28**:479.

19. Padilla HC, Lainez H, J.A., Pacheco JA. A new agent (hydrophilic fraction of polypodium leucotomos) for management of psoriasis. *Int J Dermatol* 1974;**13**:276–82.

20. Jimenez D, Naranjo R, Doblare E, Munoz C, Vargas JF. Anapsos, an antipsoriatic drug, in atopic dermatitis. *Allergol Immunopathol* 1987;**15**:185–9.

21. Vasange-Tuominen M, Perera-Ivarsson P, Shen J, Bohlin L, Rolfsen W. The fern Polypodium decumanum, used in the treatment of psoriasis, and its fatty acid constituents as inhibitors of leukotriene B4 formation, Prostaglandins, Leukotrienes Essent. *Fatty Acids* 1994;**50**:279–84.

22. Sempere JM, Rodrigo C, Campos A, Villalba JF, Diaz J. Effect of Anapsos (Polypodium leucotomos extract) on in vitro production of cytokines. *Br J Clin Pharmacol* 1997;**43**:85–9.

23. Garcia F, Pivel JP, Guerrero A, Brieva A, Martinez-Alcazar MP, Caamano-Somoza M, et al. Phenolic components and antioxidant activity of Fernblock, an aqueous extract of the aerial parts of the fern Polypodium leucotomos. *Methods Find Exp Clin Pharmacol* 2006;**28**:157–60.

24. Gombau L, Garcia F, Lahoz A, Fabre M, Roda-Navarro P, Majano P, et al. Polypodium leucotomos extract: Antioxidant activity and disposition. *Toxicol In Vitro* 2006;**20**:464–71.

25. Creasey WA. Antitumoral activity of the fern Cibotium schiedei. *Nature* 1969;**222**:1281–2.

26. Middelkamp-Hup MA, Bos JD, Rius-Diaz F, Gonzalez S, Westerhof W. Treatment of vitiligo vulgaris with narrow-band UVB and oral Polypodium leucotomos extract: A randomized double-blind placebo-controlled study. *J Eur Acad Dermatol Venereol* 2007;**21**:942–50.

27. Reyes E, Jaen P, de las Heras E, Carrion F, Alvarez-Mon M, de Eusebio E, et al. Systemic immunomodulatory effects of Polypodiumleucotomos as an adjuvant to PUVA therapy in generalized vitiligo: A pilot study. *J Dermatol Sci* 2006;**41**:213–6.

28. Alonso-Lebrero JL, Domínguez-Jiménez C, Tejedor R, Brieva A, Pivel JP. Photoprotective properties of a hydrophilic extract of the fern Polypodium leucotomos on human skin cells. *J Photochem Photobiol B* 2003;**70**:31–7.

29. Mulero M, Rodriguez-Yanes E, Nogues MR, Giralt M, Romeu M, Gonzalez S, et al. Polypodium leucotomos extract inhibits glutathione oxidation and prevents Langerhans cell depletion induced by UVB/UVA radiation in a hairless rat model. *Exp Dermatol* 2008;**17**:653–8.

30. Mitchell DL, Jen J, Cleaver JE. Sequence specificity of cyclobutane pyrimidine dimers in DNA treated with solar (ultraviolet B) radiation. *Nucleic Acids Res* 1992;**20**:225–9.

31. Emanuel P, Scheinfeld N. A review of DNA repair and possible DNA-repair adjuvants and selected natural anti-oxidants. *Dermatol Online J* 2007;**13**:10.

32. Hattori Y, Nishigori C, Tanaka T, Uchida K, Nikaido O, Osawa T, et al. 8-hydroxy-2' deoxyguanosine is increased in epidermal cells of hairless mice after chronic ultraviolet B exposure. *J Invest Dermatol* 1996;**107**:733–7.

33. Hart RW, Setlow RB, Woodhead AD. Evidence that pyrimidine dimers in DNA can give rise to tumors. *Proc Natl Acad Sci USA* 1977;**74**:5574–8.

34. Norval M, Simpson TJ, Ross JA. Urocanic acid and immunosuppression. *Photochem Photobiol* 1989;**50**:267–75.

35. Kripke ML, Cox PA, Alas LG, Yarosh DB. Pyrimidine dimers in DNA initiate systemic immunosuppression in UV-irradiated mice. *Proc Natl Acad Sci USA* 1992;**89**:7516–20.

36. Villa A, Viera MH, Amini S, Huo R, Perez O, Ruiz P, et al. Decrease of ultraviolet A light-induced "common deletion" in healthy volunteers after oral Polypodium leucotomos extract supplement in a randomized clinical trial. *J Am Acad Dermatol* 2010;**62**:511–3.

37. Black AK, Greaves MW, Hensby CN, Plummer NA, Warin AP. The effects of indomethacin on arachidonic acid and prostaglandins E2 and F2alpha levels in human skin 24 h after u.v.B and u.v.C irradiation. *Br J Clin Pharmacol* 1978;**6**:261–6.

38. Deliconstantinos G, Villiotou V, Stravrides JC. Release by ultraviolet B (u.v.B) radiation of nitric oxide (NO) from human keratinocytes: a potential role for nitric oxide in erythema production. *Br J Pharmacol* 1995;**114**:1257–65.

39. Gonzalez S, Alcaraz MV, Cuevas J, Perez M, Jaen P, Alvarez-Mon M, et al. An extract of the fern Polypodium leucotomos (Difur) modulates Th1/Th2 cytokines balance in vitro and appears to exhibit anti-angiogenic activities in vivo: Pathogenic relationships and therapeutic implications. *Anticancer Res* 2000;**20**:1567–75.

40. Janczyk A, Garcia-Lopez MA, Fernandez-Penas P, Alonso-Lebrero JL, Benedicto I, Lopez-Cabrera M, et al. A Polypodium leucotomos extract inhibits solar-simulated radiation-induced TNF-alpha and iNOS expression, transcriptional activation and apoptosis. *Exp Dermatol* 2007;**16**:823–9.

41. Prescott SM, Fitzpatrick FA. Cyclooxygenase-2 and carcinogenesis. *Biochim Biophys Acta* 2000;**1470**:69–78.

42. Simon JC, Tigelaar RE, Bergstresser PR, Edelbaum D, Cruz Jr PD. Ultraviolet B radiation converts Langerhans cells from immunogenic to tolerogenic antigen-presenting cells. Induction of specific clonal anergy in CD4+ T helper 1 cells. *J Immunol* 1991;**146**:485–91.

43. De la Fuente H, Tejedor R, Garcia-Lopez MA, Mittelbrunn M, Alonso-Lebrero JL, Sanchez-Madrid F, et al. Polypodiumleucotomos induces protection of UV-induced apoptosis in human skin cells. *J Invest Dermatol* 2005;**124**:A 121.

44. Norval M. Immunosuppression induced by ultraviolet radiation: Relevance to public health. *Bull World Health Organ* 2002;**80**:906–7.

45. Euvrard S, Kanitakis J, Pouteil-Noble C, Claudy A, Touraine JL. Skin cancers in organ transplant recipients. *Ann Transplant* 1997;**2**:28–32.

46. Capote R, Alonso-Lebrero JL, Garcia F, Brieva A, Pivel JP, Gonzalez S. Polypodium leucotomos extract inhibits trans-urocanic acid photoisomerization and photodecomposition. *J. Photochem Photobiol B* 2006;**82**:173–9.

47. Prater MR, Blaylock BL, S.D., Holladay SD. Molecular mechanisms of cis-urocanic acid and permethrin-induced alterations in cutaneous immunity. *Photodermatol Photoimmunol Photomed* 2003;**19**:287–94.

48. Tanaka K, Asamitsu K, Uranishi H, Iddamalgoda A, Ito K, Kojima H, et al. Protecting skin photoaging by NF-kappaB inhibitor. *Curr Drug Metab* 2010;**11**:431–5.

49. Kryston TB, Georgiev AB, Pissis P, Georgakilas AG. Role of oxidative stress and DNA damage in human carcinogenesis. *Mutat Res* 2011;**711**:193–201.

50. Philips N, Keller T, Hendrix C, Hamilton S, Arena R, Tuason M, et al. Regulation of the extracellular matrix remodeling by lutein in dermal fibroblasts, melanoma cells, and ultraviolet radiation exposed fibroblasts. *Arch Dermatol Res* 2007;**299**:373–9.

51. Siscovick JR, Zapolanski T, Magro C, Carrington K, Prograis S, Nussbaum M, et al. Polypodium leucotomos inhibits ultraviolet B radiation-induced immunosuppression. *Photodermatol Photoimmunol Photomed* 2008;**24**:134–41.

52. Tanew A, Radakovic S, Gonzalez S, Venturini M, Calzavara-Pinton P. Oral administration of a hydrophilic extract of Polypodium leucotomos for the prevention of polymorphic light eruption. *J Am Acad Dermatol* 2012;**66**:58–62.

Skin Cancer, Polyphenols, and Oxidative Stress

Neena Philips, Halyna Siomyk, David Bynum

School of Natural Sciences, Fairleigh Dickinson University, Teaneck, NJ, USA

Salvador Gonzalez

Dermatology Service, Memorial Sloan-Kettering Cancer Center, New York, NY, USA

List of Abbreviations

ROS Reactive oxygen species
MMP Matrix metalloproteinases
ECM Extracellular matrix
UV Ultraviolet
MAPK Mitogen activated protein kinase
NF-κB Nuclear factor-kappa beta
STAT Signal Transduction and Activation of Transcription
ERK1/2 Extracellular signal-regulated kinase 1/2
JNK c-Jun-N-terminal-kinase
AP-1 activator protein-1
MAE Michelia alba extract
COX Cyclooxygenase
EGCG Epigallocatechin-3-gallate
IL Interleukin
TGF-β Transforming growth factor-β
VEGF Vascular Endothelial Growth Factor

INTRODUCTION

The characteristics of malignancy are loss of cellular regulation and metastasis.[1] One of the major causes of skin cancer (melanoma and nonmelanoma) is the cellular accumulation of reactive oxygen species (ROS), which outbalances the cellular antioxidant system, and the subsequent oxidative damage, inflammation, activation of oxidative/inflammatory signal transduction pathways, and the degradation/remodeling of the extracellular matrix (ECM) for angiogenesis and metastasis. The predominant structural ECM proteins, collagen and elastin fibers, are remodeled by matrix metalloprotienases (MMP) and elastases. This chapter reviews the fundamental biology, the alterations, and counteraction by polyphenols in the mechanism to carcinogenesis or its prevention as (1) oxidative stress, inflammation, and

associated signal transduction pathways; and (2) ECM remodeling and associated growth factors.

OXIDATIVE STRESS, INFLAMMATION, AND ASSOCIATED SIGNAL TRANSDUCTION PATHWAYS: FUNDAMENTAL BIOLOGY, THE ALTERATION, AND COUNTERACTION BY POLYPHENOLS

Oxidative Stress and Polyphenols

The reactive oxygen species (ROS), which cause oxidative stress, include hydroxyl radicals, superoxide, and hydrogen peroxide; the counteracting cellular antioxidants include catalase, glutathione peroxidase, glutathione, ascorbate, α-tocopherol, and carotene. Phytonutrients, most of which are rich in polyphenols, have been gaining popularity for improving the cellular antioxidant score and thereby preventing oxidative stress and cancer. The structure of the phenolic compounds possesses at least one aromatic ring with one or more hydroxyl groups. These have been categorized as phenolic acids, monophenols, polyphenols, and associated substructures; and polyphenols have been classified as flavonoids (xanthohumol), anthocyanin (grapes), catechins (tea leaves), flavones (luteolin), flavonols (quercetin), isoflavones (genistein), lignans (flax seeds), proanthocyanidins (red wine), procyanidins (apple), stilbenes (resveratrol), and tannins (nuts, tea).[2] The structure as well as the number or location of the hydroxyl groups may correlate with the radical scavenging potential of polyphenols.[3–5]

Cancer
http://dx.doi.org/10.1016/B978-0-12-405205-5.00026-X

© 2014 Elsevier Inc. All rights reserved.

The cellular oxidative stress occurs with intrinsic aging, though more so from the exposure of skin to environmental pollutants, and ultraviolet (UV) radiation. With intrinsic aging there is cellular and mitochondrial DNA damage, and diminished expression of protective hormones and growth factors that strengthen the ECM.[6] The environmental pollutants include benzene, dioxins, heavy metals, and aryl or chlorinated hydrocarbons. Benzene is metabolized by skin cells and its metabolites, and copper increases oxidative stress and the expression of ECM proteolytic enzymes by dermal fibroblasts.[7,8] The UVA radiation increases cellular ROS and oxidative chain reaction.[2,9] The damage by UVB irradiation is mediated initially by reactive oxygen species, and then because of the increased reactive nitrogen species, lipid peroxidant species, and reduced counteracting catalase and glutathione levels.[10] The ROS attack the DNA, proteins, and lipids directly. The oxidative damage includes 8-oxo-7, 8-dihydro-2′-deoxyguanosine (8-oxo-dG) that generates a GC to TA transversion mutation during replication, pyrimidine dimers, carbonyl amino acid derivatives, lipid inflammatory mediators/peroxides, advanced glycation end products, calcification, and mechanical fatigue.[2,11–13] The repair of oxidative DNA damage (8-oxoG) is reduced because of the glutathione depletion in fibroblasts and melanoma cells.[14] The application of oligomeric proanthocyanidins (OPCs) as creams reduces oxidative stress in skin, while a combination of 5 phytochemicals (protandim, rich in polyphenols) potently induces cellular antioxidant enzymes for cancer prevention.[15,16] Green tea polyphenols improve DNA repair and induce the expression of tumor suppressor gene through the inhibition of histone methylation.[2,17,18]

Inflammation and Polyphenols

ROS are inducers of pro-inflammatory genes. The inflammatory mediators are released from keratinocytes, fibroblasts, tumor cells, leukocytes, and the endothelial lining of blood vessels. The mediators include the plasma mediators (bradykinin, plasmin, fibrin), lipid mediators (prostaglandins, leukotrienes, and platelet activating factor), and the inflammatory cytokines [interleukin-1 (IL-1), interleukin-6 (IL-6), and tumor necrosis factor-α (TNF-α)].[19] UV radiation directly induces inflammation, and the release of prostaglandins, histamine, and active phospholipase.[20] The UV radiation also activates the tissue leukocytes to generate prostaglandins and inflammatory cytokines.[2] A major mechanism in the anticarcinogenic effect of polyphenols is the inhibition of inflammation. In our laboratory, xanthohumol (a flavonoid from hop plant *Humulus lupulus* L. [Cannabinaceae]) and *P. leucotomos* (rich in mono- and polyphenols) have shown anticancer activities.[21–24] The anticancer effects of these polyphenols are similar to those of ascorbate or lutein.[25,26] The effects of lutein are similar to that of superoxide dismutase, indicating the inhibition of ROS as one of its mechanisms.[25] The anticancer mechanism of xanthohumol includes the inhibition of cycloxygenases/prostaglandins and reactive oxygen/ reactive nitrogen species.[28]

Signal Transduction Pathways and Polyphenols

The predominant pathways activated ROS and inflammatory cytokines for carcinogenesis include the mitogen-activated protein kinase (MAPK), the nuclear factor-kappa beta (NF-κB)/p65, and the JAK/STAT (Signal Transduction and Activation of Transcription).[1,2,11,19]

The MAPK pathway is comprised of the extracellular signal-regulated kinase 1/2 (ERK1/2), c-Jun-N-terminal-kinase (JNK), and p38 proteins. The activation of MAP kinase pathway, through the receptor tyrosine kinase, results in the activation of transcription factor activator protein-1 (AP-1) that activates expression of MMPs.[11] The JNK and p38 pathways play a major role in the UVA radiation mediated increase in AP-1 and COX-2 expression, and are targets for chemoprevention of skin cancer.[29] The inhibition of MAPK is a mechanism to the anticancer activity of polyphenols. Michelia alba extract (MAE) inhibits UVB-induced ERK and JNK kinase in the prevention of ECM degradation.[30] Resveratrol and black tea polyphenol decrease the expression of phosphorylated ERK1/2, JNK, and p38, and increase phosphorylated p53 and apoptosis in skin tumors, and thereby suppress skin carcinogenesis.[31] Grape skin polyphenols inhibit MAPK pathway to inhibit cancer cell migration and metastasis.[32]

The *NF-κB pathway* is activated by oxidative stress and inflammation through the activation of cytoplasmic I-κB kinase. Active I-κB kinase phosphorylates and degrades I-κB, the inhibitor of NF-κB (p65/p50 heterodimeric protein) transcription factor.[1] The NF-κB activation is associated with UVA and UVB radiation mediated oxidative modification of cellular membrane components.[33] The release of NF-κB, from its inhibitor (I-κB), results in the translocation of active NF-κB to the nucleus to activate the inflammatory cytokines and prostaglandins.[1] The active NF-κB also activates nitric oxide synthase, cycloxygenase, and histone acetylase.[1,33] The inhibition of NF-κB activation is a mechanism to the anticancer property of dietary polyphenols.[34] Xanthohumol inhibits NF-κB transcription factor.[35] Epigallocatechin-3-gallate (EGCG), the major polyphenolic in green tea, decreases NF-κB, inflammation, interleukin-1 β (IL-1β) secretion, and cell growth of melanoma cells.[36] The antioxidant polyphenols from olive oil (oleuropein and hydroxytyrosol) and red wine (resveratrol and quercetin) inhibit NF-κB and thereby COX-2, MMP-9, and tumor angiogenesis.[37]

The STAT pathway is activated by the inflammatory cytokines.[19] Many of the cytokines bind to their cytokine receptors on cell surfaces to activate JAK kinase that activates STAT transcription factors through phosphorylation.[1] The phosphorylation of STAT results in its dimerization and translocation into the nucleus to modulate gene expression.[1] The different cytokines activate different STAT proteins; one of its effects is the activation of $Bcl-x_L$ to prevent apoptosis, and allow for cell proliferation.[1] The polyphenols apigenin and luteolin inhibit the cytokine IL-6/STAT3 pathway and thereby endothelial cell proliferation, migration, and differentiation for angiogenesis.[38]

ECM REMODELING AND ASSOCIATED GROWTH FACTORS: FUNDAMENTAL BIOLOGY, THE ALTERATION, AND COUNTERACTION BY POLYPHENOLS

Collagen/Elastin and Polyphenols

Collagen and elastin are the structural proteins of the ECM. The deterioration/remodeling of the collagen and elastin fibers facilitates angiogenesis and metastasis, and the damaged collagen and elastin proteins serve as additional sensitizers of photo-oxidative stress.[39]

Collagen is the predominant ECM protein, central to the interstitial ECM as well as to the basement membrane. Collagens are homo or hetero trimeric triple helical proteins, composed of repeating Gly-X-Y motifs where X or Y are proline or hydroxyproline.[1] The specific properties of the different collagens are based on the lengths of the triple helical segments, interruptions to the triple helix, and amino acid modifications. There are about 28 types of collagen.[1] Collagens are classified as fibrillar collagens (types I, II, III, V), fibril associated collagens (types VI, IX), sheet forming anchoring collagens (types IV, VII, XV), transmembrane collagens (types XIII, XVII), and host defense collagens.[1] The dermal collagen fibers are formed of the type I (90%), III, V, and VII collagens.[11]

The elastin fibers that provide stretch-recoil properties to skin are composed predominantly of an elastin core (90%) surrounded by fibrillin microfibrils. Elastin is expressed as soluble hydrophobic tropoelastin, rich in proline, valine, lysine, alanine, glycine, leucine, and isoleucine, transferred to the microfibril scaffold, and cross-linked by lysyl oxidase and transglutamise.[13,40] The fibrillins are cysteine-rich, highly disulphide-bonded glycoproteins, with calcium-binding epidermal growth factor like domains.[40,41] They are secreted in pro-forms, assembled into microfibrils, and associated with other microfibril-associated glycoproteins.[40,41] The loss of proper elastin fibers occurs with the exposure of skin to UV radiation.[13] UV radiation also depletes the microfibrillar network in the epidermal-dermal layer and the dermis, which contributes to the aberrant elastic fibers.[41]

Xanthohumol dramatically stimulates the expression of types I, III, and V collagens, elastin, fibrillin-1, and fibrillin-2 in dermal fibroblasts.[21] P. leucotomos stimulates the expression of type I, III, and V collagens, and elastin in nonirradiated, UVA radiated, and UVB radiated dermal fibroblasts.[21,22,24]

Matrix Metalloproteinases/Elastases

The ECM proteolytic enzymes (MMPs/elastases) are produced by epidermal keratinocytes, dermal fibroblasts, and melanoma in the mediation of ECM remodeling and skin cancer.[7,8,21–26,42–44] The basal levels of MMPs increase with aging, and are further increased by environmental pollutants and UV radiation, resulting in the fragmentation of collagen and elastin fiber proteins for carcinogenesis.

The MMPs are zinc metalloproteinases that are secreted extracellularly in an inactive form, with the propeptide linked to the zinc in the catalytic domain. One of the mechanisms of activation of MMPs is through proteolysis of its propeptide by plasmin, which is formed from plasminogen following the secretion of plasminogen activator by tumor cells.[1,45]

The MMPs are classified on the basis of their substrate specificity, or their regulatory promoter elements. The categories based on substrate specificity are the interstitial collagenases (predominantly MMP-1) that cleave the fibrillar collagens, the gelatinases (MMP-2, 9) that cleave the basement membrane collagens and elastin, the stromelysines (MMP-3, 10) that degrade the basement membrane proteins, the membrane-type MMPs with a transmembrane domain that act on pro-MMPs and fibrillar collagens, and the other MMPs including the matrilysin (MMP-7) and metalloelastase (MMP-12) that degrade the basement membrane proteins and elastin.[45] MMPs are categorized on the basis of the presence of AP-1 or TATA sequences in the promoters into group I MMPs (MMP-1, 3, 7, 9, 10, 12, 13, 19, and 26) that contain TATA box and activator protein-1 (AP-1 site), group II MMPs (MMP-8, 11, 21) without the AP-1 site, and group III (MMP-2, 14, 28) without the TATA box and AP-1 site.[46]

The transcription factor AP-1, stimulated largely by the MAPK pathway, stimulates the transcription of several MMPs that collectively degrade the ECM, such as MMP-1, MMP-2/9, and MMP-3.[33] Further, AP-1 inhibits the transcription of type I collagen gene.[11] Hence, the damage to the ECM and tissue integrity is from the enhanced degradation of ECM by MMPs as well as the reduced expression of the structural ECM proteins.

The pro- and active forms of MMPs are inhibited by the tissue inhibitors of MMPs or TIMPs.[45,46] The genes for four of the TIMPs (1–4) have been mapped to chromosomes X, 17, 22, and 3.[47] TIMP-1 and TIMP-3 are inducible, TIMP-2 is constitutive, and TIMP-4 is restricted in its expression to certain tissues.[47] The TIMPs (1-4) bind to all of the MMPs, though TIMP-1 has preference for MMP-1 and TIMP-2 for MMP2.[45,47] The N-terminal conserved region of TIMPs is essential to binding to the active site of MMPs.[47] The remodeling of collagen and elastin, for angiogenesis, metastasis, and tissue destruction, is largely from the increased expression or activation of MMPs and reduced expression of TIMPs.[48,49]

The inhibition of expression and activity of MMPs and the stimulation of expression of TIMPs have been active targets in managing carcinogenesis. Specific MMP-1 gene silencing with siRNAs inhibits the expression of MMP-1 in melanoma cells.[23] An MMP-9 inhibitor prevents epithelial cancer cell migration and tubular network formation.[48] A human serum albumin/TIMP-2 fusion protein inhibits MMP-2 expression and thereby inhibits *in vivo* vascularization/angiogenesis.[50] Polyphenols provide a means to accomplish the inhibition of MMPs and stimulation of TIMPs. The activities of MMP-1, MMP-2, MMP-3, and MMP-9 are directly inhibited by xanthohumol and *P. leucotomos*.[21,22] *P. leucotomos* inhibits the expression of MMP-1 in nonirradiated, UVA or UVB radiated keratinocytes and fibroblasts.[24] Further, *P. leucotomos* inhibits the expression of MMP-2 while stimulating TIMP-1 and/or TIMP-2 expression in dermal fibroblasts and melanoma cells.[22] The polyphenols apigenin and luteolin inhibit the expression of MMP-2 whereas Michelia alba extract (MAE) inhibits the UVB radiation-induced expression of MMP-1, MMP-3, and MMP-9.[30,38] A synthetic flavanoid (SR13179) simultaneously inhibits cancer cell invasiveness and stimulates TIMP-2 expression.[49]

Transforming Growth Factor-β (TGF-β) and Polyphenols

TGF-β is a key regulator of cell cycle and the expression of collagen, elastin, and MMPs. Carcinogenesis is associated with altered expression of TGF-β gene, TGF-β receptors, or transcription factors (Smads). It is expressed in high levels in tumors, and mediates epithelial-mesenchymal transition, angiogenesis, and tumor invasion.[51]

There are three isoforms of *TGF-β*.[1,7,21] *TGF-β* is secreted in an inactive form, and forms a disulphide bonded dimer with a latency associated protein (LAP) as its prodomain and associates with latent *TGF-β* binding protein (LTBP).[1] *TGF-β* is activated by the dissociation of the mature *TGF-β* dimer from LAP and LTBP.[1] The active *TGF-β* binds to transmembrane serine/threonine kinase receptors on target cells.[1] There are three types

of receptors that specifically bind TGF-β; however only type I (TβRI) and type II (TβRII) receptors have intrinsic serine/threonine kinase activity whereas type III (TβRIII) receptor facilitates TGF-β binding to TβRII. Once TGF-β binding has occurred, TβRI and TβRII form a heteromeric complex and phosphorylate intracellular Smads.[1] The phosphorylated Smad 2 /3 bind Smad 4 translocate to the nucleus, and bind TGF-β regulatory elements.[1] The inhibitory Smads, Smad 6 and 7, interact with TβRI to prevent phosphorylation and subsequent activation of Smad 2/3.[1]

TGF-β has differential effects in different cell types.[42–44,52] It stimulates the expression of MMPs in keratinocytes and epithelial cells, whereas it inhibits the expression of MMPs in normal dermal fibroblasts.[42–44] TGF-β counteracts the effects of UV radiation on the expression of ECM remodeling enzymes, and inhibits the expression of MMP-1 at the protein, mRNA, and promoter levels in normal dermal fibroblasts.[43,52] TGF-β is coordinately stimulated with MMP-1 in cancer cells.[53,54] Cancer metastasis is associated with enhanced MMPs and TGF-β expression.[52–54] *P. leucotomos* inhibits the expression of TGF-β in melanoma cells.[22,23]

Vascular Endothelial Growth Factor (VEGF) and Polyphenols

VEGF is central to angiogenesis for cancer growth and metastasis, and its overexpression is associated with poor prognosis.[55] VEGF is regulated by various factors, TGF-β, hypoxia inducible factor-1, and MMPs, and binds to its receptor tyrosine kinase.[1,48] The VEGF receptor tyrosine kinase activates the mitogen activated protein kinase (MAPK) pathway, which results in the activation of several transcription factors including the serum response factor that activates c-fos gene expression for malignancy.[1] Epigallocatechin gallate (EGCG) in green tea inhibits cancer cell growth by preventing the activation of VEGF receptor.[56] This effect is similar to the VEGF antibodies, such as Avastin, which inhibit angiogenesis, tumor growth, and metastasis.[48]

CONCLUSION

Oxidative stress from the increased cellular ratio of ROS to antioxidants is the major factor in the etiology of cancer. It occurs with intrinsic aging, but more so from exposure of skin to environmental pollutants and UV radiation. The ROS directly damage the DNA, proteins, and lipids; induce inflammation; and activate the MAPK, NF-κB, and JAK/STAT pathways to activate the expression of MMPs, TGF-β, and VEGF that collectively remodel the ECM, composed primarily of collagen and elastin, for tumor growth, angiogenesis, and metastasis.

Polyphenols are increasingly playing a role in cancer prevention through the restoration of the cellular antioxidant balance; anti-inflammation; inhibition of the MAPK, NF-κB, and JAK/STAT pathways; inhibition of MMPs, TGF-β, and VEGF; and the inhibition of cancer growth, angiogenesis, and metastasis.

Acknowledgement

We would like to acknowledge Hui Jia and David Sherer, for participating in the literature search.

SUMMARY POINTS

- ROS that causes oxidative stress, which facilitates carcinogenesis, are induced with intrinsic aging and more so with exposure of skin to environmental pollutants and UV radiation, and are counteracted by polyphenols that act as antioxidants.
- ROS damage DNA, proteins, and lipids; they are counteracted and damage-repaired by polyphenols.
- ROS induce inflammation and inflammatory mediators, and are counteracted by polyphenols.
- ROS induce oxidative and inflammatory pathways such as MAPK, NF-κB, and STAT pathways that activate transcription factors that increase cell proliferation and metastasis; they are counteracted by polyphenols.
- ROS induce MMPs, TGF-β, and VEGF that remodel the ECM composed of structural collagen and elastin for tumor growth, angiogenesis, and metastasis, which is counteracted by polyphenols.
- Polyphenols and their combinations are effective in natural cancer prevention or as supplements in cancer treatment.

References

1. Lodish H, Berk A, Kaiser CA, Krieger M, Scott MP, Bretscher A, et al. *Molecular cell biology.* 6th ed. W.H Freeman and Company; 2008.
2. Nichols JA, Katiyar SK. Skin photoprotection by natural polyphenols: anti-inflammatory, antioxidant and DNA repair mechanisms. *Arch Dermatol Res* 2010;**302**:71–83.
3. Cheng LX, Tang JJ, Luo H, Jin XL, Dai F, Yang J, et al. Antioxidant and antiproliferative activities of hydroxyl-substituted Schiff bases. *Bioorg Med Chem Lett* 2010;**20**:2417–20.
4. Chen ZY, Chan PT, Ho KY, Fung KP, Wang J. Antioxidant activity of natural flavonoids is governed by number and location of their aromatic hydroxyl groups. *Chem Phys Lipids* 1996;**79**:157–63.
5. Gazák R, Sedmera P, Vrbacký M, Vostálová J, Drahota Z, Marhol P, et al. Molecular mechanisms of silybin and 2, 3-dehydrosilybin antiradical activity—Role of individual hydroxyl groups. *Free Radic Biol Med* 2009;**46**:745–58.
6. Philips N, Devaney J. Beneficial regulation of type I collagen and matrixmetalloproteinase–1 expression by estrogen, progesterone, and its combination in skin fibroblasts. *J Am Aging Assoc* 2003;**26**:59–62.

7. Philips N, Hwang H, Chauhan S, Leonardi D, Gonzalez S. Stimulation of cell proliferation, and expression of matrixmetalloproteinase-1 and interleukin-8 genes in dermal fibroblasts by copper. *Connective Tissue Res* 2010;**51**:224–9.
8. Philips N, Burchill D, O'Donoghue D, Keller T, Gonzalez S. Identification of benzene metabolites in dermal fibroblasts: Regulation of cell viability, apoptosis, lipid peroxidation, and expression of MMP-1 and elastin by benzene metabolites. Skin Pharmacol. *Physiol* 2004;**17**:147–52.
9. Krutmann J. The role of UVA rays in skin aging. *Eur J Dermatol* 2001;**11**:170–1.
10. Terra VA, Souza-Neto FP, Pereira RC, Silva TNX, Costa ACC, Luiz RCC, et al. Time-dependent reactive species formation and oxidative stress damage in the skin after UVB irradiation. *J Photochem Photobiol B: Biol* 2012;**109**:34–41.
11. Callaghan TM, Wilhelm KP. A review of ageing and an examination of clinical methods in the assessment of ageing skin. Part I: cellular and molecular perspectives of skin ageing. *Int J Cosmet Sci* 2008;**30**:313–22.
12. Briganti S, Picardo M. Antioxidant activity, lipid peroxidation and skin diseases. *What's new? J Euro Acad Dermatol Venereol* 2003;**17**:663–9.
13. Yaar M, Gilchrest BA. Photoageing: Mechanism, prevention and therapy. *Br J Dermatol* 2007;**157**:874–87.
14. Eiberger W, Volkmer B, Amouroux R, Dhérin C, Radicella JP, Epe B. Oxidative stress impairs the repair of oxidative DNA base modifications in human skin fibroblasts and melanoma cells. *DNA Repair* 2008;**7**:912–21.
15. Van Wijk EPA, Van Wijk R, Bosman S. Using ultra-weak photon emission to determine the effect of oligomeric proanthocyanidins on oxidative stress of human skin. *J Photochem Photobiol B: Biol* 2010;**98**:199–206.
16. Liu J, Gu X, Robbins D, Li G, Shi R. Protandim: A fundamentally new antioxidant approach in chemoprevention using mouse two-stage skin carcinogenesis as a model. *PloS One* 2009;**4**(4): e5284.
17. Choudhury SR, Balasubramanian S, Chew YC, Han B, Marquez VE, Eckert RL. (-)-Epigallocatechin-3-gallate and DZNep reduce polycomb protein level via a proteasome-dependent mechanism in skin cancer cells. *Carcinogenesis* 2011;**32**:1525–32.
18. Thakur VS, Gupta K, Gupta S. The chemopreventive and chemotherapeutic potentials of tea polyphenols. *Curr Pharm Biotechnol* 2012;**13**:191–9.
19. Kindt TJ, Goldsby RA, Osborne BA. *Kuby Immunology.* 6th ed. W. H. Freeman and Company; 2007.
20. Hruza LL, Pentland AP. Mechanisms of UV-induced inflammation. *Invest Dermatol* 1993;**100**:35S–41S.
21. Philips N, Samuel M, Arena R, Chen Y, Conte J, Natrajan P, et al. Direct inhibition of elastase and matrixmetalloproteinases., and stimulation of biosynthesis of fibrillar collagens, elastin and fibrillins by xanthohumol. *J Cosmet Sci* 2010;**61**:125–32.
22. Philips N, Conte J, Chen Y, Natrajan P, Taw M, Keller T, et al. Beneficial regulation of matrixmetalloproteinases and its inhibitors, fibrillar collagens and transforming growth factor-β by P. Polypodium leucotomos, directly or in dermal fibroblasts, ultraviolet radiated fibroblasts, and melanoma cells. *Arch Dermatol Res* 2009;**301**(7):487–95.
23. Philips N, Dulaj L, Upadhya T. Growth inhibitory mechanism of ascorbate and counteraction of its matrix metalloproteinases-1 and transforming growth factor-beta stimulation by gene silencing or P. leucotomos. *AntiCancer Res* 2009;**29**:3233–8.
24. Philips N, Smith J, Keller T, Gonzalez S. Predominant effects of *Polypodium leucotomos* on membrane integrity, lipid peroxidation, and expression of elastin and matrixmetalloproteinase-1 in ultraviolet radiation exposed fibroblasts, and keratinocytes. *J Dermatol Sci* 2003;**32**:1–9.

25. Philips N, Keller T, Hendrix C, Hamilton S, Arena R, Tuason M, et al. Regulation of the extracellular matrix remodeling by lutein in dermal fibroblasts, melanoma cells., and ultraviolet radiation exposed fibroblasts. *Arch Dermatol Res* 2007;**299**:373–9.

26. Philips N, Keller T, Holmes C. Reciprocal effects of ascorbate on cancer cell growth and the expression of matrix metalloproteinases and transforming growth factor-beta. *Cancer Lett* 2007;**256**:49–55.

27. Lemay M, Murray MA, Davies A, Roh-Schmidt H, Randolph R. In vitro and ex vivo cyclooxygenase inhibition by a hops extract. *Asia Pac J Clin Nutr* 2004;**13**:S110.

28. Gerhauser C, Alt A, Heiss E, Gamal-Eldeen A, Klimo K, Knauft J, et al. Cancer chemopreventive activity of Xanthohumol, a natural product derived from hop. *Mol Cancer Ther* 2002;**1**:959–69.

29. Bachelor MA, Bowden GT. UVA mediated activation of signaling pathways involved in skin tumor promotion and progression. *Semin Cancer Biol* 2004;**14**:131–8.

30. Chiang HM, Chen HC, Lin TJ, Shih IC, Wen KC, Ellis LZ, et al. Michelia alba extract attenuates UVB-induced expression of matrix metalloproteinases via MAP kinase pathway in human dermal fibroblasts. *Food Chem Toxicol* 2012;**50**:4260–9.

31. George J, Singh M, Srivastava AK, Bhui K, Roy P, Chaturvedi PK. Resveratrol and black tea polyphenol combination synergistically suppress mouse skin tumors growth by inhibition of activated MAPKs and p53. *PLoS One* 2011;**6**:e23395.

32. Sun T, Chen QY, Wu LJ, Yao XM, Sun XJ. Antitumor and antimetastatic activities of grape skin polyphenols in a murine model of breast cancer. *Food Chem Toxicol* 2012;**50**:3462–7.

33. Fisher GJ, Datta SC, Talwar HS, Wang ZQ, Varani J, Kang S, et al. Molecular basis of sun-induced premature skin ageing and retinoid antagonism. *Nature* 1996;**379**:335–9.

34. Guo W, Kong E, Meydani M. Dietary polyphenols, inflammation, and cancer. *Nutr Cancer* 2009;**61**(6):807–10.

35. Albini A, Dell'Eva R, Vené R, Ferrari N, Buhler DR, Noonan DM, et al. Mechanisms of the antiangiogenic activity by the hop flavonoid xanthohumol: NF-kappaB and Akt as targets. *FASEB J* 2006;**20**:527–9.

36. Ellis LZ, Liu W, Luo Y, Okamoto M, Qu D, Dunn JH, et al. Green tea polyphenol epigallocatechin-3-gallate suppresses melanoma growth by inhibiting inflammasome and IL-1β secretion. *Biochem Biophys Res Commun* 2011;**414**:551–6.

37. Scoditti E, Calabriso N, Massaro M, Pellegrino M, Storelli C, Martines G, et al. Mediterranean diet polyphenols reduce inflammatory angiogenesis through MMP-9 and COX-2 inhibition in human vascular endothelial cells: a potentially protective mechanism in atherosclerotic vascular disease and cancer. *Arch Biochem Biophysics* 2012;**527**:81–9.

38. Lamy S, Akla N, Ouanouki A, Lord-Dufour S, Béliveau R. Diet-derived polyphenols inhibit angiogenesis by modulating the interleukin-6/STAT3 pathway. *Exp Cell Res* 2012;**318**:1586–96.

39. Wondrak GT, Roberts MJ, Cervantes-Laurean D, Jacobson MK, Jacobson EL. Proteins of the Extracellular Matrix Are Sensitizers of Photo-oxidative Stress in Human Skin Cells. *J Investigative Dermatol* 2003;**121**:578–86.

40. Kielty CM, Sherratt MJ, Shuttleworth CA. Elastic fibres. *J Cell Sci* 2002;**115**:2817–28.

41. Watson RE, Griffiths CE, Craven NM, Shuttleworth CA, Kielty CM. Fibrillin-rich microfibrils are reduced in photoaged skin. Distribution at the dermal-epidermal junction. *J Invest Dermatol* 1999;**112**:782–7.

42. Philips N, Tuason M, Chang T, Lin Y, Tahir M, Rodriguez SG. Differential effects of ceramide on cell viability and extracellular matrix remodeling in keratinocytes and fibroblasts. *Skin Pharmacol Physiol* 2009;**22**:151–7.

43. Philips N, Keller T, Gonzalez S. TGF β like regulation of matrix metalloproteinases by anti transforming growth factor-β and anti transforming growth factor-β1 antibodies in dermal fibroblasts: implications to wound healing. *Wound Rep Regenerat* 2004;**12**:53–9.

44. Philips N. An anti TGF-β increased the expression of transforming growth factor-β, matrix metallproteinase-1, and elastin, and its effects were antagonized by ultraviolet radiation in epidermal keratinocytes. *J Dermatol Sci* 2003;**33**:177–9.

45. Herouy Y. Matrixmetalloproteinases in skin pathology. *Int J Mol Med* 2001;**7**:3–12.

46. Yan C, Boyd DD. Regulation of matrix metalloproteinase gene expression. *J Cell Physiol* 2007;**211**:19–26.

47. Verstappen J, Von den Hoff JW. Tissue Inhibitors of Metalloproteinases (TIMPs): Their Biological Functions and Involvement in Oral Diseases. *Crit Rev Oral Biol Med J Dent Res* 2006;**85**:1074–84.

48. Karroum A, Mirshahi P, Faussat A, Therwath A, Mirshahi M, Hatmi M. Tubular network formation by adriamycin-resistant MCF-7 breast cancer cells is closely linked to MMP-9 and VEGFR-2/VEGFR-3 over-expressions. *Eur J Pharmacol* 2012;**685**:1–7.

49. Waleh NS, Murphy BJ, Zaveri NT. Increase in tissue inhibitor of metalloproteinase-2 (TIMP-2). levels and inhibition of MMP-2 activity in a metastatic breast cancer cell line by an anti-invasive small molecule SR13179. *Cancer Lett* 2010;**289**:111–8.

50. Lee MS, Jung JI, Kwon SH, Lee SM, Morita K. TIMP-2 Fusion Protein with Human Serum Albumin Potentiates Anti-Angiogenesis-Mediated Inhibition of Tumor Growth by Suppressing MMP-2 Expression. *PLoS One* 2012;**7**:e35710.

51. Fuxe J, Karlsson MCI. TGF-β-induced epithelial-mesenchymal transition: a link between cancer and inflammation. *Semi Cancer Biol* 2012;**22**:455–61.

52. Philips N, Arena R, Yarlagadda S. Inhibition of ultraviolet radiation mediated extracellular matrix remodeling in fibroblasts by transforming growth factor-β. *BIOS* 2009;**80**:1–5.

53. Philips N, Tahir M, Stellatella J, Stephan K, Givant J, Zhou L, et al. Differential regulation of growth factors and matrix metalloproteinase-1 by estrogen, progesterone, and tamoxifen in normal and cancerous endometrial cells. *J Cancer Mol* 2009;**4**:169–73.

54. Philips N, McFadden K. Inhibition of transforming growth factor-beta and matrix metalloproteinases by estrogen and prolactin in breast cancer cells. *Cancer Lett* 2004;**206**:63–8.

55. Ali EM, Sheta M, El Mohsen MA. Elevated serum and tissue VEGF associated with poor outcome in breast cancer patients. *Alexandria J Med* 2011;**47**:217–24.

56. Shimizu M, Shirakami Y, Sakai H, Yasuda Y, Kubota M, Adachi S, et al. (-).-Epigallocatechin gallate inhibits growth and activation of the VEGF/VEGFR axis in human colorectal cancer cells. *Chem Biol Int* 2010;**185**:247–52.

Pterostilbene Protection and Bladder Cancer Cells

Ying-Jan Wang

Department of Environmental and Occupational Health, National Cheng Kung University Medical College, Tainan, Taiwan

Rong-Jane Chen

Department of Environmental and Occupational Health, National Cheng Kung University Medical College, Tainan, Taiwan, Graduate Institute of Clinical Medicine, Taipei Medical University, Taipei, Taiwan

List of Abbreviations

ABAP 2,2′-azo-bis(2-amidinopropane)
ABTS 2,2′-azino-bis(3-ethylbenzothiazoline-6-sulphonic acid)
AGE advanced glycation and products
Akt Protein Kinase B
AMACR alpha-methylacyl-CoA recemase
AMPK AMP-activated protein kinase
AOM azoxymethane
AP-1 activator protein 1
Atg autophagy related gene
BCG Bacille Calmette-Guérin
BLT2 leukotriene B4
Cdh1 E-cadherin
COX-2 Cyclooxygenase-2
CYP1A1 cytochrome P450, family 1, subfamily A, polypeptide 1
DMBA 7, 12 dimethylbenz[a] antracene
DPPH 2,2-diphenyl-1-picrylhydrazyl
EGF epidermal growth factor
EGFR epidermal growth factor receptor
ERK1/2 extracellular signal-regulated kinase
HIF-1α Hypoxia inducible factor 1, alpha subunit
HRG-β1 heregulin-β1
HUVEC human vascular endothelial cells
IKK IκB kinase
IL interleukin
iNOS inducible nitric oxide synthase
JAK Janus-activated kinase
LPS lipopolysaccharide
MAPK mitogen-activated protein kinases
NMI non-muscle-invasive
MMPs matrix metalloproteinases
mTOR the mammalian TOR
NFκB nuclear factor-κB
NO nitric oxide

NOX NADPH oxidase
Nrf2 NF-E2 related factor 2
oxLDL oxidized low-density lipoprotein
PAH polycyclic aromatic hydrocarbons
PGs prostaglandins
PI3K phosphatidylinositol-3-kinase
PKC protein kinase c
PPARα peroxisome proliferator-activated receptor
RNS reactive nitrogen species
ROS reactive oxygen species
SOD superoxide dismutase enzymes
Stat signal transducer and activator of transcription
TNFα tumor necrosis factor alpha
TPA 12-O-tetradecanoylphorbol-13-acetate
VEGF vascular endothelial growth factor

INTRODUCTION

Pterostilbene is a naturally occurring stilbenoid compound that originates from several natural plant sources including several types of grapes, peanuts, blueberries, and some plants widely used in traditional medicine, such as *Pterocarpus marsupium*, *Pterocarpus santalinm*, and *Vitis vinifera* leaves, as well as the stem bark of *Gauibourtia tessmanii*,[1] a tree found in Africa that is commonly used in folk medicine.[2] Pterostilbene is a dimethyl ether analog of resveratrol and has recently drawn much attention due to its health benefits as an antioxidative, anticancer, antidiabetic, and antihyperlipidemic agent.[3] The anticancer effects of pterostilbene have been shown to inhibit cancer cell growth, induce

Cancer
http://dx.doi.org/10.1016/B978-0-12-405205-5.00027-1

© 2014 Elsevier Inc. All rights reserved.

apoptosis, necrosis, and autophagy. In addition, pterostilbene inhibits adhesion, invasion, and metastasis in various cancer cells.

Recently, pterostilbene has been identified as a CYP1A1 (cytochrome P450, family 1, subfamily A, polypeptide 1) inhibitor.[1] Because cytochrome p450 was thought to play a role in carcinogenesis, pterostilbene may reduce the risk of mutagenesis and cancer. Regarding antidiabetic activities, pterostilbene has been demonstrated to lower blood glucose levels and modulate PPARα (peroxisome proliferator-activated receptor α) activation signaling pathways. Other miscellaneous healthy benefits of pterostilbene include antiaging, analgesic, and neuroprotective effects. These benefits associated with pterostilbene are mainly attributed to its antioxidant activity.[1]

Bladder cancer is the fourth most common cancer in men, and the ninth most common cancer in women.[4] The risk factors of bladder cancer include the following: inhalation of and exposure to aromatic amines and chemicals associated with the rubber and paint industry, tobacco smoke, hair dyes, chronic absorption of nitrates or arsenic in food or drinking water, bladder stones, chronic urinary tract infections or parasitic infections, the use of cyclophosphamide, and rarely, renal transplantation.[4] Bladder cancer is characterized by frequent recurrence and poor clinical outcome when tumors from noninvasive, flat, and papillary urothelial neoplasias invade to muscle or metastasize.[5] Traditional chemotherapy for advanced bladder cancer uses the combination of methotrexate, vinblastine, doxorubicin, and cisplatin (MVAC). However, this regimen is toxic, not easily tolerated, and shows only a low long-term survival rate in patients.[6]

Presently, chemotherapy for bladder cancer is focused on newer active agents, such as cisplatin-based regimens combined with gemcitabine or taxanes. Nevertheless, a cisplatin-based regimen is not suitable for all patients with advanced disease. Moreover, although patients receiving a cisplatin-based regimen demonstrated lower toxicity, they showed a similar 5-year survival rate compared with MVAC treatment. In addition, most deaths from bladder cancer are due to advanced unresectable disease, which is resistant to chemotherapy.[6] Chemoresistance in human bladder cancer cells could be associated with defects in the apoptosis machinery.[7] These findings raised the important issue of developing newer agents for bladder cancer therapy to reduce the toxicity, enhance treatment efficacy, and also increase effectiveness for treatment of resistant bladder cancer. Recently, pterostilbene is reported to induce cytotoxicity in sensitive and chemoresistant bladder cancer cells by modulating various cell death mechanisms and signaling pathways.[8] Therefore, pterostilbene may be effective in the management of bladder cancer, including advanced bladder cancer.

PTEROSTILBENE

Pterostilbene (trans-3,5-dimethoxy-4′-hydroxystilbene; $C_{16}H_{16}O_3$; MW, 256.296 g/mol) is a chemical classified as a benzylidene compound (stilbene) and is a naturally occurring phytoalexin identified in several plants of the genus *Pterocarpus* (in which *Pterocarpus marsupium* has been used for many years in the treatment of diabetes mellitus), in *Vitis vinifera* leaves, in some berries, and in grapes.[1] Pterostilbene is a naturally occurring analog of the well-studied resveratrol as shown in Figure 27.1.[9] Pterostilbene has pharmacologic properties similar to those of resveratrol. However, pterostilbene has several advantages superior to those of resveratrol due to its structure; pterostilbene has two methoxy groups and one hydroxyl group, whereas resveratrol has three hydroxyl groups. Therefore, pterostilbene is more lipophilic than resveratrol, easily increasing oral absorption and rendering a higher potential for cellular intake, and the bioavailability of pterostilbene is increased, and it has a longer half-life (105 min) than resveratrol (14 min).[10]

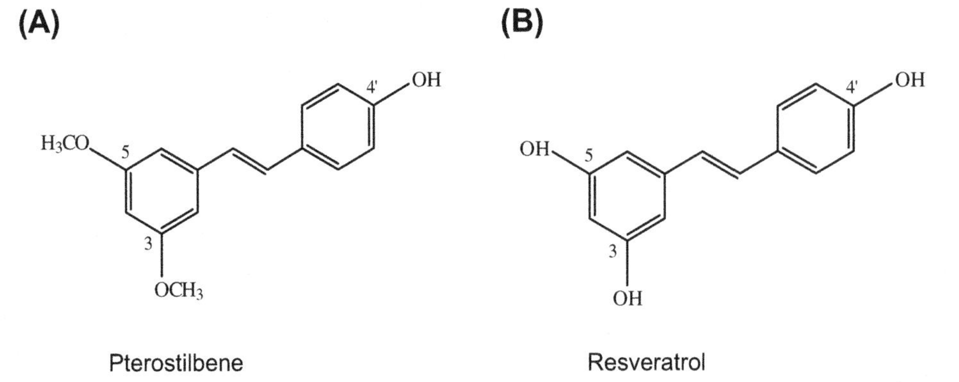

(A)

(B)

Pterostilbene   Resveratrol

FIGURE 27.1   Structure of (A) Pterostilbene and (B) Resveratrol.

Pharmacokinetics of Pterostilbene

Oral administration of pterostilbene in male CD rats for 14 days (56 or 168 mg/kg/day) showed 80% bioavailability. Pterostilbene exhibited an extensive tissue distribution with a V_{ss} value (5.3 L/kg) greater than that of total body water (~0.7 L/kg). Pterostilbene can be metabolized in mice to generate mono-demethylated and mono-hydroxylated metabolites. Pterostilbene, mono-demethylated metabolites, and mono-hydroxylated metabolites are good substrates for glucuronidation and sulfation to form related phase II metabolites. The GI (gastrointestinal) tract, liver, kidney, and urine processes high levels of phase II metabolites activities, and the GI tract may be the major site for pterostilbene metabolism.[11] Phase II metabolites in plasma are much higher than the parent compound. Using LC/APCI-MS/MS and LC/ESI/MS/MS analyses showed nine novel metabolites of pterostilbene from female C57BL/6J mice urine samples collected 24 h after administration of 200 mg/kg pterostilbene through oral gavage.[12] Therefore, glucuronidation, sulfation, demethylation, and hydroxylation are the major biotransformation processes of pterostilbene in mice. Glucuronide and sulfate conjugates were reported to be the primary metabolites of pterostilbene, and sulfate conjugation was more extensive than glucuronidation of pterostilbene.[13] These conjugates could serve as storage pools for the parent drug; however, the biological activity and significance of the metabolites is unclear.[13]

Studies in rodents and *in vitro* with normal cells showed that pterostilbene administration is nontoxic. In mice fed trans-pterostilbene for 28 days at a dose up to 3000 mg/kg/body weight/day (500 times the estimated mean human intake) showed no significant toxic effect or adverse biochemical parameter compared with control. This indicated that subchronic oral intake of a high dose of pterostilbene is safe.[14]

Oxidative Stress and Inflammation

Oxidative stress results from an imbalance between production and accumulation of ROS (reactive oxygen species) or RNS (reactive nitrogen species) and the antioxidant mechanisms.[15] Increased oxidative stress is not beneficial for cells; however, complete elimination of free radicals could also disrupt the normal function of the body because ROS signaling plays an important role in biological processes such as regulation of the processes catalyzed by protein kinases, phosphatases, and other enzymes.[16] Therefore, ROS signaling can initiate both inhibition and activation of tumor formation. For instance, previous reports have indicated that almost all tumors showed an overproduction of intracellular ROS in cancer cells compared with their normal counterparts. Conversely, ROS formation stimulated by external agents and pro-oxidants can result in apoptosis and cancer cell death.[16]

Oxidative stress and generation of inflammatory mediators are hallmarks of inflammation.[17] The link between inflammation and cancer has two paradigms. First, inflammation-driven tumorigenesis through extrinsic pathways includes ROS that damage DNA, impairs the DNA repair system, thereby initiating carcinogenesis. Second, excessive expression of pro-inflammatory mediators perturbs cell signaling pathways, promotes cell growth, angiogenesis, metastasis, and reduces susceptibility to the host immune response.[18] Central to these altered processes, diverse inflammatory mediators such as cytokines, chemokines, COX-2 (cyclooxygenase-2), PGs (prostaglandins), iNOS (inducible nitric oxide synthase), NO (nitric oxide), ROS, and AGE (advanced glycation end-products) are demonstrated as elevated or overproduced, or are abnormally activated.[18] In addition, inflammation-induced activation of various protein kinases, including JAK (Janus-activated kinase), Akt (Protein Kinase B), and MAPK (mitogen-activated protein kinases) and induced aberrant activation of transcriptional factors, such as Stat (signal transducer and activator of transcription), NFκB (nuclear factor-κB), AP-1 (activator protein 1), and HIF-1α (Hypoxia inducible factor 1, alpha subunit), have been implicated in tumor growth, angiogenesis, and metastasis.[18] Therefore, anti-inflammatory compounds may be useful for chemoprevention.[17]

Antioxidant and Anti-inflammatory Effects of Pterostilbene

Pterostilbene has been reported as an effective antioxidant and anti-inflammatory agent including the following mechanisms:

- The antioxidative effects of pterostilbene (with one hydroxyl group and two methoxy groups) were attributed to its unique structure that reduced extracellular ROS, whereas resveratrol (with three hydroxyl groups) neutralized ROS in whole blood and isolated lymphoblasts.[19] This effect allowed the use of pterostilbene to target extracellular ROS species that are responsible for tissue damage during chronic inflammation.[20]
- The free radical-scavenging property of pterostilbene has shown that pterostilbene's peroxyl radical-scavenging activity appears to be similar to that of resveratrol. Pterostilbene inhibits oxidation of ABTS (2,2'-azino-bis(3-ethylbenzothiazoline-6-sulphonic acid) indicating that pterostilbene scavenged DPPH (2,2-diphenyl-1-picrylhydrazyl) radicals and ABAP [2,2'-azo-bis(2-amidinopropane)]-derived peroxy radicals.[1]

The antioxidant ability of pterostilbene most importantly contributes to many health benefits of pterostilbene. For instance, pterostilbene prevents oxLDL (oxidized low-density lipoprotein)-induced ROS, NFκB activation, p53 accumulation, apoptosis, and decreased mitochondrial membrane potential, thereby preventing apoptosis in HUVECs (human vascular endothelial cells).[21] The antioxidative effects of pterostilbene combined with resveratrol were investigated in human erythrocytes *in vitro*. The results indicated that pterostilbene alone can protect erythrocyte membranes against lipid peroxidation and, when combined with resveratrol at a lower concentration, it induced a synergistic protective effect in lipid peroxidation; however, at a high concentration, induced additive protective effects.[22] The protective effect can partially explain the healthy benefits of these bioactive compounds when combined in the diet.

The anti-inflammatory effects of pterostilbene were studied in the *in vivo* model; pterostilbene meliorates the immediate inflammatory responses in TNFα (tumor necrosis factor alpha)-induced pancreatitis through down-regulation of Stat3 and the secretion of lipase and inflammatory cytokines IL-1β (interleukin-1 beta) and IL-6 (interleukin-6).[23] In another cell model, pterostilbene down-regulates inflammatory iNOS and COX-2 gene expression in macrophages by inhibiting NFκB activation through interfering with the activation of PI3K/Akt/IKK (IκB kinase) and MAPK.[24] Another study further indicated that pterostilbene can suppress the expression of several genes and their inflammatory products in LPS (lipopolysaccharide)-stimulated RAW264.7 macrophages and peritoneal macrophages from mice. These effects could be partly due to the diminution of NFκB activation by the proteasome, thereby suppressing activation of iNOS genes, and decreased secretion of TNFα, IL-1β, IL-6, and NO levels. This study suggested that pterostilbene may act as a proteasome inhibitor, thus indicating one of the mechanisms for its anti-inflammatory effects.[25] Taken together, these reports have an important implication for using pterostilbene toward the development of an effective anti-inflammatory agent. The antioxidant and anti-inflammatory responses regulated by pterostilbene are shown in Figure 27.2.

Chemopreventive Effects of Pterostilbene by Inhibiting Inflammatory Responses

Numerous studies have been conducted to evaluate the cancer chemopreventive potential of pterostilbene by modulating inflammatory responses. Using a mouse mammary organ cultured model, carcinogen-induced preneoplastic lesions were significantly inhibited by pterostilbene, partially due to its antioxidant

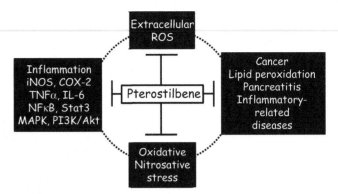

FIGURE 27.2 **The Antioxidant and Anti-Inflammatory Responses Regulated by Pterostilbene.** Pterostilbene scavenges extracellular ROS and eliminates intracellular oxidative/nitrosative stress. Pterostilbene inhibits production of inflammatory cytokines TNFα and IL-6 and inhibits the expression of iNOS, COX-2, NFκB, Stat3, MAPK, and PI3K/Akt signaling pathways, thereby reducing the risk of inflammatory-related diseases such as cancer, lipid peroxidation, and pancreatitis.

activity that scavenges peroxyl radicals and reduces singlet-oxygen-induced peroxidation.[9] In the skin carcinogenesis model, pterostilbene significantly inhibited 7, 12 dimethylbenz[a] antracene (DMBA)/TPA (12-O-tetradecanoylphorbol-13-acetate)-induced skin tumor formation by suppressing TPA-induced activation of MAPK, PI3K/Akt, and AP-1, thereby down-regulating iNOS and COX-2 expression.[26] In colon cancer, pterostilbene was reported to be more potent than resveratrol in preventing azoxymethane (AOM)-induced colon tumorigenesis, and cell proliferation through reduced gene expression of Myc, cyclin D, NFκB p65 subunit, iNOS, COX-2, VEGF (vascular endothelial growth factor), EGFR (epidermal growth factor receptor) signaling, MMPs (matrix metalloproteinases), and PI3K/Akt. Pterostilbene also was shown to induce apoptosis and Nrf2 (NF-E2 related factor 2)-mediated antioxidant signaling pathways in the mouse colon.[27,28] The other study indicated that pterostilbene reduced the colon tumor multiplicity of noninvasive adenocarcinomas, lowered proliferating cell nuclear antigen and down-regulated the expression of β-catenin and cyclin D1.[29]

Colon tumors from pterostilbene-fed animals showed reduced expression of inflammatory markers, including TNFα, IL-1β and IL-4, as well as nuclear staining for phospho-p65.[27,28] In HT-29 colon cancer cells, pterostilbene reduced the activation of NFκB, JAK, and MAPK, and the PI3K protein kinase cascade seems to be a key pathway for eliciting the anti-inflammatory action of pterostilbene.[17] These beneficial effects of pterostilbene make it a potential chemopreventive agent for inflammation-associated tumorigenesis. The mechanism of the chemopreventive effects of pterostilbene is shown in Figure 27.3.

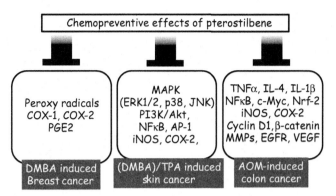

FIGURE 27.3 **The Mechanism of Chemopreventive Effects of Pterostilbene.** Chemopreventive effects of pterostilbene in DMBA-induced breast cancer, skin cancer, and AOM-induced colon cancer through inhibition of inflammatory processes.

Anticancer Mechanisms of Pterostilbene in Preclinical Studies

Anticancer effects of pterostilbene have been investigated in several cancer cell lines *in vitro*, including bladder, breast cancer, melanoma, colon, liver, and gastric cancer cells.[23] Pterostilbene possesses potent antioxidant (low dose) and pro-oxidant (high dose) properties in treated cells.[20] The pro-oxidant effects of high-dose pterostilbene may induce cancer cell death through ROS generation. Other possible anticancer mechanisms of pterostilbene include cell cycle arrest, apoptosis, necrosis, and autophagy.[23] However, the mechanistic details underlying pterostilbene protective effects against cancer cells are still unclear and are worth further exploration. The current proposed anticancer mechanisms of pterostilbene are discussed in the following section.

Pterostilbene Induces Cell Cycle Arrest in Cancer Cells

Pterostilbene can induce both cytotoxic and cytostatic mechanisms in many cancer cells. Pterostilbene was reported to block cell cycle progression at the G0/G1 phase in AGS gastric cancer, prostate cancer, pancreatic cancer, and breast cancer cells with similar mechanisms.[30] In AGS cancer cells, pterostilbene decreased the expression of cell cycle regulatory proteins, such as pRb, cyclin A, and cyclin E, and upregulated the expression of p53, p21, p27, and p16.[31] Similar results were observed in p53 wild-type LNCaP prostate cancer cells in which pterostilbene induced G0/G1 arrest by inducing p53 expression, AMPK (AMP-activated protein kinase) activation, and further upregulating p21 expression.[32] Additionally, pterostilbene inhibits androgen- and estrogen-mediated pathways that may contribute to its antiprostate cancer effect.[33] In breast cancer cells, pterostilbene induced apoptosis, cell cycle arrest through down regulation of cyclin D1 and wnt signaling as well.[30] Treatment of

pterostilbene at a lower dose induced S phase arrest in MCF-7 breast cancer and MIA PaCa-2 pancreatic cancer cells, whereas a higher dose of pterostilbene induced accumulation of cells in the G0/G1 phase. In addition, pterostilbene caused an increase in S phase cells of HL60 myeloid leukemia cells but had no significant effect on the cell cycle distribution of another leukemia cell line K562 cells.[34] Thus, the effects of pterostilbene on cell cycle distribution depend on different cancer types, and the precise mechanisms remain unclear and need to be further investigated.

Pterostilbene Induces Apoptosis in Cancer Cells

Pterostilbene was found to induce apoptosis in bladder, breast, colon, gastric, leukemia, liver, lung, melanoma, pancreas, and prostate cancers. ROS formation can result in apoptosis and cancer cell death.[16] Therefore, the pro-apoptotic effects of pterostilbene were suggested to be correlated with its pro-oxidant properties. Both intrinsic and extrinsic apoptosis pathways were identified in pterostilbene-induced apoptotic cell death. Pterostilbene increased ROS generation, which induces altered mitochondrial transmembrane potential, causing release of cytochrome-c, followed by activation of the caspase cascade-triggered mitochondria-dependent intrinsic pathways.[33,35] In several cancer cell lines, such as breast cancer, gastric cancer, leukemia, lung cancer, melanoma, and pancreatic cancer, down-regulation of Akt, phosphor-Stat3, and Bcl-2, and activation of Bax, Bad, cytochrome c, Smac/Diablo, and caspases as well as cytosolic Ca^{2+} overload were involved in the pterostilbene-induced intrinsic apoptotic pathway.[10] Pterostilbene also induced extrinsic death pathways in leukemia through Fas-mediated mechanisms.[32,34] In general, one of the important anticancer mechanisms of pterostilbene is apoptosis induction. The apoptotic mechanisms regulated by pterostilbene are shown in Figure 27.4.

Pterostilbene Inhibits Invasion and Metastasis

In addition to apoptosis, pterostilbene inhibits cancer growth by inducing cell cycle arrest as well as inhibits invasion and metastasis. Pterostilbene regulates multiple signaling pathways that control metastasis and invasion in cancer of breast, colon, liver, and melanoma.[10] In breast cancer, the dominant active form of β-catenin could reverse the growth inhibitory effects of pterostilbene, indicating that wnt signaling is important for the growth inhibitory effect of pterostilbene.[28,30] In liver cancer, Pan et al. suggested that pterostilbene protects against TPA-mediated metastasis of liver cancer via down-regulation of EGF, VEGF, EGFR, PKC (Protein kinase C), PI3K, MAPK, AP-1, and NFκB pathways followed by suppression of MMP-9 expression.[34,35] Pterostilbene could also

FIGURE 27.4 **The Mechanisms of Apoptosis Induced by Pterostilbene.** Apoptosis induction by pterostilbene occurs through activation of mitochondria-dependent intrinsic pathways through ROS generation, loss of mitochondria membrane potential, released cytochrome c and Smac/Diablo, and activation of caspases. Pterostilbene induces activation of Fas-mediated extrinsic pathways. Pterostilbene inhibits activation of Akt and Stat3, which are correlated with apoptosis induction.

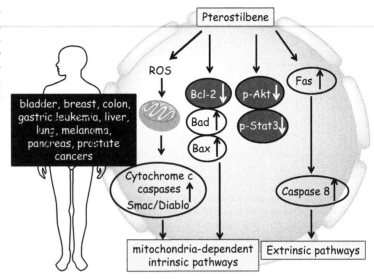

suppress HRG-β1 (heregulin-β1)-mediated cell invasion, motility, and cell transformation of MCF-7 human breast carcinoma through down-regulation of MMP-9 activity and growth inhibition.[35,36] In prostate cancer, pterostilbene inhibits expression of MMP-9 and AMACR (alpha-methylacyl-CoA recemase) and could be effective in its treatment.[36,37] In summary, these current results reveal that pterostilbene is a novel and effective antimetastatic agent that functions by partially down-regulating MMP-9-related pathways.

Pterostilbene Induces Protective Autophagy

Autophagy was first described by Christian de Duve in 1963 as a lysosome-mediated degradation process for nonessential or damaged cellular components.[37,38] Autophagy degradation process involves sequestration of parts of the cytoplasm in double-membrane vesicles (autophagosomes) that fuse with lysosomes, form autolysosomes where cytoplasmic material is hydrolyzed, and the resulting amino acids and other macromolecular precursors can be recycled.[38,39] Autophagy is ubiquitous in eukaryotic cells and is important in the development and diverse pathophysiological conditions, including providing protection against neurodegeneration, infections, and tumor development.[39,40]

Autophagy activation is the formation of the autophagosome by a series of steps, including phagophore elongation and maturation. These processes are controlled by a set of proteins called Atg (autophagy-related) proteins and require two ubiquitin-like conjugation systems: Atg5-Atg12 conjugate and Atg8 (a homolog of human LC3) lipidation and type III PI3K-Beclin1, which is negatively regulated by interaction with Bcl-2 family proteins. Signaling pathways that regulate autophagy include the PI3K/Akt/mTOR and the MEK/ERK1/2 signaling pathways.[40,41] ERK1/2 phosphorylates the Gα-interacting protein, which accelerates GTP hydrolysis

resulting in induction of autophagy.[41,42] In addition, AMPK activation then inhibits mTOR, thereby activating autophagy.[42,43] It has also been suggested that autophagy may provide an important way to prevent cancer development, limit tumor progression, and enhance the efficacy of cancer treatments.[40,41] Therefore, extensive attention has been paid to the role of autophagy in cancer development and therapy.[43,44]

Studies regarding pterostilbene induced autophagy in cancer cells remain limited. Pterostilbene was reported to induce protective autophagy in breast cancer cells suggesting that the combination of autophagy inhibitors with pterostilbene could serve as a promising strategy for the treatment of breast cancer.[45] Pterostilbene can induce a cytoprotective autophagy at a low concentration in vascular endothelial cells via a rapid elevation in intracellular Ca^{2+} concentration, subsequent AMPK activation, mTOR inhibition, which in turn helped to remove accumulated toxic oxLDL and inhibit apoptosis in HUVECs.[44,45] Therefore, it has been proposed that pterostilbene could be a potential lead compound for developing a class of autophagy regulators to treat autophagy-related diseases.

PTEROSTILBENE AND BLADDER CANCER

Bladder cancer is the fourth most common cancer in men and the ninth most common cancer in women.[45,46] Most newly diagnosed bladder cancers are non-muscle-invasive that are confined to the urothelium or the suburothelial connective tissues; the remaining are invasive to the muscularis propria. The strategy in the management of bladder cancers includes using chemotherapeutic agents, immunotherapies with Bacille Calmette-Guérin (BCG), and interferon.[45,46]

Bladder Cancer Risk Factors

The risk factors for bladder cancer include male gender, exposed to carcinogens such as tobacco smoke, aryl amines from occupational exposure, ionizing radiation, arsenic in the environment, and cyclophosphamide and phenacetin-containing analgesics. Suspected risk factors include dietary factors, chlorinated by-products in drinking water, and various types of occupational exposure that accounts for 20% of all bladder cancer cases.[45,46] Occupational factors include aromatic amines present in aniline dyes and many other industrial chemicals, chlorinated hydrocarbon (perchloroethylene) found in dry-cleaning solvents, and polycyclic aromatic hydrocarbons.[46,47] Cigarette smoking is considered to be one of the most important risk factors for urinary bladder cancer. The major carcinogens in cigarette smoke for bladder cancer development are aromatic amines that are metabolically N-hydroxylated and esterified, leading to reactive electrophiles, DNA adduct formation, and mutation. Bladder cancer follows the general concept of multistep carcinogenesis that DNA lesions of targeted cells are required for malignant transformation.[46,47] Beyond DNA damage induced by carcinogens, cocarcinogens may facilitate the development of different genetic alterations by epigenetic modification. For instance, nicotine acts as a cocarcinogen that can induce bladder cell proliferation and chemoresistance through epigenetic modification.[47,48]

Role of ROS and RNS in Linking Inflammation and Bladder Cancer

Excessive generation of ROS in bladder cancer is a consequence of environmental stress, infection, metabolic disorder, and defects in the body's detoxification pathways.[18] In addition, chronic inflammation can lead to generation of cellular oxidative or nitrosative stress, thereby contributing to carcinogenesis.[48,49] Inflammation preceding bladder tumorigenesis has been reported in a series of recent studies (Figure 27.5). One mechanism of inflammation-induced bladder oncogenesis is the induction of genomic instability leading to chromosome aberrations. Chromosome aberrations in bladder include deletion of chromosome 9, which contains a gene that encodes tuberous sclerosis complex 1 (tsc1), a protein that together with tuberous tuberin can negatively regulate mTOR (mammalian target of rapamycin).[46,47] In addition, the cdkn2a locus at 9p21, which encodes two key cell cycle regulators p14 and p16, were deleted in bladder cancer. Deletions of 17p13 and 13q14 result in inactivation of p53 and Rb1 genes[49,50] that play an important role in bladder carcinogenesis. For instance, progression-free survival is significantly shorter in patients with a p53 mutation compared with

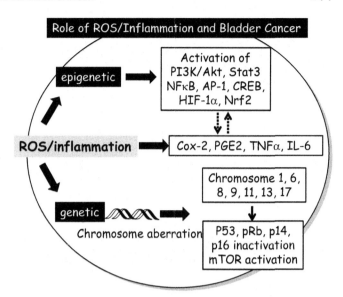

FIGURE 27.5 Inflammation and Bladder Cancer. Inflammation precedes bladder tumorigenesis through genetic and epigenetic modulation. Oxidative stress damages DNA, resulting in chromosome aberration at chromosomes 1, 6, 8, 9, 11, 13, 17, followed by reduction in expression of p53, pRb, p14, and p16, as well as increased mTOR expression. ROS and inflammation-altered epigenetic signaling pathways lead to over activation or expression of PI3K/Akt, Stat3, NFκB, AP-1, CREB, HIF-1α, and Nrf2.

patients with wild-type p53.[50,51] Alteration in the Rb1 pathway was commonly found in the development of invasive bladder cancer, and has been demonstrated to be significantly associated with tumor stage and tumor grade.[51,52] In addition, epigenetic aberrations were also identified in bladder cancers that p16INK4A was hypermethylation in 7 to 60% of bladder cancers. Methylation of Cdh1 (E-cadherin) was associated with shortened survival, and methylation of lama3, lamb3, and lamc2 increased cell adhesion, differentiation, and migration.[46,47]

Furthermore, inflammation and deregulated prostaglandin E metabolism in the tumor microenvironment promote immunosuppressive and increased infiltrating of myeloid-derived suppressor cells that is a source of multiple chemokines and cytokines, correlating with inflammation and immune dysfunction in bladder cancer.[52,53] Additionally, several inflammatory mediators, including COX-2 and PGE2, as well as transcriptional factors NFκB, Stat3, and HIF-1, mediated bladder cancer survival. The other inflammatory cytokine TNFα could stimulate the secretion of MMP-9, which contributes to bladder cancer invasion and migration.[53,54] The aberrant expression or production by the aforementioned pro-inflammatory mediators are regulated by redox-sensitive transcription factors, such as NFκB, AP-1, CREB, Stat3, HIF-1α, and Nrf2.[54,55]

General Mechanistic Alteration in Bladder Cancer

Cell signaling pathways play a critical role in tumorigenesis, and affect cell cycle control, DNA repair, cell death, carcinogen metabolism, and cell survival. In particular, a subset of mammalian target of rapamycin (mTOR) pathway alterations is shown to occur in bladder cancer. The mTOR pathway plays an important role in cell cycle progression, cell proliferation, angiogenesis, apoptosis, and energy balance as well as regulates autophagy. Activation of Akt leads to mTOR activation and subsequently phosphorylation of p70S6K and 4E-BP, therefore inactivating the proapoptotic molecule Bad and increasing cyclin D1 expression for cell cycle progression. One recent study showed that mTOR activity was associated with reduced patient survival and increased pathological stage. Therefore, mTOR inhibition by rapamycin prevents the carcinoma in situ transition to invasive bladder cancer.[55,56] Hence, in addition to rapamycin, other mechanisms by which mTOR can be targeted or alternatives to inhibit PI3K/Akt may be promising treatment strategies for bladder cancer. However, clinical studies evaluating the role of PI3K/Akt inhibition in bladder cancer are lacking, and continued investigation of the regulation of these pathways and usage of targeted inhibition in bladder cancer is necessary.

Treatment of Bladder Cancer

The overall survival (OS) after local therapy for bladder cancer is 52 to 77% in T2 disease, 40 to 64% in T3 disease, and 26 to 44% in T4 or node-positive disease.[56,57] Most patients with bladder cancer ultimately die due to metastatic disease. Therefore, metastatic disease at the time of local therapy is a major challenge. The treatment of bladder cancer uses chemotherapeutic agents including (1) neoadjuvant and adjuvant chemotherapy, (2) systemic chemotherapy for metastatic bladder cancers, and (3) novel targeted agents.[56,57] Neoadjuvant chemotherapy offers potential advantages compared with adjuvant chemotherapy and is used for early treatment of systemic metastasis as well as potential downstaging of the primary and regional disease. Cisplatin-based neoadjuvant chemotherapy provides a modest survival benefit. Systemic chemotherapy remains the standard therapy for metastatic bladder cancer patients. The standard first-line systemic chemotherapy therapy is MVAC. In Europe, CM (cisplatin, methotrexate), CMV (cisplatin, methotrexate, vinblastine), and MVEC (methotrexate, vinblastine, epirubicin, cisplatin) are also commonly used.[57,58] GC (gemcitabine, cisplatin) has a similar overall response rate, time to progression, and median survival rate but lower toxicity compared with MVAC. Therefore, GC is now considered the standard first-line

therapy for metastatic bladder cancer. However, this regimen is only suitable in fit patients. Recently, novel therapeutic agents, such as gemcitabine and taxanes, are among the most interesting therapeutic agents, but due to high toxicity and moderate responsive rate, they could not replace the GC regimen as standard therapy.[57,58] Recently, several novel targeted agents have been considered for bladder cancer therapy. However, one recent phase II clinical trial in advanced bladder cancer using a VEGF inhibitor indicated a disappointed result in that only four responses were obtained in 77 patients.[58,59] Thus, development of less toxic, more effective agents is critical, and clinical trial needs to be prioritized.

Treatment of Bladder Cancer by Pterostilbene

Mounting evidence suggests that inflammation plays an important role in bladder carcinogenesis and pathogenesis. Therefore, modulating inflammatory pathways may potentially enhance chemotherapy for bladder cancer. In patients with invasive bladder cancer, short-term treatment with a COX-2 inhibitor increased apoptosis in tumor tissue. However, COX-2 inhibitors have been associated with an increased cardiovascular risk.[52,53] Alternatively, previous reports have indicated the critical role of the NFκB-IL-6-Stat3 cascade involved in bladder carcinogenesis and chemoresistance induced by long-term nicotine exposure. Therefore, targeting NFκB and Stat3 can improve chemotherapeutic effects of bladder cancer.[7,59,60]

One recent study conducted by Chen et al. indicated that pterostilbene effectively inhibits the growth of both sensitive (T24 cells) and chemoresistant (T24R cells) bladder cancer cells by inducing apoptosis, necrosis, autophagy, and cell cycle arrest (Figure 27.6).[8] Pterostilbene demonstrated potent effects in the inhibition of cancer cell growth by down-regulating NFκB and Stat3 signaling pathways, which are both potential targets for bladder cancer therapy. Pterostilbene can induce necrosis in bladder cancer cells after a sudden stress, and then induce autophagy and apoptosis in both sensitive (T24 cells) and resistant (T24R cells) bladder cancer cells. It is suggested that T24 and T24R cells exposed to pterostilbene may not adapt to the changes, and thus, these cells undergo necrosis at an earlier exposure time, followed by autophagy and apoptosis. The antiapoptotic mitochondrial proteins Bcl-2 and Bcl-xl were significantly decreased, and expression of caspase 3 was increased by pterostilbene treatment in both cell lines. The author indicated that pterostilbene-induced bladder cancer death mainly occurred by apoptosis induction. However, whether ROS contributed to pterostilbene-induced necrosis and apoptosis in bladder cancer cells remained unclear and needs to be further investigated.

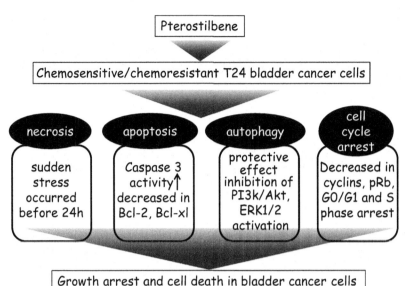

FIGURE 27.6 **Pterostilbene Protection in Bladder Cancer.** Pterostilbene induces cytotoxic and cytostatic effects in sensitive and chemoresistant human bladder cancer cells through induction of necrosis, autophagy, apoptosis, and cell cycle arrest. Pterostilbene induces caspase 3 activation, Bcl-xl and Bcl-2 inhibition, leading to apoptosis, and inhibits activation of ERK1/2 and Akt/mTOR pathways, leading to autophagy.

The author further indicated that pterostilbene induced autophagy earlier than apoptosis. Pterostilbene induced autophagy was mediated by inhibition of Akt/mTRO/p70S6K pathway and activation of the ERK1/2 pathway. However, pterostilbene-induced autophagy is not sufficient to induce cell death. In contrast, inhibition of autophagy by inhibitors may increase the sensitivity of cells to apoptosis. This indicated that pterostilbene induced protective autophagy in bladder cancer cells; therefore, cytotoxicity effects of pterostilbene may increase when in combination with autophagy inhibitors.

Pterostilbene treatment induces S phase arrest at lower concentration, whereas higher doses of pterostilbene induced G0/G1 arrest, followed by a shift to S phase arrest at 48 and 72 h. The G0/G1 arrest may be associated with the time-dependent decrease in cell cycle regulatory cyclin proteins and phosphorylation of Rb. This indicated that pterostilbene delayed cell cycle progression in both cell types. Therefore, pterostilbene-induced cell cycle arrest also contributed to the overall growth suppression in T24 and T24R cells.

Cancer cells are sensitive to apoptotic induction, but deregulation of apoptosis resulted in therapy resistance; therefore, agents that kill cancer cells through nonapoptotic pathways may circumvent this drug resistance.[60,61] Thus, there is growing interest to identify novel death mechanisms by anticancer agents. The optimal antiproliferative effects of pterostilbene in sensitive and chemoresistant human bladder cancer cells are regulated by necrosis, autophagy, apoptosis, and cell cycle arrest through ERK1/2 activation, Akt/mTOR inhibition, caspase activation, and down-regulation of cyclins. In conclusion, pterostilbene may induce cytostatic and cytotoxic effects in both chemosensitive and chemoresistant bladder cancer cells and serves as a promising agent for bladder cancer therapy.[8]

SUMMARY POINTS

- This chapter discusses pterostilbene protective effects in bladder cancer. Pterostilbene promotes anticancer effects in both sensitive and chemoresistant bladder cancer cells through necrosis, apoptosis, autophagy, and cell cycle arrest, making it a potential agent for treatment of different types of bladder cancer.

- Pterostilbene is a naturally occurring stilbenoid compound that originates from several natural plant sources including grapes, peanuts, blueberries, and some plants widely used in traditional medicine, such as *Pterocarpus marsupium*, *Pterocarpus santalinm*, and *Vitis vinifera* leaves, as well as the stem bark of *Gauibourtia tessmanii*.

- Pterostilbene is a dimethyl ether analog of resveratrol and has many health benefits such as antioxidant, anticancer, antidiabetic, and antihyperlipidemic effects. The antioxidant effects make it an effective chemopreventive agent for many cancers.

- Treatment of bladder cancer by chemotherapy has been reported in many settings. Neoadjuvant cisplatin-based combination chemotherapy has demonstrated a significant survival benefit. Gemcitabine plus cisplatin is the current standard first-line treatment of metastatic bladder cancer but only in fit patients. Other novel therapeutic agents showed high toxicity and a low responsive rate and thus could not replace gemcitabine and cisplatin as a standard therapy. Development of less toxic, more effective agents is critical, and clinical trial participation needs to be prioritized.

References

1. Remsberg CM, Yanez JA, Ohgami Y, Vega-Villa KR, Rimando AM, Davies NM. Pharmacometrics of pterostilbene: Preclinical pharmacokinetics and metabolism, anticancer, antiinflammatory, antioxidant and analgesic activity. *Phytother Res.* 2008;**22**(2):169–79.

2. Fuendjiep V, Wandji J, Tillequin F, Mulholland DA, Budzikiewicz H, Fomum ZT, et al. Chalconoid and stilbenoid glycosides from *Guibourtia tessmanii*. *Phytochemistry* 2002;**60**(8):803–6.

3. Cichocki M, Paluszczak J, Szaefer H, Piechowiak A, Rimando AM, Baer-Dubowska W. Pterostilbene is equally potent as resveratrol in inhibiting 12-O-tetradecanoylphorbol-13-acetate activated NFkappaB, AP-1, COX-2, and iNOS in mouse epidermis. *Mol Nutr Food Res* 2008;**52**(Suppl. 1):S62–70.

4. Letasiova S, Medve'ova A, Sovcikova A, Dusinska M, Volkovova K, Mosoiu C, et al. Bladder cancer, a review of the environmental risk factors. *Environ Health: A global access science source* 2012;**11**(Suppl. 1):S11.

5. Sengupta N, Siddiqui E, Mumtaz FH. Cancers of the bladder. *J R Soc Promot Health* 2004;**124**(5):228–9.

6. Dreicer R. Locally advanced and metastatic bladder cancer. *Curr Treat Opt Oncol* 2001;**2**(5):431–6.

7. Chen RJ, Ho YS, Guo HR, Wang YJ. Long-term nicotine exposure-induced chemoresistance is mediated by activation of Stat3 and downregulation of ERK1/2 via nAChR and beta-adrenoceptors in human bladder cancer cells. *Toxicol Sci* 2010;**115**(1):118–30.

8. Chen RJ, Ho CT, Wang YJ. Pterostilbene induces autophagy and apoptosis in sensitive and chemoresistant human bladder cancer cells. *Mol Nutr Food Res.* 2010;**54**(12):1819–32.

9. Rimando AM, Cuendet M, Desmarchelier C, Mehta RG, Pezzuto JM, Duke SO. Cancer chemopreventive and antioxidant activities of pterostilbene, a naturally occurring analogue of resveratrol. *J Agric Food Chem* 2002;**50**(12):3453–7.

10. McCormack D, McFadden D. Pterostilbene and cancer: Current review. *J Surg Res* 2012;**173**(2):e53–61.

11. Kuhnle G, Spencer JP, Chowrimootoo G, Schroeter H, Debnam ES, Srai SK, et al. Resveratrol is absorbed in the small intestine as resveratrol glucuronide. *Biochem Biophys Res Commun* 2000;**272**(1):212–7.

12. Shao X, Chen X, Badmaev V, Ho CT, Sang S. Structural identification of mouse urinary metabolites of pterostilbene using liquid chromatography/tandem mass spectrometry. *Rapid Commun Mass Spectrom RCM* 2010;**24**(12):1770–8.

13. Kapetanovic IM, Muzzio M, Huang Z, Thompson TN, McCormick DL. Pharmacokinetics, oral bioavailability, and metabolic profile of resveratrol and its dimethylether analog, pterostilbene, in rats. *Cancer Chemother Pharmacol* 2011;**68**(3):593–601.

14. Ruiz MJ, Fernandez M, Pico Y, Manes J, Asensi M, Carda C, et al. Dietary administration of high doses of pterostilbene and quercetin to mice is not toxic. *J Agric Food Chem* 2009;**57**(8):3180–6.

15. Da Costa LA, Badawi A, El-Sohemy A. Nutrigenetics and modulation of oxidative stress. *Ann Nutr Metab* 2012;**60**(Suppl. 3):27–36.

16. Afanas'ev I. Reactive oxygen species signaling in cancer: Comparison with aging. *Aging Dis* 2011;**2**(3):219–30.

17. Paul S, Rimando AM, Lee HJ, Ji Y, Reddy BS, Suh N. Antiinflammatory action of pterostilbene is mediated through the p38 mitogen-activated protein kinase pathway in colon cancer cells. *Cancer Prev Res (Philadelphia, Pa)* 2009;**2**(7):650–7.

18. Kundu JK, Surh YJ. Emerging avenues linking inflammation and cancer. *Free Radic Biol Med* 2012;**52**(9) 2013–37.

19. Perecko T, Jancinova V, Drabikova K, Nosal R, Harmatha J. Structure-efficiency relationship in derivatives of stilbene. Comparison of resveratrol, pinosylvin and pterostilbene. *Neuro Endocrinol Lett* 2008;**29**(5):802–5.

20. Pterostilbene. Monograph. *Altern Med Rev* 2010;**15**(2):159–63.

21. Zhang L, Zhou G, Song W, Tan X, Guo Y, Zhou B, et al. Pterostilbene protects vascular endothelial cells against oxidized low-density lipoprotein-induced apoptosis in vitro and in vivo. *Apoptosis* 2012;**17**(1):25–36.

22. Mikstacka R, Rimando AM, Ignatowicz E. Antioxidant effect of trans-resveratrol, pterostilbene, quercetin and their combinations in human erythrocytes in vitro. *Plant Foods Hum Nutr* 2010;**65**(1):57–63.

23. McCormack D, McDonald D, McFadden D. Pterostilbene ameliorates tumor necrosis factor alpha-induced pancreatitis in vitro. *J Surg Res* 2012;**178**(1):28–32.

24. Pan MH, Chang YH, Tsai ML, Lai CS, Ho SY, Badmaev V, Ho CT. Pterostilbene suppressed lipopolysaccharide-induced up-expression of iNOS and COX-2 in murine macrophages. *J Agric Food Chem* 2008;**56**(16):7502–9.

25. Qureshi AA, Guan XQ, Reis JC, Papasian CJ, Jabre S, Morrison DC, et al. Inhibition of nitric oxide and inflammatory cytokines in LPS-stimulated murine macrophages by resveratrol, a potent proteasome inhibitor. *Lipids Health Dis* 2012;**11**:76.

26. Tsai ML, Lai CS, Chang YH, Chen WJ, Ho CT, Pan MH. Pterostilbene, a natural analogue of resveratrol, potently inhibits 7,12-dimethylbenz[a]anthracene (DMBA)/12-O-tetradecanoylphorbol-13-acetate (TPA)-induced mouse skin carcinogenesis. *Food Funct* 2012;**3**(11):1185–94.

27. Chiou YS, Tsai ML, Nagabhushanam K, Wang YJ, Wu CH, Ho CT, et al. Pterostilbene is more potent than resveratrol in preventing azoxymethane (AOM)-induced colon tumorigenesis via activation of the NF-E2-related factor 2 (Nrf2)-mediated antioxidant signaling pathway. *J Agric Food Chem* 2011;**59**(6):2725–33.

28. Chiou YS, Tsai ML, Wang YJ, Cheng AC, Lai WM, Badmaev V, et al. Pterostilbene inhibits colorectal aberrant crypt foci (ACF) and colon carcinogenesis via suppression of multiple signal transduction pathways in azoxymethane-treated mice. *J Agric Food Chem* 2010;**58**(15):8833–41.

29. Paul S, DeCastro AJ, Lee HJ, Smolarek AK, So JY, Simi B, Wang CX, Zhou R, Rimando AM, Suh N. Dietary intake of pterostilbene, a constituent of blueberries, inhibits the beta-catenin/p65 downstream signaling pathway and colon carcinogenesis in rats. *Carcinogenesis* 2010;**31**(7):1272–8.

30. Wang Y, Ding L, Wang X, Zhang J, Han W, Feng L, et al. Pterostilbene simultaneously induces apoptosis, cell cycle arrest and cyto-protective autophagy in breast cancer cells. *Am J Transl Res* 2012;**4**(1):44–51.

31. Pan MH, Chang YH, Badmaev V, Nagabhushanam K, Ho CT. Pterostilbene induces apoptosis and cell cycle arrest in human gastric carcinoma cells. *J Agric Food Chem* 2007;**55**(19):7777–85.

32. Lin VC, Tsai YC, Lin JN, Fan LL, Pan MH, Ho CT, et al. Activation of AMPK by pterostilbene suppresses lipogenesis and cell-cycle progression in p53 positive and negative human prostate cancer cells. *J Agric Food Chem* 2012;**60**(25):6399–407.

33. Wang TT, Schoene NW, Kim YS, Mizuno CS, Rimando AM. Differential effects of resveratrol and its naturally occurring methylether analogs on cell cycle and apoptosis in human androgen-responsive LNCaP cancer cells. *Mol Nutr Food Res* 2010;**54**(3):335–44.

34. Tolomeo M, Grimaudo S, Di Cristina A, Roberti M, Pizzirani D, Meli M, et al. Pterostilbene and 3'-hydroxypterostilbene are effective apoptosis-inducing agents in MDR and BCR-ABL-expressing leukemia cells. *Int J Biochem Cell Biol.* 2005;**37**(8):1709–26.

35. Pan MH, Chiou YS, Chen WJ, Wang JM, Badmaev V, Ho CT. Pterostilbene inhibited tumor invasion via suppressing multiple signal transduction pathways in human hepatocellular carcinoma cells. *Carcinogenesis* 2009;**30**(7):1234–42.

36. Pan MH, Lin YT, Lin CL, Wei CS, Ho CT, Chen WJ. Suppression of Heregulin-beta1/HER2-Modulated Invasive and Aggressive Phenotype of Breast Carcinoma by Pterostilbene via Inhibition of Matrix Metalloproteinase-9, p38 Kinase Cascade and Akt Activation. *Evid Based Complement Altern Med* 2011;**2011**:562187.

37. Chakraborty A, Gupta N, Ghosh K, Roy P. In vitro evaluation of the cytotoxic, anti-proliferative and anti-oxidant properties of pterostilbene isolated from Pterocarpus marsupium. *Toxicol In Vitro* 2010;**24**(4):1215–28.

38. De Duve C, Wattiaux R. Functions of lysosomes. *Annu Rev Physiol* 1966;**28**:435–92.

39. Lum JJ, DeBerardinis RJ, Thompson CB. Autophagy in metazoans: Cell survival in the land of plenty. *Nat Rev Mol Cell Biol* 2005;**6**(6):439–48.

40. Karantza-Wadsworth V, Patel S, Kravchuk O, Chen G, Mathew R, Jin S, et al. Autophagy mitigates metabolic stress and genome damage in mammary tumorigenesis. *Genes Dev* 2007;**21**(13):1621–35.

41. Thorburn A. Apoptosis and autophagy: Regulatory connections between two supposedly different processes. *Apoptosis* 2008;**13**(1):1–9.

42. Klionsky DJ, Emr SD. Autophagy as a regulated pathway of cellular degradation. *Science* 2000;**290**(5497):1717–21.

43. Hoyer-Hansen M, Bastholm L, Szyniarowski P, Campanella M, Szabadkai G, Farkas T, et al. Control of macroautophagy by calcium, calmodulin-dependent kinase kinase-beta, and Bcl-2. *Mol Cell* 2007;**25**(2):193–205.

44. Chiu HW, Ho SY, Guo HR, Wang YJ. Combination treatment with arsenic trioxide and irradiation enhances autophagic effects in U118-MG cells through increased mitotic arrest and regulation of PI3K/Akt and ERK1/2 signaling pathways. *Autophagy* 2009;**5**(4):472–83.

45. Zhang L, Cui L, Zhou G, Jing H, Guo Y, Sun W. Pterostilbene, a natural small-molecular compound, promotes cytoprotective macroautophagy in vascular endothelial cells. *J Nutr Biochem* 2012.

46. Sexton WJ, Wiegand LR, Correa JJ, Politis C, Dickinson SI, Kang LC. Bladder cancer: A review of non-muscle invasive disease. *Cancer Control* 2010;**17**(4):256–68.

47. Volanis D, Kadiyska T, Galanis A, Delakas D, Logotheti S, Zoumpourlis V. Environmental factors and genetic susceptibility promote urinary bladder cancer. *Toxicol Lett* 2010;**193**(2):131–7.

48. Chen RJ, Chang LW, Lin P, Wang YJ. Epigenetic effects and molecular mechanisms of tumorigenesis induced by cigarette smoke: an overview. *J Oncol* 2011;**2011**:654931.

49. Poljsak B, Milisav I. The neglected significance of antioxidative stress. *Oxid Med Cell Longev* 2012;**2012**:480895.

50. Chan MW, Hui AB, Yip SK, Ng CF, Lo KW, Tong JH, et al. Progressive increase of genetic alteration in urinary bladder cancer by combined allelotyping analysis and comparative genomic hybridization. *Int J Oncol* 2009;**34**(4):963–70.

51. Ecke TH, Sachs MD, Lenk SV, Loening SA, Schlechte HH. TP53 gene mutations as an independent marker for urinary bladder cancer progression. *Int J Mol Med* 2008;**21**(5):655–61.

52. Gallucci M, Guadagni F, Marzano R, Leonardo C, Merola R, Sentinelli S, et al. Lopez Fde L, Cianciulli AM. Status of the p53, p16, RB1, and HER-2 genes and chromosomes 3, 7, 9, and 17 in advanced bladder cancer: Correlation with adjacent mucosa and pathological parameters. *J Clin Pathol* 2005;**58**(4):367–71.

53. Zhu Z, Shen Z, Xu C. Inflammatory pathways as promising targets to increase chemotherapy response in bladder cancer. *Mediators Inflamm* 2012;**2012**:528690.

54. Lee SJ, Park SS, Cho YH, Park K, Kim EJ, Jung KH, et al. Activation of matrix metalloproteinase-9 by TNF-alpha in human urinary bladder cancer HT1376 cells: The role of MAP kinase signaling pathways. *Oncol Rep* 2008;**19**(4):1007–13.

55. Leibovici D, Grossman HB, Dinney CP, Millikan RE, Lerner S, Wang Y, et al. Polymorphisms in inflammation genes and bladder cancer: from initiation to recurrence, progression, and survival. *J Clin Oncol* 2005;**23**(24):5746–56.

56. Seager CM, Puzio-Kuter AM, Patel T, Jain S, Cordon-Cardo C, Mc Kiernan J, et al. Intravesical delivery of rapamycin suppresses tumorigenesis in a mouse model of progressive bladder cancer. *Cancer Prev Res (Phila)* 2009;**2**(12):1008–14.

57. Costantini C, Millard F. Update on chemotherapy in the treatment of urothelial carcinoma. *Sci World J* 2011;**11**:1981–94.

58. Racioppi M, D'Agostino D, Totaro A, Pinto F, Sacco E, D'Addessi A, et al. Value of current chemotherapy and surgery in advanced and metastatic bladder cancer. *Urol Int* 2012;**88**(3):249–58.

59. Gallagher DJ, Milowsky MI, Gerst SR, Ishill N, Riches J, Regazzi A, et al. Phase II study of sunitinib in patients with metastatic urothelial cancer. *J Clin Oncol* 2010;**28**(8):1373–9.

60. Chen RJ, Ho YS, Guo HR, Wang YJ. Rapid activation of Stat3 and ERK1/2 by nicotine modulates cell proliferation in human bladder cancer cells. *Toxicol Sci* 2008;**104**(2):283–93.

61. Ruefli AA, Smyth MJ, Johnstone RW. HMBA induces activation of a caspase-independent cell death pathway to overcome P-glycoprotein-mediated multidrug resistance. *Blood* 2000;**95**(7): 2378–85.

Index

Note: Page numbers with "*f*" denote figures; "*t*" tables.

Color Plates

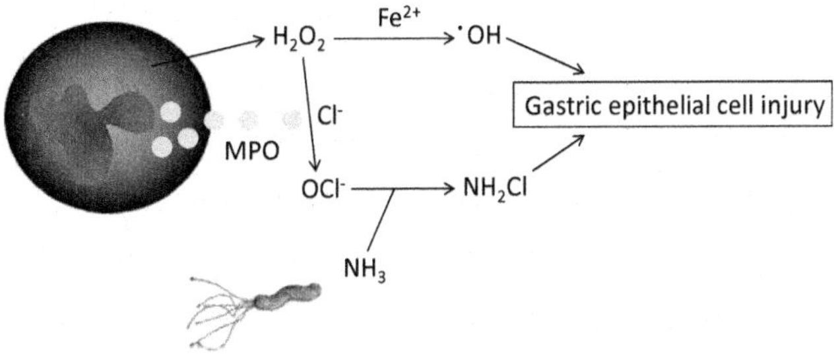

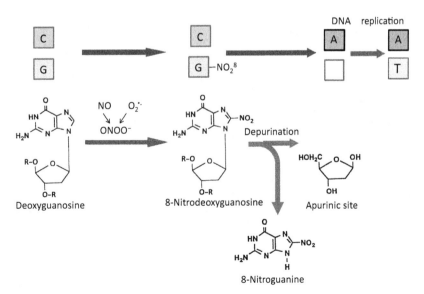

FIGURE 4.1 *Helicobacter pylori*-associated gastric oxidative injury: monochloramine pathway. MPO: myeloperoxidase.

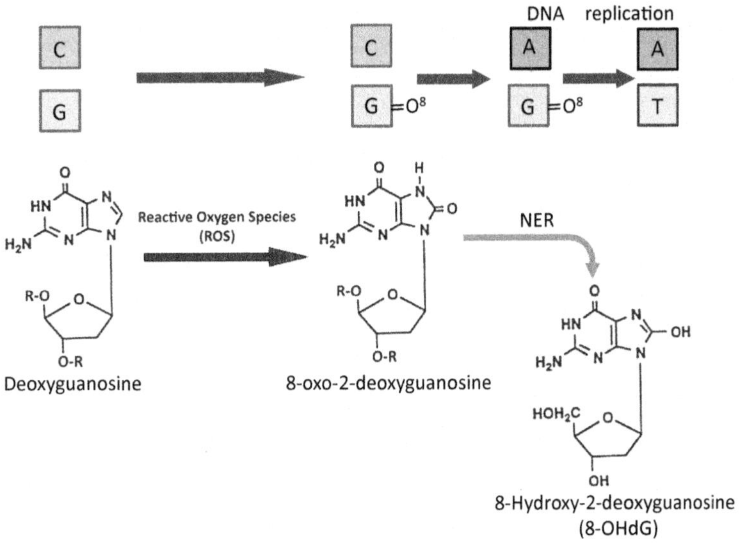

FIGURE 4.2 Point mutation induced by 8-oxo-2-deoxyguanosine via induction of the G:C → T:A transversion. NER: nucleotide excision repair enzyme.

FIGURE 4.3 Point mutation induced by 8-nitrodeoxyguaninnosine via induction of the G:C → T:A transversion.

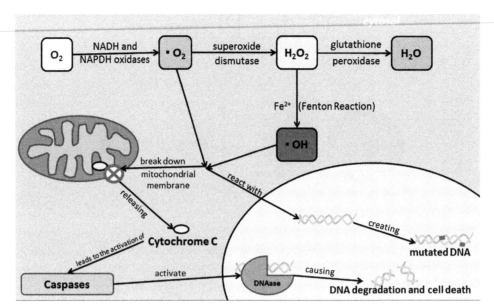

FIGURE 5.1 Oxidative Stress can Damage Critical Ovarian Cell Components. ROS are created endogenously through normal reactions within the cell. Antioxidant enzymes within the cell normally convert these ROS into more harmless substances in the cell. However, when these antioxidant mechanisms are overwhelmed, superoxide radicals are capable of breaking down the outer mitochondrial membrane, initiating a cascade that results in apoptosis. Free radicals within the cytosol can also enter the nucleus and react with DNA. If apoptosis does not occur, mutations within the DNA can be carried on to future generations of cells, potentially deactivating crucial tumor suppressor genes.

Necessary

- ROS produced by the pre-ovulatory follicle are required to induce ovulation
- ROS stimulate cell growth and proliferation
- ROS control the expression of various tumor suppressor genes
- Increased ROS levels can block angiogenesis and metastasis by destroying cancer cells

Harmful

- Follicular ROS promotes apoptosis in the ovary
- Excessive ROS can damage cellular DNA, proteins, and lipids
- ROS can cause permanent DNA damage, resulting in tumor suppressor gene mutations
- Excessive ROS can activate transcription factors that enable tumor cell survival and proliferation as well as angiogensis and metastasis

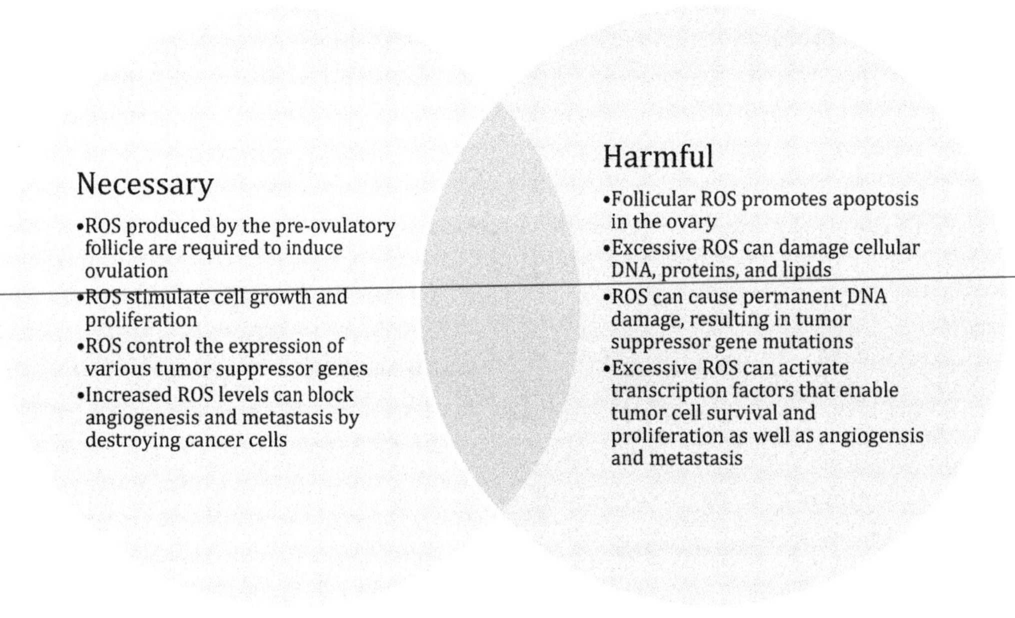

FIGURE 5.4 The Yin and Yang of Oxidative Stress—A Delicate Balance of ROS. ROS are necessary components of the cellular environment that are responsible for key regulatory functions of the cell. However, in excess, ROS can become harmful to the cell and initiate processes that often result in tumorogenesis.

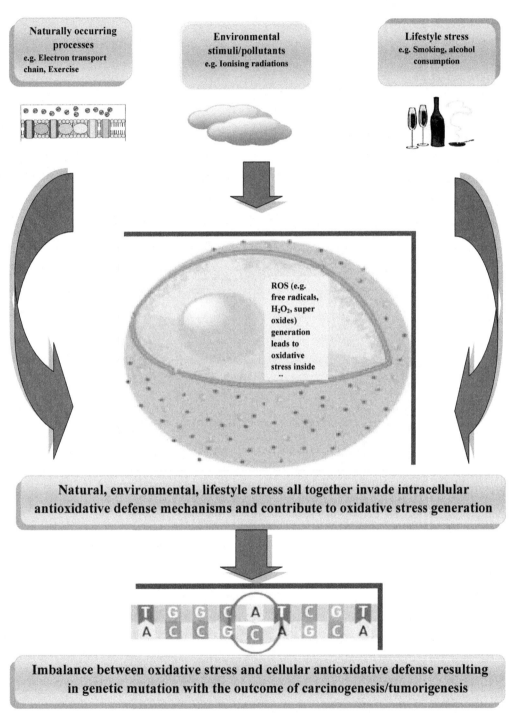

FIGURE 8.1 Oxidative stress leads to ROS generation with the ultimate outcome of carcinogenesis/tumorigenesis: natural, environmental, lifestyle stress all together contribute to oxidative stress generation, resulting in the imbalance of cellular antioxidant defense mechanisms that ultimately cause genetic mutation with the outcome of carcinogenesis.

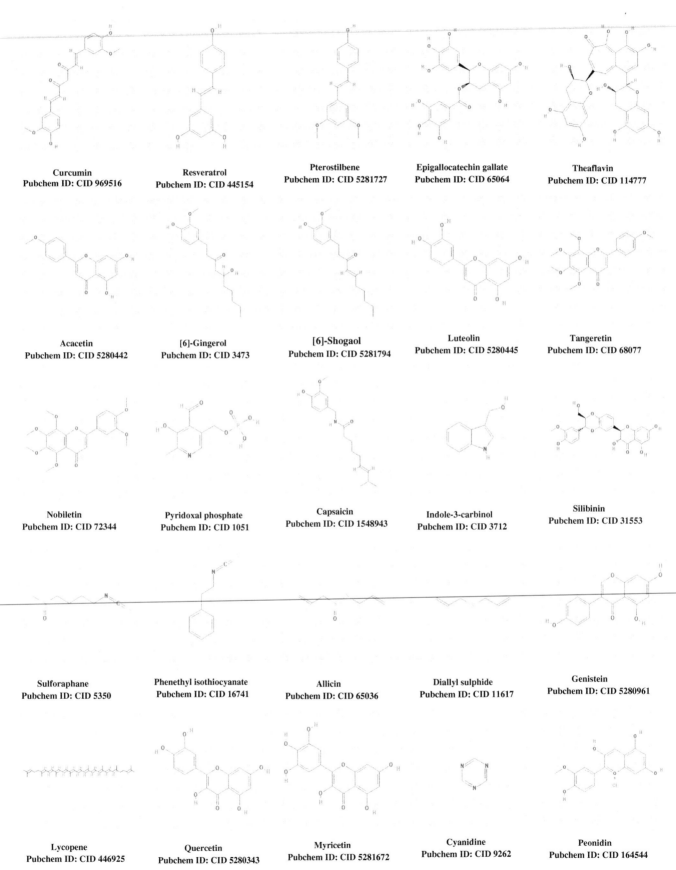

Curcumin
Pubchem ID: CID 969516

Resveratrol
Pubchem ID: CID 445154

Pterostilbene
Pubchem ID: CID 5281727

Epigallocatechin gallate
Pubchem ID: CID 65064

Theaflavin
Pubchem ID: CID 114777

Acacetin
Pubchem ID: CID 5280442

[6]-Gingerol
Pubchem ID: CID 3473

[6]-Shogaol
Pubchem ID: CID 5281794

Luteolin
Pubchem ID: CID 5280445

Tangeretin
Pubchem ID: CID 68077

Nobiletin
Pubchem ID: CID 72344

Pyridoxal phosphate
Pubchem ID: CID 1051

Capsaicin
Pubchem ID: CID 1548943

Indole-3-carbinol
Pubchem ID: CID 3712

Silibinin
Pubchem ID: CID 31553

Sulforaphane
Pubchem ID: CID 5350

Phenethyl isothiocyanate
Pubchem ID: CID 16741

Allicin
Pubchem ID: CID 65036

Diallyl sulphide
Pubchem ID: CID 11617

Genistein
Pubchem ID: CID 5280961

Lycopene
Pubchem ID: CID 446925

Quercetin
Pubchem ID: CID 5280343

Myricetin
Pubchem ID: CID 5281672

Cyanidine
Pubchem ID: CID 9262

Peonidin
Pubchem ID: CID 164544

FIGURE 8.2 Chemical structures of some important bioactive compounds used as antioxidants for chemoprevention: curcumin, resveratrol, pterostilbene, epigallocatechin gallate, theaflavin, Acacetin, [6]- gingerol, [6]- shogaol, luteolin, tangeretin, nobiletin, pyridoxal phosphate, capsaicin, indole-3-carbinol, silibinin, sulforaphane, phenethyl isothiocyanate, allicin, diallyl sulphide, genistein, lycopene, quercetin, myricetin, cyanidine, peonidin, emodin, eugenol, beta carotene, apigenin, carnosol, ellagic acid, guggulsterone.

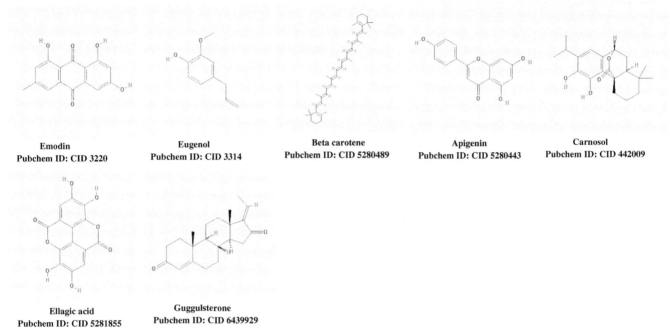

Emodin
Pubchem ID: CID 3220

Eugenol
Pubchem ID: CID 3314

Beta carotene
Pubchem ID: CID 5280489

Apigenin
Pubchem ID: CID 5280443

Carnosol
Pubchem ID: CID 442009

Ellagic acid
Pubchem ID: CID 5281855

Guggulsterone
Pubchem ID: CID 6439929

FIGURE 8.2 Cont'd

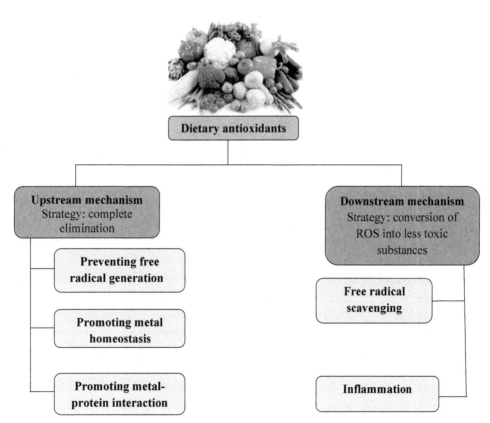

FIGURE 8.3 Therapeutic strategy of chemoprevention by dietary antioxidants: dietary antioxidants employed different upstream (e.g., preventing free radical generation, promoting metal homeostasis, and metal-protein interaction) and downstream strategies (e.g., free radical scavenging, inflammation).

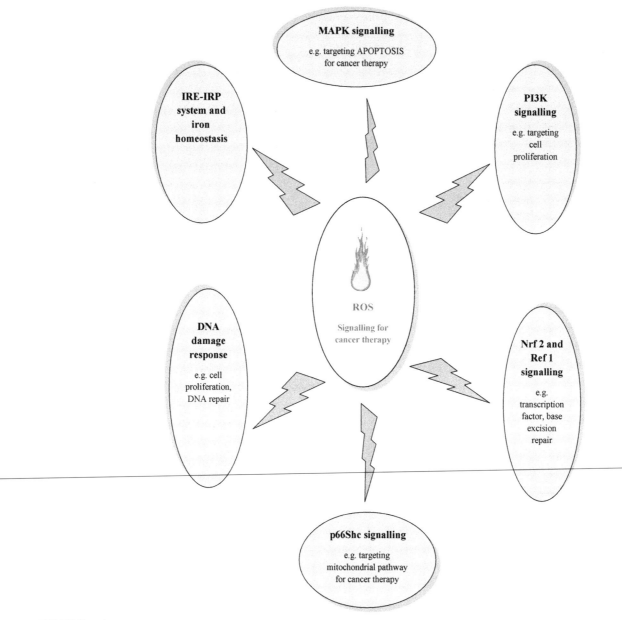

FIGURE 8.4 ROS-related signaling pathways for targeted cancer therapy include MAPK signaling, PI3K signaling, Nfr 2 and Ref 1 signaling, P66shc signaling, DNA damage response, and IRE-IRP system and iron homeostasis.

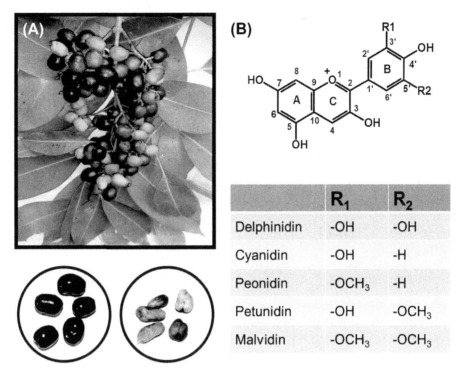

FIGURE 10.1 The photographs of Indian blackberry jamun fruits and seeds (A). Structures of various anthocyanidins in jamun fruit (B). *Photographs of jamun fruits were kindly provided by Dr. Manjeshwar S. Baliga.*

	R₁	R₂
Delphinidin	-OH	-OH
Cyanidin	-OH	-H
Peonidin	-OCH₃	-H
Petunidin	-OH	-OCH₃
Malvidin	-OCH₃	-OCH₃

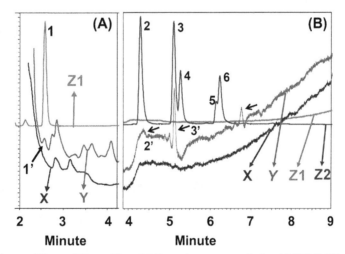

FIGURE 10.10 Detection of anthocyanidins at 520 nm (B) and 260 nm for protocatechuic acid (PCA) (A), a bioactive metabolite of cyanin in lung tissue of rats given dietary blueberry (5% w/w) for 5 weeks by HPLC-PDA. Chromatograms shown are composite: X = lung tissue from rats on control diet; Y = lung tissue from rats on blueberry diet. Reference compounds: Z1, peak 1 (PCA); Z2, reference anthocyanidins, peaks: 2 (Dp), 3 (Cy), 4 (Pt), 5 (Pe), and 6 (Mv). Solvent, gradient of 3.5% phosphoric acid in acetonitrile. Maps are cropped for presentation purposes.

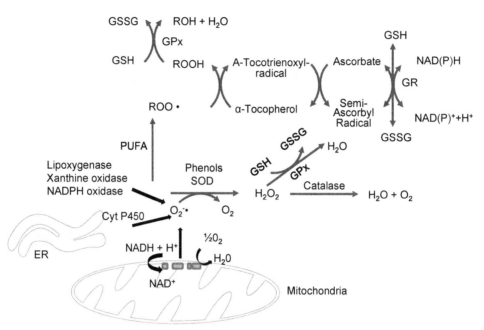

FIGURE 13.1 Generation of superoxide radical and antioxidant defense system. Superoxide radicals are generated during normal cellular metabolism. Endogenous antioxidant defense enzymes include superoxide dismutase (SOD), catalase, and glutathione peroxidase (GPx). Exogenous antioxidants are tocopherols, ascorbic acid, and phenols derived from dietary sources. *From Sung & Bae (2013) Dietary antioxidants and Rheumatoid Arthritis. In: Watson RR and Preedy VR (eds.) Bioactive Food as Dietary Interventions for Arthritis and Related Inflammatory Diseases, pp. 515–527. San Diego: Academic Press. Reprinted with permission from Elsevier.*

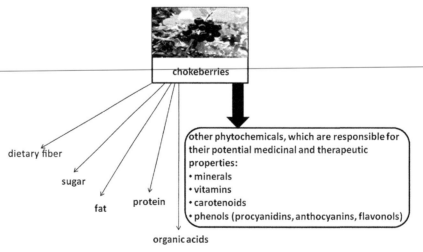

FIGURE 14.1 Chemical compounds of chokeberries, which are responsible for their potential medical and therapeutic properties [42; modified].

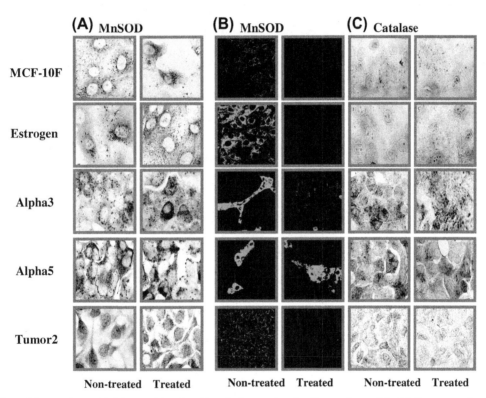

(A) MnSOD   **(B)** MnSOD   **(C)** Catalase

MCF-10F

Estrogen

Alpha3

Alpha5

Tumor2

Non-treated   Treated   Non-treated   Treated   Non-treated   Treated

FIGURE 15.2 **Effect of Curcumin on MnSOD and Catalase Protein Expression in a Breast Cancer Model.** Representative images of the effect of 15 uM curcumin on MnSOD protein expression analyzed by either (A) peroxidase or (B) immunofluorescent techniques in the MCF-10F, Estrogen, Alpha3, Alpha5, and Tumor2 cell lines. (C) Catalase protein expression analyzed by peroxidase technique in the MCF-10F, Estrogen, Alpha3, Alpha5, and Tumor2 cell lines.

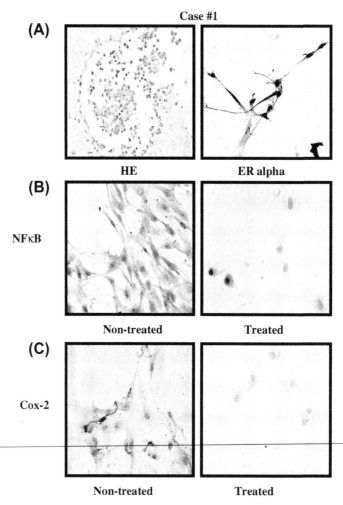

FIGURE 15.4 Effect of Curcumin on NFκB and COX-2 Protein Expression in Isolated Cells Derived from Specimens with Breast Carcinomas. Cross-section of breast specimen with noninvasive ductal carcinoma (Case #1) (Hematoxylin–eosin) (A; left side). Protein expression of isolated cells from such tissue after 3 passages of Estrogen receptor alpha (Right side); NFκB, (B) and COX-2, (C) in nontreated and treated cells with 30uM curcumin for 2 days in culture.

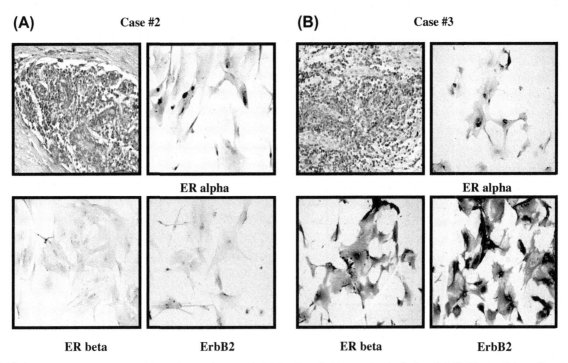

(A) Case #2

ER alpha

ER beta   ErbB2

(B) Case #3

ER alpha

ER beta   ErbB2

FIGURE 15.5 **Effect of Curcumin on ER alpha, ER beta, and ErbB-2 Protein Expression in Isolated Cells Derived from Specimens with Breast Carcinomas.** (A) Cross-section of breast specimen with invasive ductal carcinoma, grade II (Case # 2). Protein expression of positive ER alpha, ER beta, and ErbB-2 of isolated cells derived from such tissue after 10 passages and 2 days in culture in the presence of curcumin. (B) Cross-section of breast specimen with lobular carcinoma, grade III (Case #3). Protein expression of negative ER alpha, ER beta, and ErbB-2 of isolated cells derived from such tissue after 10 passages and 2 days in culture in the presence of curcumin.

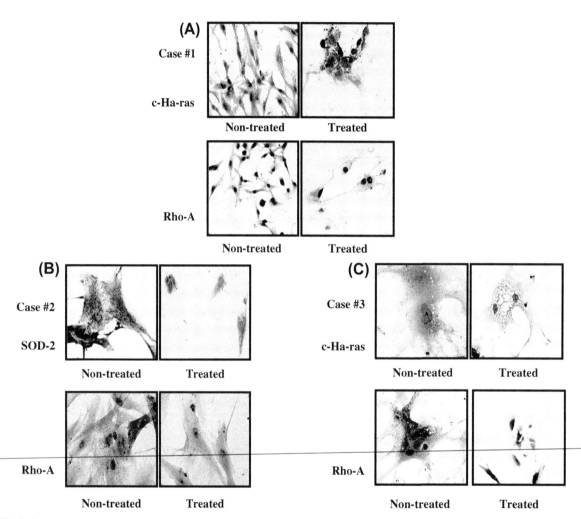

FIGURE 15.6 **Effect of Curcumin on c-Ha-ras and Rho-A and SOD-2 Protein Expression in Isolated Cells Derived from Specimens with Breast Carcinomas.** (A) Isolated cells derived from breast specimen with ductal carcinoma after 3 passages (Case #1). Protein expression of c-Ha-ras and Rho-A in nontreated and treated cells with 30 μm curcumin for 2 days in culture. (B) Isolated cells derived from breast specimen with invasive ductal carcinoma, grade II (Case #2) after 10 passages. Protein expression of SOD-2 and Rho-A in nontreated and treated with 30 μm curcumin after 10 passages and 2 days in culture. (C) Isolated cells derived from breast specimen with lobular carcinoma, grade III. Protein expression of c-Ha-ras and Rho-A in nontreated and treated (30 μm) curcumin in isolated cells derived from such tissue after 10 passages and 2 days in culture.

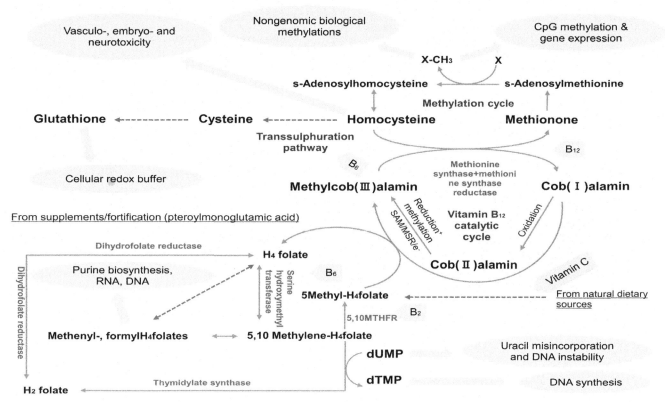

FIGURE 16.1 The schematic of folate-dependent process and the cooperation with the other vitamins in the elaboration, maintenance, and expression of DNA.

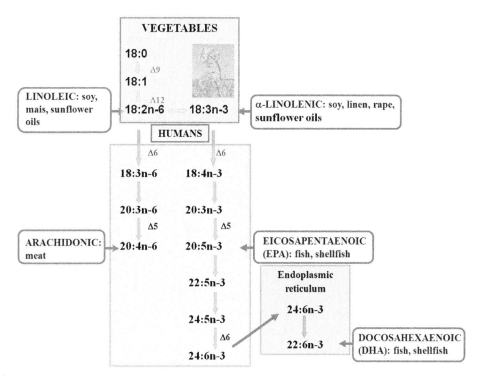

FIGURE 19.1 Synthesis of PUFAs in human beings from linoleic and α-linolenic acids taken with vegetables.

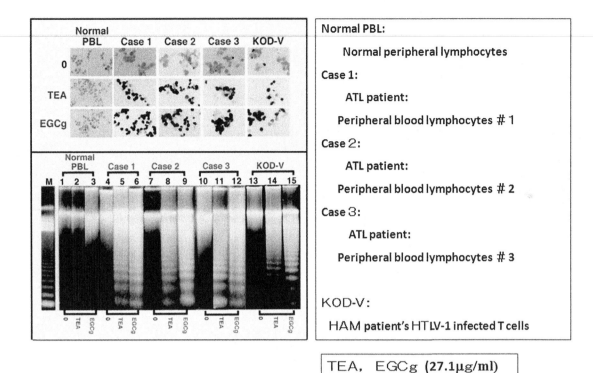

FIGURE 21.4 Apoptosis induction in ATL cells and HTLV-1-infected cells by green tea.

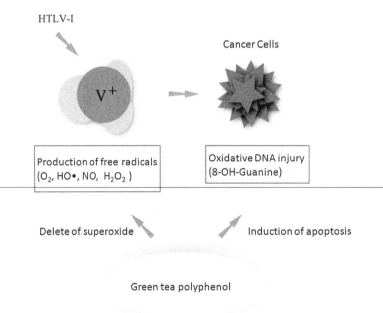

FIGURE 21.5 Scheme of interaction between green tea polyphenols and HTVL-1-infected lymphocytes.

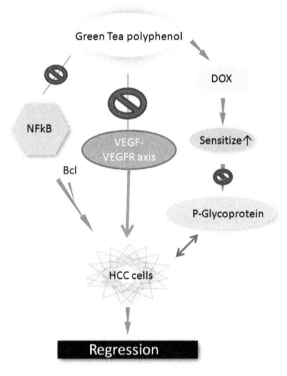

FIGURE 21.6 Scheme of interaction between green tea polyphenols and hepatocellular carcinoma (HCC). Note: NF-κB, nuclear factor κB; Bcl, B-cell lymphoma; VEGF, vascular endothelial growth factor; DOX, doxorubicin; ⊘, blockade.

Printed in the United States
By Bookmasters